Advances in Research on Pediatric Gender Affirming Healthcare

AF739422

Advances in Research on Pediatric Gender Affirming Healthcare

Edited by

Danielle N. Moyer
Department of Pediatrics, Oregon Health & Science University, Portland, OR, United States

Jess T. Guerriero
Doernbecher Gender Clinic, Oregon Health & Science University, Portland, OR, United States

Kara J. Connelly
Department of Pediatrics, Oregon Health & Science University, Portland, OR, United States

ELSEVIER

Elsevier
Radarweg 29, PO Box 211, 1000 AE Amsterdam, Netherlands
125 London Wall, London EC2Y 5AS, United Kingdom
50 Hampshire Street, 5th Floor, Cambridge, MA 02139, United States

Copyright © 2026 Elsevier Inc. All rights are reserved, including those for text and data mining, AI training, and similar technologies.

For accessibility purposes, images in electronic versions of this book are accompanied by alt text descriptions independently provided by Elsevier, generated with the aid of AI. For more information, see https://www.elsevier.com/about/accessibility or Radarweg 29, 1043 NX Amsterdam, Netherlands.

Books and Journals published by Elsevier comply with applicable product safety requirements. For any product safety concerns or queries, please contact our authorised representative, Elsevier B.V., at productsafety@elsevier.com.

Publisher's note: Elsevier takes a neutral position with respect to territorial disputes or jurisdictional claims in its published content, including in maps and institutional affiliations.

No part of this publication may be reproduced or transmitted in any form or by any means, electronic or mechanical, including photocopying, recording, or any information storage and retrieval system, without permission in writing from the publisher. Details on how to seek permission, further information about the Publisher's permissions policies and our arrangements with organizations such as the Copyright Clearance Center and the Copyright Licensing Agency, can be found at our website: www.elsevier.com/permissions.

This book and the individual contributions contained in it are protected under copyright by the Publisher (other than as may be noted herein).

Notices

Knowledge and best practice in this field are constantly changing. As new research and experience broaden our understanding, changes in research methods, professional practices, or medical treatment may become necessary.

Practitioners and researchers must always rely on their own experience and knowledge in evaluating and using any information, methods, compounds, or experiments described herein. In using such information or methods they should be mindful of their own safety and the safety of others, including parties for whom they have a professional responsibility.

To the fullest extent of the law, neither the Publisher nor the authors, contributors, or editors, assume any liability for any injury and/or damage to persons or property as a matter of products liability, negligence or otherwise, or from any use or operation of any methods, products, instructions, or ideas contained in the material herein.

ISBN: 978-0-443-29019-0

For Information on all Elsevier publications
visit our website at https://www.elsevier.com/books-and-journals

Publisher: Andre Wolff
Acquisitions Editor: Wendi Baker
Editorial Project Manager: Kshitij Jha
Production Project Manager: Gomathi Sugumar
Cover Designer: Miles Hitchen

Printed in England by CPI Group (UK) Ltd
Croydon CR0 4YY
United Kingdom

Typeset by MPS Limited, Chennai, India

Working together
to grow libraries in
developing countries
www.elsevier.com • www.bookaid.org

Contents

List of contributors ix
Foreword xi
Acknowledgments xiii

Part I
Setting the stage for healthcare

1. Putting pediatric gender affirming healthcare and research into context

Meredithe McNamara, Joseph T. Whelihan Jr, Jessica M. Bernacki and Juanita K. Hodax

Background and history 4
Sex and gender 4
Gender incongruence and gender dysphoria 4
Historical evolution of pediatric gender-affirming care 4
Structural transphobia and minority stress 5
The provision of pediatric gender-affirming healthcare 6
Standards of care and clinical guidelines 6
Current clinical practice 7
Research 9
Principles of ethical research in transgender health 10
Systematic reviews and guideline development 13
The sociopolitical climate 14
The policy climate in the United States and beyond 14
Misinformation and disinformation 14
Future directions in research and policy 15
Conclusion 15
References 15

2. Gender development and developmental pathways

Alora A. Rando and Esha Vaid

Introduction to gender identity development research 22
Importance of understanding gender identity development 22
Cisnormativity and bias in research 22
Overview of childhood gender development 23
(Trans)gender identity development 25
(Trans)gender developmental milestones 25
Common developmental pathways in transgender and gender diverse children 27
Transactional model of transgender identity development 28
Nonbinary gender identity development 28
Factors influencing gender development 29
Biological factors 29
Social and environmental factors 31
Children as active agents in their development 34
Interaction among biological, psychological, and social processes 34
Summary and future research directions 34
References 35

3. Risk and resilience among transgender and gender diverse youth

Jonathan Poquiz, Afiya Sajwani, Sabrina Gretkierewicz and Lisa M. Buckloh

Risk factors for negative health outcomes 42
Resiliency and protective factors 43
Models of minority stress and resiliency 44
Gender minority stress and resilience 44
Transgender resilience intervention model 46
Multidimensional transgender and nonbinary resilience 46
Summary and future directions for research 47
References 48

4. Overview of healthcare and barriers to access

Jess T. Guerriero and Danielle N. Moyer

Health and healthcare disparities 52
Research strengths and limitations 54

Recommendations for clinical practice and research 55
Access and barriers to pediatric gender-affirming care 55
Research strengths and limitations 58
Recommendations for clinical practice and research 58
Telehealth and electronic consultation 59
Political attacks and legal barriers 60
Summary and recommendations for future research 61
References 62

Part II
Pediatric gender-affirming healthcare

5. Holistic and interdisciplinary healthcare for transgender and gender diverse youth

T. Zachary Huit, Amanda F. Terry, Michaela Patton, Claire Coyne and Diane Chen

Interdisciplinary gender-affirming care 68
Existing gender-affirming care models 69
Initial assessment of care needs 69
Psychosocial care and supportive interventions 69
Medical and surgical interventions 70
Adjunctive programming and collaboration 70
Evidence base for holistic and interdisciplinary care 70
Social interventions 71
Medical and surgical interventions 71
Psychological assessments and interventions 71
Regional and global distinctions and barriers to care 72
Cultural context, inclusivity, and equity in gender-affirming healthcare practices 74
Recommendations for research, practice, and advocacy 74
Tailoring inclusive care approaches 74
Building social supports and community connection 75
Navigating healthcare and addressing care barriers 75
Conclusion 76
References 76

6. Pubertal suppression

Kelsey Eitel, Karine Khatchadourian and Daniel Klink

Pubertal suppression in gender-affirming care 82
Clinical outcomes 83
Efficacy 83
Mental health and psychosocial outcomes 83
Bone health 84
Height and skeletal structure 86
Body composition and cardiometabolic health 87
Brain development and cognition 88
Fertility and fertility preservation 89
Surgical considerations 89
Access, decision-making, and continuation of treatment 90
Summary and future research directions 91
References 92

7. Testosterone and estradiol

Kara J. Connelly and Danielle N. Moyer

Overview of testosterone and estradiol 98
Dosing methods 98
Physical health, body composition, and risk profiles 99
Anthropometric measurement and body mass index 99
Glucose metabolism 99
Blood pressure 100
Lipids 100
Hemoglobin and hematocrit 100
Renal function and serum creatinine 100
Hyperprolactinemia 101
Long-term health risks 101
Mental health, body satisfaction, and well-being 102
Longitudinal research methods 103
Cross-sectional research methods 105
Measurement tools 106
Summary and future research directions 107
References 108

8. Other medical and nonmedical treatments for gender dysphoria

Matthew Goldenberg

Social and legal transition steps 115
Management of the appearance of sex characteristics 115
Chest binding 115

Packing, tucking, and the use of breast and body forms 116
Voice training 117
Hair removal 118
Facilitating facial hair 118
Hormone management 119
Menstrual cycle suppression 119
Progesterone for individuals assigned male at birth 121
Antiandrogens 121
Managing aspects of other interventions 122
Managing dermatologic effects 122
Needle phobia and pill swallowing aversion 122
Conclusion 123
References 123

9. Surgical interventions in adolescence and young adulthood

Sarah K. Fadich, Rebecca Rada and Jessica A. Lee

Community acknowledgment and considerations 130
Overview of gender-affirming surgery 131
Surgical trends among adolescents and young adults 131
Considerations for age and timing of interventions 131
Individualized surgical planning 133
Psychosocial outcomes after gender-affirming surgery 133
Satisfaction and risk of regret 134
Preoperative considerations 134
Postoperative care and support 135
Surgical interventions for transgender women and nonbinary people assigned male at birth 137
Top surgery: breast augmentation 137
Bottom surgery: orchiectomy, vulvoplasty, vaginoplasty 137
Facial gender-affirming surgery: facial feminization surgery 138
Gender-affirming vocal surgery: glottoplasty 139
Gender-affirming body contouring 140
Surgical interventions for transgender men and nonbinary people assigned female at birth 140
Top surgery: mastectomy with chest reconstruction, chest reduction 140
Bottom surgery: metoidioplasty, phalloplasty, hysterectomy, salpingo-oophorectomy 141
Facial gender-affirming surgery: facial masculinization surgery 142
Gender-affirming vocal surgery: vocal fold augmentation, thyroplasty 143
Summary and research recommendations 143
Conclusion 143
References 144

10. Mental healthcare for youth and families

Ashleigh Lin, Helen Morgan, Penelope Strauss, Yael Perry and Blake S. Cavve

Mental health and well-being 150
Gender affirmation as a mental health intervention 151
Future research 152
Psychosocial interventions to improve mental health 153
Research strengths and weaknesses 154
Future research 154
Family systems 155
Mental health and well-being of parents 155
Mental health and well-being of siblings 156
Interventions for families 156
Future research 158
Conclusion 159
References 159

Part III
Additional topics of interest

11. Fertility and fertility preservation

Hayley K. Baines, Leena Nahata, Evelyn Mok-Lin and Norah M. van Mello

Overview of pediatric fertility preservation 170
Ethical considerations 171
Research methods and future directions 172
Knowledge, attitudes, beliefs, and decision-making 172
Decision-making 173
Research methods and future directions 173
Fertility implications before and after gender-affirming medical interventions 173
Animal models 174
Pubertal suppression 174
Spironolactone 174
Menstrual cycle suppression 175
Testosterone 175
Estradiol 175
Research methods and future directions 176
Outcomes of fertility preservation 176

Oocyte preservation 176
Sperm preservation 178
Research methods and future directions 178
Fertility counseling and rates of fertility preservation 179
Research methods and future directions 180
Summary and future research directions 181
References 181

12. Sexual development, health, and functioning

Theodore R. Burnes

Sex-positivity 188
Sexual identity and orientation 189
Sexual functioning 190
Sexual health and well-being 191
Implications and recommendations for research 193
Recommendation 1: center sexuality-related values 193
Recommendation 2: consider unique sexual development trajectories 194
Recommendation 3: promote sexual health and well-being 194
Recommendation 4: diversify research samples 194
Recommendation 5: reconstruct sexual safety 194
Recommendation 6: revise and empirically validate measurement tools 195
Conclusion 195
References 195

13. Long-term satisfaction and risk of regret

Josh Tiller-Ormord

Measuring patient-reported outcomes 202
Psychosocial and mental health outcomes 202
Discontinuation or reversal of medical treatment 203
Detransition 205
Reidentification 206
Satisfaction, regret, and other emotions associated with care 207
Recommendations for future research 209
Longitudinal research 209
Qualitative research 209
Increasing representation 209
Community-based studies 210
Gender as a process, not a destination 210
References 210

14. Conclusion and future research directions

Jonna Finocchio and Natalie Nokoff

Risk, resilience, and access to healthcare 213
Key research gaps 215
Future directions for research 215
Gender identity development 215
Key research gaps 216
Future directions for research 216
Mental health 216
Minority stress as context 217
Caregiver support and acceptance 217
Social, medical, and legal affirmation 218
Intervention research 218
Key research gaps 218
Future directions for research 219
Disentangling the detransition literature 219
Key research gaps 220
Future directions for research 220
Medical interventions 220
Pubertal suppression 221
Key research gaps 221
Future directions for research 222
Testosterone and estradiol 222
Key research gaps 222
Future directions for research 222
Other treatments 223
Key research gaps 223
Future directions for research 223
Surgical interventions 223
Key research gaps 224
Future directions for research 224
Sexual health 224
Key research gaps 225
Future directions for research 225
Reproductive health and fertility preservation 225
Key research gaps 226
Future directions for research 226
Conclusion 226
AI disclosure 227
References 227

Author index 237
Subject index 243

List of contributors

Hayley K. Baines Department of Pediatric Endocrinology, Oregon Health & Science University, Portland, OR, United States

Jessica M. Bernacki Division of General Internal Medicine & Health Services Research, Department of Medicine, David Geffen School of Medicine at UCLA, University of California Los Angeles (UCLA) Medicine-Pediatrics, Los Angeles, CA, United States

Lisa M. Buckloh Nemours Children's Health, Jacksonville, FL, United States; PATHProgram (PediatricAcute Telemental Health) Director of Clinical Training Division of Psychology, Psychology (Mayo Clinic), Jacksonville, FL, United States

Blake S. Cavve The Kids Research Institute Australia, Perth, WAS, Australia

Diane Chen The Potocsnak Family Division of Adolescent and Young Adult Medicine, Ann & Robert H. Lurie Children's Hospital of Chicago, Chicago, IL, United States; Pritzker Department of Psychiatry & Behavioral Health Ann & Robert H. Lurie Children's Hospital of Chicago, Chicago, IL, United States; Department of Psychiatry and Behavioral Sciences, Northwestern University Feinberg School of Medicine, Chicago, IL, United States

Kara J. Connelly Department of Pediatrics, Oregon Health & Science University, Portland, OR, United States

Claire Coyne The Potocsnak Family Division of Adolescent and Young Adult Medicine, Ann & Robert H. Lurie Children's Hospital of Chicago, Chicago, IL, United States; Pritzker Department of Psychiatry & Behavioral Health Ann & Robert H. Lurie Children's Hospital of Chicago, Chicago, IL, United States; Department of Psychiatry and Behavioral Sciences, Northwestern University Feinberg School of Medicine, Chicago, IL, United States

Kelsey Eitel Department of Pediatrics, University of Washington, Seattle, WA, United States

Sarah K. Fadich David Geffen School of Medicine, University of California, Los Angeles, CA, United States

Jonna Finocchio University of Colorado Anschutz, Aurora, CO, United States

Matthew Goldenberg Department of Adolescent Medicine, Seattle Children's Hospital, Seattle, WA, United States

Sabrina Gretkierewicz Nemours Children's Health, Jacksonville, FL, United States

Jess T. Guerriero Doernbecher Gender Clinic, Oregon Health & Science University, Portland, OR, United States

Juanita K. Hodax Pediatric Endocrinology, University of Washington, Seattle, WA, United States

T. Zachary Huit The Potocsnak Family Division of Adolescent and Young Adult Medicine, Ann & Robert H. Lurie Children's Hospital of Chicago, Chicago, IL, United States; Pritzker Department of Psychiatry & Behavioral Health Ann & Robert H. Lurie Children's Hospital of Chicago, Chicago, IL, United States; Department of Psychiatry and Behavioral Sciences, Northwestern University Feinberg School of Medicine, Chicago, IL, United States

Karine Khatchadourian Department of Pediatrics, Children's Hospital of Eastern Ontario and Children's Hospital of Eastern Ontario Research Institute, University of Ottawa, Ottawa, ON, Canada

Daniel Klink ZAS Center for Gender Care, Antwerp, Belgium; Department of Paediatric Endocrinology, Ghent University Hospital, Ghent, Belgium

Jessica A. Lee Children's Hospital Los Angeles, University of Southern California, Los Angeles, CA, United States

Ashleigh Lin Australian Research Centre in Sex, Health and Society, La Trobe University, Melbourne, VIC, Australia; School of Population and Global Health, The University of Western Australia, Perth, WAS, Australia

Meredithe McNamara Yale School of Medicine, New Haven, CT, United States

Evelyn Mok-Lin Department of Obstetrics, Gynecology and Reproductive Sciences, University of California, San Francisco, CA, United States

Helen Morgan The Kids Research Institute Australia, Perth, WAS, Australia

Danielle N. Moyer Department of Pediatrics, Oregon Health & Science University, Portland, OR, United States

Leena Nahata Department of Pediatrics, The Ohio State University College of Medicine, Columbus, OH, United States

Natalie Nokoff University of Colorado Anschutz, Aurora, CO, United States

Michaela Patton Pritzker Department of Psychiatry & Behavioral Health Ann & Robert H. Lurie Children's Hospital of Chicago, Chicago, IL, United States; Division of Endocrinology & Diabetes, Seattle Children's Hospital, Seattle, WA, United States; Embedded Behavioral Health, Division of Psychiatry and Behavioral Medicine, Seattle Children's Hospital, Seattle, WA, United States

Yael Perry The Kids Research Institute Australia, Perth, WAS, Australia

Jonathan Poquiz Department of Family Medicine and Community Health, University of Minnesota Medical School, Minneapolis, MN, United States

Theodore R. Burnes Rossier School of Education, University of Southern California, Los Angeles, CA, United States

Rebecca Rada UCLA Gender Health Program, University of California, Los Angeles, CA, United States

Alora A. Rando Division of Psychology, Department of Pediatrics, Nemours Children's Health, Wilmington, DE, United States

Afiya Sajwani Department of Psychiatry and Behavioral Sciences, Northwestern University Feinberg School of Medicine, Chicago, IL, United States

Penelope Strauss The Kids Research Institute Australia, Perth, WAS, Australia; School of Population and Global Health, The University of Western Australia, Perth, WAS, Australia

Amanda F. Terry The Potocsnak Family Division of Adolescent and Young Adult Medicine, Ann & Robert H. Lurie Children's Hospital of Chicago, Chicago, IL, United States; Pritzker Department of Psychiatry & Behavioral Health Ann & Robert H. Lurie Children's Hospital of Chicago, Chicago, IL, United States; Department of Psychiatry and Behavioral Sciences, Northwestern University Feinberg School of Medicine, Chicago, IL, United States

Josh Tiller-Ormord Randall Children's Hospital, Portland, OR, United States

Esha Vaid Graduate School of Applied and Professional Psychology, Rutgers University, New Brunswick, NJ, United States

Norah M. van Mello Department of Obstetrics & Gynaecology, Reproduction and Development, Amsterdam University Medical Center, Center of Expertise on Gender Dysphoria, Amsterdam, Netherlands

Joseph T. Whelihan Jr Division of Adolescent Medicine, Department of Pediatrics, Children's Hospital of Philadelphia, Philadelphia, PA, United States

Foreword

In 2012, I attended my first World Professional Association for Transgender Health (WPATH) Biennial Symposium in Atlanta, GA. I had seen my first transgender patient in 2007 and, by 2012, had cared for a total of five adult transgender men. I had recently started working with Arlene Lev, Licensed Clinical Social Worker (LCSW-R), and her team of gender-affirming therapists in Albany, which is a 90-minute drive from my office in Oneonta, NY. Arlene had just referred my first transfeminine patient and asked me to present on hormone therapy at the Empire Conference, a social networking event for transgender people, later that year.

"I only have five patients," I said. "I'm no expert."

"You need more training," she replied. "But you're the only doctor I know to ask. Come to the WPATH symposium. We'll be presenting the new Standards of Care, Version 7. You'll learn what you need to know."

While working with Lev, I'd drive between Albany and Oneonta, my town of 15,000, musing on the irony of transgender people from a city traveling to our rural office for care. Transgender people didn't live in my town. Main Street consisted of two blocks of small shops, many shuttered, along with a few social service offerings, none of them LGBTQ-related.

I packed my bags and boarded a plane to Atlanta.

That year, WPATH shared some events with Southern Comfort, a social networking conference attended primarily by older, Southern transgender and gender-diverse (TGD) women. I met women who'd grown up in an era steeped in stigma who desperately wanted to pass. Passing ensured respect, acceptance, employment, and safety in an intolerant world. I met professionals, some transgender and some cisgender—doctors, therapists, psychologists, researchers—as well as the allies who supported the field of transgender health care.

The field of transgender health emerged in the United States after 1952, when Christine Jorgensen returned home following her medical transition in Denmark. Her story introduced the concept of gender mutability to the American public and offered hope to transgender individuals. At that time, most clinicians were taught that transgender people suffered from a mental disorder and that their cross-gender "delusions" should not be tolerated. Jorgensen's story sparked intense cultural clashes across the nation. The lack of clinical research and guidance created a medical vacuum, soon filled with fervent public judgment.

Jorgensen, whose photo frequently graced the front pages of news outlets, was perceived as white, beautiful, heterosexual, healthy, and well-adjusted. Within the dominant American culture, she seemed normal. Her story captured the baffled public's attention. When Harry Benjamin stepped forward to provide medical care for Jorgensen, the two of them forged the path for the field of transgender medicine in the United States. Trans individuals across the United States began seeking medical care in droves, though often from uninformed physicians who deemed them insane. Even Benjamin lacked evidence-based treatment guidelines, but he believed that transgender individuals deserved compassion, respect, and medical gender alignment.

The seventh edition of the WPATH Standards of Care, presented at the 2012 conference in Atlanta, testified to the evolution of health care that aimed to affirm rather than pathologize TGD individuals. This event took place 60 years after Christine Jorgensen's return to the United States. I attended the plenary session on new treatments for transgender adolescents, where the speaker described studies from the Netherlands that used gonadotropin-releasing hormones to prevent irreversible pubertal changes, followed by hormone therapy to induce a puberty aligned with a youth's experienced gender. Dr. Norman Spack introduced these protocols at his new pediatric gender clinic in Boston.

This therapy boggled my mind. I listened agape. Patients would never ask me for this care. This was work for the experts, pediatric endocrinologists in tertiary hospitals, not family doctors in small towns. I doodled in the margins of my program.

The middle-aged transwoman sitting next to me sighed. "I missed this," she said. "I could have looked like myself instead of like this." A tear rolled down her powdered cheek. I compared her masculine features with the feminine face on the slide. I touched her hand. What could I say?

The week I returned home, Lev sent two new referrals—a prepubertal trans girl, Tessa, and a boy, Skylar. "I know it's too early for blockers," she said. "But they may need them soon. These kids have been consistent, persistent, and insistent in their gender identities from a young age. Just meet with the families. Do well-child checks."

"I've never cared for trans youth," I reminded her.

"You know more than anyone else in the region," she said. "And these families can't afford to travel for care."

I followed Tessa and Skylar until puberty hit, when they both experienced crippling dysphoria. We discussed care options, and their families requested pubertal blockers followed by hormone therapy. I reviewed the literature and implemented the appropriate interventions. I watched, breathless, as Tessa's and Skylar's bodies transformed into young adult forms that aligned with their sense of self. Tessa sang soprano in the choir and beamed in her school photo, while Skylar joined the wrestling team and started a Gay-Straight Alliance at school. Their parents relaxed in the chairs in my office, expressing pride in their children's accomplishments rather than fear for their future. Over the following years, our practice became a hub for adults and families seeking gender-affirming care for TGD youth. To my surprise, many patients came from my small town and the surrounding rural area. It was now safe to come out.

I was not alone in this clinical experience. Families began seeking care for their TGD youth nationwide, and clinicians adopted and adapted the Dutch protocols, astonished by the results. However, we all faced questions that we could not answer. Did the blossoming we witnessed in individual youth translate to a larger cohort? Was there a significant improvement in mental health or a decrease in suicidal ideation? Did blockers increase the risk of osteoporosis later in life? Could an 11-year-old make an informed choice about the loss of fertility? What about long-term satisfaction and the risk of regret? The research did not exist.

Since 2012, the field has exploded with studies supporting gender-affirming care for youth. All major medical societies endorse this care, and clinicians, now following evidence-based protocols, have established hundreds of gender clinics for youth nationwide. The existence of this book—a volume summarizing this body of work—demonstrates the progress made in the field of transgender healthcare for adolescents. We now have more answers to questions.

However, we now face new threats. Trump's executive orders undermine care for TGD youth while disregarding the research that demonstrates its effectiveness. As of this writing, 26 states have banned gender-affirming care for TGD adolescents, and the administration threatens further action against clinicians and families in protected states. The words on the pages that follow provide clinical information and strengthen our fight to protect the care we know saves many lives. Please read them. Stand strong. Take care of your patients.

Carolyn Wolf-Gould

Acknowledgments

The team of editors would like to take this opportunity to extend gratitude to a number of people who helped make this project possible. When we first embarked on this project, the state of medical care and research for transgender and gender diverse (TGD) youth was drastically different than it is at the time of publication. TGD youth continue to be incredibly creative and resilient, even as the current sociopolitical climate demands more from them. First and foremost, this book is dedicated to the fierce TGD youth and their families who trust clinicians with their care, participate in research, and stand on the front lines to protect themselves and their loved ones. Without you, this book would not be possible.

We would also like to thank our contributors. These chapters were a labor of love from some of the most passionate and compassionate minds in the field. As care and research have been impacted internationally, this group worked tirelessly to ensure that we had a product that will preserve existing research and serve to advocate for continued advances in the field. We know that many of you are early in your careers, clinician-researchers, or working in geographical areas that have fewer protections for care, and that involvement in this project added to an already heavy workload. Thank you for lending your time and expertise and for your determination and persistence throughout this process.

Additionally, care provision and research for TGD youth would not be possible without the countless medical practitioners, behavioral health clinicians, and researchers working behind the scenes. Your work is not unnoticed, and we hope that you can celebrate this publication as a reflection of your passion and dedication to this field.

Finally, we would like to thank our friends and families who have been the cheerleaders for this project. Your support and encouragement made it so that we could persevere in the face of growing global challenges, and your excitement to see the project come to fruition continues to remind us of the importance of this work.

Part I

Setting the stage for healthcare

Chapter 1

Putting pediatric gender affirming healthcare and research into context

Meredithe McNamara[1], Joseph T. Whelihan Jr[2], Jessica M. Bernacki[3], and Juanita K. Hodax[4]
[1]*Yale School of Medicine, New Haven, CT, United States,* [2]*Division of Adolescent Medicine, Department of Pediatrics, Children's Hospital of Philadelphia, Philadelphia, PA, United States,* [3]*Division of General Internal Medicine & Health Services Research, Department of Medicine, David Geffen School of Medicine at UCLA, University of California Los Angeles (UCLA) Medicine-Pediatrics, Los Angeles, CA, United States,* [4]*Pediatric Endocrinology, University of Washington, Seattle, WA, United States*

Learning objectives

- Review the history of transgender and gender diverse youth and their healthcare.
- Describe the role of structural transphobia and minority stress on the lives and healthcare of transgender youth.
- Summarize modern advancements in research on pediatric gender-affirming care and review common research strategies.
- Introduce recent sociopolitical developments that will profoundly impact research and clinical care for transgender young people into the future.

Key terms and definitions

- **Transgender and gender diverse** is an umbrella term used to describe individuals whose gender identity does not align with the sex they were assigned at birth. Transgender is sometimes used interchangeably with this term. Gender diverse, used on its own, typically refers to diverse identities that do not fall into binary transgender categories.
- **Sex assigned at birth** refers to the designation of male or female by a person of authority at the time of birth, also sometimes referred to as sex designated at birth, sex assumed at birth, or birth registered sex, depending on context and geographical region.
- **Gender identity** is a social and legal concept that includes norms, expectations, and behaviors that exist on a continuum. Gender identity refers to an internal sense of gender, which may or may not align with the sex assigned at birth.
- **Pediatric gender-affirming care** is an individualized, holistic, and developmentally informed framework of care for transgender and gender diverse youth. Components of this care include social and legal affirmation, behavioral healthcare, support and resources, and medical care, among other individually tailored interventions.

People of all ages may experience a difference between their sex assigned at birth and their gender identity. The term "transgender and gender diverse" (TGD) may describe people with this experience. TGD people benefit from compassionate, multidisciplinary care delivered by clinicians who are informed in matters of gender diversity due to the unique health needs of this population. This care, inclusive of medical interventions, nonmedical options, and mental health supports, is often termed "gender-affirming care." It is guided by internationally recognized standards that are in turn informed by a continually growing body of evidence generated from research that is inclusive of diverse voices, and uses a range of recruitment, measurement, and analytic strategies (Hembree et al., 2017; Coleman et al., 2022).

Advances in Research on Pediatric Gender Affirming Healthcare. DOI: https://doi.org/10.1016/B978-0-443-29019-0.00009-9
© 2026 Elsevier Inc. All rights are reserved, including those for text and data mining, AI training, and similar technologies.

The clinical care and its supporting research have developed within a historical context of discrimination, disinformation, and structural transphobia that continue to impose limitations on this field. Despite these challenges the field of pediatric gender-affirming care has evolved to offer holistic, interdisciplinary models, grounded in evidence, that allow young people to thrive. Pediatric clinicians and researchers are well-positioned to continue building on research advancements to overcome ongoing challenges and continue to improve the field of gender-affirming care for TGD young people.

Background and history

Sex and gender

Sex is a biological construct that encompasses chromosomes, endogenous hormone profiles, and physical traits, including internal and external reproductive anatomy. Sex assignment typically occurs at birth and is based on visual inspection of external genitalia, often informed by available prenatal testing, such as cell-free DNA screening.

Gender identity refers to an individual's internal sense of self as it relates to gender and is shaped by an interplay of biological, psychological, and social factors (Karamanis et al., 2022). Awareness and expression of gender identity are influenced by social and cultural contexts, including societal norms, stigma toward nonbinary identities, and cultural recognition of genders outside the binary framework (Turban and Ehrensaft, 2018). Gender expression describes the outward presentation of gender through behavior, clothing, and appearance, which is communicated by the individual and interpreted by others.

Gender incongruence and gender dysphoria

In the *DSM-5-TR* from the American Psychiatric Association, gender dysphoria is defined as a marked and persistent incongruence between an individual's experienced or expressed gender and the sex assigned at birth that is accompanied by clinically significant distress or impairment in social, occupational, or other important areas of functioning. This diagnosis is explicitly focused on distress, rather than identity. The *International Classification of Diseases-11* from the World Health Organization uses the term gender incongruence, classifying it within conditions related to sexual health rather than mental disorders, and defines it as a marked and persistent incongruence between experienced gender and assigned sex, without requiring distress or dysfunction (Harrison et al., 2021). This reflects a shift to reduce pathologization while still articulating the importance of access to care.

Historical evolution of pediatric gender-affirming care

Transgender youth have sought and received medically affirming interventions since at least the 1930s (Giles-Petersen, 2018). Gender-affirming hormones were first officially provided to youth in healthcare settings in the 1960s, on an extremely selective basis (Newman, 1970). More often, youth were subjected to conversion practices that attempted to force gender identity to match the norms expected of sex assigned at birth (Bradley and Zucker, 1990). These practices have been shown by empirical evidence to be both ineffective and harmful (Craig et al., 2017; Green et al., 2020).

Building off care for transgender adults, physicians in the Netherlands introduced a protocol for youth in the 1980s, commonly known as the Dutch Protocol (de Vries and Cohen-Kettenis, 2012; Gooren and Delemarre-van de Waal, 1996). The Center for Expertise in Gender Dysphoria in the Netherlands became an early clinical research hub, and findings from this group showed improvements in depression, anxiety, and suicidality following medical treatment of gender dysphoria (e.g., puberty blockers and hormones) (de Vries et al., 2014; De Vries et al., 2011). The success of the Dutch Protocol led other clinicians to open up similar dedicated clinics. The first known multidisciplinary gender care clinic for adolescents opened in 2007 in Boston (Edwards-Leeper and Spack, 2012). Since then, multidisciplinary gender clinics have emerged in many US states and other countries.

The gender-affirmative model

The progression in approaches to caring for TGD youth has been influenced by an evolution in understanding of gender identity, gender dysphoria, and gender incongruence, and the context of minority stress and structural transphobia. The gender-affirmative model acknowledges the risk of interfering with the authentic expression of gender identity and recommends that youth be allowed to live as the gender that most aligns with who they are, without restriction, rejection, or punishment (Hidalgo et al., 2013). The gender-affirmative model is grounded in three key facts: (1) gender diversity is not a psychological disorder, and that cooccurring pathology is often the result of external stressors; (2) gender identity development involves a complex interaction of biological, social, and cultural factors; and (3) gender presentations vary across cultures, and identity may be fluid, and not necessarily binary or linear.

The gender-affirmative model is grounded in established principles of pediatric and adolescent medicine, mental healthcare, and bioethics, including individualized assessment, developmentally appropriate care, and shared decision-making. Its clinical components—such as careful psychosocial support, treatment of mental health conditions, and, when indicated, stepwise medical interventions—are supported by a growing body of observational studies, systematic reviews, and clinical consensus statements showing associations with improved psychosocial functioning and reduced distress for many TGD youth. The model emphasizes ongoing evaluation of gender embodiment goals, discussions about whether a more reversible intervention can meet goals, recognition that treatment plans can be adjusted to address evolutions in gender identity or expression, and alignment with existing standards for managing other chronic or stigmatized conditions in pediatric care.

Structural transphobia and minority stress

Structural transphobia refers to the systemic devaluation of TGD people through laws, policies, and cultural norms that create and reinforce barriers to inclusion and equality. Within the minority stress framework, these structural conditions give rise to chronic, identity-specific stressors—such as anticipation of rejection, identity concealment for safety, exposure to stigma and threats of violence, institutional invalidation, and negative dominant cultural narratives—that are repeatedly encountered across domains of life. Literature has explored the specific role of minority stress for TGD youth, demonstrating that it occurs at the individual, interpersonal, and structural level (Hendricks and Testa, 2012; Testa et al., 2015).

At the individual level, individuals may experience internalized transphobia, reaction sensitivity, and concealment. Internalized transphobia refers to the internalization of negative societal attitudes toward TGD people. Reaction sensitivity refers to the psychological process by which some individuals develop anticipatory anxiety around interpersonal rejection due to prior lived or witnessed experiences of antitransgender bias. Concealment refers to efforts to hide a gender minority status to evade interpersonal and structural biases (Hatzenbuehler and Pachankis, 2016). Individuals may also experience interpersonal forms of minority stress, including prejudice (a biased emotion or evaluation), discrimination (an action that advantages or disadvantages), or stereotyping (a belief that characterizes TGD people). These manifest as bullying and other forms of victimization, social (including parental) rejection, abuse, and microaggressions (Hatzenbuehler and Pachankis, 2016).

As structural transphobia and minority stress mutually reinforce each other, external harms can become internalized through experiences such as involuntary disclosure of gender identity, loss of healthcare, employment discrimination, cancellation of affirming cultural events, bullying, and harassment. This sustained and cumulative stress burden is associated with higher rates of anxiety, depression, suicidality, social isolation, and barriers to healthcare among TGD people (Hunter et al., 2021). Critically, the minority stress model demonstrates that these health disparities are socially produced rather than inherent to transgender identity itself.

Globally, examples of structural transphobia include criminalization of gender expression, denial of legal gender recognition, barriers to gender-congruent identity documents, exclusion from health systems, and restrictions on education, employment, housing, and asylum protections. In countries that do not criminalize authentic gender expression, administrative requirements—such as mandatory medical interventions and psychiatric diagnoses—can function as structural barriers that erode autonomy and human rights. These systems arbitrate whose identities are considered legitimate and worthy of protection.

In the United States, TGD people of all ages also face extensive housing and workplace discrimination, contributing to disproportionate rates of homelessness, unemployment, and poverty (Kattari et al., 2015). TGD people are four times more likely to experience violent victimization, and over-policing and incarceration are also common (Flores et al., 2021). Concurrent with rising anti-lesbian, gay, bisexual, transgender, or queer (LGBTQ) legislation, TGD youth in the United States now report more suicidal ideation and more frequent suicide attempts (Lee et al., 2024). Of the 1.4 million youth experiencing homelessness in the United States, 40% identify as LGBTQ, and TGD youth are over-represented in foster care (Baams et al., 2019).

While modern clinical practice has improved and now predominantly takes a gender-affirmative approach to care, systemic barriers to accessing pediatric gender-affirming healthcare remain. Wait times are increasingly long, likely due to increasing referrals and patient volume. Some studies have shown wait times of 9 to 10 months for an initial hormone appointment (Tordoff et al., 2023; Lawson et al., 2024). Longer wait times are associated with self-harm and suicidal thoughts, and disproportionately impact youth from families with lower income levels, Medicaid insurance, and a lack of family support (Tordoff et al., 2023; Lawson et al., 2024). Models for responding to these long wait times are in development (Russell et al., 2022; Dahlgren Allen et al., 2021).

Due to several restrictive state-level policies in the United States, gender-affirming care for youth is less available in large swaths of the country (Borah et al., 2023). In addition to the barriers of finding a gender clinic or gender-affirming primary care provider, waiting for and traveling to an appointment, and insurance coverage are barriers to care for many families (Gridley et al., 2016). Many insurance plans do not provide comprehensive coverage of gender-affirming care, and some states are intentionally excluding coverage with new legislation. Without insurance coverage, many of these costly medications are not feasible for most families to afford out of pocket. Structural transphobia, minority stress, historical pathologizing of gender identity, and current barriers to accessing high-quality, affirming care all form the context in which clinical care and modern clinical research occur, and these systemic issues should be taken into account when providing care or interpreting the results of research in this area. The ongoing development of evidence-based care for TGD youth should include adaptations of evidence-based practices that consider the impact of minority stress (Coyne et al., 2023; Delozier et al., 2020).

Mitigators of structural transphobia and minority stress

Viewing the well-being of TGD people through a structural lens underscores that disparities are not inherent to transgender identities themselves. Rather, these disparities are produced by social, legal, and political systems—and are therefore amenable to change through multilateral evidence-informed policy that fosters environmental affirmation. Minority stress is measurable and clinically actionable. Reducing stigma while strengthening resilience and affirmation is central to improving health outcomes for TGD youth. Environments that affirm transgender identities through legal recognition, nondiscrimination protections, and access to evidence-based healthcare can reduce minority stress and improve health outcomes.

Studies examining access to accurate identity documents (e.g., name and gender marker changes) consistently find associations with improved mental health, reduced psychological distress, and greater safety in daily life. A large US cohort study found that implementation of state-level health insurance nondiscrimination policies (protecting gender identity) was associated with significant decreases in suicidality among TGD individuals in the years following policy enactment, suggesting that such protections may reduce gender minority stress and improve mental health outcomes (McDowell et al., 2020). Legal affirmation reduces misgendering, harassment, and institutional barriers that contribute to stress and avoidance of services.

Population-level studies show that jurisdictions with explicit protections for TGD people in healthcare, employment, and public accommodations are associated with better self-reported health outcomes and reduced disparities. Comparative research demonstrates that TGD people living in more affirming legal and policy environments report better health and well-being than those in more restrictive contexts.

There is also a growing body of research demonstrating the positive impacts of affirmation for youth. For instance, measurable improvements in mental health are observed when a TGD young person's chosen name is used, as well as gender-appropriate pronouns (Russell et al., 2018). Youth affirmed in their educational environments are less likely to miss school and be targeted negatively by peers (Kosciw et al., 2014), and youth allowed to participate in extracurricular activities with the peers of their choosing demonstrate better mental health outcomes (LaRocca et al., 2023).

The provision of pediatric gender-affirming healthcare

Multidisciplinary gender clinics are typically comprised of medical providers (pediatric endocrinologists, adolescent medicine specialists, or primary care providers) and mental health providers (psychologists or social workers). Some teams may also include care navigators, nurses, speech pathologists, surgeons, and gynecologists. Because there are many disciplines relevant to gender-affirming care, there is often care coordination and communication between specialty gender clinics and other health systems. Primary care providers who have specific training and experience with gender healthcare play an important role in decreasing barriers to access and improving overall healthcare for TGD communities (Wylie et al., 2016). Critically, primary care providers can provide support for social transition and menstrual suppression, as well as overall trauma-informed and culturally appropriate healthcare for TGD young people (Hodax et al., 2023).

Standards of care and clinical guidelines

The two prevailing guidelines for gender-affirming medical care are published by the World Professional Association for Transgender Health (WPATH) (Coleman et al., 2022) and the Endocrine Society (Hembree et al., 2017). WPATH, which is the leading expert medical association in the field of transgender health, issued the first guidelines for pediatric and adolescent care in 1998. The most recent WPATH guidelines are the Standards of Care, Version 8 (SOC8), published in 2022.

WPATH is an international, multidisciplinary association whose mission is to "promote evidence-based care, research, education, public policy, and respect in transgender health" (Coleman et al., 2022). The WPATH SOC8 covers a comprehensive landscape of gender-affirming care throughout the lifespan and across diverse identities. In addition to recommendations for medical and surgical care the SOC8 provides guidance on sexual and reproductive healthcare, voice, mental health, education, and institutional environments, among others.

The SOC8 was developed by a multidisciplinary team of clinicians and researchers who conducted systematic reviews to inform the drafting of the recommendation statements. The Delphi process was followed for approval of recommended statements, and each statement required at least 75% approval by the SOC8 committee members. The chapter authors then graded each statement using Grading of Recommendations, Assessment, Development, and Evaluation (GRADE). Before publication the SOC8 draft was distributed to WPATH members for comments, which were then reviewed and incorporated into the final version.

The Endocrine Society guidelines were developed by nine subject matter experts selected by the Endocrine Society leadership and published in 2017 (Hembree et al., 2017). Unlike the WPATH SOC8, which are broad in scope, the Endocrine Society guidelines are more focused on the diagnosis of gender dysphoria, defining gender-affirming medical care and offering medical guidance on its provision. The guidelines offer criteria for starting gender-affirming medical care in adolescents and adults and provide details about the benefits and risks of pubertal suppression and gender-affirming hormones. These guidelines provide recommendations for dosing of gender-affirming hormones, including guidance on puberty induction with hormones in adolescents.

In addition to these clinical care guidelines, several organizations have published position statements in support of gender-affirming care. These position statements do not provide specific guidelines for medical care but affirm the importance of gender-affirming care as a treatment for gender dysphoria and support the existing research findings demonstrating long-term benefits of access to gender-affirming care. One of the most well-known position statements for gender-affirming care in adolescents was published by the American Academy of Pediatrics (AAP) in 2018 and reaffirmed in 2023 (Rafferty et al., 2018). The AAP Position Statement outlines the social, legal, medical, mental health, and surgical aspects of gender-affirming care, the health disparities TGD youth face, and the value of family and community support.

Current clinical practice

Gender-affirming care for youth is best understood as an individualized, developmentally informed framework rather than a single intervention. See Fig. 1.1 for a visual metaphor describing pediatric gender-affirming care. This section introduces the core components of youth gender care—social and legal affirmation, behavioral healthcare, informed consent and assent, and medical care—highlighting how each element plays a distinct yet interconnected role in supporting young

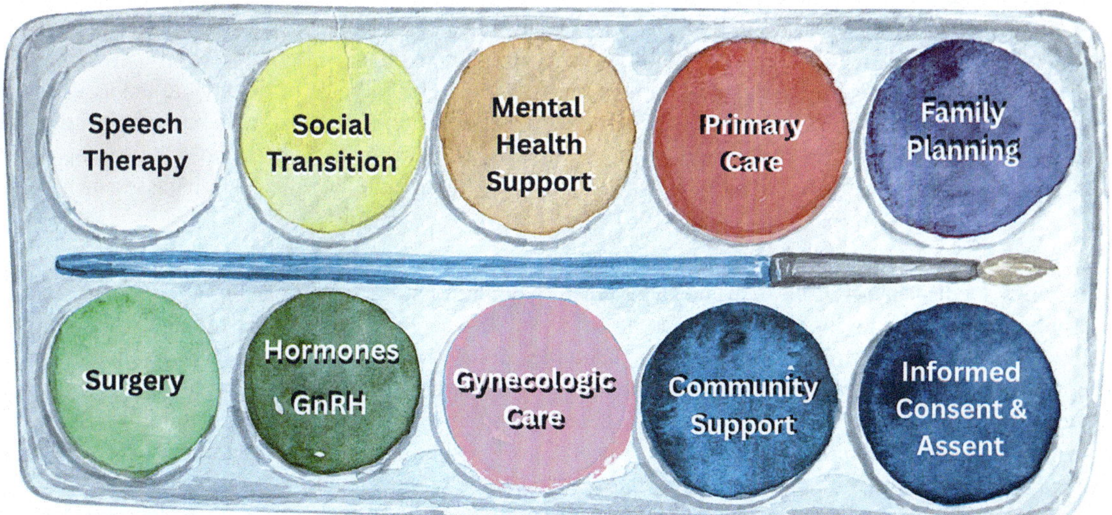

FIGURE 1.1 **Gender-affirming care: a visual metaphor.** A schematic describing pediatric gender-affirming care.

people's health and well-being. Together, these domains reflect an approach that centers the lived experiences of youth, family, and community contexts, and ethical clinical practice while remaining responsive to ongoing social and legal challenges.

Social and legal affirmation

For many patients, changes in gender expression are an important step in gender affirmation and may be more easily accessible than medical care. Social gender affirmation may include changing name and pronouns, hairstyle, and clothing choices, speech therapy to change voice pitch and communication, using a chest binder to flatten chest tissue, or tucking to create a smooth contour in the genital region (Rafferty et al., 2018). Social gender affirmation can be a helpful step in exploring gender identity. However, some individuals do not feel comfortable making changes in gender expression until they have had some physical changes from gender-affirming medical care to align their physical appearance with their gender identity. Historically, in many domains, social gender affirmation was a requirement before starting gender-affirming hormones or having gender-affirming surgery. Contemporary guidelines on gender-affirming care describe the importance of an individualized approach that centers the patient's values, physical safety, and preferences (Hembree et al., 2017; Coleman et al., 2022).

Legal gender affirmation involves changing one's name and/or gender marker on official documents to reflect gender identity. For many individuals, this can help to decrease confusion or the need for explanation when the gender marker on identification documents does not match gender expression or appearance. It can also decrease misgendering or being called the incorrect name at places like school or work, potentially leading to increased feelings of safety in public. Some challenges to legal affirmation include cost, administrative burdens, and, in some regions, legal restrictions to doing so.

Behavioral healthcare

Behavioral health providers experienced in supporting TGD youth can serve many purposes, including offering biopsychosocial assessment to facilitate medical care, parent and caregiver support, and gender-affirmative therapy to support gender identity exploration, address the impacts of minority stress and stigma, and provide evidence-based interventions for specific behavioral health concerns (Buckloh et al., 2022; Cousino et al., 2014; Spencer et al., 2021). Youth presenting to multidisciplinary gender clinics commonly interact with behavioral health providers, though to varying degrees (Chen et al., 2016; Cousino et al., 2014; Coyne et al., 2023). Many programs require youth to meet with a behavioral health provider as part of the intake process, including a biopsychosocial assessment. Other programs may not require a behavioral health provider to meet with a patient if the patient has an established relationship with a community-based provider experienced in working with TGD youth. Clinical consultation with this provider may also take place as part of establishing care with a gender program.

The role of behavioral health providers in gender-affirming care models has shifted over time, with current models seeking to balance providing optimal support and minimizing unnecessary barriers to care that have been historically discussed as gatekeeping efforts (Inwards-Breland et al., 2021). Whereas TGD youth were previously required to undergo an explicit course of therapy before transition, the current WPATH SOC8 acknowledges that this may not be indicated for all youth. Current guidelines recommend that youth undergo a multidisciplinary biopsychosocial assessment before the initiation of medical care to establish a diagnosis of gender dysphoria, clarify goals of treatment, and assess capacity for informed assent or consent. The SOC8 indicates that any experienced clinician, medical, or behavioral health professional can conduct this assessment, while acknowledging that behavioral health providers often have the appropriate training, time to spend with families, and previously established rapport to most effectively provide this assessment.

Informed consent and assent

Gender-affirming medical interventions are indicated when they align with a young person's gender embodiment goals and the young person is able to give informed consent, or assent with parental consent (depending on age and geographic region). Informed assent and consent form a process that varies in length and structure, based on individual development and understanding, clarity of gender embodiment goals, family dynamics, and overall individualized care.

Informed consent includes discussion of the risks, benefits, and alternatives of an intervention, the likely impact on physical functioning and psychological adjustment, expectations for timing and effectiveness, methods of administration, dosing and monitoring, considerations for fertility and reproductive choices, and expectations for satisfaction and the possibility of regret. It is also beneficial to discuss the possibility that goals of care may evolve over time. Existing research shows that most adolescents have the capacity to make informed decisions and provide consent for gender-affirming medical care (Clark and Virani, 2021; Ravindranath et al., 2024). While the neurocognitive processes involved in decision-making strengthen during adolescence and early adulthood, the capacity to access these skills often relies on contextual factors that facilitate or hinder thorough reasoning. Medical decision-making occurs most often within a context that promotes slowly paced, mature decision-making (Steinberg, 2013).

According to one study that compared clinician assessment with the standardized MacArthur Competence Assessment Tool for Treatment, approximately 90% of adolescents aged 10–18 years old demonstrated competence to provide informed consent on their own (Vrouenraets et al., 2021). While this research suggests that adolescents are able to provide informed consent, most often they do so with the support of their parents, who also contribute to thoughtful decision-making. Qualitative research of TGD adolescents, their parents, and healthcare providers further supports the ability these young people have to meaningfully participate in the informed consent process and to make thoughtful decisions, particularly with the support of accepting parents (Daley et al., 2019; Clark and Virani, 2021).

Shared decision-making involves the combined input of clinicians who bring their professional experience and clinical evidence, patients who bring their own values and treatment preferences, and parents and caregivers who bring their values as well as their care and concern for their child. The four primary components of shared decision-making are as follows: (1) clarifying the decision and the relevant individuals involved; (2) describing the options, including doing nothing, as well as the potential benefits and risks; (3) eliciting the values, beliefs, and attitudes of those involved; and (4) making a clear decision and identifying unresolved decisional needs (Stiggelbout et al., 2015).

Shared decision-making can be achieved without formal tools, although many formal tools have been created for various medical conditions and treatments to facilitate a standardized approach. To date, one formal decision aid tool has been developed specifically for pediatric gender-affirming care (Salvetti et al., 2025). The tool was designed based on professional standards, expert consensus, and a youth advisory board, and it was reviewed and refined by a multidisciplinary expert panel. It offers key terms and definitions, sexual and reproductive health information, medication options (including no treatment), and community resources, such as support groups, voice therapists, and fertility specialists.

Medical and mental health supports

Gender-affirming care for youth is a developmentally tailored, individualized, evidence-based approach that supports a young person's physical, psychological, and social well-being by affirming their gender identity. Care often begins with social transition, which may include use of a chosen name and pronouns, gender-affirming clothing or hairstyles, and support within family, school, and peer environments. These interventions are fully reversible and are commonly associated with improvements in gender dysphoria and psychological distress (Durwood et al., 2017; Olson et al., 2015). The majority of prepubertal children who socially transition, when indicated, experience persistence of their gender identity into adolescence and beyond and demonstrate favorable mental health outcomes (Olson et al., 2016, 2022).

Medical interventions do not occur before pubertal onset, and only if desired and indicated for individual patients. For TGD adolescents who experience distress with pubertal development (sexual maturity rating ≥ 2), pubertal suppression—most commonly using gonadotropin-releasing hormone agonists (GnRHa)—may be considered following a comprehensive multidisciplinary assessment and informed consent process involving legal guardians. GnRHa pauses pubertal progression and is fully reversible, with typical pubertal development resuming if treatment is discontinued. These medications may be administered via long-acting implants or periodic injections. When GnRHa is unavailable or not clinically indicated, alternative medications may be used to suppress menses or mitigate androgen effects.

With ongoing assessment and informed consent, gender-affirming hormones such as estrogen or testosterone may be initiated in older adolescents who desire the development of secondary sex characteristics aligned with their gender identity. Timely access to gender-affirming care during puberty may reduce the later need for surgical interventions to address body incongruence. In rare cases, selected older adolescents may benefit from gender-affirming surgical interventions (Dai et al., 2024). Across all stages, care is guided by international, evidence-based clinical guidelines and emphasizes individualized decision-making, caregiver involvement, and attention to the broader social context, recognizing that psychosocial support is foundational to youth well-being.

Research

Gender dysphoria and gender incongruence are researched using a range of methods, including structured clinical assessments aligned with diagnostic criteria, validated self-report measures of dysphoria and body-related distress, and longitudinal designs that examine developmental trajectories, mental health outcomes, and responses to social and medical interventions. Contemporary research distinguishes between gender diversity as a normal aspect of human variation and dysphoria or incongruence as context-dependent experiences shaped by developmental, biological, and social factors.

Early research related to gender diversity in youth was conducted largely within the context of gender identity change efforts (i.e., conversion therapy) and as a result focused heavily on the permanence or transience of gender identity, expression, and behaviors that did not align with what is stereotypically expected based on sex assigned at birth. As understanding and acceptance of gender diversity has progressed, contemporary research has shifted focus toward ethical research questions and less biased methodology. Advances in research have led to a better understanding of the population of TGD youth, factors that contribute to gender identity development, risk factors for negative health outcomes, and resilience. Research on pediatric gender-affirming care specifically has significantly grown over the last decade and has provided a wealth of information about barriers to accessing care, interest and desire for different aspects of gender-affirming care, the long-term mental health benefits of gender-affirming interventions, and youth and parent satisfaction with care. This body of research serves to inform and improve clinical practice and to establish a foundation of knowledge on which future research can build.

Principles of ethical research in transgender health

Historically, research in transgender health has not consistently used best practices, as gender diversity and having a transgender identity were once considered pathological. Even now, some research in this field represents TGD people as a homogenous population with minimal variability in gender identity, ethnicity, and sexuality. A bioethically sound research lens considers the multidimensionality and nuance of this population, with participation in research being confidential and not required for receipt of care.

Those who care for transgender people and conduct research on their health should assess their own positionality (i.e., how social and cultural identities, experiences, and power dynamics influence perspectives and behaviors). It should also account for intersectionality, meaning how diverse aspects of people's identities and lives intersect and affect their health outcomes. There is an inherent power imbalance between researchers and clinicians and the TGD people they study and care for, which can impact trust and communication. These power imbalances can be addressed by creating inclusive, community-based approaches to research where TGD peoples' values, preferences, and priorities are centered.

TGD youth have described their recommendations for researchers, including accepting diverse gender experiences, acknowledging the spectrum and fluidity of gender, reflecting on bias, avoiding pathologizing of gender, committing to learning more about gender, and engaging TGD youth in research when it is about them (Dixon et al., 2022). These results emphasize the importance of grounding research in a meaningful collaboration with community stakeholders and including TGD individuals in the design and execution of research studies. Best practices in community-based research include acknowledging and defining the community of TGD youth and families to be studied, engaging them in all aspects of the research process, addressing power dynamics and any tensions related to data collection tools, developing safety plans to address suicidality disclosures, and differentiating the role of academics as researchers versus advocates (Katz-Wise et al., 2019).

Epidemiology of gender diversity

Evaluating the naturalistic incidence of gender diversity in youth and observing gender in youth are essential aspects of research in youth gender care. Health systems, schools, and social safety net services must have a robust understanding of the epidemiology of TGD youth and their demographic characteristics to better support them. This information is also needed to close intersectionality-driven disparities, because TGD people of color are far more likely to experience violence and socioeconomic disadvantage. Population estimates of TGD youth are difficult to ascertain and dependent on the setting of data collection, the methods used to ask about gender diversity, and whether answers could be influenced by parental support or involvement. The number of young people who identify as TGD has primarily been assessed through either school-based self-report surveys or health record data, largely collected in North America and Western Europe (Coleman et al., 2022).

The Centers for Disease Control use the Youth Risk Behavior Survey (YRBS) to assess various aspects of adolescent and young adult populations in the United States. Based on the 2017 and 2019 versions of the YRBS the Williams Institute estimates that 1.4% or approximately 300,000 youth between the ages of 13 and 17 years old identify as transgender (Herman et al., 2022). According to the 2023 YRBS, 3.3% of US high school students identified as transgender and 2.2% identified as questioning their gender identity (Suarez et al., 2024), an increase from 1.8% and 1.6%, respectively, in 2019 when the gender identity question was optional (Miller-Jacobs et al., 2023). Using a modified version of the YRBS that employed a two-step method of assessing gender identity (i.e., "what was your sex assigned at birth" and "what best describes your gender identity?"), 9.2% of high schoolers in an urban US setting identified as TGD in 2019 (Kidd et al., 2021). When measured this way, TGD youth are slightly less likely to report being white compared to the overall US population, and the racial/ethnic distribution appears to be otherwise quite similar to population estimates (Herman et al., 2022).

One comparative study found that 0.9% of patients in health system data had a gender dysphoria diagnosis compared to 1.9% of youth who responded to the YRBS (Kahn et al., 2024). The difference was more pronounced for individuals who had a male sex in their health record and for those who identified as Asian, Black/African American, and Hispanic/Latino. These studies suggest that youth with gender dysphoria, and particularly non-white youth, are underrepresented in health system data. While some TGD youth may not want medical interventions, these data suggest that many TGD youth, and particularly those of intersecting identities, may not be accessing health systems or providers to understand gender-affirming medical options.

Over the past decade an increasing number of TGD youth have presented to health settings to learn about medical and nonmedical options for supporting their gender identity and expression. Several factors likely contribute to higher numbers of TGD youth presenting for care. It is important to note that the positive developments in improved access to care over the past several decades are also accompanied by more recent political attacks and limitations to care. Increased awareness and visibility of transgender people, via public figures and media representation, have likely made it easier for people to understand and articulate gender identity earlier in life. This increased awareness also impacts parents, educators, and healthcare providers, who may work to create more supportive environments for TGD youth. Educational campaigns and resources from LGBTQ organizations have improved public knowledge about TGD identities and available healthcare options.

Evaluating the impacts of gender-affirming medical treatments

Gender-affirming medical treatments, including GnRHa and hormone therapy, have long been used for other medical indications. Puberty blockers are a well-known treatment for precocious puberty in youth and for prostate cancer in adults, and the safety and efficacy have been studied at length for those indications. Testosterone and estrogen are both used regularly in adolescents and adults for delayed puberty and for the treatment of hypogonadism. As a result, foundational knowledge of these treatments came from research in these other areas, including expected changes, safety, and potential risks. In the last several decades, research on these treatments specifically in the context of gender-affirming care for youth has grown.

Outcome data on pediatric gender-affirming care largely come from observational studies that either follow TGD youth over time as they receive specific aspects of care or cross-sectional comparisons of groups of youth who have and have not received an intervention. As most research has emerged from clinics providing specialty care, retrospective chart review has also been used to supplement research methodology when a prospective research design is not possible. The earliest research data on the outcomes following pediatric gender-affirming care were produced from programs with embedded behavioral healthcare and protracted assessment periods (De Vries et al., 2011). However, as research in this area has expanded, results have remained consistent across studies from programs with a variety of models and practices. Large, multicenter, prospective, longitudinal cohort studies are ongoing, including TransYouth Care in the United States (Olson-Kennedy et al., 2019), Trans Youth CAN! in Canada (Lawson et al., 2024) and the European Network for the Investigation of Gender Incongruence (ENIGI) (Boogers et al., 2024).

While recruitment and observation of TGD adolescents receiving care in a specialty clinic reflect the most common methodology for studying gender-affirming care, there may be benefits to community-based studies and recruitment methods. For example, a community-based study may be more generalizable to a variety of care provision models if study participants are receiving care across different clinical sites. One US pilot study demonstrated the feasibility of using social media as a recruitment strategy, emphasizing the potential need to target subgroups in recruitment and retention strategies to maximize representation across diverse identities (Holt et al., 2024). There is one large prospective cohort study in the United States that used a community-based sample, the Trans Youth Project (TYP) (Olson et al., 2015). While the initial goals of the TYP were not specifically related to gender-affirming medical care, the cohort that was recruited is now beginning to access care, and therefore outcomes can be evaluated.

Measurement and outcomes

Research outcomes in pediatric gender-affirming care typically pertain to (1) the experience of distress before or in the face of barriers to accessing treatment, (2) body changes occurring in a linear, often binary, direction, (3) improvements in distress following treatment, and (4) satisfaction with treatment (Oosthoek et al., 2024).

Mental health outcomes, including depression, anxiety, and suicidality, are often studied to assess the impact of gender-affirming care. These are typically measured using standardized assessment tools, with some primarily used in clinical settings and others used more in prospective standardized research. Brief screening measures such as the Patient Health Questionnaire-9 (PHQ-9) for depression (Kroenke et al., 2001) and the Generalized Anxiety Disorder-7 (GAD-7) for anxiety (Spitzer et al., 2006) are often used clinically, and these data are therefore more widely available in clinical settings and for retrospective research. Other studies may use large composite measures of mental health, such as the Youth Self-Report (Achenbach, 2009), or mental health proxy measures, such as diagnoses, medications, and treatment utilization. More recently, there have also been shifts toward measuring positive mental health outcomes, such as quality of life, subjective well-being, and psychosocial functioning. Improvements in mental health appear to occur when youth with gender dysphoria feel better about their bodies (Chen et al., 2023).

Given the increased risk of suicidal ideation and self-harm among TGD youth, this is also often used as a benchmark for improved mental health and well-being. Some measures, such as question #9 in the PHQ-9, indirectly assess for suicidal ideation, asking if a patient has had "*thoughts that you would be better off dead, or of hurting yourself*" in the last 2 weeks (Kroenke et al., 2001). There are also measures such as the Ask Suicide-Screening Questions (ASQs) that ask direct questions about suicidal ideation (Horowitz et al., 2012).

The implicit goals of gender-affirming care are to relieve gender dysphoria and to achieve appearance congruence, or the alignment between physical body and gender identity. While appearance congruence was historically considered from a binary lens of masculine or feminine, current validated assessment tools consider individual embodiment goals. Table 1.1 describes validated measurement tools for appearance congruence, body satisfaction, and gender embodiment.

Early measures of gender dysphoria, appearance congruence, and body satisfaction, such as the Utrecht Gender Dysphoria Scale (UGDS) (Cohen-Kettenis and Van Goozen, 1997) and Body Image Scale (BIS) (Lindgren and Pauly, 1975), are less favored now due to their use of binary and noninclusive language. Both measures have since been updated

TABLE 1.1 Validated measurement tools for appearance congruence, body satisfaction, and gender embodiment.

Measurement tool	Outcome measured	Ages	Typical uses
Transgender Congruence Scale (TCS) (Kozee et al., 2012)	Gender congruence between identity and appearance; acceptance of gender identity	Adolescents and adults (used primarily in adults, sometimes older adolescents)	Useful for assessing psychosocial outcomes and changes over time in relation to social or medical affirmation
Trans Youth CAN! Gender Distress Scale (Bauer et al., 2021).	Gender-related distress (social and body-related), inclusive of nonbinary identities	Children and adolescents	Developmentally appropriate; youth-informed; captures distress beyond binary or diagnostic frameworks
Trans Youth CAN! Gender Positivity Scale (Bauer et al., 2021)	Positive gender experiences (pride, comfort, affirmation)	Children and adolescents	Strengths-based complement to distress measures; highlights resilience and well-being rather than pathology
Gender Preoccupation and Stability Questionnaire (GPSQ) (Donaghy et al., 2024)	Gender-related preoccupation; perceived stability of gender identity	Adolescents (13+) and adults	Assesses cognitive and temporal aspects of gender experience; useful in longitudinal research
Gender Preoccupation and Stability Questionnaire—2nd Edition (GPSQ-2) (Bowman et al., 2024)	Refined assessment of gender preoccupation and identity stability	Adolescents and adults	Improved psychometric properties and clarity; more appropriate for adolescent samples than the earlier version
Gender Self-Report (GSR) (Strang et al., 2023)	Broad gender-related experiences using inclusive, community-centered language	Adolescents and adults	Flexible, nonpathologizing; useful in research settings aiming to capture diverse gender experiences
GENDER-Q (Kaur et al., 2024)	Health-related quality of life, body image, social functioning, care experiences	Adults; pediatric validation underway	Comprehensive, designed to assess outcomes of gender-affirming care and patient-centered quality-of-life domains

in an attempt to be more inclusive, resulting in the Body Image Scale—Gender Spectrum and the UGDS—Gender Spectrum (McGuire and Berg, 2020). The Gender Congruence and Life Satisfaction Scale (GCLS) is a measure of appearance congruence and body satisfaction following gender-affirming care (Jones et al., 2019).

The Transgender Congruence Scale (TCS) is a 12-item self-report measure that assesses the degree to which TGD individuals feel authentic and comfortable with their gender identity and external appearance (Kozee et al., 2012). It yields a total score and two subscale scores—*Appearance Congruence* (reflecting how well an individual's outward presentation matches their internal sense of gender) and *Gender Identity Acceptance* (capturing comfort with gender identity)—using a 5-point scale. Higher scores indicate greater congruence between inner gender identity and outward expression, and the scale has demonstrated good internal consistency and validity across diverse settings, making it useful for both research and clinical assessment of gender identity experiences.

The Trans Youth CAN! Gender Distress Scale (TYC-GDS) was developed as part of the Trans Youth CAN! cohort to measure distress related to gender incongruence in youth, including those with nonbinary identities (Bauer et al., 2021). It adapts traditional gender dysphoria constructs to be more inclusive and developmentally appropriate by capturing distress related both to social aspects of gender (e.g., discomfort being misgendered) and to body-related aspects (e.g., discomfort with bodily features). Items were informed by feedback from TGD youth and are scored on a 5-point scale, with higher scores indicating greater gender-related distress. This scale helps clarify nuances in youth experiences of distress that may not be reflected in measures designed primarily for adults. Complementing the distress scale the Trans Youth CAN! Gender Positivity Scale (TYC-GPS) was designed to assess positive experiences and feelings related to gender, such as pride, comfort, and enjoyment of gender expression (Bauer et al., 2021).

The original Gender Preoccupation and Stability Questionnaire (GPSQ) and its revised *2nd Edition* (GPSQ-2) are 14-item self-report measures designed to assess the intensity of gender-related preoccupation and the stability of gender identity over time—constructs closely associated with gender dysphoria (Donaghy et al., 2024). The GPSQ-2 was updated to improve relevance and comprehension for individuals aged 13 years and older, addressing earlier validity concerns and making the tool more applicable in adolescent populations. Items probe persistent thoughts about gender and the consistency of an individual's sense of gender, and psychometric testing has supported the reliability and structural validity of the GPSQ-2 in diverse age samples.

The Gender Self-Report (GSR) is a self-report instrument developed to provide a broad, community-centered assessment of gender-related experiences without relying on complex clinical jargon (Strang et al., 2023). It allows respondents to describe their gender identity and experiences in ways that are meaningful and affirming, aiming to capture a range of gender-related psychological phenomena.

The GENDER-Q is a comprehensive patient-reported outcome measure (PROM) originally developed for adults undergoing gender-affirming care (Kaur et al., 2024). It includes multiple independently functioning scales that assess aspects of health-related quality of life, body experiences, social functioning, voice and communication, and care experiences. An adaptation for youth (*GENDER-Q Youth*) is currently under development, reflecting an international effort to ensure that PROMs are grounded in what matters most to TGD youth themselves. By centering youth voices in item development and validation, this instrument aims to provide rigorous, patient-centered assessments of care outcomes that can inform both clinical practice and research.

In addition, there is recent research focusing on regret, discontinuation of treatment, and birth sex reidentification. These are highly nuanced concepts that are best understood in the context of the individual, rather than applied as a critique of gender-affirming care for youth. Someone who regrets receiving care may continue to identify as transgender, someone who stops medications may not experience regret, and someone who no longer identifies as transgender may not regret receiving medical care. It is rare that an individual would later determine that they are not transgender after receiving medications or experience significant regret (Cavve et al., 2024; Olson et al., 2024). However, for those who do feel negatively about having received gender-affirming care, clinicians should be prepared to support them, necessitating continued research on these outcomes despite the low occurrence.

Systematic reviews and guideline development

A systematic review is a structured, methodologically rigorous approach to identifying, evaluating, and synthesizing all relevant research on a clearly defined question. It follows meticulously predefined protocols for literature search, study selection, quality appraisal, and data synthesis to minimize bias and increase transparency. Systematic reviews are generally considered more rigorous and reliable than individual studies or expert opinion. A well-conducted systematic review reports on the strength and consistency of evidence and describes the degree to which evidence is robust, limited, or uncertain. These conclusions allow expert guideline panels to base recommendations on a balanced assessment of benefits, harms, and evidence quality, and to clearly justify the confidence placed in each recommendation. In contentious or rapidly evolving fields, well-conducted systematic reviews are especially critical for preventing the mischaracterization of evidence and for ensuring that guidelines reflect the best available science rather than political, ideological, or anecdotal influences. Systematic reviews are not intended to be the sole determinant of medical decision-making, but rather inform guideline developers and shared decision-making processes.

Systematic reviews in pediatric gender care play a critical role by synthesizing the totality of available evidence to inform clinical decision-making in a highly politicized area of medicine. When conducted rigorously and transparently, they help counter selective citation and ideologically motivated claims, support beneficence and nonmaleficence in care for children and adolescents, and clarify where evidence is strong versus where genuine uncertainty remains (Dopp et al., 2025; LaFleur et al., 2025).

The sociopolitical climate

The heightened visibility of pediatric gender-affirming care has expanded public understanding that supporting TGD youth is fundamentally a matter of child health, family well-being, and health equity. Despite political scrutiny, pediatric gender-affirming care continues to be supported by a broad and durable consensus among leading medical and mental health organizations, reflecting the strength of its underlying clinical and scientific foundations (Lavalley et al., 2025). The current sociopolitical climate has prompted unprecedented engagement from clinicians, researchers, policymakers, families, and youth themselves. Across many jurisdictions, healthcare systems are innovating to protect continuity of care for TGD youth, demonstrating resilience and a sustained commitment to evidence-based practice. Globally, youth voices and lived experiences are increasingly centered in research, policy discussions, and care models, which signals a shift toward more participatory, rights-based approaches to this care.

Nevertheless, the gap between medical consensus and political rhetoric widens. Clinical questions that were once addressed within the privacy of the patient–clinician relationship have been reframed as matters of ideological debate, with youth gender-affirming care becoming a proxy battleground for broader cultural conflicts over gender, parental authority, and the role of government in healthcare. As a result, policy decisions are increasingly driven by rhetoric, selective use of evidence, and moral panic rather than by the standards of pediatric care. This creates an environment of legal uncertainty, heightened scrutiny, and fear for young people, their families, and their clinicians.

The policy climate in the United States and beyond

Policies impacting pediatric gender-affirming care in the United States and globally vary widely, reflecting divergent legal frameworks, political climates, and approaches to medical regulation, rather than a clear medical consensus. In the United States the regulation of medical care is considered to be a function of each state. Regarding gender-affirming care, this is now producing a fragmented landscape in which some states explicitly protect access to gender-affirming care for minors while others restrict, prohibit, and/or criminalize aspects of that care through statutory bans, felony laws, professional sanctions, or limitations on public funding. Several states have also passed legislation to safeguard the care.

Internationally, approaches are similarly heterogeneous. Some countries that allow gender-affirming care embed it within publicly funded health systems and established clinical pathways (Brezin et al., 2024), while others impose age-based restrictions or confine care within research environments only.

Misinformation and disinformation

Disinformation is a term used to describe deliberate use of false or misleading information or the deliberate omission of correct information from information dissemination processes, whereas misinformation does not necessarily assume the intent of its source (Pérez-Escolar et al., 2023). These bans and legislative processes through which they were deliberated relied on false information about gender identity, the safety, evidence, and standard practices of gender-affirming medical care, and the medical authorities that support this care (McNamara et al., 2024). Bans on gender-affirming care constitute disinformation because they rely on mischaracterized or outdated research, discredited sources, omission of relevant findings, and blatant falsehoods (Boulware et al., 2022). Such tactics resemble "spin science" such as what has been weaponized by tobacco, oil, sugar, and other industries to dupe government and skirt environmental and public health protections (Goldberg and Vandenberg, 2021). State-sponsored disinformation about TGD identities and gender-affirming care contributes to a well-described pattern of "institutional and informational erasure" of transgender people (Bauer et al., 2009) and potentiates structural transphobia and minority stress.

The hostile sociopolitical climate surrounding gender-affirming care for youth has impacted related clinical research and its funding. In some jurisdictions, legislative restrictions and the threat of legal or professional consequences have made it difficult for clinicians and institutions to initiate or continue research involving transgender adolescents, even when studies are observational, ethically approved, and clinically necessary. In 2025, over half of all US federally funded grants in transgender health were terminated, impacting translational scientific and health equity research alike (Miller et al., 2026). The result is a paradox in which policymakers demand stronger evidence while simultaneously fostering conditions that impede research needed to generate it—ultimately slowing scientific progress and constraining the ability of clinicians, researchers, and policymakers to make scientifically informed decisions.

Future directions in research and policy

Future research in gender-affirming care for youth should continue to prioritize multicenter longitudinal studies that follow diverse cohorts of young people across adolescence and into adulthood. Much of the existing evidence demonstrates short- to medium-term improvements in gender dysphoria for youth, yet policymakers and clinicians alike would benefit from more robust data on long-term physical, psychological, and social outcomes across different care pathways. These studies should include those with diverse ethnic and racial identities, socioeconomic backgrounds, and geographic backgrounds, and be attentive to evolving clinical practice. In addition to traditional clinical outcomes, research should examine broader indicators of self-determined well-being and resilience in the context of stigma and structural barriers. Such longitudinal approaches are essential not only for refining clinical guidelines but also for countering the misuse of scientific information to justify restrictive laws and policies.

Studies should also continue to explore how young people evaluate different aspects of care, including decision-making processes, autonomy, psychosocial support, and alignment with their sense of self. This is especially critical for understanding how best to serve youth with nonbinary identities, whose goals and experiences may not align with traditional, binary models of transition or care. Research must move beyond framing nonbinary identities as deviations from a norm and instead investigate flexible, individualized approaches that center young people's own definitions of well-being and success. Developing and validating measures that capture these complexities will be key to ensuring that future models of youth gender care are responsive, developmentally tailored, affirming, and evidence-informed.

Regarding policy, ongoing regulatory and legal challenges will shape the future of care provision in the United States and abroad. The current US presidential administration has issued executive orders and regulatory proposals aimed at restricting pediatric gender-affirming care. Federal agencies that once protected the well-being of TGD people, including the Department of Health and Human Services and the Center for Medicaid Services, are now advancing rules that would cut reimbursement and participation in federal health programs for hospitals and clinics providing gender-affirming care to youth. These federal actions have triggered legal challenges from states and civil rights groups, with some court rulings temporarily blocking parts of the funding restrictions. This reflects a significant departure from earlier federal stances under the previous US presidential administration, which had been more supportive of transgender rights and access to medically guided care, and illustrates how current federal policy is actively working to limit youth gender-affirming healthcare at the national level.

Conclusion

Gender-affirming medical care for TGD youth is multidisciplinary, individualized, and evidence-based. Those who desire this care are supported by a consenting legal guardian, and when medical care is indicated, its receipt is associated with improvements in gender dysphoria, psychosocial well-being, depression, anxiety, and suicidality. The field faces unprecedented, politicized challenges that are driven by scientific disinformation. In the face of this hostility the research continues to advance and offer increasingly sophisticated analyses that continue to replicate evidence of its benefits.

References

Achenbach, T.M., 2009. Achenbach System of Empirically Based Assessment (ASEBA): Development, Findings, Theory, and Applications. University of Vermont, Research Center of Children, Youth & Families.

Baams, L., Wilson, B.D.M., Russell, S.T., 2019. LGBTQ youth in unstable housing and foster care. Pediatrics 143 (3). Available at https://doi.org/10.1542/peds.2017–4211.

Bauer, G.R., Hammond, R., Travers, R., Kaay, M., Hohenadel, K.M., Boyce, M., 2009. I Don't think this is theoretical; this is our lives": how erasure impacts healthcare for transgender people. Journal of the Association of Nurses in AIDS Care 20 (5), 348–361. Available at https://doi.org/10.1016/j.jana.2009.07.004.

Bauer, G., Churchill, S., Feder, J., Gillis, S., Gotovac, L., Holmes, S., et al., 2021. Trans Youth CAN! Gender Distress Scale (TYC-GDS). Trans Youth CAN! Research Team, London, Canada.

Boogers, L.S., Wiepjes, C.M., Staphorsius, A.S., Klink, D.T., Ciancia, S., Romani, A., et al., 2024. A European network for the investigation of gender incongruence in adolescents. Journal of Sexual Medicine 21 (4), 350–356. Available at https://doi.org/10.1093/jsxmed/qdae014.

Borah, L., Zebib, L., Sanders, H.M., Lane, M., Stroumsa, D., Chung, K.C., 2023. State restrictions and geographic access to gender-affirming care for transgender youth. JAMA 330 (4), 375–378. Available at https://doi.org/10.1001/jama.2023.11299.

Boulware, S., Kamody, R., Kuper, L., McNamara, M., Olezeski, C., Szilagyi, N., et al., 2022. Biased Science: The Texas and Alabama Measures Criminalizing Medical Treatment for Transgender Children and Adolescents Rely on Inaccurate and Misleading Scientific Claims. Yale Law School, Public Law Research Paper. Available at 10.2139/ssrn.4102374.

Bowman, S.J., Hakeem, A., Demant, D., McAloon, J., Wootton, B.M., 2024. Assessing gender dysphoria: development and validation of the gender preoccupation and stability questionnaire–2nd edition (GPSQ-2). Journal of Homosexuality 71 (3), 666–690. Available at https://doi.org/10.1080/00918369.2022.2132440.

Bradley, S.J., Zucker, K.J., 1990. Gender identity disorder and psychosexual problems in children and adolescents. Canadian Journal of Psychiatry 35 (6), 477–486. Available at https://doi.org/10.1177/070674379003500603.

Brezin, F., Busiah, K., Leroy, C., Fiot, E., Bensignor, C., Amouroux, C., et al., 2024. Endocrine management of transgender adolescents: expert consensus of the french society of pediatric endocrinology and diabetology working group. Archives de Pediatrie. Available at https://doi.org/10.1016/j.arcped.2024.08.003.

Buckloh, L.M., Poquiz, J.L., Alioto, A., Moyer, D.N., Axelrad, M.E., 2022. Best practices in working with parents and caregivers of transgender and gender diverse youth. Clinical Practice in Pediatric Psychology 10 (3), 325–335. Available at https://doi.org/10.1037/cpp0000442.

Cavve, B.S., Bickendorf, X., Ball, J., Saunders, L.A., Thomas, C.S., Strauss, P., et al., 2024. Reidentification with birth-registered sex in a Western Australian pediatric gender clinic cohort. JAMA Pediatrics 178 (5), 446–453. Available at https://doi.org/10.1001/jamapediatrics.2024.0077.

Chen, D., Hidalgo, M.A., Leibowitz, S., Leininger, J., Simons, L., Finlayson, C., et al., 2016. Multidisciplinary care for gender-diverse youth: a narrative review and unique model of gender-affirming care. Transgender Health 1 (1), 117–123. Available at https://doi.org/10.1089/Trgh.2016.0009.

Chen, D., Berona, J., Chan, Y.-M., Ehrensaft, D., Garofalo, R., Hidalgo, M.A., et al., 2023. Psychosocial functioning in transgender youth after 2 years of hormones. New England Journal of Medicine 388 (3), 240–250. Available at https://doi.org/10.1056/NEJMoa2206297.

Clark, B.A., Virani, A., 2021. This wasn't a split-second decision": an empirical ethical analysis of transgender youth capacity, rights, and authority to consent to hormone therapy. Journal of Bioethical Inquiry 18 (1), 151–164. Available at https://doi.org/10.1007/s11673-020-10086-9.

Cohen-Kettenis, P.T., Van Goozen, S.H.M., 1997. Sex reassignment of adolescent transsexuals: a follow-up study. Journal of the American Academy of Child and Adolescent Psychiatry 36 (2), 263–271. Available at https://doi.org/10.1097/00004583-199702000-00017.

Coleman, E., Radix, A.E., Bouman, W.P., Brown, G.R., de Vries, A.L.C., Deutsch, M.B., et al., 2022. Standards of care for the health of transgender and gender diverse people, version 8. International journal of transgender health 23 (Suppl 1). Available at https://doi.org/10.1080/26895269.2022.2100644.

Cousino, M.K., Davis, A., Ng, H., Stancin, T., 2014. An emerging opportunity for pediatric psychologists: our role in a multidisciplinary clinic for youth with gender dysphoria. Clinical Practice in Pediatric Psychology 2 (4), 400–411. Available at https://doi.org/10.1037/cpp0000077.

Coyne, C.A., Yuodsnukis, B.T., Chen, D., 2023. Gender dysphoria: optimizing healthcare for transgender and gender diverse youth with a multidisciplinary approach. Neuropsychiatric Disease and Treatment 19, 479–493. Available at https://doi.org/10.2147/NDT.S359979.

Craig, S.L., Austin, A., Rashidi, M., Adams, M., 2017. Fighting for survival: the experiences of lesbian, gay, bisexual, transgender, and questioning students in religious colleges and universities. Journal of Gay and Lesbian Social Services 29 (1), 1–24. Available at https://doi.org/10.1080/10538720.2016.1260512.

Dahlgren Allen, S., Tollit, M.A., McDougall, R., Eade, D., Hoq, M., Pang, K.C., 2021. A waitlist intervention for transgender young people and psychosocial outcomes. Pediatrics 148 (2), 0031–4005. Available at https://doi.org/10.1542/peds.2020–042762.

Dai, D., Charlton, B.M., Boskey, E.R., Hughes, L.D., Hughto, J.M.W., Orav, E.J., et al., 2024. Prevalence of gender-affirming surgical procedures among minors and adults in the US. JAMA Network Open 7 (6), e2418814. Available at https://doi.org/10.1001/jamanetworkopen.2024.18814.

Daley, T., Grossoehme, D., McGuire, J.K., Corathers, Conard, L.A., Lipstein, E.A., 2019. I couldn't see a downside": decision-making about gender-affirming hormone therapy. Journal of Adolescent Health 65 (2), 274–279. Available at https://doi.org/10.1016/j.jadohealth.2019.02.018.

De Vries, A.L.C., Steensma, T.D., Doreleijers, T.A.H., Cohen-Kettenis, P.T., 2011. Puberty suppression in adolescents with gender identity disorder: a prospective follow-up study. Blackwell Journal of Sexual Medicine 8 (8), 2276–2283. Available at https://doi.org/10.1111/j.1743-6109.2010.01943.x.

Delozier, A.M., Kamody, R.C., Rodgers, S., Chen, D., 2020. Health disparities in transgender and gender expansive adolescents: a topical review from a minority stress framework. Journal of Pediatric Psychology 45 (8), 842–847. Available at https://doi.org/10.1093/jpepsy/jsaa040.

Dixon, M., Hawke, L.D., Relihan, J., Darnay, K., Henderson, J.L., 2022. Let's talk gender: ten things transgender and nonbinary youth want all researchers to know. Journal of the American Academy of Child and Adolescent Psychiatry 61 (8), 960–964. Available at https://doi.org/10.1016/j.jaac.2021.12.010.

Donaghy, O.J.E., Cobham, V., Lin, A., 2024. Screening adolescent transgender-related distress: gender preoccupation and stability questionnaire demonstrates excellent criterion validity with multi-disciplinary, pediatric gender specialist assessment. International Journal of Transgender Health. Available at https://doi.org/10.1080/26895269.2024.2378378.

Dopp, A.R., Peipert, A., Buss, J., Jesús-Romero, R.D., Palmer, K., Lorenzo-Luaces, L., 2025. Interventions for gender dysphoria and related health problems in transgender and gender-expansive youth: a systematic review of benefits and risks to inform practice, policy, and research. Rand health quarterly 12 (2).

Durwood, L., McLaughlin, K.A., Olson, K.R., 2017. Mental health and self-worth in socially transitioned transgender youth. Journal of the American Academy of Child and Adolescent Psychiatry 56 (2). Available at https://doi.org/10.1016/j.jaac.2016.10.016.

Edwards-Leeper, L., Spack, N.P., 2012. Psychological evaluation and medical treatment of transgender youth in an interdisciplinary "gender management service" (GeMS) in a major pediatric center. Journal of Homosexuality 59 (3), 321–336. Available at https://doi.org/10.1080/00918369.2012.653302.

Flores, A.R., Meyer, I.H., Langton, L., Herman, J.L., 2021. Gender identity disparities in criminal victimization: national crime victimization survey, 2017-2018. American journal of public health 111 (4), 726–729. Available at https://doi.org/10.2105/AJPH.2020.306099.

Giles-Petersen, J., 2018. History of the Transgender Child. University of Minnesota Press, Minneapolis, MN.

Goldberg, R.F., Vandenberg, L.N., 2021. The science of spin: targeted strategies to manufacture doubt with detrimental effects on environmental and public health. Environmental Health: A Global Access Science Source 20 (1). Available at https://doi.org/10.1186/s12940-021-00723-0.

Gooren, L., Delemarre-van de Waal, H., 1996. The feasibility of endocrine interventions in juvenile transsexuals. Journal of Psychology & Human Sexuality 8 (4), 69–74. Available at https://doi.org/10.1300/j056v08n04_05.

Green, A.E., Price-Feeney, M., Dorison, S.H., Pick, C.J., 2020. Self-reported conversion efforts and suicidality among us lgbtq youths and young adults, 2018. American Journal of Public Health 110 (8), 1221–1227. Available at https://doi.org/10.2105/AJPH.2020.305701.

Gridley, S.J., Crouch, J.M., Evans, Y., Eng, W., Antoon, E., Lyapustina, M., et al., 2016. Youth and caregiver perspectives on barriers to gender-affirming healthcare for transgender youth. Journal of Adolescent Health 59 (3), 254–261. Available at https://doi.org/10.1016/j.jadohealth.2016.03.017.

Harrison, J.E., Weber, S., Jakob, R., Chute, C.G., 2021. ICD-11: an international classification of diseases for the twenty-first century. BMC Medical Informatics and Decision Making 21. Available at https://doi.org/10.1186/s12911-021-01534-6.

Hatzenbuehler, M.L., Pachankis, J.E., 2016. Stigma and minority stress as social determinants of health among lesbian, gay, bisexual, and transgender youth: research evidence and clinical implications. Pediatric Clinics of North America 63 (6), 985–997.

Hembree, W.C., Cohen-Kettenis, P.T., Gooren, L., Hannema, S.E., Meyer, W.J., Murad, M.H., et al., 2017. Endocrine treatment of gender-dysphoric/gender-incongruent persons: an endocrine society clinical practice guideline. Endocrine practice: official journal of the American College of Endocrinology and the American Association of Clinical Endocrinologists 23 (12). Available at https://doi.org/10.4158/1934-2403-23.12.1437.

Hendricks, M.L., Testa, R.J., 2012. A conceptual framework for clinical work with transgender and gender nonconforming clients: An adaptation of the minority stress model. Professional Psychology: Research and Practice 43 (5), 460–467. Available at https://doi.org/10.1037/a0029597.

Herman, J., Flores, A., O'Neill, K., 2022. How many adults and youth identify as transgender in the United States. https://williamsinstitute.law.ucla.edu/wp-content/uploads/Trans-Pop-Update-Jun-2022.pdf

Hidalgo, M.A., Ehrensaft, D., Tishelman, A.C., Clark, L.F., Garofalo, R., Rosenthal, S.M., et al., 2013. The gender affirmative model: what we know and what we aim to learn. Human Development 56 (5), 285–290. Available at https://doi.org/10.1159/000355235.

Hodax, J.K., Crouch, J.M., Sethness, J.L., Loren, D., Kahn, N.F., Asante, P.G., et al., 2023. Strategies for providing gender-affirming care for adolescents in the primary care setting. Pediatric Annals 52 (12), e442. Available at https://doi.org/10.3928/19382359-20231016-04.

Holt, M.K., Ranker, L.R., Kraus, A.D., Xuan, Z., Greif Green, J., Katz-Wise, S.L., et al., 2024. Recruitment of a U.S. Nation-Wide sample of transgender and gender diverse youth. Journal of Adolescent Health 75 (6), 969–973. Available at https://doi.org/10.1016/j.jadohealth.2024.07.021.

Horowitz, L.M., Bridge, J.A., Teach, S.J., Ballard, E., Klima, J., Rosenstein, D.L., et al., 2012. Ask suicide-screening questions (ASQ): a brief instrument for the pediatric emergency department. Archives of Pediatrics and Adolescent Medicine 166 (12), 1170–1176. Available at https://doi.org/10.1001/archpediatrics.2012.1276.

Hunter, J., Butler, C., Cooper, K., 2021. Gender minority stress in trans and gender diverse adolescents and young people. Clinical child psychology and psychiatry 26 (4), 1182–1195. Available at https://doi.org/10.1177/13591045211033187.

Inwards-Breland, D.J., Karrington, B., Sequeira, G.M., 2021. Access to care for transgender and nonbinary youth: ponder this, innumerable barriers exist. JAMA Pediatrics 175 (11), 1112–1114. Available at https://doi.org/10.1001/jamapediatrics.2021.2068.

Jones, B.A., Bouman, W.P., Haycraft, E., Arcelus, J., 2019. The gender congruence and life satisfaction scale (GCLS): development and validation of a scale to measure outcomes from transgender health services. The International Journal of Transgenderism 20 (1), 63–80. Available at https://doi.org/10.1080/15532739.2018.1453425.

Kahn, N.F., Sequeira, G.M., Asante, P.G., Kidd, K.M., Coker, T.R., Christakis, D.A., et al., 2024. Estimating transgender and gender-diverse youth populations in health systems and survey data. Pediatrics 153 (6). Available at https://doi.org/10.1542/peds.2023-065197.

Karamanis, G., Karalexi, M., White, R., Frisell, T., Isaksson, J., Skalkidou, A., et al., 2022. Gender dysphoria in twins: a register-based population study. Scientific Reports 12 (1). Available at https://doi.org/10.1038/s41598-022-17749-0.

Kattari, S.K., Whitfield, D.L., Walls, E., Langenderfer-Magruder, L., Ramos, D., 2015. Poli uals. Journal of the Society for Social Work and Research 7 (3), 427–447. Available at https://doi.org/10.1086/686920.

Katz-Wise, S.L., Pullen Sansfaçon, A., Bogart, L.M., Rosal, M.C., Ehrensaft, D., Goldman, R.E., Austin, B., 2019. Lessons from a community-based participatory research study with transgender and gender nonconforming youth and their families. Action Research 17 (2), 186–207. Available at https://doi.org/10.1177/1476750318818875.

Kaur, M.N., Morrison, S.D., Kennedy, S.L., van de Grift, T.C., Højgaard, A., Jacobsen, A.L., et al., 2024. International study to develop a patient-reported outcome measure to evaluate outcomes of gender-affirming care - the GENDER-Q. Journal of Patient-Reported Outcomes 8 (1). Available at https://doi.org/10.1186/s41687-024-00785-x.

Kidd, K.M., Sequeira, G.M., Douglas, C., Paglisotti, T., Inwards-Breland, D.J., Miller, E., et al., 2021. Prevalence of gender-diverse youth in an urban school district. Pediatrics 147 (6). Available at https://doi.org/10.1542/peds.2020-049823.

Kosciw, J.G., Greytak, E.A., Palmer, N.A., M.J., 2014. The 2013 National School Climate Survey: The Experiences of Lesbian, Gay, Bisexual, and Transgender Youth in Our Nation's Schools. GLSEN, New York.

Kozee, H.B., Tylka, T.L., Bauerband, L.A., 2012. Measuring transgender individuals' comfort with gender identity and appearance: development and validation of the transgender congruence scale. Psychology of Women Quarterly 36 (2), 179–196. Available at https://doi.org/10.1177/0361684312442161.

Kroenke, K., Spitzer, R.L., Williams, J.B., 2001. The PHQ-9: validity of a brief depression severity measure. Journal of General Internal Medicine 16 (9), 606–613.

LaFleur, J., Heath, L., Gonzalez, V., 2025. Gender-affirming medical treatments for pediatric patients with gender dysphoria: A report of the University of Utah College of Pharmacy Drug Regimen Review Center (DRRC). Available from: https://le.utah.gov/AgencyRP/reportingDetail.jsp?rid=636.

LaRocca, D., James, K.A., Rosenberg, S., Ma, M., Brooks-Russell, A., 2023. Team sports participation, depression, and suicidal ideation in lesbian, gay, bisexual, transgender, and questioning adolescents. Psychology in the Schools 60 (4), 902–911. Available at https://doi.org/10.1002/pits.22792.

Lavalley, M.N., Eble, D.J., Kuzon, W.M., Peters, B.R., Skokan, A.J., Ettinger, R.E., et al., 2025. Making a statement: positions of professional medical organizations towards gender-affirming care. Annals of Surgery 281 (2), 219–220. Available at https://doi.org/10.1097/SLA.0000000000006342.

Lawson, M.L., Gotovac, S., Couch, B., Gale, L., Vandermorris, A., Ghosh, S., et al., 2024. Pathways to care for adolescents attending a first hormone appointment at Canadian gender affirming medical clinics: a cross-sectional analysis from the trans youth CAN! study. Journal of Adolescent Health 74 (1), 140–147. Available at https://doi.org/10.1016/j.jadohealth.2023.07.021.

Lee, W.Y., Hobbs, J.N., Hobaica, S., DeChants, J.P., Price, M.N., Nath, R., 2024. State-level anti-transgender laws increase past-year suicide attempts among transgender and non-binary young people in the USA. Nature Human Behaviour 8 (11), 2096–2106. Available at https://doi.org/10.1038/s41562-024-01979-5.

Lindgren, T.W., Pauly, I.B., 1975. A body image scale for evaluating transsexuals. Archives of Sexual Behavior 4 (6), 639–656. Available at https://doi.org/10.1007/BF01544272.

McDowell, A., Raifman, J., Progovac, A.M., Rose, S., 2020. Association of nondiscrimination policies with mental health among gender minority individuals. JAMA Psychiatry 77 (9), 952–958. Available at https://doi.org/10.1001/jamapsychiatry.2020.0770.

McGuire, J.K., Berg, D., Catalpa, J.M., Morrow, Q.J., Fish, J.N., Nic Rider, G., Steensma, T., Cohen-Kettenis, P.T., Spencer, K., 2020. Utrecht Gender Dysphoria Scale - Gender Spectrum (UGDS-GS): Construct validity among transgender, nonbinary, and LGBQ samples. Int J Transgend Health 21 (2), 194–208. Available at 10.1080/26895269.2020.1723460.

McNamara, M., McLamore, Q., Meade, N., Olgun, M., Robinson, H., Alstott, A., 2024. A thematic analysis of disinformation in gender-affirming healthcare bans in the United States. Social Science & Medicine 1982, 351.

Miller, A.S., Liu, M., Graham, R., Berlin, N.L., Carruthers, K.H., Keuroghlian, A.S., 2026. Termination of gender-affirming care grants at the national institutes of health. JAMA Pediatrics. Available at https://doi.org/10.1001/jamapediatrics.2025.4653.

Miller-Jacobs, C., Operario, D., Hughto, J.M.W., 2023. State-level policies and health outcomes in U.S. transgender adolescents: findings from the 2019 youth risk behavior survey. LGBT Health 10 (6), 447–455. Available at https://doi.org/10.1089/lgbt.2022.0247.

Newman, L.E., 1970. Transsexualism in adolescence: problems in evaluation and treatment. Archives of General Psychiatry 23 (2), 112–121. Available at https://doi.org/10.1001/archpsyc.1970.01750020016003.

Olson, K.R., Key, A.C., Eaton, N.R., 2015. Gender cognition in transgender children. Psychological Science 26 (4), 467–474. Available at https://doi.org/10.1177/0956797614568156.

Olson, K.R., Durwood, L., DeMeules, M., McLaughlin, K.A., 2016. Mental health of transgender children who are supported in their identities. Pediatrics 137 (3). Available at https://doi.org/10.1542/peds.2015-3223.

Olson, K.R., Durwood, L., Horton, R., Gallagher, N.M., Devor, A., 2022. Gender identity 5 years after social transition. Pediatrics 150 (2). Available at https://doi.org/10.1542/peds.2021-056082.

Olson, K.R., Raber, G.F., Gallagher, N.M., 2024. Levels of satisfaction and regret with gender-affirming medical care in adolescence. JAMA Pediatrics 178 (12), 1354–1361. Available at https://doi.org/10.1001/jamapediatrics.2024.4527.

Olson-Kennedy, J., Chan, Y.-M., Rosenthal, S., Hidalgo, M.A., Chen, D., Clark, L., Chen, D., et al., 2019. Creating the trans youth research network: a collaborative research endeavor. Transgender Health 4 (1), 304–312. Available at https://doi.org/10.1089/trgh.2019.0024.

Oosthoek, E.D., Stanwich, S., Gerritse, K., Doyle, D.M., de Vries, A.L.C., 2024. Gender-affirming medical treatment for adolescents: a critical reflection on "effective" treatment outcomes. BMC Medical Ethics 25 (1). Available at https://doi.org/10.1186/s12910-024-01143-8.

Pérez-Escolar, M., Lilleker, D., Tapia-Frade, A., 2023. A systematic literature review of the phenomenon of disinformation and misinformation. Media and Communication 11 (2), 76–87. Available at https://doi.org/10.17645/mac.v11i2.6453.

Rafferty, J., Yogman, M., Baum, R., Gambon, T.B., Lavin, A., Mattson, G., et al., 2018. Ensuring comprehensive care and support for transgender and gender-diverse children and adolescents. Pediatrics 142 (4). Available at https://doi.org/10.1542/peds.2018-2162.

Ravindranath, O., Perica, M.I., Parr, A.C., Ojha, A., McKeon, S.D., Montano, G., et al., 2024. Adolescent neurocognitive development and decision-making abilities regarding gender-affirming care. Developmental Cognitive Neuroscience 67. Available at https://doi.org/10.1016/j.dcn.2024.101351.

Russell, S.T., Pollitt, A.M., Li, G., Grossman, A.H., 2018. Chosen name use is linked to reduced depressive symptoms, suicidal ideation, and suicidal behavior among transgender youth. The Journal of Adolescent Health: Official Publication of the Society for Adolescent Medicine 63 (4), 503–505.
Russell, M.R., Rogers, R.L., Rosenthal, S.M., Lee, J.Y., 2022. Increasing access to care for transgender/gender diverse youth using telehealth: a quality improvement project. Telemedicine and e-Health 28 (6), 847–857. Available at https://doi.org/10.1089/tmj.2021.0268.
Salvetti, B., Westphaln, K.K., Pike, N.A., 2025. Web-based gender affirming treatment decision aid for youth and caregivers: a pilot study. Journal of the American Association of Nurse Practitioners 37 (7), 379–385. Available at https://doi.org/10.1097/JXX.0000000000001138.
Spencer, K.G., Berg, D.R., Bradford, N.J., Vencill, J.A., Tellawi, G., Rider, G.N., 2021. The gender-affirmative life span approach: a developmental model for clinical work with transgender and gender-diverse children, adolescents, and adults. Psychotherapy 58 (1), 37–49. Available at https://doi.org/10.1037/pst0000363.
Spitzer, R.L., Kroenke, K., Williams, J.B.W., Löwe, B., 2006. A brief measure for assessing generalized anxiety disorder: the GAD-7. Archives of Internal Medicine 166 (10), 1092–1097.
Steinberg, L., 2013. Does recent research on adolescent brain development inform the mature minor doctrine? Journal of Medicine and Philosophy (United Kingdom) 38 (3), 256–267. Available at https://doi.org/10.1093/jmp/jht017.
Stiggelbout, Pieterse, De Haes, 2015. Shared decision making: concepts, evidence, and practice. Patient Education and Counseling 98 (10), 1172–1179. Available at https://doi.org/10.1016/j.pec.2015.06.022.
Strang, J.F., Wallace, G.L., Michaelson, J.J., Fischbach, A.L., Thomas, T.R., Jack, A., et al., 2023. The gender self-report: a multidimensional gender characterization tool for gender-diverse and cisgender youth and adults. American Psychologist 78 (7), 886–900. Available at https://doi.org/10.1037/amp0001117.
Suarez, N.A., Trujillo, L., McKinnon, I.I., Mack, K.A., Lyons, B., Robin, L., et al., 2024. Disparities in school connectedness, unstable housing, experiences of violence, mental health, and suicidal thoughts and behaviors among transgender and cisgender high school students - youth risk behavior survey, united states, 2023. MMWR Supplements 73 (4), 50–58. Available at https://doi.org/10.15585/mmwr.su7304a6.
Testa, R.J., Habarth, J., Peta, J., Balsam, K., Bockting, W., 2015. Development of the gender minority stress and resilience measure. Psychology of Sexual Orientation and Gender Diversity 2 (1), 65–77. Available at https://doi.org/10.1037/sgd0000081.
Tordoff, D.M., Sequeira, G.M., Shook, A.G., Williams, F., Hayden, L., Kasenic, A., et al., 2023. Factors associated with time to receiving gender-affirming hormones and puberty blockers at a pediatric clinic serving transgender and nonbinary youth. Transgender Health 8 (5), 420–428. Available at https://doi.org/10.1089/trgh.2021.0116.
Turban, J.L., Ehrensaft, D., 2018. Research review: gender identity in youth: treatment paradigms and controversies. Journal of Child Psychology and Psychiatry and Allied Disciplines 59 (12), 1228–1243. Available at https://doi.org/10.1111/jcpp.12833.
de Vries, A.L.C., Cohen-Kettenis, P.T., 2012. Clinical management of gender dysphoria in children and adolescents: the Dutch approach. Journal of Homosexuality 59 (3), 301–320. Available at https://doi.org/10.1080/00918369.2012.653300.
de Vries, A.L.C., McGuire, J.K., Steensma, T.D., Wagenaar, E.C.F., Doreleijers, T.A.H., Cohen-Kettenis, P.T., 2014. Young adult psychological outcome after puberty suppression and gender reassignment. Pediatrics 134 (4), 696–704. Available at https://doi.org/10.1542/peds.2013-2958.
Vrouenraets, L.J.J.J., de Vries, A.L.C., de Vries, M.C., van der Miesen, A.I.R., Hein, I.M., 2021. Assessing medical decision-making competence in transgender youth. Pediatrics 148 (6). Available at https://doi.org/10.1542/peds.2020-049643.
Wylie, K., Knudson, G., Khan, S.I., Bonierbale, M., Watanyusakul, S., Baral, S., 2016. Serving transgender people: clinical care considerations and service delivery models in transgender health. The Lancet 388 (10042), 401–411. Available at https://doi.org/10.1016/S0140-6736(16)00682-6.

Chapter 2

Gender development and developmental pathways

Alora A. Rando[1] and Esha Vaid[2]
[1]*Division of Psychology, Department of Pediatrics, Nemours Children's Health, Wilmington, DE, United States,* [2]*Graduate School of Applied and Professional Psychology, Rutgers University, New Brunswick, NJ, United States*

Learning objectives

- Understand the biological, psychological, social, and cultural factors influencing gender development, including how they interact to shape gender identity.
- Evaluate research on young transgender and gender-nonconforming children, focusing on developmental considerations and challenges.
- Summarize research methodologies used to investigate and understand childhood gender identity development.
- Identify and address limitations in gender identity research, many of which are born of methodologies based on a binary understanding of gender, and make suggestions for future research.

Key terms and definitions

- Sex assigned at birth refers to a designation of female or male (or occasionally unknown) made at birth, often determined by a physician and typically based on the appearance of the genitals. The term *assigned gender* is sometimes used rather than the term *sex assigned at birth* in the context of early childhood gender development and cognition. The terms female and male are used in this chapter to indicate sex assignment, sex-based traits (such as in animal models), or gender-stereotyped behaviors and characteristics.
- Gender identity refers to a personal sense of gender, which may or may not correspond with the sex assigned at birth. This is distinct from sexual or romantic orientation. The terms boy, girl, woman, and man are used in this chapter to indicate gender identity, in addition to a variety of other terms (e.g., nonbinary, bigender, and others).
- Gender expression is the external display of gender through clothing, haircut, voice, and other forms of presentation.
- Cisgender refers to individuals whose gender identity aligns with the sex assigned at birth. Cisnormative refers to the societal assumption that cisgender identities are the default, while gender nonconformity is abnormal.
- Gender nonconforming refers to behaviors and interests that fall outside of established stereotypes that are based on sex assigned at birth.
- Transgender and gender diverse is an umbrella term for individuals whose gender identity differs from the sex assigned at birth. Transgender is sometimes used when referring to research participants who specifically identify as binary women or men.
- Nonbinary is an umbrella term for individuals whose gender identity does not fit within the traditional binary of man and woman, such as genderfluid, genderqueer, agender, and bigender, among others. Nonbinary can also be used as a specific identity term.

Over the past few decades, understanding of gender identity development in childhood has evolved significantly, including among transgender and gender diverse (TGD) youth. Research to date has demonstrated the complex, dynamic,

Advances in Research on Pediatric Gender Affirming Healthcare. DOI: https://doi.org/10.1016/B978-0-443-29019-0.00016-6
© 2026 Elsevier Inc. All rights are reserved, including those for text and data mining, AI training, and similar technologies.

and evolving nature of childhood gender identity development. Rather than prescribing what children's gender identity development *should* look like, the clinical research summarized here is intended to help readers understand how best to support children as their gender identity develops and grows (Newhook et al., 2018).

Introduction to gender identity development research

The prevalence of TGD children globally is difficult to determine. According to the 2023 US Youth Risk Behavior Survey, 3.3% of a nationally representative sample of high school students endorsed a transgender identity and 2.2% identified as questioning their gender identity (Suarez et al., 2024). A systematic review of studies across New Zealand, the Netherlands, and the United States found that self- and parent-reported transgender identity ranged from 1.2% to 2.7%, while self- and parent-reported gender variance or gender nonconformity were higher, ranging from 2.5% to 8.4% (Zhang et al., 2020).

Emerging research has begun to describe the prevalence of nonbinary identities and other identities outside the gender binary among youth (Cohen-Kettenis, 2015; McGuire et al., 2019; Twist and de Graaf, 2019). In a large study of lesbian, gay, bisexual, and transgender (LGBT) youth (N=10,030), when asked if they considered themselves "male, female, transgender, or other gender," 925 youth (9%) selected one of the latter two options. Of these gender-expansive youth, nearly twice as many opted to write in a gender identity (6%) as youth who selected "transgender" (3.2%) (Baum et al., 2013). Prevalence estimates are impacted by the ever-growing and evolving glossary of terms related to TGD identities and experiences, with the recognition that identity terms may resonate differently with each individual (e.g., not all nonbinary individuals understand themselves as transgender).

Importance of understanding gender identity development

Although visibility and acceptance of TGD identities have increased over the last several years, TGD youth continue to face persecution and discrimination. In addition to widespread social discourse and media campaigns that stigmatize TGD identities and challenge the right of TGD youth to exist, a great number of anti-TGD laws, executive orders, and bills have been proposed and passed in the past few years—with particular focus on targeting transgender youth and adolescents—in the United States and Europe (Khonina and Salway, 2025; Salinas-Quiroz et al., 2025). Discrimination at the political, institutional, and interpersonal levels prevents TGD youth from using appropriate bathrooms; accessing healthcare; participating in sports; and accessing tools necessary for legal, medical, and social transition. Ongoing discrimination limits TGD youth from having basic needs met and prevents them from participating safely in their communities. There is a clear need for healthcare and research resources dedicated to understanding gender identity development in childhood to meet the needs of TGD youth.

Cisnormativity and bias in research

Research on gender identity development is limited by the pervasive cisnormativity in the operationalization of gender identity and related constructs and, in turn, the framing of research questions. Since the 1950s, developmental science has devoted significant attention to studying gender development patterns exhibited by one particular type of child: those who demonstrate the modal pattern of gender development. Research on developmental pathways has historically presupposed alignment between sex assigned at birth and gender identity, and TGD children have therefore falsely been assumed to be cisgender. Increasingly, investigations into gender identity have begun to capture the development and experiences of TGD children. However, an overemphasis on binary gender identities, at the exclusion of nonbinary identities, continues to be a limitation (Goodey, 2021).

Research approaches rooted in gender-expansive frameworks have become more widespread in recent years. However, contemporary research faces the unfortunate conundrum that investigating the gender development of TGD children in terms of peer, activity, clothing, and toy preferences inherently involves labeling certain preferences as feminine or masculine, which may reify and perpetuate stereotypes. As such, developmental constructs such as "gender-typical development" and "gender stability" are meaningful only within a cisnormative framework and do not reflect the range of experiences of gender diverse youth (Salinas-Quiroz et al., 2025). Even efforts to broaden the lens of gender development typically borrow frameworks originally developed for understanding the experience of cisgender individuals (Gülgöz et al., 2019; Olson et al., 2022).

Any attempt to describe and synthesize the literature on gender development in childhood faces this same predicament: to describe and critique the existing literature, it is necessary to reference notions rooted in cisnormativity. Given concerns about reification of cisnormativity and binary understandings of gender, it is important to clarify that descriptors such as "gender-nonconforming" and "gender-stereotyped" are meant to signify what is *perceived by members of the child's culture as masculine or feminine*, rather than intending to imply that any object, preference, or behavior is inherently gendered (Gülgöz et al., 2019; Fast and Olson, 2018; Olson et al., 2022).

Early research on gender-nonconforming youth largely emerged from clinical contexts intended for corrective intervention (i.e., conversion therapy), resulting in an over-focus on the "stability" of gender and predicting gender identity outcomes. From this research the term "desistance" was used to describe early childhood gender nonconformity or gender dysphoria that was no longer present in adolescence or adulthood (Drummond et al., 2008; Wallien and Cohen-Kettenis, 2008). Due to the theoretical, ethical, and methodological limitations, minimal utility, and problematic implications of this body of research, the term desistance is no longer considered appropriate for research or clinical work (Karrington, 2022; Olson, 2016; Olson et al., 2022). Any future research on the stability of gender identity in young children or adolescents should consider this historical context during research design, provide a clear, hypothesis-driven rationale for examining this outcome, and ideally use community-based participatory research methods to ensure community-driven priorities and ethical interpretation of results.

Overview of childhood gender development

What is known about gender development milestones and processes in childhood is based primarily on research conducted among presumed cisgender samples. This literature is supplemented by emerging research on young TGD children, largely derived from the TransYouth Project (TYP), a large-scale, longitudinal study following over 300 socially transitioned binary transgender children (henceforth: transgender), as well as two comparison cohorts composed of (1) cisgender siblings of the transgender participants (henceforth: sibling) and (2) unrelated, cisgender age- and gender-matched controls (henceforth: control) from the United States and Canada (Olson and Gülgöz, 2018). A cohort of gender-nonconforming children was also recruited in parallel to the primary study, and more recent data from the TYP are beginning to clarify the experiences of youth whose identities fall outside of the binary (Rae et al., 2019). A full description of the research methods and measurement tools used in the TYP can be found in the original articles (Gülgöz et al., 2019; Durwood et al., 2017; Olson et al., 2016).

Ages 0–3 years

Children begin to demonstrate awareness of gender categories and to apply binary gender categories and stereotyped preferences to themselves and others early in life. As young as 6 months, infants can differentiate between cisgender male and female voices and faces, suggesting an awareness of gender as a category by which the world around them can be organized (Fagan, 1976; Miller, 1983). Between 6 and 12 months, infants begin to demonstrate awareness of congruence between different gender-stereotyped traits—they are able to pair gender-stereotyped male and female voices with gender-stereotyped male and female faces (Poulin-Dubois et al., 1994).

As knowledge acquisition accelerates, children exhibit increasingly sophisticated differentiation by binary gender categories and begin to apply these categories to themselves. By 1 year, infants exhibit an understanding of their own assigned gender categorization (Rhodes and Baron, 2019). Between 18 and 24 months, children begin to apply gender labels to themselves and others (Zosuls et al., 2009). Verbalization of their own gender identity becomes more consistent and frequent between 2 and 3 years, and most children reliably label their assigned gender by their third birthday (Leinbach and Fagot, 1986).

In the second year of life, gender becomes a more salient influence on behavior, preferences, and expectations of others. By 18 months, many children demonstrate a preference for toys that are stereotypically associated with their assigned gender (Solomon, 2016). By 2 years, children exhibit a preference for people who share their assigned gender (Serbin et al., 2001). Children between 2 and 3 years demonstrate increasing knowledge of gendered stereotypes and associate binary gender labels with faces, toys, and activities (Cooley and Killen, 2015; Rhodes and Baron, 2019). This body of research is limited by a lack of longitudinal data about gender identity development over time. It is unknown whether the children in these samples continued to identify with their assigned gender or whether findings generalize to infants or toddlers who later are gender nonconforming as children and/or realize TGD identities in childhood or adolescence.

Early childhood

Between the ages of 3 and 5 years, gender becomes a highly influential motivator of preferences and behaviors. Preschool-age children use gender to guide their toy choices and express greater interest in gender-associated objects (Serbin et al., 2001; Zosuls et al., 2009). They begin identifying as more similar to peers of the same gender and display a strong preference for same-gender peers (Martin et al., 2017). Children use gender to inform expectations of others and tend to negatively evaluate children who do not conform to gender stereotypes (Blakemore et al., 2013; Eaton et al., 1981; Martin and Fabes, 2001; Signorella et al., 1993).

By 5 years, many children have established gender constancy, or an understanding of gender as a stable aspect of identity, across time (e.g., from childhood to adulthood), and some have established gender constancy across context (e.g., when playing with stereotypically masculine toys vs stereotypically feminine toys) (Martin et al., 2002). The notion of gender constancy is inherently biased by the idea that gender is, in fact, stable and has therefore not historically taken gender diversity or fluidity into account.

Results from the TYP provide evidence that transgender children do not differ from their same-gender peers in how gender informs their attitudes and choices. Specifically, transgender children aged 3–5 years did not differ from their same-gender peers in their likelihood to self-categorize their gender identity, nor on measures of gender-stereotyped behavior, toy, peer, and clothing preferences, or beliefs. Transgender children and their cisgender peers also did not differ in gender constancy across contexts. This was true when thinking about themselves and when thinking about others. There were some differences related to gender constancy across time. While all children were equally likely to believe their gender would remain the same into adulthood, transgender children were less likely to believe that their gender at birth was the same as their current gender identity. Transgender children and their siblings were both less likely than cisgender controls to believe that other people's gender is stable over time. Rather than suggesting that transgender children or their siblings are not establishing gender constancy at the same age as their peers, these findings highlight that exposure to diverse identities may lead to more flexibility with regard to beliefs about gender. This study was among the first to show that gender identity development in young transgender children is very similar to that of cisgender children (Fast and Olson, 2018).

Emerging research suggests that it is between the ages of 3 and 5 years that many nonbinary individuals begin to recognize discomfort with the gender norms associated with their assigned gender, prompting them to question their gender identity (Waagen, 2022). This research is retrospective, and there is no research to date on very young children expressing gender identity or preferences outside of binary stereotypes.

Middle childhood

Middle childhood is a time of change and increasing variability, and research on this developmental period has largely focused on understanding changes over time. There tends to be more variability across the spectrum of gender conformity versus nonconformity with regard to identity, similarity, and preferences in middle compared to early childhood. In general, children who are more nonconforming tend to stay that way over time, while children who are more conforming are more likely to change over time (Golombok et al., 2008, 2012). In other words, by the end of middle childhood, fewer children are as rigidly gender-conforming as compared to earlier childhood.

Transgender children in the TYP generally showed similar levels of gender conformity to their cisgender peers, with some differences emerging over time (from approximately 7 to 9.5 years of age) (Hässler et al., 2022). Transgender children showed consistency over time in gender identity (i.e., how they described their gender identity and pronouns) and similarity (i.e., how similar they were to other kids of the same gender), whereas children in the cisgender comparison group showed a small increase in gender conformity over time. Transgender children exhibited decreased gender-conforming preferences over time, whereas cisgender children did not. Finally, differences emerged between binary gender identities: transgender girls and cisgender girls demonstrated a decrease in gender-stereotypical preferences and conformity with gender identity over time, while boys in both groups remained similarly conforming.

Adolescence

Adolescence is a period of life characterized by overall identity development and consolidation. There is significantly less research on gender development in adolescence compared to other aspects of identity and compared to gender development during childhood (Steensma et al., 2013).

Gender stereotyping peaks around 6 years and then steadily declines, but stereotypes persist into adolescence (Alfieri et al., 1996), and some adolescents and adults continue to evaluate gender nonconformity negatively (Rudman and Fairchild, 2004). Gender essentialism, or the idea that gender is fixed, innate, and biologically determined, peaks in middle childhood before beginning to decline (Taylor, 1996). Adolescents are more likely to believe that social factors impact gender identity in addition to biological factors compared to younger children. For some, gender essentialism persists strongly through adolescence and into adulthood. Essentialism tends to be held more toward stereotypically masculine people, and essentialist beliefs are more likely to be held by stereotypically masculine individuals (Smiler and Gelman, 2008).

Transgender adolescents in the TYP demonstrated similar levels of gender stereotyping to other cisgender adolescents and at lower levels than their cisgender parents, showing a similar developmental trend toward adulthood (deMayo et al., 2022a).

Binary transgender adolescents tend to think of their specific gender (boy or girl) as more important to their identity than either their cisgender or nonbinary peers (Wittlin et al., 2025b).

Overall, research has demonstrated that gender development in transgender children and adolescents occurs remarkably similarly to that of cisgender children, with a few key differences. Transgender youth and, to some extent, their cisgender siblings tend to be less rigid and more flexible in their gender-related preferences and beliefs. The TYP has pioneered the movement to examine similar gender development processes among TGD youth. By recruiting binary, socially transitioned transgender youth, this body of research has been able to control for identity differences that would not otherwise be possible, but it nonetheless limits generalizability to other gender diverse children and provides little information on the dynamic and fluid nature of gender identity development.

(Trans)gender identity development

The milestones in gender development observed in the general population rely heavily on environmental feedback and reinforcement of expected gender norms. When gender identity does not align with expectations based on sex assigned at birth, these feedback and reinforcement mechanisms often are not available, and may even be confusing and contradictory to young children. As a result the timing of gender development and pathways toward developmental milestones among TGD individuals are highly variable. Gender identity models have therefore shifted away from the historical emphasis on gender identity outcomes (e.g., onset, stability) and toward a conceptualization of gender identity development as a *process* (Katz-Wise et al., 2017).

(Trans)gender developmental milestones

Milestones in the gender development of TGD individuals have been primarily studied in two ways. First, TGD adults have been surveyed retrospectively about their early experiences, such as with the US Transgender Survey, which captures thousands of people each year (James et al., 2016; Turban et al., 2023). Another large retrospective survey assessed milestones among TGD individuals aged 16–73 years to evaluate generational differences specifically (Puckett et al., 2022). Gender development milestones for TGD individuals may include the following: awareness that gender does not align with sex assigned at birth, identifying as transgender, nonbinary, or another TGD identity, living as affirmed gender, and accessing gender-affirming medical interventions (Puckett et al., 2022).

Retrospective data from TGD adults are limited by the possibility of cohort effects (e.g., the childhood experiences of TGD adults in the last two decades may differ from those of contemporary TGD children). Furthermore, younger children who hold TGD identities may not be concurrently navigating sexual orientation development; they may be less aware of violating gender norms, and they often have parents who exert strong influence on access to support and resources (Katz-Wise et al., 2017). Finally, minimal representation of nonbinary identities in adult research limits generalizability to TGD youth who endorse gender identities outside of the binary at a much higher rate than adults (Baum et al., 2013; James et al., 2016; Kuper et al., 2018; Puckett et al., 2022). Despite these limitations, retrospective data are helpful for reaching community samples, including those who may not have had support in childhood and adolescence.

The second way in which early developmental milestones have been studied has been among clinic-referred youth. These studies have been able to explore differences between those presenting for gender-affirming medical care at younger ages and stages of puberty compared to older adolescents and those in advanced stages of puberty (Chen et al., 2021; Sorbara et al., 2020). While these methods avoid cohort effects, outcomes are not necessarily generalizable to youth who lack access to care. There is also evidence that nonbinary youth are underrepresented in clinic-referred samples (Kuper et al., 2019). The TYP is the only community sample that has been able to look at milestones related to social and medical transition, but the sample consists explicitly of youth who recognized and disclosed a transgender identity early in life and who, like the samples of clinic-referred youth, are more likely to have strong parental support.

Overall, reaching gender development milestones is associated with better mental health (Puckett et al., 2022), especially at younger ages (Chen et al., 2021; Sorbara et al., 2020; Topaz et al., 2024). For young people, parents and caregivers play a significant role in whether milestones can be met—even recognition of identity may be more easily achieved in a home that provides safety for and acceptance of gender exploration. One research group conducted semistructured interviews about the barriers to meeting gender development milestones and used the results to conduct a survey on the same topic (Sorbara et al., 2021). According to the results some parents question the validity of their child's diverse gender identity, primarily due to a perceived lack of gender diverse expression and/or worry that external factors are playing a role. Family religious affiliation and discomfort discussing gender with family were also barriers, particularly to disclosure of identity.

Realization and disclosure

According to the 2015 US Transgender Survey, 60% of TGD adults recognized a difference between their gender identity and sex assigned at birth by 10 years of age and 96% by 22 years, suggesting a large majority of TGD individuals begin to understand their gender during childhood and adolescence (James et al., 2016; Turban et al., 2023). Generationally, while individuals in the Boomer cohort overall report having met their other milestones later than those in Generation X, Millennial, and Generation Z cohorts, all groups tend to report realization of gender identity before age 15 years (Puckett et al., 2022).

Clinic-referred youth similarly report early realization, with one study finding an average age of recognition of 8 years (Olson et al., 2015). Those seeking medical care at earlier ages and pubertal stages report earlier realization than those seeking care later. The average age of disclosure in a cohort of youth seeking pubertal suppression earlier in puberty was 6 years, while the average age of realization in an older cohort of youth seeking hormones was 10 years (Chen et al., 2021). This study found no differences in the age of realization across sex assigned at birth. In another study the average age of realization in youth 14 years and younger was approximately 6 years, while the average age in those 15 years and older was 9 years (Sorbara et al., 2020).

One analysis of the US Transgender Survey found that transgender women were more likely to report realization in childhood, while nonbinary individuals were more likely to report realization in adolescence (Turban et al., 2023). Clinic-referred nonbinary youth also report a later age of realization than their binary transgender peers (Topaz et al., 2024). Generationally, individuals in Boomer and Generation X cohorts are more likely to identify as transgender women, while those in Millennial and Generation Z cohorts are more likely to report nonbinary identities (Puckett et al., 2022). These findings may be related to representation. Transgender women have had more historical media representation than transgender men or nonbinary people, and nonbinary identities are not discussed as much in general discourse compared to binary identities.

Young people exhibit a high degree of variability in both the timing of their initial disclosure of gender identity and to whom they choose to make this disclosure (e.g., friend, caregiver, sibling) (Kuper et al., 2019). According to the US Transgender Survey the median age of telling someone about being TGD was 20 years among those who had realized before age 10, and the median number of years between realization and disclosure was 14 years (Turban et al., 2023). These results likely represent some cohort effects, and the time between realization and disclosure may be getting smaller with greater acceptance of gender diversity. These results nonetheless highlight the fact that many young people have recognized a difference in their gender identity long before telling anyone.

In one study of clinic-referred youth the average age of realization was 8 years, while the average age of telling others was 16 years, suggesting that there still exists a gap between realization and disclosure even among young people with parental support (Olson et al., 2015). In another study, those presenting for medical care at younger ages also reported disclosure at younger ages, but there was no difference across age regarding the length of time between realization and disclosure (Sorbara et al., 2020). TGD youth also report a longer time between recognition and disclosure than their parents perceive (Sorbara et al., 2021). Higher socioeconomic status has been found to be associated with a shorter period between realization and disclosure (Topaz et al., 2024).

Social transition

Once young people come out, there is often a sense of urgency to pursue social and/or medical transition, having overcome the hurdles of labeling their experience and fear of rejection to disclose their identity (Medico et al., 2020). While transgender women tend to report earlier age of recognition of gender, they conversely report later ages for social or medical transition (Puckett et al., 2022). In clinic-referred samples of youth, those assigned male at birth are significantly less likely to be living full-time in their affirmed gender than those assigned female at birth (Chen et al., 2021; Olson et al., 2015). They are also more likely to change appearance simultaneously with name and pronouns, whereas those assigned female at birth often change appearance before name and pronouns (Kuper et al., 2019). These results may be related to higher stigmatization of feminine characteristics among people perceived as male compared to masculine characteristics among people perceived as female.

One study from the TYP conducted semistructured interviews with parents and caregivers of transgender children and gender-nonconforming children about their social transition experiences (Olson et al., 2019). For transgender children who underwent early social transition, parents described the decision as being driven largely by the child. Fewer parents cited a notable behavior change or the advice of a professional as reasons for pursuing social transition, and very few cited the media, support groups, or other people. In a separate survey, these parents also reported having discussed the possibility of retransition.

Decisions not to pursue social transition were also largely driven by the child, either because it was never brought up or because the child specifically declined. Parent resistance to transition was rarely brought up in this study, likely due to a predominantly supportive group of participating parents. Most families of gender-nonconforming children had discussed the possibility of transition, suggesting that discussions alone do not necessarily lead to transition. Gender-nonconforming children from the TYP who later went on to socially transition had improved levels of anxiety and depression compared to before transition (Durwood et al., 2017).

Medical transition

Transgender youth from the TYP had low levels of anxiety and depression at baseline, which continued throughout medical treatment, supporting both social and medical transition as protective factors (Wittlin et al., 2025a). Furthermore, while cisgender girls had worsening mental health into adolescence and cisgender boys had improving mental health, the transgender adolescents remained consistent and fell between the other two groups.

Timing of presentation for medical care is associated with differences in mental health, with those presenting at younger ages and in earlier stages of puberty having better mental health outcomes (Chen et al., 2021; Sorbara et al., 2020; Topaz et al., 2024). Those presenting for medical care at younger ages were more likely to initially approach their primary care doctor about gender, while those presenting at older ages were more likely to initially approach a mental health provider (Sorbara et al., 2021). Higher socioeconomic status is associated with earlier age of accessing medical care and a shorter time between disclosure of identity and referral for care. Those who access care younger are also more likely to be living with both parents, suggesting that access to care may be impacted by families who are not on the same page and may be less supportive (Arnoldussen et al., 2023). Among those who present later for care, there also tends to be a larger discrepancy between youth and their caregivers regarding the importance of medical care.

Some studies have found a greater proportion of youth assigned female at birth access care at older ages compared to younger ages (Arnoldussen et al., 2023), but other studies have found no differences across sex assigned at birth (Topaz et al., 2024). In a clinic sample a smaller proportion of nonbinary adolescents (62%) desired medical interventions than binary transgender adolescents (95%), although it was nonetheless the majority (Herrmann et al., 2024). In a community sample, nonbinary adolescents and young adults were less likely to have accessed medical care or to have plans to access care compared to their binary transgender peers (Todd et al., 2019).

Common developmental pathways in transgender and gender diverse children

While the timing of transgender developmental milestones is highly varied, patterns of realization, disclosure, and affirmation do follow some common pathways. It is well-established that some children who explore gender identity and/or expression in childhood do not go on to identify as TGD in adolescence. While desistance is no longer considered an appropriate or useful concept, information about the experiences of gender among young people seeking clinical care can be found in this body of literature, separate from the flawed interpretations about rates of desistance (Newhook et al., 2018). For gender-nonconforming children who do not go on to identify as TGD in adolescence, 10–13 years appears to be a critical time for realization (Olson et al., 2022; Steensma et al., 2011). Reasons for identifying with sex assigned at birth include puberty changes, growing understanding of anatomy, and sexual interests (Steensma et al., 2011).

For adolescents who do identify as TGD, qualitative research among clinic-referred youth has revealed three common pathways (Medico et al., 2020; Sansfaçon et al., 2020). Qualitative interviews were conducted with 36 youth aged 9–17 years and their parents in Canada (Sansfaçon et al., 2020) and 10 youth aged 9–21 years and their parents in Switzerland (Medico et al., 2020). Across both studies, thematic analyses revealed three similar trajectories: (1) early dissonance paired with early affirmation (the affirmed child); (2) early dissonance paired with delayed transition (the silent child); and (3) later-emerging experience of gender dysphoria (the neutral child).

The first pathway describes children who have early realization and disclosure, often followed by social transition and earlier access to gender-affirming medical interventions of interest. Participants in the TYP represent a good example of youth following this pathway. The second pathway describes children who have an early realization of their gender as different from the sex assigned at birth, but for a variety of reasons do not disclose this identity until the onset of puberty. For example, a child who identifies as TGD but is not experiencing distress may decide that the fear of potential parental rejection outweighs the potential benefits of disclosing identity. With the onset of puberty and changes to secondary sex characteristics, gender dysphoria and a desire to seek intervention may begin to outweigh the fear of rejection.

The third pathway describes adolescents who both recognize and disclose their gender identity after the onset of puberty. In these cases, puberty and secondary sex characteristics likely play a role in facilitating realization by being perceived as incongruent with the internal sense of gender and potentially experienced as distressing. For the latter two pathways, as with children who explore gender identity in childhood but do not go on to identify as TGD in adolescence, puberty is a critical time for identity consolidation. The second pathway (early realization and delayed disclosure) appears to be the most common pathway (Medico et al., 2020), consistent with quantitative research on gender development milestones.

Transactional model of transgender identity development

The Transactional Model of Transgender Identity Development provides a comprehensive understanding of TGD identity development as a dynamic, transactional process between youth and their caregivers, such that each mutually influences the other's experiences and understanding of gender identity (Katz-Wise et al., 2017). To explore all factors that contribute to identity development, semistructured interviews were conducted with transgender and gender-nonconforming youth and caregivers recruited from the community, allowing for a broad range of experiences to be captured. Youth with highly unsupportive families are still underrepresented, given barriers to participation in that context. The results of the study demonstrated an array of factors that influence identity development among TGD youth, including broad sociocultural influences, individual biological influences such as puberty, and the bidirectional influence of family, community, and access to resources.

Central to this model is the notion that identity development is not isolated but instead occurs in relation to caregiver attitudes and responses. Specifically, caregiver reactions, support, and acceptance of a child's gender expression and identity shape identity development, while the experiences and expressed identities of the child influence caregiver understanding and acceptance. This framework moves away from linear, stage-based views of identity formation, instead conceptualizing development as occurring through transactional pathways. These pathways are shaped by caregiver responses to early gender nonconformity, youth disclosure, and changing family dynamics.

According to this model, gender identity development is impacted by internal and external challenges, including self-acceptance, peer relationships, societal stigma, gender-related stress, and decision-making about social or medical transition. The model also addresses caregiver internal and interpersonal processes that impact youth well-being and resilience, including managing expectations, learning about gender diversity, and seeking social support. The degree of caregiver support can significantly impact the mental health and resilience of TGD youth, influencing their capacity to navigate their identity development in positive ways. Finally, the model acknowledges that contextual and cultural factors shape these pathways by influencing experiences of both youth and caregivers. Such factors include race, ethnicity, socioeconomic status, and community attitudes and beliefs about gender identity.

The model challenges traditional, stage-based frameworks, as well as unhelpful concepts like "onset" and "persistence," by suggesting that identity development is fluid and shaped by a combination of youth experiences and caregiver influences. This model highlights that identity development among TGD youth is not an isolated process, but rather intertwined with family dynamics, and thereby facilitates a more holistic understanding of TGD identity development in youth. The model underscores the importance of family-centered approaches in supporting TGD youth and encouraging supportive family environments for positive youth outcomes.

Nonbinary gender identity development

While less research has focused on nonbinary identity development specifically, models that focus on nonbinary gender identity development tend to be nonlinear and flexible, to account for various trajectories and fluidity (Matsuno and Budge, 2017; Ward and Lucas, 2024). According to one such framework that has been proposed to move beyond the limitations of cisnormative assumptions to characterize gender development in nonbinary children, gender socialization and exposure to gender identities are key aspects of development (Salinas-Quiroz and Sweder, 2023). Often between the ages of 4–8 years, children expand their understanding of gender as a spectrum by exposure to additional contexts (e.g., school and media). Rigid and binary gender norms decrease exploration, while learning about a wide range of gender expression may increase perceptions of safety to explore gender identity, particularly for nonbinary youth.

Nonbinary youth with a gender expression that conforms to binary expectations may be less subject to victimization than those who demonstrate greater gender nonconformity, but they may also receive less validation from allies (Diamond, 2020). Using Photovoice, a methodology to collect photographs of lived experiences, nonbinary emerging adults were found to experience feelings of invisibility (Cosgrove, 2021). Invisibility and erasure of nonbinary identities often resulted in barriers to developing interpersonal connections and reduced safety, especially in public spaces (e.g., binary bathroom or gendered activities). More research is needed to clarify the specific experiences of youth who identify outside of the gender binary, whether there are similarities or differences in reaching gender development milestones, and what additional supports may be needed to promote healthy identity development across the gender spectrum.

Factors influencing gender development

There is strong empirical support to suggest that the development of gender is influenced by a complex interplay of biological, psychological, and social factors, and emerging research supports a similar interplay of factors in the development of TGD identities.

Biological factors

Research has long emphasized the role of genetics and hormones in gender identity and gender-linked traits. Twin studies and genetic testing have been used to investigate genetic influences on gender identity, expression, and the heritability of TGD identities specifically. Androgens and estrogens, which are thought to interact with social factors in gender development, have also been studied among children with conditions like congenital adrenal hyperplasia (CAH) and complete androgen insensitivity syndrome (CAIS), as well as those exposed to typical and atypical prenatal hormone levels. Research on biological factors influencing gender development has historically been limited by methodologies predicated on cisnormative, binary models of gender, a reliance on sex and gender stereotypes, and a failure to consider the influence of the social interpretation of biological differences.

Heritability

Research with monozygotic and dizygotic twins indicates that both genetic and shared environmental factors may contribute to gender development (Coolidge et al., 2002; Elizabeth and Green, 1984; Heylens et al., 2012). Early research explored gender-stereotyped behavior among twins presumed to be cisgender, while more recent research has incorporated TGD identities specifically. Among 702 pairs of presumed cisgender twins aged 4–12 years, parents of assigned male twins rated their gender-typed interests as the most similar, followed by assigned female twins and lastly female–male pairs (Elizabeth and Green, 1984). More variability in behavior and preferences was observed among female twins and dizygotic twins compared to monozygotic and male pairs.

Using a 6-item, parent-report measure based on the DSM-IV criteria for gender identity disorder among parents of 162 pairs of monozygotic and dizygotic twins aged 4–17 years, a genetic component accounted for 62% of the variance in different scores, supporting the potential heritability of gender incongruence and associated distress (Coolidge et al., 2002). A systematic review of case studies of twins in which at least one twin was diagnosed with gender identity disorder found 37.5% concordance among monozygotic twins assigned female at birth and 40.0% concordance among monozygotic twins assigned male at birth, while there was zero concordance among dizygotic twins (Heylens et al., 2012). These studies rely on the assumption that monozygotic and dizygotic twins experience equally similar environments. Because monozygotic twins may be treated more similarly than dizygotic twins in terms of gendered expectations and responses to nonconformity, it is difficult to differentiate genetic and environmental influences in these cases (Hood et al., 2010).

While a single case study bears limited generalizability, it is notable that there is one documented case of concordance for The Diagnostic Statistical Manual of Mental Disorders, Fifth Edition (DSM-5) diagnosis of gender dysphoria among three monozygotic triplets, all of whom went on to socially and medically transition (Kauffman et al., 2022). The use of Diagnostic Statistical Manual of Mental Disorders, Fourth Edition (DSM-IV) criteria for gender identity disorder in the majority of these studies limits the ability to generalize the results to the TGD population as it is understood today. Furthermore, the DSM diagnostic criteria for both gender identity disorder and gender dysphoria require the presence of distress or impairment, and more research is needed to understand the heritability along a more expansive spectrum of gender diverse identities and expressions.

The role of genetics has also been researched through DNA sequencing. Among 94 TGD individuals and 94 cisgender individuals in Spain, two single-nucleotide polymorphism differences were observed between the groups, specifically at genes related to metabotropic glutamate receptors (Fernández et al., 2024). Animal models have similarly demonstrated a relationship between these genetic differences and sex-stereotyped behaviors.

Prenatal hormonal exposure

Hormone levels during gestation have been researched in nonhuman mammals and in human children who were exposed to typical as well as atypical variations in hormone levels (Hines et al., 2015). The presence or absence of androgens has been proposed to play a role in shaping gender identity and gender-stereotyped preferences (Hines, 2011). One mechanism for the relationship between prenatal androgen exposure and variations in gender identity and behaviors may be neural organization (Hines, 2011). For most births, testosterone is higher for assigned males than females between 8 and 24 weeks of gestation and the first 3 months after birth (Hines et al., 2015). Sexual differentiation of the genitals occurs in the first 2 months of pregnancy, while differentiation of the brain occurs in the second half of pregnancy (Bao and Swaab, 2011; Nguyen et al., 2019). It is now well-known that brains are neither "female" nor "male," but a complex mosaic of features common to males, females, or both (Brown et al., 2020; Joel et al., 2015; Nguyen et al., 2019). However, there are some differences that are helpful for understanding gender development.

In animal models, females treated with prenatal androgens display more male-stereotyped behavior after birth, while males treated with prenatal antiandrogens display more female-stereotyped behavior, including play, sexual behaviors, and aggression (Hines et al., 2015; McCarthy et al., 2012). Further supporting the role of neural organization, animal models have also identified sexually dimorphic neural structures that differ in size in male versus female animals and change in size in opposite directions for male animals compared to female animals when hormones are manipulated (Hines, 2011).

Some aspects of neural development experience organization prenatally, while activation does not occur until puberty. The size of the anteroventral paraventricular nucleus appears to be determined prenatally, but volume size changes do not occur until puberty (Hines, 2011). This observation, that neural organization occurs in utero but with activation in puberty, may begin to explain why some TGD youth do not experience gender incongruence or dysphoria until the onset of puberty. Animal models have also shown that the level of manipulation of hormones affects subsequent structural changes, suggesting that androgen exposure likely plays a part in individual differences within sex in addition to differences between male and female animals. Because the effects of prenatal hormones on neural differences can differ across species, these results cannot be automatically generalized to humans (Hines, 2011).

Due to the ethical limitations of prenatal hormone manipulations in humans, most human research comes from populations with atypical prenatal hormone environments (Hines, 2011; Hines et al., 2015). For example, individuals with CAIS have XY chromosomes and testes but are unable to process androgens, leading to female primary and secondary sex characteristics. The majority of individuals with CAIS are assigned female at birth and grow up to identify as women. Additionally, this population tends to display more female- than male-stereotyped behaviors (Hines et al., 2015). For example, scores on a Gender/Sex and Activities Questionnaire did not significantly differ between adult women with CAIS and cisgender women without CAIS, and both groups were significantly different from cisgender men on several aspects of gender-stereotyped preferences and activities (Hines et al., 2003). Because most women with CAIS are assigned female at birth, it is difficult to differentiate the influence of prenatal hormone exposure from the influence of gender socialization.

While research consistently shows that assigned female children with CAH, who experience higher levels of androgen exposure prenatally, tend to demonstrate more gender-nonconforming behaviors and interests in childhood, the literature is mixed when it comes to gender identity development. Individuals assigned female at birth with CAH tend to display more gender-nonconforming play and preferences in childhood compared to the general female population (Hines, 2011; Hines et al., 2015). In one research study in which presumed cisgender children 4–11 years were randomized into conditions and taught that gender-neutral toys held gendered properties, females with CAH demonstrated a much lower preference for the gender-stereotyped toys compared to assigned males with CAH or to children without CAH (Hines et al., 2016). Some studies show slightly elevated rates of TGD identity and/or gender dysphoria (5.2%) among assigned female children with versus without CAH (Dessens et al., 2005), while other studies found no differences in TGD identities and even that gender identity and satisfaction with being female were stronger in assigned female children with compared to without CAH (Endendijk et al., 2016b; Meyer-Bahlburg et al., 2004).

Another strategy for researching prenatal hormone exposure has been to measure blood samples during pregnancy to evaluate the impact of normal variation in hormone exposure (Hines et al., 2015). This line of research has shown some evidence that normal variability leads to differences in behavior and preferences (Hines et al., 2002), but this overall body of research is more mixed (Hines et al., 2015). Using parent reports of child behaviors after having collected prenatal blood samples showed that normal variations in hormone levels appeared to be related to behavioral differences among assigned females, but not among assigned males (Hines et al., 2002). Children of mothers who were prescribed prenatal hormones that increased or decreased the presence of androgens are also more likely to display less conforming gender-stereotyped behaviors and preferences (Hines et al., 2015; Hines, 2011). Finally, there is a small, limited amount of research evidence that hormone levels during mini-puberty (the surge of estrogen or testosterone in the first few months of life) and the onset of puberty in adolescence may also impact development, although these impacts are less well-understood (Hines et al., 2015; Hines, 2011).

Transgender neural development

Some studies have shown differences in neural structure and functionality specific to TGD individuals that are not clearly tied to specific genetic or hormonal influences. Some of the earliest research on the determinants of transgender identity employed posthumous brain dissections and found differences in specific brain regions. For example, transgender women were found to have a smaller bed nucleus of the stria terminalis compared to cisgender men (independent of gender-affirming hormone use), which is more similar to cisgender women (Zhou et al., 1995). This is another brain region particularly relevant to puberty and further evidence that while puberty is unlikely to change established identity, it can be a clarifying time both biologically and socially.

In neuroimaging studies, transgender adults who have not initiated hormones have demonstrated more similar structural and activation patterns to their same-gender cisgender peers, although there is also conflicting evidence in this regard (Nguyen et al., 2019). In a study of early neural lateralization, transgender women and men both demonstrated more balanced language lateralization, in a manner more similar to cisgender women and less similar to cisgender men, who tend to show stronger one-sided lateralization (Govier et al., 2010). In contrast, in a large study of (N=2165) adolescents, neuroimaging data revealed few differences between teens who endorsed TGD identities versus those who did not (Xerxa et al., 2023). The one exception was among teens assigned male at birth, for whom there was more cortical thickness in the left inferior temporal gyrus among TGD participants. However, this study did not control for individual differences, including hormone use, and generalization is therefore limited.

Gender-affirming medical interventions

Historically, researchers, clinicians, and caregivers have expected or assumed a degree of gender stability in a TGD identity before encouraging youth to engage in steps to socially transition and an even greater degree of stability before allowing youth to pursue gender-affirming medical interventions. Recently, however, researchers have emphasized that social and medical transitions play an important role in gender identity development and consolidation (Ashley, 2019; Ehrensaft, 2017).

Interventions such as gender-affirming hormones may influence brain and body composition in a manner that shapes gender development and identity. Largely conducted among TGD adults, cross-sectional studies have found that adults taking gender-affirming hormones show some structural differences more similar to cisgender adults of the same gender, but also that some areas are more similar among transgender men and women compared to cisgender men and women, suggesting some differences may be more related to factors specific to TGD identities, such as gender dysphoria or experiences of minority stress. Longitudinal studies have more consistently shown structural changes in the direction of same-gender cisgender peers. The literature on functional activation, as opposed to structure, is much more mixed, with few differences consistently identified (Nguyen et al., 2019).

Social and environmental factors

An individual's social and cultural environment influences gender development through various agents, including parents, peers, media, and school. These are the avenues through which children learn to assign genders to people, objects, and behaviors (deMayo et al., 2022b). These influences vary in salience throughout childhood (e.g., parent influence is strongest in infancy and early childhood and later superseded by peer influence). Social cognitive theorists posit three avenues of influence: (1) modeling, which involves exposure to gender-linked information portrayed by models in the environment, (2) enactive experience, which involves the child discerning gender-linked information based on the evaluative social reactions resulting from their actions, and (3) direct instruction (Bussey and Bandura, 1999).

Importantly, while these factors influence gender development and gender-related beliefs and experiences, no research has supported the idea that gender identity can be voluntarily changed or influenced via intentional intervention. Conversely, a large body of research has shown that such attempts to alter gender identity are both ineffective and harmful (APA, 2021). Research on parenting behaviors or parent–child relationship factors as a *cause* of gender identity has similarly been problematic and has not generated any supportive results (Steensma et al., 2013). While social factors undoubtedly influence gender development, TGD children by their very existence demonstrate that social and environmental factors alone cannot fully account for gender identity. Across historical periods, youth have articulated TGD identities even as they are raised in cisnormative contexts wherein gender-nonconforming behavior and TGD identities are discouraged and constrained (deMayo et al., 2022b). While parents, peers, and other social factors do not appear to influence identity formation itself, it is undeniable that supportive social relationships play an important role in healthy gender identity development, gender expression, and self-acceptance among TGD young people (Doyle, 2022).

Parents as sources of social influence

Parents are perhaps the most influential agents of gender socialization in children and serve as the primary influencers of gender socialization in early childhood, specifically. Most parents initiate the process of gender socialization even before their child is born, through activities such as choosing gender-specific names and nursery colors (Bussey, 2014). There are mixed findings related to the impact of gender bias on parental perceptions of the behavior and emotions of infants. These studies present adults with pictures, videos, or animations of infants and randomize them to believe the baby is assigned female or male, to measure differential behaviors toward the infants (Araujo et al., 2022; Condry and Condry, 1976; Plant et al., 2000). While some studies have demonstrated that adults describe crying infants differently based on the perceived sex of the infant, such that they are more likely to attribute emotions like fear to females and anger to males (Araujo et al., 2022; Condry and Condry, 1976), other research has failed to replicate these findings (Plant et al., 2000).

Research on parental influence on gender in early childhood relies primarily on observational methods, often recording parent–child interactions and coding various behaviors, such as differential control strategies and conversation content (Endendijk et al., 2016a; Fivush et al., 2000; Gelman et al., 2004). Between ages 3 and 5 years, parents may continue to enforce gender norms by encouraging gender-stereotyped play (Bussey, 2014), generalizing traits based on gender while reading stories to children (Gelman et al., 2004), and using more emotion-based language and context with children assigned female compared to male (Fivush et al., 2000). In contrast a meta-analysis found no differences in the use of control strategies toward children assigned female compared to male, possibly suggesting changes over time or a lesser role of gender in control strategies compared to other aspects of parenting (Endendijk et al., 2016a).

Beyond encouraging, modeling, and direct and indirect teaching of gender-*conforming* behavior, parents also provide influence via enactive experiences and direct instruction when children exhibit gender-nonconforming behavior. Parental attitudes and behaviors toward gender nonconformity appear to influence the gender-nonconforming behavior of adolescents, with notable differences in parental attitudes about gender nonconformity based on the assigned gender of the child. Research has found that parents are generally less accepting of gender nonconformity in children assigned male compared to female, particularly when the parents hold more traditional beliefs (Spivey et al., 2018). Studies have also shown that adolescents perceive pressure from parents to conform to gender-stereotypic behavior (Cook et al., 2019; Jackson et al., 2021), and parental attempts to alter an adolescent's behaviors are a stronger predictor of adolescent gender expression than a parent's own gender conformity and attitudes about gender conformity (Lan and Isacoff, 2022).

Certain parental responses to gender nonconformity appear to be harmful to the psychosocial adjustment of children (D'Augelli et al., 2006; Ryan et al., 2010). One study that showed parents two vignettes, one of a gender-conforming boy and another of a gender-nonconforming boy, found greater concern among parents for the future of the nonconforming boy (Jacques et al., 2022). These findings suggest that parents may discourage gender-nonconforming behavior out of protective concern for their child's future, even as this parental response itself may exert negative effects on a child's development. Unconditional positive regard from parents of gender-nonconforming children is associated with the most positive outcomes, fostering an open, safe environment for healthy exploration and development of gender identity (Seager van Dyk et al., 2021). Regarding TGD youth specifically, parental reactions toward a child's gender expression and identity typically encompass many stages, with many parents exhibiting a trajectory from personal emotionality toward acceptance and child-focused advocacy (Abreu et al., 2019).

In complement to research investigating presumed cisgender parental reactions and behavior related to child gender nonconformity, research has also investigated the influence of parental TGD identities and of family structures that do not conform to societal gender stereotypes (e.g., families with two same-gender parents) on children's gender development. Although a significant proportion of TGD adults are parents, and many childless TGD people endorse interest in becoming parents, literature on TGD parents is relatively limited (Riskind and Tornello, 2022). Transgender parents might model less gender-conforming behaviors and sometimes forgo labeling a child's gender identity, allowing more possibilities for children's gender expression and identity development (Riskind and Tornello, 2022). Young children with a TGD sibling or parent appear to hold less stereotypical gender beliefs and are more accepting of gender nonconformity (Olson and Enright, 2018). Overall, children with at least one nonbinary or binary transgender parent generally show typical gender development, though they may be more gender-expansive in their play and interests compared to normed samples, particularly for children assigned male at birth (Riskind and Tornello, 2022). Nonbinary and binary transgender parents may be more likely to take a child-centered, developmentally aware approach to gender, challenging common gender assumptions (McGuire et al., 2016).

Peers and other sources of social influence

Peers begin to play a significant role in gender socialization as early as preschool by reinforcing gender-typed play and criticizing gender-nonconforming behavior, also known as "gender policing," with cisgender boys being particularly unforgiving of gender-nonconforming play (Brown et al., 2020; Xiao et al., 2019). The more often young children play with same-gender peers, the more their behavior aligns with gender stereotypes (Martin and Fabes, 2001). Peer groups play a crucial role in gender development during elementary school years. Single-gender peer groups are more common in elementary school and reinforce gender norms. In contrast, research suggests that mixed-gender groups are less stereotyped and may provide more support and flexibility for nonbinary children to express their gender identity (Waagen, 2022; Salinas-Quiroz et al., 2025).

Overall, research suggests that peers likely play little to no role in gender identity development, but they likely do play a role in behaviors and experiences of gender. In one longitudinal study of seventh and eighth graders, several dimensions of gender identity and socialization were measured over time to examine the role of peer influence (Kornienko et al., 2016). Students who participated in the study selected up to 10 peers that they considered friends, self-identified their own gender identity, and completed measures of gender typicality (how similar they felt to those of the same gender), gender contentedness (how positive they felt to their own gender group), felt pressure (how much pressure they felt from peers to be gender-conforming), and intergroup bias (how negative they felt toward the other gender group). The results of the study found that peer influence played a role in between-gender dimensions (felt pressure and intergroup bias) but not within-gender dimension factors (gender typicality and gender connectedness). In other words, how negatively students felt about the other gender and how much pressure they felt to be gender-conforming was influenced by their peers, but how gender-conforming they actually felt and how positively they felt about their own gender was not influenced by peers.

Supportive peer relationships, including those accessed through social media, play an important role in healthy identity development and expression for TGD youth specifically. Use of social media has been found to be a helpful contributor to TGD identity development by allowing young people to try out different gender expressions, often anonymously, exposing them to diverse identities and sometimes validating their own experiences, and by being a source of information on gender identity and gender transition options (Berger et al., 2022). A large national survey of US adolescents found that, in contrast to their cisgender peers who demonstrated a correlation between negative body image and social media use, more social media use among TGD adolescents was associated with more positive body image (Allen et al., 2021). Qualitative interviews with TGD adolescents seeking medical care found similar themes, including the importance of finding community online, accessing information on identity and coming out, and seeing other gender diverse young people who are happy and thriving, thereby giving TGD youth hope for themselves and their futures (Adkins et al., 2018).

Media, including television programs, books, and movies, also play a role in shaping gender stereotypes internalized by children (Brown et al., 2020). Historically, media portrayals have reinforced traditional gender roles. However, some content intended for youth audiences has shifted to include more diverse and inclusive representations of gender. Media consumed by children exerts an influence on their understanding of gender roles and expectations (Durkin and Nugent, 1998; Stever, 2017).

Teachers and school environments also contribute to gender socialization by reinforcing gender norms through curriculum choices and classroom interactions. Schools can either perpetuate traditional gender roles or promote a more inclusive understanding of gender diversity (Eccles, 1987). A research review found that many educators demonstrate clear gender biases, perpetuate heteronormativity, and lack appropriate knowledge about TGD identities (Martin et al., 2025). In contrast, positive modeling of gender nonconformity in a child's environment can positively influence gender development (Salinas-Quiroz et al., 2025).

Cultural factors

Ideas about whether behaviors, objects, and characteristics are masculine or feminine, neither masculine nor feminine, or both masculine and feminine are informed by culture and differ across cultures (Boskey, 2014). There are also differences across cultures with regard to how gender is defined, differential treatment on the basis of gender, and the rigidity or flexibility with which gender stereotypes are upheld (Brown et al., 2020). Some cultures have long histories of recognizing gender identities outside of the binary, while white Westernized cultures have historically punished gender nonconformity. Hijras in India, Fa'afafine in Samoa, and Two Spirit people in North America are examples of identities that do not conform to binary gender identities or roles but were accepted and, in some cases, celebrated within their cultural context (Brown et al., 2020). As a result of historical colonization and intersectional oppression, Two Spirit and other indigenous gender diverse identities have been largely excluded from research on gender and gender identity (Angelino and Bell, 2023; Veldhuis et al., 2024). One unique study asked parents of transgender youth in New Zealand to draw pictures about their experiences, the results of which found that a cultural shift toward decolonization is playing a role in the identity development of young people (de Bres and Morrison-Young, 2024).

A small body of research has examined Two Spirit identity development, specifically. Two Spirit identities, which do not directly correspond to Westernized understandings of gender and sexuality, are diverse, fluid, and nonlinear, and the use of and specific meaning of this identity term may differ regionally and/or with regard to specific tribal teachings

(Angelino and Bell, 2023; Thomas et al., 2022). Two Spirit identity formation is also often interwoven with the development of ethnic and spiritual identity (McCoy et al., 2024). One mixed-methods study employed a Two-Eyed Seeing framework to capture stories about coming out and Two Spirit identity development, the findings of which demonstrate both similarities and differences to other conceptualizations of gender identity development (Sylliboy, 2022). For example, the roles of gender dysphoria and social support in identity formation parallel findings in other research, whereas an emphasis on cultural development and spirituality is distinct and in need of further prioritization within identity development research. A 2021 review of all published articles related to Two Spirit identities found that the majority adopted a disease perspective framework, very few focused on adolescents and none on children, and none evaluated identity development in youth (Thomas et al., 2022). The authors called for future research focusing on protective factors and strengths associated with Two Spirit identities, as well as the role of cultural practices in health outcomes.

Children as active agents in their development

Some research on gender identity development may be interpreted to suggest that external biological and social forces act *upon* children and adolescents, and that young people are passive recipients of these influences. However, although no research has demonstrated that individuals can intentionally change their own identity (APA, 2021), children and adolescents do exercise agency through exploration of gender, gender expression, and the eventual consolidation and acceptance of gender identity. This active role is evident in children's engagement with and interpretation of gendered information from the environment, choices of friends and community, and internal pressure around gender identity and conformity. Longitudinal research among the general population of adolescents has shown that felt pressure from self to be gender-conforming is more impactful than perceived pressure from parents or peers (Cook et al., 2019; Jackson et al., 2021).

Recent research on TGD children specifically has also helped to shift the understanding of children as passive recipients of gender socialization to children as active agents who seek out and make sense of information related to gender. Studies examining gender development from the perspective of TGD youth have demonstrated the interaction of internal processes with external and social elements (e.g., family adjustment and support, sociocultural influences, and societal discourse and norms) (Katz-Wise et al., 2017; Kuper et al., 2018).

Interaction among biological, psychological, and social processes

Each child is an active participant in interactions between nature and nurture, which take place within and are influenced by the given cultural contexts. These transactional relationships between nature, nurture, and culture have been likened to a "gender web" (Ehrensaft, 2017). The threads of this web include primary and secondary sex characteristics, hormones, chromosomes, psychological and cognitive processes and schemas, socialization processes (family, school, community), and cultural factors (values, policies, laws, ethics, praxis, theory) (Ehrensaft, 2014). Children are actively engaged in negotiating and weaving these threads over time. No existing models or research fully explain how children develop gender identity or why some children gravitate toward gender-conforming versus gender-nonconforming expression or realize a TGD identity (deMayo et al., 2022b). What is clear, however, is the myriad benefits when children and adolescents are trusted to explore, know, and articulate their own gender. When a particular theory of development fails to account for such diversity in gender across young people, the flaw is with the theory—not with the young people themselves (Brown et al., 2020).

Summary and future research directions

Broad population research has the potential to clarify much about gender development but has historically been limited by cisnormative assumptions. Future research would greatly benefit from longitudinal designs capable of capturing the diverse, fluid, and potentially nonlinear trajectories of gender expression and gender identity over time, to clarify developmental nuances across diverse gender identities. Similarly, research focused on TGD youth has often relied on binary conceptualizations of gender, excluding or erasing nonbinary identities (Riskind and Tornello, 2022). Advancing the field will require research frameworks that explicitly include, rather than exclude, the full spectrum of gender expression and identities (deMayo et al., 2022a). Particularly in population-based research on the biological bases of gender differences, but also in research more broadly, it is recommended that theory-grounded, a priori hypotheses guide research design and implementation (Hines et al., 2015).

This literature has also given limited attention to intersecting identities, including intersections of gender with race, ethnicity, and socioeconomic status, and relatively little is known about the role of culture in gender identity development.

Multiinstitutional and cross-disciplinary collaborations are needed to deepen understanding of the experiences of TGD young people across various dimensions of identity and cultural contexts (deMayo et al., 2022a).

Research has relied heavily on adult retrospective reports and samples of youth referred to gender specialty clinics, limiting generalizability to the broader population of TGD youth. Young people who are not in supportive environments or who do not feel safe disclosing their TGD identity are largely unrepresented in the literature. Although it is important to acknowledge these limitations, ensuring the safety and privacy of TGD youth is ethically imperative, even as doing so constrains the scope of potential research.

Research on gender identity development has also faced persistent measurement challenges. Measurement tools used in broad population-based studies have generally not been validated in TGD populations, and even within TGD-focused research, evolving terminology and conceptual frameworks of gender diversity require frequent updates to measurement tools, limiting direct comparisons across studies. Inconsistent use of terms like "gender dysphoria" and "gender diversity" can further contribute to misinterpretation of study samples (Pullen Sansfaçon et al., 2020).

Both self-report and parent-report measures present distinct limitations. When parents are asked to report on their child's gender identity, the result may not accurately reflect how a child understands their own identity. Although self-report is likely more accurate, young children, especially, may have difficulty responding to questions about gender or articulating their gender identity. Finally, measurement tools have not been used consistently across research specialties or areas of interest. It is recommended that standardized measurements relevant to multiple subfields be developed to work toward integration across subspecialties and comparison across studies (deMayo et al., 2022a).

Finally, advancing research on gender and gender identity development in childhood requires increasing attention to and inclusion of TGD identities and intersecting marginalized identities. The overrepresentation of cisgender researchers operating within cisnormative systems has constrained progress in this area of research (Gill-Peterson, 2017; Salinas-Quiroz and Sweder, 2023). Increased training, mentorship, and structural support for TGD researchers—particularly those with intersecting marginalized identities—represents one pathway toward addressing inequities that shape research agendas and outcomes. Community-driven research priorities, depathologizing frameworks, and conceptualizations of gender as a spectrum rather than a binary are essential to advance ethical, rigorous, and informative research on childhood gender development (deMayo et al., 2022a,b). Ultimately, understanding gender development requires a multifaceted approach that integrates biological, social, and cultural influences. Addressing the gaps in current research, particularly through more inclusive and validated methodologies, is essential for advancing our understanding and support of diverse gender identities in youth.

References

Abreu, R.L., Rosenkrantz, D.E., Ryser-Oatman, J.T., Rostosky, S.S., Riggle, E.D.B., 2019. Parental reactions to transgender and gender diverse children: a literature review. Journal of GLBT Family Studies 15 (5), 461–485. Available at https://doi.org/10.1080/1550428X.2019.1656132.

Adkins, V., Masters, E., Shumer, D., Selkie, E., 2018. Exploring transgender adolescents' use of social media for support and health information seeking. Journal of Adolescent Health 62 (2), S44. Available at https://doi.org/10.1016/j.jadohealth.2017.11.087.

Alfieri, T., Ruble, D.N., Higgins, E.T., 1996. Gender stereotypes during adolescence: developmental changes and the transition to junior high school. Developmental Psychology 32 (6), 1129–1137. Available at https://doi.org/10.1037/0012-1649.32.6.1129.

Allen, B.J., Stratman, Z.E., Kerr, B.R., Zhao, Q., Moreno, M.A., 2021. Associations between psychosocial measures and digital media use among transgender youth: cross-sectional study. JMIR Pediatrics and Parenting 4 (3). Available at https://doi.org/10.2196/25801.

Angelino, A.C., Bell, S., 2023. Rethinking gender diversity terminology: a call to include native two-spirit and LGBTQ1 youth. Pediatrics 152 (4). Available at https://doi.org/10.1542/peds.2023-062093.

APA, 2021. APA Resolution on Gender Identity Change Efforts. APA. Available at https://www.apa.org/about/policy/resolution-gender-identity-change-efforts.pdf.

Araujo, V., Schaffer, D., Costa, A.B., Musse, S.R., 2022. Towards virtual humans without gender stereotyped visual features. Proceedings - SIGGRAPH Asia 2022: Technical Communications. Available at https://doi.org/10.1145/3550340.3564232.

Arnoldussen, M., de Rooy, F.B.B., de Vries, A.L.C., van der Miesen, A.I.R., Popma, A., Steensma, T.D., 2023. Demographics and gender-related measures in younger and older adolescents presenting to a gender service. European Child and Adolescent Psychiatry 32 (12), 2537–2546. Available at https://doi.org/10.1007/s00787-022-02082-8.

Ashley, F., 2019. Thinking an ethics of gender exploration: against delaying transition for transgender and gender creative youth. Clinical Child Psychology and Psychiatry 24 (2), 223–236. Available at https://doi.org/10.1177/1359104519836462.

Bao, A.M., Swaab, D.F., 2011. Sexual differentiation of the human brain: relation to gender identity, sexual orientation and neuropsychiatric disorders. Frontiers in Neuroendocrinology 32 (2), 214–226. Available at https://doi.org/10.1016/j.yfrne.2011.02.007.

Baum, J., Brill, S., Brown, J., Delpercio, A., Kahn, E., Kennedy, L., et al., 2013. Supporting and caring for our gender expansive youth.

Berger, M.N., Taba, M., Marino, J.L., Lim, M.S.C., Skinner, S.R., 2022. Social media use and health and well-being of lesbian, gay, bisexual, transgender, and queer youth: systematic review. Journal of Medical Internet Research 24 (9). Available at https://doi.org/10.2196/38449.

Blakemore, J.E.O., Berenbaum, S.A., Liben, L.S., 2013. Gender Development. Psychology Press.
Boskey, E.R., 2014. Understanding transgender identity development in childhood and adolescence. American Journal of Sexuality Education 9 (4), 445–463. Available at https://doi.org/10.1080/15546128.2014.973131.
de Bres, J., Morrison-Young, I., 2024. Intersectional perspectives of parents of transgender children in Aotearoa (New Zealand). International Journal of Transgender Health 25 (3), 584–601. Available at https://doi.org/10.1080/26895269.2024.2316689.
Brown, C.S., Biefeld, S.D., Tam, M.J., 2020. Gender in Childhood. Cambridge University Press. Available at https://doi.org/10.1017/9781108874281.
Bussey, K., Bandura, A., 1999. Social cognitive theory of gender development and differentiation. Psychological Review 106 (4), 676–713. Available at https://doi.org/10.1037/0033-295X.106.4.676.
Bussey, K., 2014. In: Ryan, M., Branscombe, N. (Eds.), The SAGE Handbook of Gender and Psychology Gender Development. SAGE Publications, Ltd.
Chen, D., Abrams, M., Clark, L., Ehrensaft, D., Tishelman, A.C., Chan, Y.M., et al., 2021. Psychosocial characteristics of transgender youth seeking gender-affirming medical treatment: baseline findings from the trans youth care study. Journal of Adolescent Health 68 (6), 1104–1111. Available at https://doi.org/10.1016/j.jadohealth.2020.07.033.
Cohen-Kettenis, P.T., 2015. Adolescents with gender dysphoria. Best Practice & Research Clinical Endocrinology & Metabolism 29 (3), 485–495. Available at https://doi.org/10.1016/j.beem.2015.01.004.
Condry, J., Condry, S., 1976. Sex differences: a study of the eye of die beholder. Child Development 47 (3), 812–819. Available at https://doi.org/10.1111/j.1467-8624.1976.tb02248.x.
Cook, R.E., Nielson, M.G., Martin, C.L., DeLay, D., 2019. Early adolescent gender development: the differential effects of felt pressure from parents, peers, and the self. Journal of Youth and Adolescence 48 (10), 1912–1923. Available at https://doi.org/10.1007/s10964-019-01122-y.
Cooley, S., Killen, M., 2015. Children's evaluations of resource allocation in the context of group norms. Developmental Psychology 51 (4), 554–563. Available at https://doi.org/10.1037/a0038796.
Coolidge, F.L., Thede, L.L., Young, S.E., 2002. The heritability of gender identity disorder in a child and adolescent twin sample. Behavior Genetics 32 (4), 251–257. Available at https://doi.org/10.1023/A:1019724712983.
Cosgrove, D., 2021. I am allowed to be myself": a photovoice exploration of non-binary identity development and meaning-making. Journal of Gay and Lesbian Social Services 33 (1), 78–102. Available at https://doi.org/10.1080/10538720.2020.1850385.
D'Augelli, A.R., Grossman, A.H., Starks, M.T., 2006. Childhood gender atypicality, victimization, and PTSD among lesbian, gay, and bisexual youth. Journal of Interpersonal Violence 21 (11), 1462–1482.
deMayo, B., Kahn-Samuelson, S., Olson, K.R., 2022a. Endorsement of gender stereotypes in gender diverse and cisgender adolescents and their parents. PLoS ONE 17 (6). Available at https://doi.org/10.1371/journal.pone.0269784.
deMayo, B.E., Jordan, A.E., Olson, K.R., 2022b. Gender development in gender diverse children. Annual Review of Developmental Psychology 4, 207–229. Available at https://doi.org/10.1146/annurev-devpsych-121020-034014.
Dessens, A.B., Slijper, F.M.E., Drop, S.L.S., 2005. Gender dysphoria and gender change in chromosomal females with congenital adrenal hyperplasia. Archives of Sexual Behavior 34 (4), 389–397. Available at https://doi.org/10.1007/s10508-005-4338-5.
Diamond, L.M., 2020. Gender fluidity and nonbinary gender identities among children and adolescents. Child Development Perspectives 14 (2), 110–115. Available at https://doi.org/10.1111/cdep.12366.
Doyle, D.M., 2022. Transgender identity: development, management and affirmation. Current Opinion in Psychology 48, 101467. Available at https://doi.org/10.1016/j.copsyc.2022.101467.
Drummond, K.D., Bradley, S.J., Peterson-Badali, M., Zucker, K.J., 2008. A follow-up study of girls with gender identity disorder. Developmental Psychology 44 (1), 34–45. Available at https://doi.org/10.1037/0012-1649.44.1.34.
Durkin, K., Nugent, B., 1998. Kindergarten children's gender-role expectations for television actors. Sex Roles 38 (5-6), 387–402. Available at https://doi.org/10.1023/a:1018705805012.
Durwood, L., McLaughlin, K.A., Olson, K.R., 2017. Mental health and self-worth in socially transitioned transgender youth. Journal of the American Academy of Child and Adolescent Psychiatry 56 (2), 116. Available at https://doi.org/10.1016/j.jaac.2016.10.016.
Eaton, W.O., von Bargen, D., Keats, J.G., 1981. Gender understanding and dimensions of preschooler toy choice: sex stereotype versus activity level. Canadian Journal of Behavioural Science / Revue canadienne des sciences du comportement 13 (3), 203–209. Available at https://doi.org/10.1037/h0081181.
Eccles, J.S., 1987. Gender roles and women's achievement-related decisions. Psychology of Women Quarterly 11 (2), 135–172. Available at https://doi.org/10.1111/j.1471-6402.1987.tb00781.x.
Ehrensaft, 2017. Gender nonconforming youth: current perspectives. Adolescent Health, Medicine and Therapeutics 8, 57–67. Available at https://doi.org/10.2147/AHMT.S110859.
Ehrensaft, D., 2014. Treating Transgender Children and Adolescents From Gender Identity Disorder to Gender Identity Creativity: True Gender Self Child Therapy. Routledge.
Elizabeth, P.H., Green, R., 1984. Childhood sex-role behaviors: similarities and differences in twins. Acta geneticae medicae et gemellologiae: twin research 33 (2), 173–179. Available at https://doi.org/10.1017/s0001566000007200.
Endendijk, J.J., Groeneveld, M.G., Bakermans-Kranenburg, M.J., Mesman, J., 2016a. Gender-differentiated parenting revisited: meta-analysis reveals very few differences in parental control of boys and girls. PLoS ONE 11 (7). Available at https://doi.org/10.1371/journal.pone.0159193.
Endendijk, J.J., Beltz, A.M., McHale, S.M., Bryk, K., Berenbaum, S.A., 2016b. Linking prenatal androgens to gender-related attitudes, identity, and activities: evidence from girls with congenital adrenal hyperplasia. Archives of Sexual Behavior 45 (7), 1807–1815. Available at https://doi.org/10.1007/s10508-016-0693-7.

Fagan, J., 1976. Infants' recognition of invariant features of faces. Child Development 47 (3), 627–638. Available at https://doi.org/10.2307/1128177, https://doi.org/10.2307/1128177 Available from.

Fast, A.A., Olson, K.R., 2018. Gender development in transgender preschool children. Child Development 89 (2), 620–637. Available at https://doi.org/10.1111/cdev.12758.

Fernández, R., Ramírez, K., Lorente-Bermúdez, R., Gómez-Gil, E., Mora, M., Guillamon, A., et al., 2024. Analysis of single nucleotide polymorphisms of the metabotropic glutamate receptors in a transgender population. Frontiers in Endocrinology 15. Available at https://doi.org/10.3389/fendo.2024.1382861.

Fivush, R., Brotman, M.A., Buckner, J.P., Goodman, S.H., 2000. Gender differences in parent-child emotion narratives. Sex Roles 42 (3-4), 233–253. Available at https://doi.org/10.1023/a:1007091207068.

Gelman, S.A., Taylor, M.G., Nguyen, S.P., Leaper, C., Bigler, R.S., 2004. Mother-child conversations about gender: understanding the acquisition of essentialist beliefs. Monographs of the Society for Research in Child Development 69 (1).

Gill-Peterson, J., 2017. Histories of the Transgender Child. University of Minnesota Press.

Golombok, S., Rust, J., Zervoulis, K., Croudace, T., Golding, J., Hines, M., 2008. Developmental Trajectories of sex-typed behavior in boys and girls: a longitudinal general population study of children aged 2.5–8 years. Child Development 79 (5), 1583–1593. Available at https://doi.org/10.1111/j.1467-8624.2008.01207.x.

Golombok, S., Rust, J., Zervoulis, K., Golding, J., Hines, M., 2012. Continuity in sex-typed behavior from preschool to adolescence: a longitudinal population study of boys and girls aged 3–13 years. Archives of Sexual Behavior 41 (3), 591–597. Available at https://doi.org/10.1007/s10508-011-9784-7.

Goodey, C., 2021. Development: The History of a Psychological Concept. Cambridge University Press. Available at https://doi.org/10.1017/9781108980845.

Govier, E., Diamond, M., Wolowiec, T., Slade, C., 2010. Dichotic listening, handedness, brain organization, and transsexuality. International Journal of Transgenderism 12 (3), 144–154. Available at https://doi.org/10.1080/15532739.2010.514219.

Gülgöz, S., Glazier, J.J., Enright, E.A., Alonso, D.J., Durwood, L.J., Fast, A.A., et al., 2019. Similarity in transgender and cisgender children's gender development. Proceedings of the National Academy of Sciences of the United States of America 116 (49), 24480–24485. Available at https://doi.org/10.1073/pnas.1909367116.

Hässler, T., Glazier, J.J., Olson, K.R., 2022. Consistency of gender identity and preferences across time: an exploration among cisgender and transgender children. Developmental Psychology 58 (11), 2184–2196. Available at https://doi.org/10.1037/dev0001419.

Herrmann, L., Barkmann, C., Bindt, C., Fahrenkrug, S., Breu, F., Grebe, J., et al., 2024. Binary and non-binary gender identities, internalizing problems, and treatment wishes among adolescents referred to a gender identity clinic in Germany. Archives of Sexual Behavior 53 (1), 91–106. Available at https://doi.org/10.1007/s10508-023-02674-8.

Heylens, G., De Cuypere, G., Zucker, K.J., Schelfaut, C., Elaut, E., Vanden Bossche, H., et al., 2012. Gender identity disorder in twins: a review of the case report literature. Journal of Sexual Medicine 9 (3), 751–757. Available at https://doi.org/10.1111/j.1743-6109.2011.02567.x.

Hines, M., Golombok, S., Rust, J., Johnston, K.J., Golding, J., 2002. Testosterone during pregnancy and gender role behavior of preschool children: a longitudinal, population study. Child Development 73 (6), 1678–1687. Available at https://doi.org/10.1111/1467-8624.00498.

Hines, M., Ahmed, S.F., Hughes, I.A., 2003. Psychological outcomes and gender-related development in complete androgen insensitivity syndrome. Archives of Sexual Behavior 32 (2), 93–101. Available at https://doi.org/10.1023/A:1022492106974.

Hines, M., 2011. Gender development and the human brain. Annual Review of Neuroscience 34 (1), 69–88. Available at https://doi.org/10.1146/annurev-neuro-061010-113654.

Hines, M., Constantinescu, M., Spencer, D., 2015. Early androgen exposure and human gender development. Biology of Sex Differences 6 (1). Available at https://doi.org/10.1186/s13293-015-0022-1.

Hines, M., Pasterski, V., Spencer, D., Neufeld, S., Patalay, P., Hindmarsh, P.C., et al., 2016. Prenatal androgen exposure alters girls' responses to information indicating gender-appropriate behaviour. Philosophical Transactions of the Royal Society B: Biological Sciences 371 (1688). Available at https://doi.org/10.1098/rstb.2015.0125.

Hood, K.E., Halpern, C.T., Greenberg, G., Lerner, R.M., 2010. Genetic research in psychiatry and psychology. Handbook of Developmental Science, Behavior, and Genetics. Wiley, pp. 557–625. Available at https://doi.org/10.1002/9781444327632.ch18.

Jackson, E.F., Bussey, K., Myers, E., 2021. Encouraging gender conformity or sanctioning nonconformity? Felt pressure from parents, peers, and the self. Journal of Youth and Adolescence 50 (4), 613–627. Available at https://doi.org/10.1007/s10964-020-01387-8.

Jacques, K.P., Feinstein, B.A., Darling, A.K., Humphreys, K.L., 2022. An analogue study investigating differential parenting of gender conforming and nonconforming boys. Archives of Sexual Behavior 51 (7), 3569–3581. Available at https://doi.org/10.1007/s10508-022-02388-3.

James, S.E., Herman, J.L., Rankin, S., Keisling, M., Mottet, L., Anafi, M., 2016. The Report of the 2015 U.S. Transgender Survey. National Center for Transgender Equality, Washington, DC.

Joel, D., Berman, Z., Tavor, I., Wexler, N., Gaber, O., Stein, Y., et al., 2015. Sex beyond the genitalia: the human brain mosaic. Proceedings of the National Academy of Sciences of the United States of America 112 (50), 15468–15473. Available at https://doi.org/10.1073/pnas.1509654112.

Karrington, B., 2022. Defining desistance: exploring desistance in transgender and gender expansive youth through systematic literature review. Transgender Health 7 (3), 189–212. Available at https://doi.org/10.1089/trgh.2020.0129.

Katz-Wise, Budge, Fugate, Flanagan, Touloumtzis, Rood, et al., 2017. Transactional pathways of transgender identity development in transgender and gender nonconforming youth and caregivers from the trans youth family study. The international journal of transgenderism 18 (3), 243–263. Available at https://doi.org/10.1080/15532739.2017.1304312.

Kauffman, R.P., Guerra, C., Thompson, C.M., Stark, A., 2022. Concordance for gender dysphoria in genetic female monozygotic (identical) triplets. Archives of Sexual Behavior 51 (7), 3647–3651. Available at https://doi.org/10.1007/s10508-022-02409-1.

Khonina, M., Salway, T., 2025. The rise of anti-trans laws and the role of public health advocacy. Canadian Journal of Public Health 116 (1), 97–99. Available at https://doi.org/10.17269/s41997-024-00942-1.

Kornienko, O., Santos, C.E., Martin, C.L., Granger, K.L., 2016. Peer influence on gender identity development in adolescence. Developmental Psychology 52 (10), 1578–1592. Available at https://doi.org/10.1037/dev0000200.

Kuper, L.E., Wright, L., Mustanski, B., 2018. Gender identity development among transgender and gender nonconforming emerging adults: an intersectional approach. International Journal of Transgenderism 19 (4), 436–455. Available at https://doi.org/10.1080/15532739.2018.1443869.

Kuper, L.E., Lindley, L., Lopez, X., 2019. Exploring the gender development histories of children and adolescents presenting for gender affirming medical care. Clinical Practice in Pediatric Psychology 7 (3), 217–228. Available at https://doi.org/10.1037/cpp0000290.

Lan, M., Isacoff, N., 2022. Parents' attitudes predict adolescent gender expression. Psychology 13 (2), 301–312. Available at https://doi.org/10.4236/psych.2022.132018.

Leinbach, M.D., Fagot, B.I., 1986. Acquisition of gender labels: a test for toddlers. Sex Roles 15 (11-12), 655–666. Available at https://doi.org/10.1007/BF00288221.

Martin, C.L., Fabes, R.A., 2001. The stability and consequences of young children's same-sex peer interactions. Developmental Psychology 37 (3), 431–446. Available at https://doi.org/10.1037/0012-1649.37.3.431.

Martin, C.L., Ruble, D.N., Szkrybalo, J., 2002. Cognitive theories of early gender development. Psychological Bulletin 128 (6), 903–933. Available at https://doi.org/10.1037/0033-2909.128.6.903.

Martin, C.L., Andrews, N.C.Z., England, D., Zosuls, K., Ruble, D.N., 2017. A dual identity approach for conceptualizing and measuring children's gender identity. Child Development 88 (1), 167–182. Available at https://doi.org/10.1111/cdev.12568.

Martin, D.J., Pettit, S.K., Stacki, S.L., Smith, K.W., Caskey, M.M., 2025. Understanding young adolescent identity and experiences through internal dimensions: a scoping review. Education Sciences 15 (2). Available at https://doi.org/10.3390/educsci15020253.

Matsuno, E., Budge, S.L., 2017. Non-binary/genderqueer identities: a critical review of the literature. Current Sexual Health Reports 9 (3), 116–120. Available at https://doi.org/10.1007/s11930-017-0111-8.

McCarthy, M.M., Arnold, A.P., Ball, G.F., Blaustein, J.D., de Vries, G.J., 2012. Sex differences in the brain: the not so inconvenient truth. Journal of Neuroscience 32 (7), 2241–2247. Available at https://doi.org/10.1523/JNEUROSCI.5372-11.2012.

McCoy, R.T., O'Brien, R.M., Bongar, B., 2024. Encyclopedia of sexual psychology and behavior. Two-Spirit Identity. Available at https://doi.org/10.1007/978-3-031-08956-5_2495-1.

McGuire, J.K., Kuvalanka, K.A., Catalpa, J.M., Toomey, R.B., 2016. Transfamily theory: how the presence of trans* family members informs gender development in families. Journal of Family Theory and Review 8 (1), 60–73. Available at https://doi.org/10.1111/jftr.12125.

McGuire, J.K., Rider, G.N., Catalpa, J.M., Steensma, T.D., Cohen-Kettenis, P.T., Berg, D.R., 2019. Utrecht Gender Dysphoria Scale – Gender Spectrum (UDGS-GS). In: Milhausen, R.R., Sakaluk, J.K., Fisher, T.D., Davis, C.M., Yarber, W.L. (Eds.), Handbook of Sexuality-Related Measures. Routledge, New York. Available at https://doi.org/10.4324/9781315183169.

Medico, D., Pullen Sansfaçon, A., Zufferey, A., Galantino, G., Bosom, M., Suerich-Gulick, F., 2020. Pathways to gender affirmation in trans youth: a qualitative and participative study with youth and their parents. Clinical Child Psychology and Psychiatry 25 (4), 1002–1014. Available at https://doi.org/10.1177/1359104520938427.

Meyer-Bahlburg, H.F.L., Dolezal, C., Baker, S.W., Carlson, A.D., Obeid, J.S., New, M.I., 2004. Prenatal androgenization affects gender-related behavior but not gender identity in 5-12-year-old girls with congenital adrenal hyperplasia. Archives of Sexual Behavior 33 (2), 97–104. Available at https://doi.org/10.1023/B:ASEB.0000014324.25718.51.

Miller, C.L., 1983. Developmental changes in male/female voice classification by infants. Infant Behavior and Development 6 (2-3), 313–330. Available at https://doi.org/10.1016/S0163-6383(83)80040-X.

Nguyen, H.B., Loughead, J., Lipner, E., Hantsoo, L., Kornfield, S.L., Epperson, C.N., 2019. What has sex got to do with it? The role of hormones in the transgender brain. Neuropsychopharmacology 44 (1), 22–37. Available at https://doi.org/10.1038/s41386-018-0140-7.

Olson, K.R., 2016. Prepubescent transgender children: what we do and do not know. Journal of the American Academy of Child and Adolescent Psychiatry 55 (3), 155. Available at https://doi.org/10.1016/j.jaac.2015.11.015.

Olson, K.R., Enright, E.A., 2018. Do transgender children (gender) stereotype less than their peers and siblings? Developmental Science 21 (4). Available at https://doi.org/10.1111/desc.12606.

Olson, K.R., Gülgöz, S., 2018. Early findings from the TransYouth project: gender development in transgender children. Child Development Perspectives 12 (2), 93–97. Available at https://doi.org/10.1111/cdep.12268.

Olson, J., Schrager, S.M., Belzer, M., Simons, L.K., Clark, L.F., 2015. Baseline physiologic and psychosocial characteristics of transgender youth seeking care for gender dysphoria. Journal of Adolescent Health 57 (4), 374–380. Available at https://doi.org/10.1016/j.jadohealth.2015.04.027.

Olson, K.R., Durwood, L., Demeules, M., McLaughlin, K.A., 2016. Mental health of transgender children who are supported in their identities. Pediatrics 137 (3). Available at https://doi.org/10.1542/peds.2015-3223.

Olson, K.R., Blotner, C., Alonso, D., Lewis, K., Edwards, D., Durwood, L., 2019. Family discussions of early childhood social transitions. Clinical Practice in Pediatric Psychology 7 (3), 229–240. Available at https://doi.org/10.1037/cpp0000289.

Olson, K.R., Durwood, L., Horton, R., Gallagher, N.M., Devor, A., 2022. Gender identity 5 years after social transition. Pediatrics 150 (2). Available at https://doi.org/10.1542/peds.2021-056082.

Plant, E.A., Hyde, J.S., Keltner, D., Devine, P.G., 2000. The gender stereotyping of emotions. Psychology of Women Quarterly 24 (1), 81–92. Available at https://doi.org/10.1111/j.1471-6402.2000.tb01024.x.

Poulin-Dubois, D., Serbin, L.A., Kenyon, B., Derbyshire, A., 1994. Infants' intermodal knowledge about gender. Developmental Psychology 30 (3), 436–442. Available at https://doi.org/10.1037/0012-1649.30.3.436.

Puckett, J.A., Tornello, S., Mustanski, B., Newcomb, M.E., 2022. Gender variations, generational effects, and mental health of transgender people in relation to timing and status of gender identity milestones. Psychology of Sexual Orientation and Gender Diversity 9 (2), 165–178. Available at https://doi.org/10.1037/sgd0000391.

Pullen Sansfaçon, A., Medico, D., Suerich-Gulick, F., Temple Newhook, J., 2020. I knew that I wasn't cis, I knew that, but I didn't know exactly": gender identity development, expression and affirmation in youth who access gender affirming medical care. International Journal of Transgender Health 21 (3), 307–320. Available at https://doi.org/10.1080/26895269.2020.1756551.

Rae, J.R., Gülgöz, S., Durwood, L., DeMeules, M., Lowe, R., Lindquist, G., et al., 2019. Predicting early-childhood gender transitions. Psychological Science 30 (5), 669–681. Available at https://doi.org/10.1177/0956797619830649.

Rhodes, M., Baron, A., 2019. The development of social categorization. Annual Review of Developmental Psychology 1 (1), 359–386. Available at https://doi.org/10.1146/annurev-devpsych-121318-084824.

Riskind, R.G., Tornello, S.L., 2022. I think it's too early to know": gender identity labels and gender expression of young children with nonbinary or binary transgender parents. Frontiers in Psychology 13. Available at https://doi.org/10.3389/fpsyg.2022.916088.

Rudman, L.A., Fairchild, K., 2004. Reactions to counterstereotypic behavior: the role of backlash in cultural stereotype maintenance. Journal of Personality and Social Psychology 87 (2), 157–176. Available at https://doi.org/10.1037/0022-3514.87.2.157.

Ryan, C., Russell, S.T., Huebner, D., Diaz, R., Sanchez, J., 2010. Family acceptance in adolescence and the health of LGBT young adults. Journal of child and adolescent psychiatric nursing: official publication of the Association of Child and Adolescent Psychiatric Nurses, Inc. 23 (4), 205–213. Available at https://doi.org/10.1111/j.1744-6171.2010.00246.x.

Salinas-Quiroz, F., Aral, T., Hillekens, J., Hölscher, S., Demos, J., 2025. You're free from just a girl or a boy": nonbinary children's understanding of their gender. International Journal of Transgender Health 26 (2), 378–395. Available at https://doi.org/10.1080/26895269.2024.2351470.

Salinas-Quiroz, F., Sweder, N., 2023. Authentic gender development in non-binary children. Frontiers in Sociology 8. Available at https://doi.org/10.3389/fsoc.2023.1177766.

Seager van Dyk, I., Shao, J., Sohn, L., Smiley, P.A., Olson, K.R., Borelli, J.L., 2021. Responding to children's diverse gender expression: validation of a parent-report measure of gender-related conditional regard. Journal of GLBT Family Studies 17 (5), 482–500. Available at https://doi.org/10.1080/1550428X.2021.1931615.

Serbin, L.A., Poulin-Dubois, D., Colburne, K.A., Sen, M.G., Eichestedt, J.A., 2001. Gender stereotyping in infancy: visual preferences for and knowledge of gender-stereortyped toys in the second year. International Journal of Behavioral Development 25 (1), 7–15. Available at https://doi.org/10.1080/01650250042000078.

Signorella, M.L., Bigler, R.S., Liben, L.S., 1993. Developmental differences in children's gender schemata about others: a smeta-analytic review. Developmental Review 13 (2), 147–183. Available at https://doi.org/10.1006/drev.1993.1007.

Smiler, A.P., Gelman, S.A., 2008. Determinants of gender essentialism in college students. Sex Roles 58 (11-12), 864–874. Available at https://doi.org/10.1007/s11199-008-9402-x.

Solomon, J., 2016. Gender identity and expression in the early childhood classroom: influences on development within sociocultural contexts. YC Young Children. 71 (3), 61–72.

Sorbara, J.C., Chiniara, L.N., Thompson, S., Palmert, M.R., 2020. Mental health and timing of gender-affirming care. Pediatrics 146 (4). Available at https://doi.org/10.1542/peds.2019-3600.

Sorbara, J.C., Ngo, H.L., Palmert, M.R., 2021. Factors associated with age of presentation to gender-affirming medical care. Pediatrics 147 (4). Available at https://doi.org/10.1542/peds.2020-026674.

Spivey, L.A., Huebner, D.M., Diamond, L.M., 2018. Parent responses to childhood gender nonconformity: effects of parent and child characteristics. Psychology of Sexual Orientation and Gender Diversity 5 (3), 360–370. Available at https://doi.org/10.1037/sgd0000279.

Steensma, T.D., Biemond, R., De Boer, F., Cohen-Kettenis, P.T., 2011. Desisting and persisting gender dysphoria after childhood: a qualitative follow-up study. Clinical Child Psychology and Psychiatry 16 (4), 499–516. Available at https://doi.org/10.1177/1359104510378303.

Steensma, T.D., Kreukels, B.P.C., de Vries, A.L.C., Cohen-Kettenis, P.T., 2013. Gender identity development in adolescence. Hormones and Behavior 64 (2), 288–297. Available at https://doi.org/10.1016/j.yhbeh.2013.02.020.

Stever, G.S., 2017. Evolutionary theory and reactions to mass media: understanding parasocial attachment. Psychology of Popular Media Culture 6 (2), 95–102. Available at https://doi.org/10.1037/ppm0000116.

Suarez, N.A., Trujillo, L., McKinnon, I.I., Mack, K.A., Lyons, B., Robin, L., et al., 2024. Disparities in school connectedness, unstable housing, experiences of violence, mental health, and suicidal thoughts and behaviors among transgender and cisgender high school students—youth risk behavior survey, United States, 2023. MMWR Supplements 73 (4), 50–58. Available at https://doi.org/10.15585/mmwr.su7304a6.

Sylliboy, J.R., 2022. Coming out is part of the life cycle: a qualitative study using two-eyed seeing to understand a two-spirits coming out process. Global Public Health 17 (10), 2428–2446. Available at https://doi.org/10.1080/17441692.2021.1993953.

Taylor, M.G., 1996. The development of children's beliefs about social and biological aspects of gender differences. Child Development 67 (4), 1555–1571. Available at https://doi.org/10.1111/j.1467-8624.1996.tb01814.x.

Temple Newhook, J., Pyne, J., Winters, K., Feder, S., Holmes, C., Tosh, J., et al., 2018. A critical commentary on follow-up studies and "desistance" theories about transgender and gender-nonconforming children. International Journal of Transgenderism 19 (2), 212–224. Available at https://doi.org/10.1080/15532739.2018.1456390.

Thomas, M., McCoy, T., Jeffries, I., Haverkate, R., Naswood, E., Leston, J., et al., 2022. Native American two spirit and LGBTQ health: a systematic review of the literature. Journal of Gay and Lesbian Mental Health 26 (4), 367–402. Available at https://doi.org/10.1080/19359705.2021.1913462.

Todd, K., Peitzmeier, S.M., Kattari, S.K., Miller-Perusse, M., Sharma, A., Stephenson, R., 2019. Demographic and behavioral profiles of nonbinary and binary transgender youth. Transgender Health 4 (1), 254–261. Available at https://doi.org/10.1089/trgh.2018.0068.

Topaz, E., Perl, L., Raphael, I., Sheppes, T., Erez, G., Israeli, G., et al., 2024. Mental health and timing of gender-related events among transgender and gender-diverse children and adolescents seeking gender-affirming consultation and care. Psychiatry Research 342, 116175. Available at https://doi.org/10.1016/j.psychres.2024.116175.

Turban, J.L., Dolotina, B., Freitag, T.M., King, D., Keuroghlian, A.S., 2023. Age of realization and disclosure of gender identity among transgender adults. Journal of Adolescent Health 72 (6), 852–859. Available at https://doi.org/10.1016/j.jadohealth.2023.01.023.

Twist, J., de Graaf, N.M., 2019. Gender diversity and non-binary presentations in young people attending the United Kingdom's National Gender Identity Development Service. Clinical Child Psychology and Psychiatry 24 (2), 277–290. Available at https://doi.org/10.1177/1359104518804311.

Veldhuis, C.B., Cascalheira, C.J., Delucio, K., Budge, S.L., Matsuno, E., Huynh, K., et al., 2024. Sexual orientation and gender diversity research manuscript writing guide. Psychology of Sexual Orientation and Gender Diversity 11 (3), 365–396. Available at https://doi.org/10.1037/sgd0000722.

Waagen, T., 2022. The University of North Dakota Non-binary gender identity development: a qualitative study. Theses and Dissertations. 4384. https://commons.und.edu/theses/4384

Wallien, M.S.C., Cohen-Kettenis, P.T., 2008. Psychosexual outcome of gender-dysphoric children. Journal of the American Academy of Child and Adolescent Psychiatry 47 (12), 1413–1423. Available at https://doi.org/10.1097/CHI.0b013e31818956b9.

Ward, L., Lucas, S., 2024. You're trying to put yourself in boxes, which doesn't work": exploring non-binary youth's gender identity development using feminist relational discourse analysis. Journal of Gender Studies 33 (5), 658–672. Available at https://doi.org/10.1080/09589236.2023.2172557.

Wittlin, N.M., Gallagher, N.M., Atwood, S., Olson, K.R., 2025a. Mental health during medical transition in a US and Canadian sample of early socially transitioned transgender youth. Journal of Adolescent Health 76 (2), 228–237. Available at https://doi.org/10.1016/j.jadohealth.2024.10.023.

Wittlin, N.M., Gallagher, N.M., Olson, K.R., 2025b. Gender identity importance in cisgender and gender diverse adolescents in the US and Canada. British Journal of Developmental Psychology 43 (2), 250–268. Available at https://doi.org/10.1111/bjdp.12485.

Xerxa, Y., White, T., Busa, S., Trasande, L., Hillegers, M.H.J., Jaddoe, V.W., et al., 2023. Gender diversity and brain morphology among adolescents. JAMA Network Open 6 (5). Available at https://doi.org/10.1001/jamanetworkopen.2023.13139 E2313139.

Xiao, S.X., Cook, R.E., Martin, C.L., Nielson, M.G., 2019. Characteristics of preschool gender enforcers and peers who associate with them. Sex Roles 81 (11-12), 671–685. Available at https://doi.org/10.1007/s11199-019-01026-y.

Zhang, Q., Goodman, M., Adams, N., Corneil, T., Hashemi, L., Kreukels, B., et al., 2020. Epidemiological considerations in transgender health: a systematic review with focus on higher quality data. International Journal of Transgender Health 21 (2), 125–137. Available at https://doi.org/10.1080/26895269.2020.1753136.

Zhou, J.N., Hofman, M.A., Gooren, L.J.G., Swaab, D.F., 1995. A sex difference in the human brain and its relation to transsexuality. Nature 378 (6552), 68–70. Available at https://doi.org/10.1038/378068a0.

Zosuls, K.M., Ruble, D.N., Tamis-LeMonda, C.S., Shrout, P.E., Bornstein, M.H., Greulich, F.K., 2009. The acquisition of gender labels in infancy: implications for gender-typed play. Developmental Psychology 45 (3), 688–701. Available at https://doi.org/10.1037/a0014053.

Chapter 3

Risk and resilience among transgender and gender diverse youth

Jonathan Poquiz[1], **Afiya Sajwani**[2], **Sabrina Gretkierewicz**[3], **and Lisa M. Buckloh**[3,4]

[1]*Department of Family Medicine and Community Health, University of Minnesota Medical School, Minneapolis, MN, United States,* [2]*Department of Psychiatry and Behavioral Sciences, Northwestern University Feinberg School of Medicine, Chicago, IL, United States,* [3]*Nemours Children's Health, Jacksonville, FL, United States,* [4]*PATHProgram (PediatricAcute Telemental Health) Director of Clinical Training Division of Psychology, Psychology (Mayo Clinic), Jacksonville, FL, United States*

Learning objectives

- Review the established research evidence on health disparities and risk factors among transgender and gender diverse (TGD) youth.
- Define resilience, review the assessment and application of resilience in research, and synthesize the existing research on resilience factors among TGD youth.
- Introduce theoretical frameworks that either incorporate or center on resilience, and describe measures of resilience that are currently validated and utilized in research.
- Identify future directions for research on risk and resiliency factors within this population.

Key terms and definitions

- **Health disparities** refer to statistical differences in the physical health, mental health, or healthcare access of members of a specific group of people compared to the majority population.
- **Risk factors** refer to contexts, conditions, and individual characteristics that are associated with an increased likelihood of negative outcomes.
- **Minority stress** is a theory that explains negative health outcomes based on the experiences of chronic stressors that impact people of marginalized groups.
- **Resilience** refers to assets and resources available at the individual, interpersonal, and societal levels that moderate the negative effects of adversity on individual health and functioning. Resilience factors boost resilience and are also called protective factors.

Risk and resilience factors among transgender and gender diverse (TGD) youth have been increasingly emphasized over the past decade. The emerging evidence and understanding of health disparities experienced by TGD youth have proliferated and provided opportunities to identify, explain, and address barriers to health equity for this population. Despite the growing body of research and understanding of risk factors impacting negative health outcomes, there remains a dearth of research emphasizing resilience, including theoretical frameworks to ground research questions and applied research methodologies.

Advances in Research on Pediatric Gender Affirming Healthcare. DOI: https://doi.org/10.1016/B978-0-443-29019-0.00004-X
© 2026 Elsevier Inc. All rights are reserved, including those for text and data mining, AI training, and similar technologies.

Risk factors for negative health outcomes

Consistent and established research has identified that TGD youth are at increased risk for negative mental and physical health outcomes. TGD youth often experience elevated anxiety (Chodzen et al., 2019; Moyer et al., 2019), are at risk of disordered eating and eating disorder diagnoses (Linsenmeyer et al., 2021; Bishop et al., 2020), and report significant depressive symptoms and suicidality (Fontanari et al., 2020; Tüzün et al., 2022; Vance et al., 2021; Veale et al., 2017; Wittlin et al., 2023). While some research on mental health disparities utilizes community samples (Fontanari et al., 2020; Veale et al., 2017), much of what is known about the mental health of TGD youth comes from clinic-based populations, often using baseline measurements before initiation of gender-affirming medical treatments (Chodzen et al., 2019; Linsenmeyer et al., 2021; Moyer et al., 2019). Because TGD youth who are able to access gender-affirming care often have other protective factors, such as parental support, these studies may underestimate mental health disparities in the larger population.

Research has similarly identified physical health disparities for TGD youth (Rider et al., 2018), including increased risk of experiencing pain (Boerner et al., 2023) and obesity (Bishop et al., 2020; Linsenmeyer et al., 2021). Difficulties maintaining a healthy weight may be negatively influenced by food insecurity, barriers to participating in physical education or active extracurricular activities, and increased risks of disordered eating (Bishop et al., 2020). Barriers to medical care also increase the risk of poor health outcomes (Delozier et al., 2020; Rider et al., 2018). In a population-based study, preventative health was utilized less in TGD youth compared to cisgender peers, while TGD youth reported less optimal general health status (Rider et al., 2018). Barriers to preventative health and healthcare utilization may include prior negative experiences, stigma, anticipated discrimination, and fears about disclosed information being shared with parents (Call et al., 2021).

A growing body of research has consistently shown that these disparities in physical and mental health are largely due to minority stress factors (Delozier et al., 2020; Wittlin et al., 2023). For example, a large population-based survey study found that increased risks for depression and suicidality are partially explained by experiences of discrimination and victimization related to gender identity (Price-Feeney et al., 2020). The theory and model of minority stress is commonly associated with Meyer's introduction and application to sexual minority populations (Meyer, 2003). Meyer's model of minority stress provided a foundation to understand how distal stressors (or external stress factors, e.g., discrimination, prejudice) and proximal stressors (or internal stress factors, e.g., identity concealment, internalized homophobia) are associated with mental health outcomes. Meyer's model of minority stress was later adapted to TGD populations and extended to address physical health outcomes as well (Hendricks and Testa, 2012).

Several systematic reviews have demonstrated which risk factors most consistently predict negative health outcomes, although these reviews do not distinguish between minority stress factors specifically and other risk factors that may not be associated with TGD identity (O'Shea et al., 2025; Patrick, 2020; Tankersley et al., 2021). Risk factors that most likely reflect distal minority stress include discrimination, victimization, bullying, healthcare stigma, and social isolation. Those that most likely reflect proximal minority stress include negative self-concept and internalized stigma. Other risk factors that may fall under minority stress or may be unrelated to identity include low self-esteem and negative relationships with family and peers. Endorsing a nonbinary identity is also a unique risk factor for negative health outcomes relative to individuals who identify as transgender male or female. Finally, research among adults has identified that several specific adverse childhood experiences (e.g., abuse, neglect, and other household challenges) predict health disparities in adulthood, particularly among those with intersecting identities (Feil et al., 2023; King et al., 2024; Strenth et al., 2024).

While the research evidence for minority stress and other risk factors for negative health outcomes is well-established, less is known about protective factors and resiliency. In the fields of social and health sciences the focus of research often centers on the identification of risk factors that contribute to poor health outcomes to elucidate pathways for interventions. However, identification and understanding of associations between risk factors and health outcomes can only go so far in achieving effective and efficacious outcomes. While still recognizing the importance of risk factors and health disparities, researchers are called to extend their research questions to identify and understand resilience factors, or factors that buffer against or moderate risk. Only in the past few years has emerging research begun to define resiliency and identify protective factors among TGD youth.

Resiliency and protective factors

Resilience is broadly defined as the ability to "bounce back" in the face of adversity. Some researchers have conceptualized resilience as a personal trait, while others have defined resilience as a process or an outcome (Fletcher and Sarkar, 2013). Resilience is understood to moderate the effects of adversity on individual functioning (Troy et al., 2023), and resilience factors refer to personal qualities or underlying experiences that promote resilience (Meyer, 2015). There is ongoing discourse and differing opinions about what constitutes resilience in the current literature and how to measure it (Troy et al., 2023; Zolkoski and Bullock, 2012). The majority of this research has focused on individuals in North America and Europe who have experienced early hardships, such as poverty, violence, substance use, family conflict, and serious physical or mental illness (Werner, 2005).

In recent years, there has been increased interest in understanding the experiences of historically marginalized groups facing stigma-based stress through the lens of resilience. However, many contemporary conceptualizations of resilience, which often focus on individual-level traits, are inadequate in explaining the experiences of historically marginalized groups. Regarding resilience as an individual-level trait or process does not consider power, privilege, and systems of oppression, which disproportionately confer stigma-based stress onto marginalized groups (Suslovic and Lett, 2024) and consequently put the onus of overcoming hardships on the individual (McLean et al., 2023).

Within the context of structural oppression an individual-focused definition of resilience asks marginalized groups to "be resilient" to survive. Meanwhile, the underlying systems of oppression are not challenged and are, in fact, reinforced. An individual-level conceptualization of resilience also assumes that the experienced adversity or stressor is temporary and time-bound (McLean et al., 2023). Such a conceptualization does not take into account the chronic and frequent nature of stigma-based stressors faced by marginalized groups. For these reasons, resilience is defined here as assets and resources available at the individual, interpersonal, and societal levels to mitigate the negative effects of adversity on health and functioning.

Given the limitations of historical frameworks, it is unsurprising that resilience remains an understudied topic among TGD youth. A scoping review of published studies between 2010 and 2020 found that only 17 qualitative studies and 33 quantitative studies assessed resilience among TGD populations of any age (Puckett et al., 2024). Only 5 out of 33 quantitative studies assessed whether resilience factors protected TGD individuals against the effects of stigma-based stress (Puckett et al., 2024), and results from these studies are mixed and contradictory (Bockting et al., 2013; Scandurra et al., 2017). Moreover, only 11 out of 33 quantitative studies included study participants under the age of 18 years (Puckett et al., 2024), highlighting the significant dearth of resilience research among TGD children and adolescents.

According to several systematic reviews of existing research, three external supports have consistently been identified as being most robust in promoting resiliency and protecting against negative health outcomes: (1) social support, especially from primary parents and caregivers, (2) affirming environments, and (3) access to gender-affirming medical care for those who need it (O'Shea et al., 2025; Patrick, 2020; Tankersley et al., 2021). Each of these requires external support and resources, and research into internal or individual factors is only recently emerging.

Social supports are among the most commonly investigated resilience factors in TGD youth. Strong family relationships, particularly with parents and caregivers, are consistently associated with positive outcomes (Call et al., 2021; Brown et al., 2020; Weinhardt et al., 2019; Veale et al., 2017). Positive parental relationships are associated with improved self-esteem, behavior, and mental health (Gibson et al., 2021; Johns et al., 2018; Tüzün et al., 2022). A large survey study of US high school students also found that perceived connectedness and positive relationships with parents predicted increased utilization of primary care and dental care in TGD youth but not cisgender youth, highlighting the distinct potential for parental support to mitigate health disparities in this population (Taliaferro et al., 2019).

While research has consistently demonstrated the associations between parental or general family support in promoting resilience, more research is needed to examine whether there are associations between support from other family members (e.g., siblings, extended family, chosen or "found" family) and well-being. In addition to family, another important interpersonal resilience factor is the role of friendships and peer support. Peer support has been identified as a protective factor against mental health concerns and perceived isolation, while also promoting improved self-esteem (Hidalgo et al., 2019; Johns et al., 2018). A mixed-method study using a survey and focus groups further found that peer support may promote improved well-being, perceived connectedness, and having a sense of purpose (Weinhardt et al., 2019).

Affirming and supportive community and social contexts further promote resilience among TGD youth. Supportive school environments are associated with better mental health (Colvin et al., 2019) and functioning within and outside of the school (Johns et al., 2018). In addition to teachers, other staff at school (i.e., school nurses and counselors) may protect against increased risks for TGD youth, who tend to have more frequent school nurse visits and increased risk of bullying (Rider et al., 2018). Research has varied on the positive benefits of LGBTQ+ affinity groups within schools, sometimes referred to as gay-straight alliances (GSA), suggesting benefits may be limited if other school supports are not present (Colvin et al., 2019). One study that surveyed both students and GSA advisors found that higher advisor self-efficacy to address TGD issues was associated with lower depression among TGD students (Andrzejewski et al., 2023). Another study using daily diaries found that TGD youth expressed more hope following meetings in which there was greater advisor responsiveness (Poteat et al., 2023), suggesting that the benefits of these groups may be dependent on the behaviors of the advisors within the groups. Outside the school environment, community support (e.g., neighbors, political environment) may foster feelings of belonging for TGD youth, and in turn, may increase the sources of affirming social support that enhance well-being (Wittlin et al., 2023).

In addition to environments being generally safe and accepting, research has consistently shown that affirming name and pronoun use across multiple environments is associated with better mental health outcomes (Patrick, 2020; Tankersley et al., 2021). In a community sample of TGD youth, use of chosen name by others across multiple contexts (i.e., school, home, peer activities) was associated with significantly lower depression, suicidal ideation, and suicidal behaviors, and depressive symptoms were lowest among youth who had their chosen name used across all contexts (Russell et al., 2018). Another large community-based study found that not all TGD youth feel strongly about disclosing a chosen name, but among those who do, use of a chosen name is significantly associated with improved mental health (Pollitt et al., 2021).

Access to gender-affirming care is also an important factor of resilience for some TGD youth, whose transition process includes medical intervention to affirm identity (Patrick, 2020; O'Shea et al., 2025). Research on the protective impact of access to gender-affirming care has largely been studied through longitudinal cohort studies that follow youth over time before and after initiating gender-affirming medical interventions, as well as cross-sectional studies examining the experiences of youth who have access to care compared to those who face barriers. Research suggests that TGD youth with access to affirming care have better mental health outcomes relative to TGD youth who experience barriers to care (Call et al., 2021), and mental health tends to improve over time with the use of gender-affirming medical interventions (Chen et al., 2023).

Internal factors of resilience, or characteristics intrinsic to the individual, are relatively under-researched in comparison. From the research available on internal resilience factors, self-esteem and a positive view of identity are protective factors for TGD youth (Johns et al., 2018). Additionally, positive personal identity predicts increased utilization of gender-affirming medical care by TGD youth (Goldenberg et al., 2020). A large survey of US students found that internal resiliency factors, such as feeling in control of life and future and accepting others who are different, were significantly associated with lower emotional distress (McPherson et al., 2023). Another large UK-based survey identified several specific personal strengths among TGD youth relative to cisgender youth, including creativity, originality, and helping others (Waterhouse et al., 2025).

Models of minority stress and resiliency

Although the literature on resilience and resiliency factors among TGD youth is still evolving, theoretical frameworks and models have been proposed and adapted to provide a foundation for future research in this area. These models were initially adapted and developed with samples of TGD adults, with recent research beginning to apply and assess these models in samples of TGD youth and adolescents. Although this research is newly emerging, these models have provided a foundation and opportunity for further growth of research in the field of risk and resiliency in TGD youth.

Gender minority stress and resilience

The gender minority stress and resilience (GMSR) model was proposed as an extension to the original minority stress model (Hendricks and Testa, 2012; Meyer, 2003). Minority stress specifically refers to experiences that are as follows: (1) unique to a minoritized group, (2) chronic, and (3) social in nature, whether interpersonally with another individual or in interactions with institutions and organizational systems (Meyer, 2003). The model was originally developed based on existing theories of stress, specifically incorporating literature on the chronic and unique stressors of lesbian, gay, and bisexual individuals. The minority stress model as a whole used pathways to demonstrate how health disparities result from distal (external) and proximal (internal) minority stressors and how individual coping and community resources can buffer this effect.

Like the original minority stress model, GMSR describes pathways between distal factors (i.e., discrimination, rejection, victimization, and identity nonaffirmation) and proximal factors (i.e., negative expectations for the future, internalized transphobia, and identity concealment) to health outcomes (i.e., mental and physical health). In this model, discrimination, rejection, and victimization are specifically related to gender, such as rejection by a parent after coming out or healthcare discrimination based on TGD identity. Negative expectations refer to the anticipation of negative future events as a result of being TGD or disclosing a TGD identity. Internalized transphobia involves directing negative feelings inward based on external discourse and stigma, and it can include wishing not to have a TGD identity or feeling depressed or embarrassed when thinking about gender identity. Importantly, proximal factors are directly caused by distal factors rather than being inherent to individuals.

In addition to stressors, resiliency factors (i.e., community connectedness, identity pride) are included in the GMSR model to emphasize the importance of identifying factors to buffer and mitigate against risk factors (Hendricks and Testa, 2012). Community connectedness refers to a sense of belonging when interacting with others in the TGD community, as well as feeling a positive, shared experience with other TGD individuals. In contrast to internalized transphobia, identity pride refers to a positive internal view of gender identity, including comfort talking about identity and pride in holding that identity. A GMSR measure was specifically developed based on this theoretical understanding and refined through psychometric analysis (Testa et al., 2017). To date the majority of research using this model has been conducted among samples of TGD adults (Sarno et al., 2020; Testa et al., 2017). Published research using the GMSR as a theoretical framework with samples of TGD youth is minimal but growing.

The GMSR model was adapted for adolescents to include family support as a resiliency factor (Coyne et al., 2020). The explicit inclusion of family support is based on existing research emphasizing the protective role of family support in minimizing risk for negative mental health outcomes (Brown et al., 2020; Call et al., 2021). The GMSR measure has also been adapted for use with adolescents (GMSR-A) with good psychometric properties (Hidalgo et al., 2019). The GMSR-A is a 59-item measure across nine subscales of minority stress and resilience, consistent with the dimensions included in the model, and the measure can be found within the supplemental materials of the original article (Hidalgo et al., 2019). One strength of research in this area has been a mix of clinic-based and community samples, allowing for generalization across a broader range of TGD youth and potential comparison between the two groups, as those with access to a specialty gender clinic may have more protective factors and less exposure to some forms of minority stress.

One cross-sectional, clinic-based study found that TGD youth with higher levels of internalized transphobia (a proximal stressor) were significantly more likely to meet diagnostic criteria for generalized anxiety disorder and major depressive disorder (Chodzen et al., 2019). Another study of youth seen in a specialty gender clinic looked at specific aspects of the GMSR model and found that internalized stigma, in particular, was associated with worse anxiety and depression, and that gender identity pride moderated this relationship (Conn et al., 2023). While most research in this area has used self-report measures of perceived health outcomes, one study measured biomarkers of stress in youth aged 9–20 years within a specialty clinic and found that gender minority stressors were associated with greater inflammation (McQuillan et al., 2021).

Cross-sectional, community-based studies have found similar results. One study of TGD adolescents and young adults (aged 16–25 years) in the community found that higher levels of experiencing minority stressors were associated with higher levels of anxiety and depression (Hunter et al., 2021). This study also found differences based on sex assigned at birth, such that the relationship between minority stress and poorer mental health was stronger among those assigned female at birth, while the relationship between resilience and better mental health was stronger among those assigned male at birth. A community-based study of TGD youth aged 15–24 years further clarified how different variables in the model relate to each other, suggesting that resilience specifically buffers against the impact of distal stress on psychological distress while proximal stressors tend to weaken the benefits of resilience (Miller-Perusse et al., 2024).

The GMSR provides a synthesized framework for research that includes both risk and resilience factors. While the GMSR framework provides an opportunity for complex statistical modeling to examine multiple pathways that can be moderated and mediated by variables of interest, existing research has primarily focused on cross-sectional and unidirectional associations between distal and proximal factors and health outcomes. The published research with TGD youth also remains primarily cross-sectional and unidirectional. Another gap in the literature is the dearth of studies utilizing GMSR resiliency factors. Stress and resilience do not occur in separate silos. To better understand the unique interactions between risk and resilience among TGD youth, research questions and methodology focused on the intersection of the two are needed. Gender identity similarly does not exist in a silo, and rather, intersects with other identities that individuals hold. More research is needed that incorporates an intersectional perspective to understand how a cumulative, intersectional experience of stress and oppression due to multiple marginalized identities impacts health outcomes.

Transgender resilience intervention model

The transgender resilience intervention model (TRIM) provides an expanded conceptualization specifically of resilience factors for TGD individuals (Matsuno and Israel, 2018). The TRIM builds on the GMSR model and synthesizes existing psychological literature on TGD health, positing that resilience factors at the individual and group levels can moderate the negative effects of minority stressors. This model hypothesizes that group-level resilience factors can moderate against the effects of distal stressors, while individual-level resilience factors can moderate against the effects of proximal stressors.

The TRIM purports that group-level resilience factors promote the development of individual-level resilience factors. Group-level resilience factors include having positive role models in the community, community belonging, family acceptance, social support, and activism on issues relevant to TGD individuals. Individual-level resilience factors include hope, self-acceptance and identity pride, self-definition, self-worth, and gender transition. Finally, the TRIM proposes community, group, and individual-level interventions to bolster resilience among TGD individuals. The TRIM offers a synthesized understanding of the current empirical literature on resilience factors among TGD individuals. Although the model as a whole has not been empirically evaluated, there is existing evidence for several pathways proposed by the TRIM. For example, having online or offline role models and participating in TGD-related activism—both group-level resilience factors proposed by the model—have been found to facilitate resilience and promote access to resources among TGD individuals (Craig et al., 2015; Singh, 2013; Singh and McKleroy, 2011).

Although the TRIM provides a helpful conceptualization of minority stress and resilience from a strengths-based perspective, there remain theoretical and empirical limitations. For example, the model does not include resilience factors beyond the individual and group levels (e.g., at the family or sociopolitical levels), and it does not include intersectional experiences of resilience among TGD individuals from other historically marginalized identities (e.g., TGD individuals of color or disabled TGD individuals). Furthermore, the model does not take into account resiliency factors specific to young people, and to date, there is no research on the TRIM and its applications among TGD youth.

Multidimensional transgender and nonbinary resilience

The Multidimensional Transgender and Nonbinary Resilience (MTNBR) framework was developed to address the limitations of the TRIM, including the development of a specific MTNBR scale that can be used in research (Puckett et al., 2025). This model of TGD people's resilience draws on Bronfenbrenner's socio-ecological model (Bronfenbrenner, 1979), Crenshaw's framework of intersectionality (Crenshaw, 1991), and the lived experiences of TGD individuals. The socio-ecological model asserts that development is influenced by individual-level factors which are nested within broader systems, for example, systems of family, neighborhood, political, legal, etc. (Bronfenbrenner, 1979). The framework of intersectionality posits that people with multiple marginalized identities are disadvantaged by interlocking systems of oppression along the axes of race, ethnicity, gender identity, sexual orientation, ability status, social class, religion, and others (Crenshaw, 1991).

Based on this theoretical foundation the MTNBR proposes that resilience for TGD individuals can be understood at five levels: individual, interpersonal, community, societal, and intersectional (Puckett et al., 2025). The specific resilience factors included under each level were drawn from qualitative interviews and focus groups conducted with TGD adults in consultation with a community advisory board made up of TGD adults from diverse backgrounds (Clements et al., 2025; Matsuno et al., 2022; Puckett et al., 2024; Puckett et al., 2019; Pulice-Farrow et al., 2019).

Individual-level factors include the following: developing motivation and agency, pride or positive self-image, cognitive strategies, basic needs, self-definition, hope, and gender euphoria. Interpersonal level factors include the following: familial social support, instrumental social support, and acceptance and affirmation of gender by others. Community-level factors include the following: a general sense of community connectedness, tangible connection with the TGD community, and having role models. Societal-level factors include the following: critical consciousness, self-advocacy and social activism, and access to gender-affirming healthcare. Finally, intersectional level factors include the following: pride in TGD as well as racial and ethnic minority identities, community connections with TGD people from shared cultural backgrounds, and recovering historical memory.

Based on this model the MTNBR scale was developed and psychometrically evaluated among TGD adults (Clements et al., 2025; Puckett et al., 2025). The scale includes 24 items across all four resiliency levels, and respondents are asked to rate how strongly they agree with each item on a 5-point scale from strongly disagree (1) to strongly agree (5). The full measure is included in the supplemental materials of the evaluation article (Puckett et al., 2025).

This multidimensional model of resilience furthers the conceptualization of TGD minority stress and resilience theory by broadening the understanding of TGD resilience across multiple levels beyond individual factors alone. The model also emphasizes the varied experiences of TGD individuals with multiple marginalized identities at all levels, which is a significant limitation of prior models. Efforts to empirically test and measure this model are currently underway, and results are forthcoming. Like the TRIM, resiliency factors specific to TGD youth are not accounted for in the model, and further research and refinement of the model are needed to understand its applicability across the developmental spectrum.

Summary and future directions for research

TGD youth are at increased risk of negative physical and mental health outcomes compared to their cisgender peers, and existing research strongly supports a minority stress framework for understanding these health disparities. Minority stress refers to chronic, identity-specific stressors that occur as a result of being a member of a minoritized group, such as discrimination (external) or identity concealment (internal). While the research on health disparities and minority stress is well-established, less is known about the unique contributions of resilience and protective factors that lessen the impact of adversity on health. Research that takes into account the risk and resiliency factors specific to youth and considers the specific outcomes and methodology most relevant to young people is needed.

Based on the literature summarized here the primary recommendations for future research in this area include the following: (1) moving beyond research questions based on risk factors and health disparities alone to incorporating resiliency and protective factors, (2) synthesizing existing research on risk and resilience to inform more complex statistical modeling to predict outcomes, (3) using community-based participatory methods involving TGD youth and their caregivers to ensure research that is beneficial to the community, and (4) meeting young people where they are by using social media and other online resources in recruitment, methodology, and dissemination of research.

There is a need for researchers in the field to move beyond solely identifying risk and disparities, and rather, identify and emphasize resiliency. Intentionally shifting the emphasis from risk to resilience not only moves the field forward in developing interventions at individual and structural levels, but it also shifts broader perspectives, demonstrating that the lives of TGD youth are filled with resilience, strength, and joy. Newer models of minority stress and resilience that emphasize protective factors are emerging, but they have not yet been empirically applied to TGD youth. Measurement tools are needed that take into consideration the particular risk and resilience factors specific to youth, such as school-based resiliency factors and the unique role of parents and caregivers in access or restriction to gender-affirming care and social affirmation.

In addition to increasing focus on resiliency, synthesizing existing research on risk and resilience is needed to inform more complex statistical modeling to predict outcomes. While there are a small number of published reviews on risk and resilience factors in youth (O'Shea et al., 2025; Patrick, 2020; Tankersley et al., 2021), no review of existing research takes into account the intersection of the two or the unique ways in which each contributes to health disparities and well-being. A synthesis of existing research that relates to specific models of minority stress may also help to validate the models and identify limitations and future directions.

To understand the complex, interactive associations between risk and resilience, a diverse and robust research methodology is needed. Larger sample sizes and longitudinal research designs are needed to better understand pathways from risk to resilience to health outcomes suggested within existing models. Outside of quantitative methods, qualitative methods and mixed-methods research designs would further contextualize data collected by incorporating findings gathered through quantitative methods to provide a robust interpretation and understanding of the specific risk and resilience experiences of TGD youth. Including the voices of TGD youth in research development and implementation is imperative, particularly for researchers and research teams who may not be comprised of individuals who share similar identity(ies) or lived experiences. Community-based participatory research designs provide researchers with opportunities to partner with community organizations, community members, and individuals with lived experience to develop their research programs and ultimately inform research questions based on the needs of the community.

Finally, researchers are encouraged to take into account the unique contexts in which TGD youth exist at all stages of research design, recruitment, implementation, and dissemination. For example, many TGD youth are developing and aging in the era of the Internet, and researchers are behooved to broaden their training and perspective of implementing research strategies in the context of social media (e.g., recruitment), online methods (e.g., administering online surveys), and creative dissemination (e.g., publishing results in laypeople's terminology on social media). In the current sociopolitical climate, there are also important considerations to ensure safety for TGD participants in research studies. This may include considerations for how to recruit TGD youth who may not have yet disclosed their identity to their family, or who typically are not represented in existing research studies (e.g., rural TGD youth, TGD youth who are not seeking gender-affirming medical care). There are also important considerations and discussions to have with study participants on how results of the study will be used—will it inform clinical care? Will it inform efforts to enact change at a policy level? Will it uplift the lived experiences of TGD youth with intersectional identities? Having this discussion with youth and their consenting adult(s) ensures that the research will directly benefit the community.

While there is still research to be done to understand risk factors and disparities experienced by TGD youth, there is a larger need to begin identifying and understanding the strengths and resiliency of this population. Parents, family members, friends, teachers, and clinicians are well aware of the creativity, resiliency, and joy that TGD young people embody every day, and a more nuanced understanding of their experiences will emerge with the incorporation of this reality into research conducted on their behalf. Enabling the perspective and appreciation of both risk and resiliency can create a more comprehensive and complex approach to developing interventions to improve the lives of TGD youth.

References

Andrzejewski, J., Salgin, L., Poteat, V.P., Yoshikawa, H., Calzo, J.P., 2023. Gender-sexuality alliance advisors' self-efficacy to address transgender issues: an interpersonal protective factor for transgender student depression. LGBT Health 10 (4), 296–305. Available at https://doi.org/10.1089/lgbt.2022.0060.

Bishop, A., Overcash, F., McGuire, J., Reicks, M., 2020. Diet and physical activity behaviors among adolescent transgender students: school survey results. Journal of Adolescent Health 66 (4), 484–490. Available at https://doi.org/10.1016/j.jadohealth.2019.10.026.

Bockting, W.O., Miner, M.H., Swinburne Romine, R.E., Hamilton, A., Coleman, E., 2013. Stigma, mental health, and resilience in an online sample of the US transgender population. American Journal of Public Health 103 (5), 943–951. Available at https://doi.org/10.2105/AJPH.2013.301241.

Boerner, K.E., Harrison, L.E., Battison, E.A.J., Murphy, C., Wilson, A.C., 2023. Topical review: acute and chronic pain experiences in transgender and gender-diverse youth. Journal of Pediatric Psychology 48 (12), 984–991. Available at https://doi.org/10.1093/jpepsy/jsad075.

Bronfenbrenner, U., 1979. The Ecology of Human Development: Experiments by Nature and Design. Harvard university press.

Brown, C., Porta, C.M., Eisenberg, M.E., McMorris, B.J., Sieving, R.E., 2020. Family relationships and the health and well-being of transgender and gender-diverse youth: a critical review. LGBT Health 7 (8), 407–419. Available at https://doi.org/10.1089/lgbt.2019.0200.

Call, D.C., Challa, M., Telingator, C.J., 2021. Providing affirmative care to transgender and gender diverse youth: disparities, interventions, and outcomes. Current Psychiatry Reports 23 (6). Available at https://doi.org/10.1007/s11920-021-01245-9.

Chen, D., Berona, J., Chan, Y.M., Ehrensaft, D., Garofalo, R., Hidalgo, M.A., et al., 2023. Psychosocial functioning in transgender youth after 2 years of hormones. New England Journal of Medicine 388 (3), 240–250. Available at https://doi.org/10.1056/NEJMoa2206297.

Chodzen, G., Hidalgo, M.A., Chen, D., Garofalo, R., 2019. Minority stress factors associated with depression and anxiety among transgender and gender-nonconforming youth. Journal of Adolescent Health 64 (4), 467–471. Available at https://doi.org/10.1016/j.jadohealth.2018.07.006.

Clements, Z.A., Puckett, J.A., Huynh, K.D., Matsuno, E., Galupo, M.P., 2025. Transforming queer mixed methods research to be trans and nonbinary-inclusive. Journal of Mixed Methods Research 19 (4), 359–383. Available at https://doi.org/10.1177/15586898251333464.

Colvin, S., Egan, J.E., Coulter, R.W.S., 2019. School climate & sexual and gender minority adolescent mental health. Journal of Youth and Adolescence 48 (10), 1938–1951. Available at https://doi.org/10.1007/s10964-019-01108-w.

Conn, B.M., Chen, D., Olson-Kennedy, J., Chan, Y.M., Ehrensaft, D., Garofalo, R., et al., 2023. High internalized transphobia and low gender identity pride are associated with depression symptoms among transgender and gender-diverse youth. Journal of Adolescent Health 72 (6), 877–884. Available at https://doi.org/10.1016/j.jadohealth.2023.02.036.

Coyne, C.A., Poquiz, J.L., Janssen, A., Chen, D., 2020. Evidence-based psychological practice for transgender and non-binary youth: defining the need, framework for treatment adaptation, and future directions. Evidence-Based Practice in Child and Adolescent Mental Health 5 (3), 340–353. Available at https://doi.org/10.1080/23794925.2020.1765433.

Craig, S.L., McInroy, L., McCready, L.T., Alaggia, R., 2015. Media: a catalyst for resilience in lesbian, gay, bisexual, transgender, and queer youth. Journal of LGBT Youth 12 (3), 254–275. Available at https://doi.org/10.1080/19361653.2015.1040193.

Crenshaw, K., 1991. Mapping the margins: identity politics, intersectionality, and violence against women. Stanford Law Review 43 (6).

Delozier, A.M., Kamody, R.C., Rodgers, S., Chen, D., 2020. Health disparities in transgender and gender expansive adolescents: a topical review from a minority stress framework. Journal of Pediatric Psychology 45 (8), 842–847. Available at https://doi.org/10.1093/jpepsy/jsaa040.

Feil, K., Riedl, D., Böttcher, B., Fuchs, M., Kapelari, K., Gräßer, S., et al., 2023. Higher prevalence of adverse childhood experiences in transgender than in cisgender individuals: results from a single-center observational study. Journal of Clinical Medicine 12 (13). Available at https://doi.org/10.3390/jcm12134501.

Fletcher, D., Sarkar, M., 2013. Psychological resilience. European Psychologist 18 (1), 12–23. Available at https://doi.org/10.1027/1016-9040/a000124.

Fontanari, A.M.V., Vilanova, F., Schneider, M.A., Chinazzo, I., Soll, B.M.H., Schwarz, K., et al., 2020. Gender affirmation is associated with transgender and gender nonbinary youth mental health improvement. LGBT Health 7 (5), 237–247. Available at https://doi.org/10.1089/lgbt.2019.0046.

Gibson, D.J., Glazier, J.J., Olson, K.R., 2021. Evaluation of anxiety and depression in a community sample of transgender youth. JAMA Network Open 4 (4). Available at https://doi.org/10.1001/jamanetworkopen.2021.4739.

Goldenberg, T., Kahle, E.M., Stephenson, R., 2020. Stigma, resilience, and healthcare use among transgender and other gender diverse youth in the united states. Transgender Health 5 (3), 173–181. Available at https://doi.org/10.1089/trgh.2019.0074.

Hendricks, M.L., Testa, R.J., 2012. A conceptual framework for clinical work with transgender and gender nonconforming clients: an adaptation of the minority stress model. Professional Psychology: Research and Practice 43 (5), 460–467. Available at https://doi.org/10.1037/a0029597.

Hidalgo, M.A., Petras, H., Chen, D., Chodzen, G., 2019. The gender minority stress and resilience measure: psychometric validity of an adolescent extension. Clinical Practice in Pediatric Psychology 7 (3), 278–290. Available at https://doi.org/10.1037/cpp0000297.

Hunter, J., Butler, C., Cooper, K., 2021. Gender minority stress in trans and gender diverse adolescents and young people. Clinical Child Psychology and Psychiatry 26 (4), 1182–1195. Available at https://doi.org/10.1177/13591045211033187.

Johns, M.M., Beltran, O., Armstrong, H.L., Jayne, P.E., Barrios, L.C., 2018. Protective factors among transgender and gender variant youth: a systematic review by socioecological level. Journal of Primary Prevention 39 (3), 263–301. Available at https://doi.org/10.1007/s10935-018-0508-9.

King, W.M., Fleischer, N.L., Operario, D., Chatters, L.M., Gamarel, K.E., 2024. Inequities in the distribution of adverse childhood experiences and their association with health among transgender people of color. Child Abuse and Neglect 149. Available at https://doi.org/10.1016/j.chiabu.2024.106654.

Linsenmeyer, W.R., Katz, I.M., Reed, J.L., Giedinghagen, A.M., Lewis, C.B., Garwood, S.K., 2021. Disordered eating, food insecurity, and weight status among transgender and gender nonbinary youth and young adults: a cross-sectional study using a nutrition screening protocol. LGBT Health 8 (5), 359–366. Available at https://doi.org/10.1089/lgbt.2020.0308.

Matsuno, E., Bricker, N.L., Collazo, E.N., Mohr, R., Balsam, K.F., 2022. "The default is just going to be getting misgendered": minority stress experiences among nonbinary adults. Psychology of Sexual Orientation and Gender Diversity 11 (2), 202–214. Available at https://doi.org/10.1037/sgd0000607.

Matsuno, E., Israel, T., 2018. Psychological interventions promoting resilience among transgender individuals: transgender resilience intervention model (TRIM). The Counseling Psychologist 46 (5), 632–655. Available at https://doi.org/10.1177/0011000018787261.

McLean, K.C., Fish, J., Rogers, L.O., Syed, M., 2024. Integrating systems of power and privilege in the study of resilience. American Psychologist 79 (8), 999–1011. Available at https://doi.org/10.1037/amp0001260.

McPherson, L.E., Svetaz, M.V., Martin, C.L., Miller, K.K., Eisenberg, M.E., 2023. Latine transgender/gender diverse adolescents: protective factors against emotional distress. LGBT Health 10 (3), 220–227. Available at https://doi.org/10.1089/lgbt.2022.0068.

McQuillan, M.T., Kuhns, L.M., Miller, A.A., McDade, T., Garofalo, R., 2021. Gender minority stress, support, and inflammation in transgender and gender-nonconforming youth. Transgender Health 6 (2), 91–100. Available at https://doi.org/10.1089/trgh.2020.0019.

Meyer, I.H., 2003. Prejudice, social stress, and mental health in lesbian, gay, and bisexual populations: conceptual issues and research evidence. Psychological Bulletin 129 (5), 674–697. Available at https://doi.org/10.1037/0033-2909.129.5.674.

Meyer, I.H., 2015. Resilience in the study of minority stress and health of sexual and gender minorities. Psychology of Sexual Orientation and Gender Diversity 2 (3), 209–213. Available at https://doi.org/10.1037/sgd0000132.

Miller-Perusse, M., Horvath, K.J., Kahle, E., Stephenson, R., 2024. Gender minority stress, resilience, and psychological distress: the role of resilience among transgender and gender diverse youth. Transgender Health 9 (4), 307–316. Available at https://doi.org/10.1089/trgh.2022.0117.

Moyer, D.N., Connelly, K.J., Holley, A.L., 2019. Using the PHQ-9 and GAD-7 to screen for acute distress in transgender youth: findings from a pediatric endocrinology clinic. Journal of Pediatric Endocrinology and Metabolism 32 (1), 71–74. Available at https://doi.org/10.1515/jpem-2018-0408.

O'Shea, J., Jenkins, R., Nicholls, D., Downs, J., Hudson, L.D., 2025. Prevalence, severity and risk factors for mental disorders among sexual and gender minority young people: a systematic review of systematic reviews and meta-analyses. European Child and Adolescent Psychiatry 34 (3), 959–982. Available at https://doi.org/10.1007/s00787-024-02552-1.

Patrick, K.L., 2020. Transgender identity and mental health in adolescence: A scoping review. medRxiv, Canada medRxiv. doi: 10.1101/2020.08.20.20178897.

Pollitt, A.M., Ioverno, S., Russell, S.T., Li, G., Grossman, A.H., 2021. Predictors and mental health benefits of chosen name use among transgender youth. Youth & Society 53 (2), 320–341. Available at https://doi.org/10.1177/0044118x19855898.

Poteat, V.P., Calzo, J.P., Sherwood, S.H., Marx, R.A., O'Brien, M.D., Dangora, A., et al., 2023. Gender-sexuality alliance meeting experiences predict weekly variation in hope among LGBTQ+ youth. Child Development 94 (4), e215. Available at https://doi.org/10.1111/cdev.13923.

Price-Feeney, M., Green, A.E., Dorison, S., 2020. Understanding the mental health of transgender and nonbinary youth. Journal of Adolescent Health 66 (6), 684–690. Available at https://doi.org/10.1016/j.jadohealth.2019.11.314.

Puckett, J.A., Matsuno, E., Huynh, K.D., Clements, Z., Galupo, M.P., 2025. Development and validation of the multidimensional transgender and nonbinary resilience scale. International Journal of Transgender Health. Available at https://doi.org/10.1080/26895269.2025.2483220.

Puckett, J.A., Matsuno, E., Dyar, C., Mustanski, B., Newcomb, M.E., 2019. Mental health and resilience in transgender individuals: what type of support makes a difference? Journal of Family Psychology 33 (8), 954–964. Available at https://doi.org/10.1037/fam0000561.

Puckett, J.A., Huit, T.Z., Hope, D.A., Mocarski, R., Lash, B.R., Walker, T., et al., 2024. Transgender and gender-diverse people's experiences of minority stress, mental health, and resilience in relation to perceptions of sociopolitical contexts. Transgender Health 9 (1), 14–23.

Puckett, J.A., Domínguez, S., Matsuno, E., 2024. Measures of resilience: do they reflect the experiences of transgender individuals? Transgender Health 9 (1), 1–13. Available at https://doi.org/10.1089/trgh.2022.0015.

Pulice-Farrow, L., Bravo, A., Galupo, M.P., 2019. "Your gender is valid": microaffirmations in the romantic relationships of transgender individuals. Journal of LGBT Issues in Counseling 13 (1), 45–66. Available at https://doi.org/10.1080/15538605.2019.1565799.

Rider, G.N., Mcmorris, B.J., Gower, A.L., Coleman, E., Eisenberg, M.E., 2018. Health and care utilization of transgender and gender nonconforming youth: a population-based study. Pediatrics 141 (3). Available at https://doi.org/10.1542/peds.2017-1683.

Russell, S.T., Pollitt, A.M., Li, G., Grossman, A.H., 2018. Chosen name use is linked to reduced depressive symptoms, suicidal ideation, and suicidal behavior among transgender youth. Journal of Adolescent Health 63 (4), 503–505. Available at https://doi.org/10.1016/j.jadohealth.2018.02.003.

Sarno, E.L., Newcomb, M.E., Mustanski, B., 2020. Rumination longitudinally mediates the association of minority stress and depression in sexual and gender minority individuals. Journal of Abnormal Psychology 129 (4), 355–363. Available at https://doi.org/10.1037/abn0000508.

Scandurra, C., Amodeo, A.L., Valerio, P., Bochicchio, V., Frost, D.M., 2017. Minority stress, resilience, and mental health: a study of italian transgender people. Journal of Social Issues 73 (3), 563–585. Available at https://doi.org/10.1111/josi.12232.

Singh, A.A., 2013. Transgender youth of color and resilience: negotiating oppression and finding support. Sex Roles 68 (11-12), 690–702. Available at https://doi.org/10.1007/s11199-012-0149-z.

Singh, A.A., McKleroy, V.S., 2011. "Just getting out of bed is a revolutionary act": The resilience of transgender people of color who have survived traumatic life events. Traumatology 17 (2), 34–44. Available at https://doi.org/10.1177/1534765610369261.

Strenth, C.R., Smith, M., Gonzalez, L., Grant, A., Thakur, B., Levy Kamugisha, E.I., 2024. Mediational pathways exploring the link between adverse childhood experiences and physical health in a transgender population. Child Abuse and Neglect 149. Available at https://doi.org/10.1016/j.chiabu.2024.106678.

Suslovic, B., Lett, E., 2024. Resilience is an adverse event: a critical discussion of resilience theory in health services research and public health. Community Health Equity Research & Policy 44 (3), 339–343. Available at https://doi.org/10.1177/2752535x231159721.

Taliaferro, L.A., Harder, B.M., Lampe, N.M., Carter, S.K., Rider, G.N., Eisenberg, M.E., 2019. Social connectedness factors that facilitate use of healthcare services: comparison of transgender and gender nonconforming and cisgender adolescents. Journal of Pediatrics 211, 172–178. Available at https://doi.org/10.1016/j.jpeds.2019.04.024.

Tankersley, A.P., Grafsky, E.L., Dike, J., Jones, R.T., 2021. Risk and resilience factors for mental health among transgender and gender nonconforming (TGNC) youth: a systematic review. Clinical Child and Family Psychology Review 24 (2), 183–206. Available at https://doi.org/10.1007/s10567-021-00344-6.

Testa, R.J., Michaels, M.S., Bliss, W., Rogers, M.L., Balsam, K.F., Joiner, T., 2017. Suicidal ideation in transgender people: gender minority stress and interpersonal theory factors. Journal of Abnormal Psychology 126 (1), 125–136. Available at https://doi.org/10.1037/abn0000234.

Troy, A.S., Willroth, E.C., Shallcross, A.J., Giuliani, N.R., Gross, J.J., Mauss, I.B., 2023. Psychological resilience: an affect-regulation framework. Annual Review of Psychology 74, 547–576. Available at https://doi.org/10.1146/annurev-psych-020122-041854.

Tüzün, Z., Başar, K., Akgül, S., 2022. Social connectedness matters: depression and anxiety in transgender youth during the COVID-19 pandemic. The Journal of Sexual Medicine 19 (4), 650–660. Available at https://doi.org/10.1016/j.jsxm.2022.01.522.

Vance, S.R., Boyer, C.B., Glidden, D.V., Sevelius, J., 2021. Mental health and psychosocial risk and protective factors among black and latinx transgender youth compared with peers. JAMA Network Open 4 (3). Available at https://doi.org/10.1001/jamanetworkopen.2021.3256.

Veale, J.F., Watson, R.J., Peter, T., Saewyc, E.M., 2017. Mental health disparities among Canadian transgender youth. Journal of Adolescent Health 60 (1), 44–49. Available at https://doi.org/10.1016/j.jadohealth.2016.09.014.

Waterhouse, P., Burkill, S., Fenaughty, J., Lucassen, M., 2025. The strengths of sexual- and gender-minoritized adolescents: results from the millennium cohort study. Journal of Adolescent Health 77 (3), 532–539. Available at https://doi.org/10.1016/j.jadohealth.2025.05.009.

Weinhardt, L.S., Xie, H., Wesp, L.M., Murray, J.R., Apchemengich, I., Kioko, D., et al., 2019. The role of family, friend, and significant other support in well-being among transgender and non-binary youth. Journal of GLBT Family Studies 15 (4), 311–325. Available at https://doi.org/10.1080/1550428X.2018.1522606.

Werner, E.E., 2005. Resilience in Children, Families, and Communities: Linking Context to Practice and Policyresilience Research: Past, Present, and Future. Springer.

Wittlin, N.M., Kuper, L.E., Olson, K.R., 2023. Mental health of transgender and gender diverse youth. Annual Review of Clinical Psychology 19, 207–232. Available at https://doi.org/10.1146/annurev-clinpsy-072220-020326.

Zolkoski, S.M., Bullock, L.M., 2012. Resilience in children and youth: a review. Children and Youth Services Review 34 (12), 2295–2303. Available at https://doi.org/10.1016/j.childyouth.2012.08.009.

Chapter 4

Overview of healthcare and barriers to access

Jess T. Guerriero[1] and Danielle N. Moyer[2]
[1]*Doernbecher Gender Clinic, Oregon Health & Science University, Portland, OR, United States,* [2]*Department of Pediatrics, Oregon Health & Science University, Portland, OR, United States*

Learning objectives

- Examine the individual, interpersonal, and systemic factors that contribute to experiences of minority stress, health inequities, and barriers to accessing pediatric healthcare.
- Summarize the existing research on access to healthcare overall, primary care, mental healthcare, and specialty gender-affirming care for transgender and gender diverse youth.
- Highlight unique considerations for nonbinary youth, those with intersecting minoritized identities, and young people living in rural geographic areas.
- Review the strengths and limitations of existing research and make recommendations for future research on healthcare access and barriers among transgender and gender diverse youth.

Key terms and definitions

- **Health disparities**, sometimes also referred to as health inequities, refer to systematic, preventable health differences that disproportionately affect specific populations.
- **Access** is a broadly defined term that refers to the ability to find, utilize, and benefit from timely and appropriate healthcare without excessive financial burden to achieve optimal health and wellness.
- **Barriers** are a broadly defined term that refers to the individual, interpersonal, and systemic factors that contribute to inequities in health and healthcare access.
- **Telehealth**, sometimes also referred to as telemedicine, refers to the use of digital technology in the provision and monitoring of healthcare.

Transgender and gender diverse (TGD) people of all ages encounter systemic barriers to engaging with healthcare. There are similarly inherent barriers and facilitators to pediatric healthcare. At the intersection of age and gender, TGD youth face unique challenges to accessing preventive, routine, and specialty care, which is growing increasingly more difficult with the impacts of the current sociopolitical context. These barriers to healthcare access contribute directly to the health disparities faced by TGD youth, including being at risk for poorer physical and mental health compared with the general population (Clark et al., 2018b; Rider et al., 2018). Health disparities refer to systematic and preventable health differences that disproportionately affect some populations, and these disparities are even greater for TGD youth who identify as nonbinary and those experiencing intersecting forms of oppression (Clark et al., 2018a; Goldenberg et al., 2019; Rusow et al., 2022). There are individual, interpersonal, and systemic healthcare barriers and facilitators that interact and impact access to care, experiences with healthcare, and overall health outcomes among TGD youth.

Advances in Research on Pediatric Gender Affirming Healthcare. DOI: https://doi.org/10.1016/B978-0-443-29019-0.00014-2
© 2026 Elsevier Inc. All rights are reserved, including those for text and data mining, AI training, and similar technologies.

Health and healthcare disparities

The family circumstances of each young person, regardless of gender identity, impact the foundation upon which access to medical care and care utilization patterns develop. Socioeconomic deprivation, or disadvantages in income, social class, employment, and schooling, is directly linked to health disparities and associated barriers. A family's total of socioeconomic determinants, rather than income alone, is associated with access to pediatric healthcare. Even in countries with universal, state-sponsored healthcare, where it may be believed that this structure is an equalizer, youth from families with higher social standing and lower socioeconomic deprivation are more likely to attend preventive and follow-up visits, while conversely less likely to utilize the emergency room (Brennan et al., 2025).

According to one study of providers in the United Kingdom's nationalized healthcare system, barriers for general pediatric populations can be broken down into two main categories: economic and accessibility (Brennan et al., 2025). According to the results of qualitative interviews, economic factors include more than simply disparities in incomes for each family. There are economic costs associated with transportation and travel, food insecurity that can be compounded by having to acquire extra food supply for long travel and inpatient stays, lost income due to travel time, increased childcare costs if other children are in the family and cannot accompany them to care, unexpected costs related to improving housing conditions to support long-term health, and medical equipment and prescriptions. Accessibility barriers refer to those associated with language and translation needs, health literacy, especially as it relates to navigating complex health systems, cultural considerations, including differences in the definition of health and illness, who the health authority is, and how illness should be treated, and movement toward a reliance on telehealth and other forms of digital care that may reduce likelihood of engagement due to lack of digital resources or safety concerns for vulnerable groups.

Similar economic barriers were identified in a cross-sectional study at a US urban outpatient clinic, primarily serving those impacted by socioeconomic deprivation (Shi et al., 2020). The caregivers in this study also identified additional barriers, including long wait times, lack of reminders, frequent changes in primary care providers, and lack of clarity on the importance of well-child checks versus acute care visits to treat a specific condition. Family relocation and the need to find new care providers, as well as reduced proximity to medical resources, were also cited as barriers to continuity of care. Caregivers described the collective impact of life stressors, gaps in insurance coverage, and feeling disrespected by providers as further reasons for difficulty with follow-through.

In addition to parent and guardian perspectives on barriers to healthcare, the results of this study also revealed factors that can facilitate engagement with providers and adherence to treatment plans. The provider–family relationship was one facilitator for access and engagement with healthcare, with the longevity of the relationship being an additional facilitating factor. Caregiver attitudes toward supportive health behaviors, including attending both preventative and acute care appointments, were another facilitator, with caregivers who recognized the benefits of preventative visits, even if not immediately apparent, being more likely to follow through with treatment plans and associated visits. Systemic and institutional practices like reminder calls, predictable wait times, and clear scheduling systems were also factors that facilitated family engagement and follow-through with children's healthcare providers (Shi et al., 2020).

TGD youth face similar overall barriers to accessing care as the general pediatric population, but there are additional contributing factors as a result of belonging to a community that is underserved and more vulnerable than the general population. Research in the general pediatric population has shown that youth, in general, are less likely to seek out primary or preventative care if the type of care they are seeking is stigmatized or if they are part of a marginalized group (Samargia et al., 2006; Williams and Chapman, 2011). Research on TGD youth specifically shows lower rates of accessing preventative, primary care, and mental healthcare (Clark et al., 2018b; Rider et al., 2018; Rusow et al., 2022).

A cross-sectional study from Canada examined the care-seeking behaviors of TGD youth in primary care and behavioral health environments (Clark et al., 2018b). The study was based on data from the Canadian Trans Youth Health Survey, which surveys TGD youth aged 14–25 years approximately every 5 years. Results revealed several themes about barriers to care, including concerns about parental involvement, access issues (i.e., clinic hours, location of clinic, and costs), and fear of a provider's reaction, potentially due to past negative experiences. Overall, TGD youth reported poorer mental and physical health outcomes than their cisgender peers and were more likely to avoid seeking mental healthcare over physical healthcare. However, provider awareness of a youth's identity and comfort level associated with discussing transgender identity was positively correlated with improved physical and mental health outcomes.

Data from the 2016 Minnesota Student Survey of 9th- and 11th-grade students reflected similar findings (Rider et al., 2018; Taliaferro et al., 2019). In one analysis of the data, TGD youth were nearly twice as likely to report poorer physical health status and nearly three times as likely to report long-term mental health concerns, compared with their cisgender peers (Rider et al., 2018). These youth also experienced more frequent illness-related school absences and higher utilization of the nurses' offices. In contrast, they were less likely to access primary care and preventative health services compared with their cisgender peers. TGD youth whose gender expression was more nonconforming or less binary were more likely to report poorer health status, although there were no significant differences in utilization of health services between youth from different identities or expressions.

In another analysis of this survey data, parent connectedness played an important role in the disparities between TGD and cisgender youth (Taliaferro et al., 2019). Parent connectedness refers to feeling cared for by parents and caregivers as well as being able to talk to them about problems. According to the results of the survey, parent connectedness was associated with utilization of primary care and dental care for TGD youth but not their cisgender peers. In other words, the actual relationship between TGD youth and their primary caregivers can impact their healthcare access in ways that are less likely to occur in the general population (Taliaferro et al., 2019). Results of the prior analysis of the survey also emphasized that the relationship and role of providers can improve health outcomes and take specific actions to increase access and utilization, including being open and receptive to being the point of first disclosure, facilitating discussions about identity and care needs with important caregivers, and offering resources or referrals to other care options (Rider et al., 2018).

Understanding barriers from an intersectional lens is also critical for healthcare delivery. The needs of TGD youth of color have historically been conflated with the experiences of white TGD youth, and there is limited data available on the experiences of TGD youth of color (Rusow et al., 2022). Black TGD youth have been particularly underrepresented in current literature (Goldenberg et al., 2019). Data from the US Trans Youth of Color Study showed that a larger percentage (26%) of TGD adolescents and young adults of color felt they had no place to go for healthcare compared with their cisgender peers (Rusow et al., 2022).

The Affirming Voices for Action project is a mixed-method study aimed at recruiting diverse TGD communities across 14 US cities (Goldenberg et al., 2019). A secondary analysis of the data, including Black TGD adolescents and young adults, showed that nonuse of preventative, primary care was associated with gender nonaffirmation (e.g., being misgendered or being asked inappropriate questions about gender). Furthermore, gender nonaffirmation moderated the relationship between anticipated stigma and healthcare use. Specifically, for those with a history of gender affirmation within healthcare, anticipated stigma did not appear to be a barrier to care utilization. However, for those with a history of nonaffirmation, there was a positive association between anticipated stigma and primary care nonuse.

TGD youth who do not identify with the binary notions of male and female similarly face unique hurdles compared with their binary TGD peers. Nonbinary youth may have a higher degree of mistrust toward providers, frustration with the lack of appropriate care options to affirm identity, and fear that their identity or expression will be judged. These factors can lead to delays in accessing both preventative and specialty care, which in turn, impact overall health outcomes (Rider et al., 2018). Data from the Canadian Trans Youth Health Survey have shown that nonbinary adolescents are more likely to forego primary healthcare than their binary TGD peers, although in this study, this outcome was more pronounced for young adults compared with adolescents (Clark et al., 2018a). This difference was not apparent when accessing specialty gender-affirming care, suggesting that parental presence may be a protective factor for accessing primary but not specialty care for nonbinary young people. Nonbinary youth of color also report a higher degree of concern for their health compared with binary peers, and while they desire mental health support more often than their binary peers, they are less likely to have access to appropriate services (Rusow et al., 2022).

In addition to research on population-level healthcare access disparities, qualitative interviews have been used to further clarify from patients and providers the barriers and facilitators to creating supportive, affirming healthcare environments (Silveira et al., 2025). Barriers to access could be framed along the following categories: approachability, acceptability, availability, affordability, and appropriateness (Pastor Bravo et al., 2024). These themes emerged from findings from a community-based sample of TGD adolescents and young adults in Spain and Sweden and relied on semistructured interviews to gather perspectives. Additional studies reflected similar themes around barriers and facilitators in accessing care.

A mixed-methods study with TGD adolescents and young adults in the United States utilized surveys and semistructured interviews to gather perspectives on positive and negative aspects of various types of medical care (Yeh et al., 2025). The study examined experiences in inpatient medical units, inpatient behavioral health units, emergency rooms, primary care, specialty clinics not providing gender-affirming care, radiology, occupational and physical therapy, laboratories, speech and language therapy, and surgical services. Participants identified several contributing factors to positive experiences across settings, including staff having prior exposure to work with TGD youth, healthcare team members with shared LGBTQ+ identities, parent advocacy in care, and seeing improvements in the care environments over time. Participants also described areas that needed improvement to increase positive experiences, including providing staff with additional training on best practices for using pronouns, terminology, and examination techniques, accurate display of affirmed name and pronouns in the medical record and on forms and stickers, and artwork in the physical environment, as well as patient resources, that were more inclusive of varied LGBTQ+ identities.

Additional research continued to elicit TGD youth and young adult perspectives, while also incorporating provider views on barriers and facilitators. One study, using semistructured interviews, sought to specifically understand experiences in inpatient psychiatric care, from both a provider and TGD adolescent patient view (Acosta et al., 2019). Patients and providers identified the characteristics of supportive environments in their responses. These characteristics included wide-ranging respect for all gender identities, consistently using affirming names and pronouns, and commitment to acknowledging when mistakes were made around the use of affirmed names and pronouns. Environments that did not feel as supportive were associated with a lack of consistent use of affirming identifiers, challenges related to the limitations within medical records and the need to use birth-assigned name in the record, on bracelets, and attendance rosters, and a perceived lack of training on cultural competency and responsiveness when caring for TGD adolescents.

Another research group obtained further input from TGD adolescent patients and providers from the same clinic, seeking to draw out information about how to characterize affirming versus nonaffirming environments (Eisenberg et al., 2020; Rider et al., 2019). Semistructured interviews with nurses and physicians working with TGD adolescents were conducted to identify perceived barriers to care and affirming environments. This provider group indicated that additional training on gender-affirming environments and more resources to implement those recommendations were needed, in addition to individual discomfort with topics related to gender and asking about gender identity, especially fears about losing patient trust and the potential for offending patients (Rider et al., 2019). From the patient experience, TGD adolescents reported that being asked about their name and pronouns helped to create support in the environment, but that they also felt providers needed more training to better meet their needs (Eisenberg et al., 2020).

Research strengths and limitations

Patterns of healthcare utilization among TGD youth have primarily been studied through large-scale national and local surveys. Overall, the results of this research suggest that TGD youth experience unique health and healthcare disparities compared with their cisgender peers of the same age and compared with TGD adults, and that these disparities are even larger for nonbinary youth and those with additional intersecting identities. Qualitative research has also been used to clarify the direct perspectives of TGD youth and healthcare providers on the factors that facilitate or hinder creating affirming and welcoming healthcare environments. Results of this research demonstrate that systemic changes are often needed, such as widespread provider and staff training on working with TGD youth, changes to medical records and other documentation that use nonaffirming language and birth-assigned names, and environmental changes to décor and available resources.

Research in this area is generally limited to patient experiences in singular healthcare systems or countries, and more national and international collaborations are needed to clarify regional similarities and differences, and cross-cultural considerations. A large proportion of this research is also limited to the United States alone. Particular strengths of this body of research include explicit efforts by some research groups to recruit diverse TGD communities, utilization of both quantitative and qualitative research methods, and the use of national surveys with large sample sizes.

The first national Canadian population survey addressing barriers and facilitators for TGD youth seeking primary care was able to decrease potential researcher bias while potentially increasing credibility and comprehension by using a strong mixed-method research design (Clark et al., 2018b). However, the representativeness and generalizability of the study sample are difficult to estimate given the absence of reliable data on the total TGD youth population in Canada. Although varied recruitment strategies were utilized to minimize a skewed sample from healthcare environments, there is still concern that the results reflect the experiences of youth with overall higher access to care.

Similarly, the first large US population survey to examine health status, care-seeking behaviors, and the role of perceived gender expression in limiting or facilitating care for TGD adolescents used the 2016 Minnesota Student Survey of over 2000 high school students (Rider et al., 2018). The research design and ability to gather information from students across the state differed from prior studies that relied on convenience sampling with less generalizability. The study also filled a gap in knowledge about the experiences of access to healthcare among TGD youth by including varied measurement strategies to assess outcomes and care-seeking behaviors. However, the design of some questions, especially regarding biological sex and lack of appropriate options for reporting gender identity, may have influenced the way students answered these questions, and the data may not be truly representative of their experiences. Furthermore, the questions about gender identity were framed based on how students felt their peers saw them rather than their internal sense of self. Finally, gathering information in school environments may have resulted in missed or skewed data due to student absences or perceived safety in the school environment.

Recommendations for clinical practice and research

While the base of information regarding health outcomes, care-seeking behaviors, and health disparities for TGD youth continues to evolve, and more information is needed to understand where potential improvements can be made, there are existing studies that give researchers and providers a general understanding of the experiences of TGD youth in healthcare. To address pediatric health disparities overall, clinicians, hospital administrators, and policy makers need to address the systemic barriers to healthcare, including economic barriers, system accessibility, parent and caregiver health behaviors and health literacy, and social stability (Brennan et al., 2025; Shi et al., 2020). For TGD youth specifically, improvements are needed at the individual, family, interpersonal, and institutional levels (Clark et al., 2018b).

Primary care providers have the opportunity to be important protective factors for health equity among TGD youth, but this requires competency, understanding, and openness (Clark et al., 2018b). To improve healthcare access, clinicians and staff need training and resources related to TGD youth, and hospital systems need to address appropriate use of gender identity and chosen name in medical records and on forms and labels (Silveira et al., 2025). Simple environmental improvements, such as LGBTQ+ art and décor, can instantly signal to patients and families that a healthcare environment is affirming and welcoming (Yeh et al., 2025). Other recommendations for improving general healthcare access include the use of telehealth, professional development trainings that focus on specialty gender-affirming care, and updating forms and patient education materials to use gender inclusive language (Clark et al., 2018b).

Future research on the experiences with healthcare among TGD youth should continue to recruit larger sample sizes across diverse communities and various geographic locations, evaluate changes made to training curricula, and incorporate an intersectional lens (Clark et al., 2018b). Utilizing robust and varied research designs, including mixed-method designs, as well as longitudinal approaches that evaluate specific interventions for improving access and that assess changes over time, is also needed (Rusow et al., 2022). To understand the needs and behaviors of specific subgroups of TGD youth, specialized research design and recruitment strategies will be needed, including TGD youth of color, those from rural geographic areas, youth with disabilities, and those living outside of North America and Europe.

Access and barriers to pediatric gender-affirming care

Access to high-quality pediatric gender-affirming care is a protective factor against physical and mental health disparities among TGD youth. Unfortunately, many barriers to accessing specialty gender-affirming care exist. Research on access and barriers to gender-affirming care for youth has relied on both qualitative and quantitative methods. Large retrospective chart review studies have been used to evaluate differential access to care based on demographic factors, such as age, gender identity, race and ethnicity, geographic location, and insurance. Individual interviews, focus groups, and surveys have been further utilized to clarify patient, caregiver, and provider perspectives on the most impactful barriers to accessing this care.

A small sample size of youth and parents in the United States was recruited to participate in focus groups and semistructured interviews, aiming to understand barriers to gender-affirming care from both parent and patient perspectives (Breland et al., 2016). Participants noted that they had difficulty finding providers who would offer gender-affirming care, and even if they located providers, they still faced challenges accessing puberty blockers and hormone therapy. Feedback also included provider practices as barriers, including being misgendered and being made to feel uncomfortable in interactions with providers and staff. There was also a lack of clarity in the steps needed to progress in care and a lack of care coordination between multiple care providers who might be involved in a patient's care plan.

Similarly, a separate sample of youth and parents in the United States was able to provide feedback about barriers to care through either an online survey, focus groups, or individual interviews (Gridley et al., 2016). Results reflected similar themes, including concerns about delays and barriers to medications, lack of affirming providers willing to offer gender-affirming care, being misgendered by clinic providers and staff, and the lack of coordination of care and clarity in navigating clinic processes. In addition, these participants noted gatekeeping practices and insurance coverage issues as other important barriers to consider addressing.

Parent perspectives from across 41 states in the United States were gathered through a single-question online survey (Kidd et al., 2023). Parents responded to the question, and responses were analyzed through inductive thematic analysis. Overall, parents described feeling that gender-affirming care offers numerous positive effects for TGD youth. Nonetheless, the parents in this study continued to reiterate the level of difficulty present in finding a knowledgeable provider willing to offer gender-affirming care, often citing the location of clinics (mostly academic medical centers in large urban areas) as a barrier. Parents also identified a lack of provider cultural competency, and at times blatant discrimination, as serving to prevent positive experiences in care or the ability to progress toward care goals. Some participants even endorsed providers accusing them of child abuse for affirming their child's identity and supporting the pursuit of gender-affirming care. Finally, insurance coverage, lack of consistency in clinical protocols, and long wait times were among the other identified barriers.

An online survey attempted to gather perspectives across the lifespan (Puckett et al., 2018). Respondents aged 16–73 years provided feedback on a variety of experiences with gender-affirming care. The majority of the sample were receiving hormone therapy, while a smaller proportion had accessed surgical care. Responses about barriers fell into several categories, many of which were identified in other studies: concerns about finances and insurance, lack of adequate availability of affirming providers, and lack of adequate provider training. Survey respondents also cited worries and fears about bias and discrimination, sometimes rooted in direct negative experiences in care, the need for parental consent, and the presence of co-occurring medical concerns, as barriers to access.

An additional online survey brought in provider opinions on provider comfort in care delivery and the barriers to providing affirming care for TGD youth (Vance et al., 2015). Survey participants were recruited through the Pediatric Endocrine Society and the Society for Adolescent Health and Medicine membership. Roughly two-thirds of the respondents had ever provided care to a TGD young person, and in terms of confidence levels in delivering gender-affirming care, roughly 62% of the respondents felt comfortable with offering care, while just under half said they felt confident. When asked what might increase levels of comfort and confidence, providers who responded to this survey identified barriers around training, collaboration with other professionals (i.e., mental health providers to complete comprehensive evaluations), unfamiliarity with working with TGD youth, and insurance coverage.

A qualitative study from Ireland used semistructured interviews with community samples of TGD young adults, parents, and providers to understand experiences with overall barriers, opinions on the assessment process, and identified positive aspects of seeking care (Kearns et al., 2024). All three cohorts noted a severe lack of both medical and mental health gender-affirming youth services in Ireland, with some families stating that mental health providers will not see a youth with a primary gender dysphoria diagnosis because they do not consider dysphoria to be a mental health concern. Relatedly, some primary care providers refused referrals to gender services because they did not feel it was a legitimate concern.

Youth and families felt that the requirements for care were routinely changing, but that there was a lack of transparency about what would be needed. They also reported that they needed to travel to other countries or overseas to receive care, especially for surgery and for younger youth. An additional barrier that was identified was the need for assessment. Youth very clearly stated they were over-assessed and wanted movement toward informed consent, although providers were not in favor of this change, and parents expressed mixed feelings. Some parents felt judged in the assessment process, and youth felt traumatized by both the questions asked and the interactions with providers. Providers recognized that improvements were needed in the way assessments were approached. Parents overall feared that mental health comorbidities would prevent their youth from accessing care, and both youth and parents felt that the requirement for social transition before medical care was sometimes prohibitive (Kearns et al., 2024).

This study also looked at perceptions of possible facilitators to care (Kearns et al., 2024). Youth and parents felt that the nationalized health system was a facilitating factor in accessing care, stating that they did not have to worry about costs or denials of care in the same way that youth in other countries may need to. Though assessments and requirements for social transition were cited as barriers by youth and some parents, both youth and parents felt that opportunities to explore identity before starting medications were beneficial. Youth and parents also felt that online support spaces and other physical support groups were helpful in providing information about care navigation strategies, and that provider interface with these spaces offered an opportunity for building trust.

Research on utilization patterns and demographic disparities, which has largely relied on retrospective chart review, has found that access to gender-affirming care is even more limited for adolescents compared with adults, nonbinary youth compared with their binary transgender peers, and TGD youth with intersecting minoritized identities (Connelly et al., 2019; Clark et al., 2018a; Kahn et al., 2024; Locke et al., 2024).

Compared with TGD adults, children and adolescents inherently need at least one supportive adult to help them access specialty healthcare. Furthermore, TGD adolescents often experience a higher degree of gatekeeping to gender-affirming medical care than adults, with prerequisites to care that are neither consistently applied nor based on specific research outcomes. These prerequisites may include protracted assessments or treatment with a mental health provider before receiving a referral to a specialty clinic, requirements for length of time since identity realization and disclosure to others, or requirements for the involvement of all legal guardians, even in contexts where one guardian is uninvolved and/or unsupportive. A large retrospective chart review of TGD adolescents and adults from a US military health system repository (N=3066) found longer time to receiving hormones for youth aged 14–18 years compared with those aged 19–22 years (Locke et al., 2024).

Similarly, TGD youth who do not identify with the binary categories of male or female experience more barriers to accessing gender-affirming care than their binary transgender peers (Clark et al., 2018a). In this study, trans and genderqueer youth across Canada completed an online survey to understand health outcomes and access for identity subgroups within the larger TGD youth populations. Youth were recruited using multiple strategies, including social media, emails, clinic contacts, word of mouth, and referrals from other youth and providers. Compared with binary transgender youth, nonbinary youth in this study more often did not have a primary care doctor, and if they did, reported less likelihood that their doctor knew about their identity. Additionally, while nonbinary youth may not seek hormone therapy as frequently as binary youth, when hormone therapy aligns with embodiment goals, nonbinary youth were less likely than binary peers to obtain it. Barriers to initiating hormone therapy might be individual (i.e., existing options do not match embodiment goals, high rates of indecision about pursuing hormones), but they are also sometimes systemic: location and availability of clinics, lack of decision-making tools specific to nonbinary experiences, and requirements to get referrals from primary care providers. Referrals specifically become a barrier when providers are dismissive of nonbinary youths' identities and clinical needs. Affirming medical care also feels out of reach when providers lack information about the range of existing options (Clark et al., 2018a).

Similar to the research on access to general medical care, intersecting identities can further impact TGD youths' access to gender-affirming care. TGD youth of color face challenges related to both diagnosis and treatment for gender dysphoria. The retrospective chart review in the military health system found that Black youth had to endure longer wait times from the first clinical contact to the initiation of hormone therapy, compared with white youth in the study (Locke et al., 2024). Another retrospective chart review sought to examine how intersecting identities compound to further impact access to care by analyzing demographic data for 2444 youth aged 13–17 years, with listed gender identity or pronouns differing from what was assigned at birth (Kahn et al., 2024). Black TGD youth in this study were less likely to receive a gender dysphoria diagnosis and less likely to receive desired gender-affirming care compared with the white youth. Asian youth were less likely to receive a gender dysphoria diagnosis compared with the white and Hispanic youth in this study, although no difference in access to gender-affirming care was noted. Hispanic youth were more likely to receive a gender dysphoria diagnosis than Black and Asian youth but were only more likely to access care compared with Black youth. However, if the primary spoken language was Spanish, then Hispanic youth were less likely than the other youth to receive a diagnosis and access care.

Beyond race, ethnicity, and gender identity, demographic groups' experiences with access can be influenced by geographic location, socioeconomic status, and types of insurance coverage. A mixed-methods study utilized an online survey, focus groups, and semistructured individual interviews to gather a baseline understanding of the needs of TGD youth and young adults in Oregon. About 52% of the survey sample identified as being from nonurban areas, and 77% of the survey respondents indicated that they had never sought medical care for gender affirmation. Among the barriers cited, the youth and young adults noted lack of parental support as the most cited barrier to care, but also indicated that lack of insurance coverage for gender-affirming care, cost, and distance from specialty services were other frequently faced barriers (Connelly et al., 2019). The retrospective demographic chart review similarly identified state insurance coverage (compared with commercial, employer-based coverage) as a predictor of lower access to gender-affirming care (Kahn et al., 2024). State insurance coverage is typically associated with lower socioeconomic status and may be associated with additional barriers related to finances, including travel and transportation to clinics.

In contrast, another US retrospective chart review with a smaller sample size from a singular clinic found that there appeared to be no difference in state or commercial insurance coverage for pubertal suppression medications (Nahata et al., 2017). Out of the 79 charts identified in the review, 27 of those patients desired pubertal suppression medications, and only 8 had insurance coverage. Of the 19 others without coverage, one decided to self-pay, and there was no significant relationship between the type of insurance (state Medicaid compared with a commercial plan) and denials for these medications. Age and gender also did not have a significant relationship when determining the likelihood of coverage or denials. A total of 40 of the 79 patients identified eventually were prescribed hormone therapy, but less information was available about insurance coverage for this intervention, potentially suggesting that insurance plays less of a role in facilitating access to these medications, as they are generally more affordable out-of-pocket.

While insurance coverage could potentially be a facilitating factor in accessing gender-affirming care, there are still steps between when a prescription is written and authorized and when a medication is actually obtained by a patient. Interactions at a pharmacy represent both opportunities for access and the potential for additional barriers. A study group in the United States implemented snowball and purposive sampling methods to gather TGD individuals aged 18–29 years for semistructured focus groups, with the intention of understanding their experiences at community pharmacies (Johnson et al., 2025). Although the recruitment was for adult participants, they were asked to reflect upon past experiences with pharmacies if they accessed care as adolescents.

Overall, pharmacies and pharmacists presented additional barriers for TGD community members attempting to access gender-affirming medications. The focus group participants identified key themes that influenced their perception of pharmacists as barriers: accessibility and difficulties in interpersonal interactions. Participants were concerned about the financial costs and insurance coverage for these medications, but found that the most significant obstacle was obtaining supplies needed to administer medications (i.e., needles and syringes). When ordered supplies were not available, participants reported that the pharmacists did not routinely offer alternatives or substitutes, and they had to choose between reusing past supplies or not administering their medically necessary treatment. Participants also felt that pharmacists lacked training in TGD cultural competence, resulting in interactions where they were deadnamed, questioned about their identity, questioned about the purpose of the medication and whether it was appropriately prescribed, and had their identity disclosed in a public space, calling into question safety concerns (Johnson et al., 2025).

Research strengths and limitations

Overall, a mix of research methods is needed, including within individual studies. Most studies on barriers to care have been retrospective chart reviews aiming to identify demographic disparities in large samples of TGD youth. Retrospective chart reviews are important for adding to the foundation of knowledge, especially for groups that have previously been underrepresented in research (Kahn et al., 2024). However, because chart review data is static and relies on documentation, there may be inaccuracies about gender identity, pronouns, and types of treatment sought. Chart reviews can identify relevant information to address research gaps on which groups are experiencing barriers to care, but cannot provide insight into how those barriers could be addressed. Qualitative methods, including interviews, focus groups, and surveys, utilizing multiple recruitment strategies to obtain diverse samples, are needed to clarify perspectives on barriers from key stakeholders. The current research has been successful in incorporating mixed-methods approaches, in addition to recruiting participants from both clinic and community populations.

Available studies have largely been based in the United States and may not be as generalizable, especially given the repeated citation of insurance coverage as a major barrier and other cultural considerations. In systems with nationalized healthcare, coverage may not be as much of a barrier or may present difficulties in different ways. Sample sizes, especially in qualitative studies, have been relatively small, and expanding sample sizes will facilitate responses that are more representative of all experiences. As most studies have relied at least partially on recruitment within clinic populations, there is some concern that perspectives from participants who experience the highest number of barriers and have the most limited contact with healthcare environments are missing. Additionally, further insight from participants who exist at multiple intersecting identities is needed.

Recommendations for clinical practice and research

Accessing gender-affirming medical care can be a precursor to increased engagement with the larger healthcare system, which is significant given the overall disparities that exist in both care utilization and long-term physical and mental health outcomes for TGD youth compared with their cisgender peers. Additionally, the majority of TGD youth who want medical support for their embodiment goals are not able to access appropriate gender-affirming care, adding another obstacle for affirming engagement with healthcare.

Regarding pharmacists, there is a clear need to increase cultural competency around working with TGD patients (Johnson et al., 2025). Study participants indicated that pharmacists need increased education about the importance of gender-affirming medications, and that there would be benefits to having pharmacists who are from TGD communities with a base level of understanding of the importance of these medications. Participants also noted that having more information about the role of pharmacists in the dispensing and administration of gender-affirming medications could increase health literacy for patients and improve navigation and access. Ongoing evaluation of pharmacy program curricula and curricula revisions was also recommended.

Feedback and findings from multiple surveys, interviews, and focus groups repeatedly identified cultural competency and the need for increased provider and staff training as key areas for improvement in clinical care. One set of focus group participants indicated that having a colocated multidisciplinary clinic for gender care would be beneficial for decreasing barriers, and advocated for TGD cultural competency for all staff and providers associated with that clinic (Breland et al., 2016). In addition to the formation of multidisciplinary clinics and required provider and staff education, study participants offered specific recommendations related to staffing patterns and clinical program design. Clear written processes for clinicians and care maps for families, designated care coordinators within gender clinics, development of culturally competent policies for documentation, and clinical practices that allow youth to access hormone therapy around the pubertal stage of their developmentally matched peers were all offered as recommendations to improve access and experience in care environments (Gridley et al., 2016).

Parents of TGD youth have identified the need for cultural competency education and the development of clear protocols regarding the collection and use of affirmed names and pronouns, both interpersonally and in the medical record (Kidd et al., 2023). Educational efforts aimed at primary care providers were highlighted as particularly important, as increased knowledge and ability in primary care environments may increase access for families that do not have a specialty clinic easily available. Revision of care protocols and prerequisites also needs to be evaluated to ensure those that are in place allow for individualized care plans, reducing delays and other barriers. Given the current political environment, families also noted the need for providers to have awareness about updated local and national policies and to be willing to advocate against restrictive policies that create additional barriers.

Future research would benefit from the creation and validation of standardized measurement tools, as well as standardized research protocols. By standardizing measurement tools and protocols, future studies can compare data regionally and internationally and evaluate changes in trends over time. Research on the development and evaluation of interventions implemented to reduce barriers and improve access is needed. Community needs assessments and community-based participatory research strategies should be utilized to clarify TGD youth and family priorities for future studies.

Telehealth and electronic consultation

The availability of telemedicine and telehealth visits has grown exponentially in the last few years, facilitated largely by the need created by the COVID-19 pandemic. Research related to the provision of gender-affirming care via telehealth has primarily used retrospective chart review, with a smaller number of qualitative and mixed-methods studies. Telehealth has the potential to reduce many structural barriers to care, such as stigma and discrimination within large health systems, geographic and transportation barriers, and difficulty navigating time off from work and school (Rodriguez and O'Neil, 2025).

Although telehealth appears to improve access for many families, this is not unequivocally true and depends largely on the specific barriers that a particular family faces when accessing care. For example, a large chart review study of TGD adolescent and young adult patients (N = 2908) found that telehealth visits were associated with a lower no-show rate, and this was especially true for those living in rural compared with urban areas and for adolescents compared with young adults (DeGuzman et al., 2024). On the contrary, another smaller chart review study (N = 49) found that telehealth patients waited longer for the initial prescription of testosterone or estradiol compared with TGD adolescents and young adults who attended in person (Foo et al., 2024). Another retrospective chart review specifically evaluated virtual testosterone injection training versus in-person training among TGD adolescents and young adults assigned female at birth (Nightingale et al., 2024). Results revealed no significant differences in injection-related questions, time to first injection, or clinical levels of testosterone at 3 or 6 months following initiation.

One mixed-method study included a retrospective chart review of TGD youth aged 7–18 years before and after the use of telehealth was implemented in 2020, as well as a qualitative survey of youth and parents who attended a telehealth visit (Hedrick et al., 2022). New patient medical visits were reviewed across three time points, including the 4 months before the 2020 modified operations and implementation of telehealth, the 4 months immediately following the onset of modified operations, and the following 4 months. Patient demographics, including age, gender identity, geographic area (rural or urban), and number of caregivers attending the initial visit, remained relatively stable across the three time points.

Thematic analysis of survey responses revealed that telehealth improved access and comfort for some but created additional barriers or discomfort for others. For example, telehealth reduced transportation barriers, but for families without consistent internet bandwidth or who would have benefited from a blood lab draw or physical examination, telehealth created more problems. Similarly, some families perceived the telehealth visits as less stressful than in-person visits, but others found that telehealth made it difficult to connect and form relationships with providers in the same way that they would be able to in person (Hedrick et al., 2022).

In another qualitative study, semistructured interviews with TGD adolescents and young adults similarly revealed that while telehealth is often not the preferred option, it is nonetheless recognized as a practical option when needed (Inwards-Breland et al., 2025). Advantages identified in this study were convenience, lower stress, and accessibility for older adolescents and young adults whose parents may not be supportive enough to facilitate attending in-person visits. Disadvantages included technical issues such as poor connectivity, less feedback from providers on the progress of physical changes from medical interventions, discomfort related to being seen on video, and, in some households, a lack of privacy from parents or other family members.

One of the clearest facilitators of improved access to gender-affirming care is availability within primary care, and many young people prefer to receive gender-affirming care from their primary care providers (Sequeria et al., 2023). However, a lack of knowledge and confidence among primary care doctors to provide pediatric gender-affirming care remains a barrier. One novel strategy for improving access is the electronic consultation (eConsult) framework through which primary care doctors can deliver gender-affirming care in direct consultation with specialty providers through structured electronic communications. In a study evaluating a new eConsult model, the content of 38 eConsults and the results of 26 provider surveys supported the feasibility of this framework (Sequeira et al., 2024). Primary care providers reported high satisfaction with and usability of the electronic program, increased confidence in providing gender-affirming care to youth, and, in some cases (19%), that the eConsult eliminated the need for a referral to specialty care.

Overall, the research on telehealth and eConsult models demonstrates the utility of these healthcare strategies to reduce barriers for TGD youth and their families, but there are also potential limitations (Rodriguez and O'Neil, 2025). The primary advantages of telehealth appear to be reduced transportation burden and avoiding large health systems in which TGD youth may experience overt stigma or more subtle microaggressions. On the contrary, lack of consistent broadband internet access, difficulty forming an emotional connection with providers, and a lack of privacy from other family members during visits have been identified as disadvantages (Rodriguez and O'Neil, 2025). Between providers, the eConsult framework appears to be especially useful for increasing healthcare access by allowing gender-affirming care provision through primary care (Sequeira et al., 2024). Research on telehealth and eConsult models is limited to US studies and has largely relied on retrospective chart review, although qualitative and mixed-method studies have helped to clarify the nuances of patient, caregiver, and provider experiences.

Political attacks and legal barriers

In the last few years alone, pediatric gender-affirming care has been limited in some European countries and legally banned in many US states based on deliberate disinformation and pseudoscientific arguments circulated by explicitly anti-LGBTQ+ religious and political groups (Keuroghlian and Radix, 2024; Preston and Lopez, 2024). Justified by pseudoscientific disinformation and under the guise of protecting children, these restrictions demonstrate their own insincerity by making explicit exceptions for similar procedures on presumed cisgender children at much younger ages (Mar et al., 2025). Because these political attacks are relatively new, research on the impacts is only beginning to emerge, and primarily from the United States, where there are full bans in some regions.

The most immediate and overt impacts of these political attacks are the direct effects on the accessibility of services. One study calculated changes in geographic access and drive times to pediatric gender-affirming care clinics in the United States as of May 2023, when there were bans in 20 states (Borah et al., 2023). A total of 271 clinics were identified, of which 70 (25.8%) were inactive due to legal bans, affecting an estimated 30% of TGD adolescents. Estimated driving times to the nearest clinic doubled after restrictions were in place. Families living more than 1 hour from a clinic increased from 27.2% to 50.0%, and those living more than a day's drive increased from 1.4% to 25.3% following restrictions. Families in Florida, Texas, and Utah were the most impacted. These results are estimates based on national population estimates of the number of TGD youth across regions and advertised or reported clinics. Therefore they may underestimate the full impact, as families may not be aware of their closest clinic and may need to travel further to one they know.

The impacts on accessibility are not restricted to the states in which there are bans, with secondary impacts on access across the nation. Clinics in states without legal restrictions on care, which may already have long wait times, are at risk of being further overwhelmed by patients traveling from other states. Qualitative interviews with providers have demonstrated these negative impacts on the clinics and clinicians who provide this care, including threats of violence, exhaustion, emotional burnout, and moral injury (Hughes et al., 2026). In US states without restrictions, including those with explicit protections, providers are overwhelmed by increasing demands by families moving to their states for care (Goldberg and Redfield, 2025).

General fear and intimidation are further causing clinics to decrease advertising and remove otherwise helpful information from their websites, even in states without bans and with legal protections. A review of 149 US pediatric hospitals revealed that the percentage of hospital websites having any information on gender-affirming care significantly decreased from 70% to 58% of websites (Band et al., 2025). While the rates of removal of information from websites were more significant in states with legal bans, states without bans nonetheless saw decreases as well. Removal of information occurred most often in the Southern region of the United States and from the websites of religious hospitals compared with academic and secular institutions. While website information is only a proxy measure of actual access to care, this study demonstrates that the availability of information for families is an added impact of legal bans in addition to serving as a proxy for access changes.

In addition to direct impacts on access to gender-affirming care, legislation that targets TGD people affects the mental health and well-being of TGD young people. By estimating population-level suicide attempts during a period before enacted legislation and a period after, one study showed an increase in suicide attempts among TGD youth, as well as an increase in the total number of youth who attempted suicide, as a direct outcome of state-level antitransgender laws (Lee et al., 2024). Many of those who are enacting these laws, especially those banning pediatric gender-affirming care specifically argue that allowing young people to exist as their authentic selves is the cause of their mental health concerns. This study clearly demonstrates the opposite—these lawmakers are causing the problems they disingenuously purport to want to remedy.

Qualitative research has further expanded upon the direct experiences of TGD youth and their family members who are impacted by legislative attacks. For TGD youth accessing gender-affirming healthcare and their family members, laws that restrict this care impact not only direct access to care, but also have several ripple effects with significant implications, including impacts to individual well-being, changing of the social ethos to be more hostile toward these families, and general upheaval of the lives of all family members (Brandon-Friedman et al., 2025). Schools and communities have become less safe, families are forced to uproot their lives and move to states with access to care, and both of these consequences contribute to growing isolation among families who only want to protect the physical and emotional safety of their children.

A survey of parents recruited from the community (i.e., not necessarily receiving care in a specialty gender clinic) showed that more than 10% of families with a TGD youth have moved states because of enacted legislation (Kidd et al., 2025). Results of the survey demonstrated several additional themes, including financial stress, negative impacts on mental health, travel burden, increased experiences of stigma and discrimination, and generally increased fear. Families felt abandoned by the medical providers and institutions that they previously trusted. Another survey of parents and caregivers showed that there is a strong consensus that parents want legislators to reconsider these politically driven initiatives, emphasizing that the health and healthcare of TGD youth should not be a political issue and that these efforts are the direct cause of increased discrimination and violence (Abreu et al., 2022).

Importantly, these laws are not keeping parents and families from fighting for their TGD children; youth and families are more empowered than ever to fight these discriminatory practices and stand up for the rights and dignity of all (Kidd et al., 2025). Clinicians who work with TGD youth and families are similarly engaging in increased advocacy and activism (Hughes et al., 2026).

While this area of research has grown out of necessity rather than a preexisting interest, it is notable that this body of literature emerged quickly and with a range of strong methodological approaches. The rapid and robust approach to this research has clearly demonstrated the negative impacts of these discriminatory legislative efforts on not only TGD youth themselves, but their family members, the clinicians who provide this care, and all people who care about justice and human rights. The primary limitation of this area of research is that it is largely US-based, and less is known about the impacts of restrictions that are not complete bans, such as those in some European countries. International collaborations and standardized protocols are needed to explore the differential impact of various levels of restriction and protection. Finally, researchers are encouraged to further explore the positive impacts of engaging in advocacy and activism, and the resilience of TGD youth and the people who care for them in the face of these political attacks.

Summary and recommendations for future research

TGD youth and their families experience health inequities, including increased risk of physical and mental health concerns, structural discrimination and violence, and barriers to preventative, routine, and specialty care. While TGD adults also experience health disparities, young people have the added layer of requiring support from adults to access and utilize the resources available to them. Youth who experience intersecting sources of oppression, such as racism, ableism, or poverty, as well as those who identify outside of the binary categories of transgender male and female, experience greater disparities and health inequity. Recently, increasing political attacks have contributed to growing stigma, discrimination, and violence in addition to direct impacts on access to general and specialty healthcare.

Research to date has explored the individual, interpersonal, and structural barriers to primary care, specialty care, and gender-affirming care specifically. Specialty gender-affirming care is a protective factor against physical and mental health disparities, but many barriers exist. One of the most prominent and pervasive barriers is a lack of provider knowledge and experience with TGD patients, healthcare environments that are not affirming, and overt discrimination within healthcare institutions. Other barriers include a lack of parental support, financial and geographic limitations, and systemic gatekeeping of care. Specifically, prerequisites for accessing gender-affirming care are inconsistently applied, and limited research has evaluated the utility or appropriateness of these prerequisites. More recently, emerging research includes the use of telehealth as a facilitator to healthcare access and the impacts of restrictive legislation on health disparities and access to healthcare.

A strength of existing research in this area includes the use of robust and varied methodology, including qualitative research, population-level surveys, and large-scale reviews of medical records and health systems data. Primary limitations include the difficulty inherent in recruiting participants who experience a high level of barriers to care, relatively fewer research initiatives in this area from outside of the United States, and a dearth of community-based participatory research methods. Future research should expand the focus on facilitators to care and interventions to improve access, especially for TGD youth experiencing intersecting areas of oppression. Standardized measurement tools and research protocols are needed to compare regional and international differences, as well as changes over time. Community-based participatory research is also needed to identify the priorities of TGD youth and their families with regard to healthcare access and barriers.

References

Abreu, R.L., Sostre, J.P., Gonzalez, K.A., Lockett, G.M., Matsuno, E., Mosley, D.V., 2022. Impact of gender-affirming care bans on transgender and gender diverse youth: parental figures' perspective. Journal of Family Psychology 36 (5), 643–652. Available at https://doi.org/10.1037/fam0000987.

Acosta, W., Qayyum, Z., Turban, J.L., van Schalkwyk, G.I., 2019. Identify, engage, understand: supporting transgender youth in an inpatient psychiatric hospital. Psychiatric Quarterly 90 (3), 601–612. Available at https://doi.org/10.1007/s11126-019-09653-0.

Band, I.C., Dubois, B., Ghofranian, A., Nietsch, K.S., Lee, J.A., Friedenthal, J., et al., 2025. An analysis of gender-affirming care offerings on United States pediatric hospital websites: exploring the impact of state legislative bans. LGBT Health 12 (6), 458–467. Available at https://doi.org/10.1089/lgbt.2024.0214.

Borah, L., Zebib, L., Sanders, H.M., Lane, M., Stroumsa, D., Chung, K.C., 2023. State restrictions and geographic access to gender-affirming care for transgender youth. JAMA: the Journal of the American Medical Association 330 (4), 375–378. Available at https://doi.org/10.1001/jama.2023.11299.

Brandon-Friedman, R.A., Tabb, A., Imburgia, T.M., Swafford, T.R., Fortenberry, J.D., Canada, M., et al., 2025. Perspectives of gender-diverse youth and caregivers facing gender-affirming medical intervention bans. LGBT Health 12 (2), 108–115. Available at https://doi.org/10.1089/lgbt.2024.0079.

Breland, D.J., Crouch, J.M., Evans, Y., Gridley, S.J., Eng, W., Antoon, E.A., et al., 2016. Understanding barriers to healthcare for transgender youth. Journal of Adolescent Health 58 (2), S3. Available at https://doi.org/10.1016/j.jadohealth.2015.10.023.

Brennan, L., Brewster, L., Lunn, J., Isba, R., 2025. Barriers to accessing pediatric health care: a focus on health equity in the UK. Pediatrics 156 (Supplement 1). Available at https://doi.org/10.1542/peds.2025–070739e.

Clark, B.A., Veale, J.F., Townsend, M., Frohard-Dourlent, H., Saewyc, E., 2018a. Non-binary youth: access to gender-affirming primary health care. International Journal of Transgenderism 19 (2), 158–169. Available at https://doi.org/10.1080/15532739.2017.1394954.

Clark, B.A., Veale, J.F., Greyson, D., Saewyc, E., 2018b. Primary care access and foregone care: a survey of transgender adolescents and young adults. Family Practice 35 (3), 302–306. Available at https://doi.org/10.1093/fampra/cmx112.

Connelly, K.J., St. Aurin, C.G., Guerriero, J.T., Penkin, A., 2019. Survey of the Needs of Transgender Youth in Oregon. https://digitalcollections.library.oregon.gov/nodes/view/138519.

DeGuzman, P.B., Lyons, G.R., Azar, F.N., Kimble, A., Huang, G., Rheuban, K., et al., 2024. Impact of telemedicine on access to care for rural transgender and gender-diverse youth. Journal of Pediatrics 267. Available at https://doi.org/10.1016/j.jpeds.2024.113911.

Eisenberg, M.E., McMorris, B.J., Rider, G.N., Gower, A.L., Coleman, E., 2020. "It's kind of hard to go to the doctor's office if you're hated there." A call for gender-affirming care from transgender and gender diverse adolescents in the United States. Health and Social Care in the Community 28 (3), 1082–1089. Available at https://doi.org/10.1111/hsc.12941.

Foo, M.M., Roberts, Z., Roden, R.C., 2024. Use of telemedicine services is associated with longer ime to tinitiation of hormone therapy in transgender teens and young adults. Telemedicine and e-Health. Available at https://doi.org/10.1089/tmj.2024.0302.

Goldberg, A.E., Redfield, E., 2025. The Experiences of Gender-Affirming Care Providers in States Without Laws Restricting Access to Care. https://williamsinstitute.law.ucla.edu/publications/experiences-gac-providers/

Goldenberg, T., Jadwin-Cakmak, L., Popoff, E., Reisner, S.L., Campbell, B.A., Harper, G.W., 2019. Stigma, gender affirmation, and primary healthcare use among black transgender youth. Journal of Adolescent Health 65 (4), 483–490. Available at https://doi.org/10.1016/j.jadohealth.2019.04.029.

Gridley, S.J., Crouch, J.M., Evans, Y., Eng, W., Antoon, E., Lyapustina, M., et al., 2016. Youth and caregiver perspectives on barriers to gender-affirming health care for transgender youth. Journal of Adolescent Health 59 (3), 254–261. Available at https://doi.org/10.1016/j.jadohealth.2016.03.017.

Hedrick, H.R., Glover, N.T., Guerriero, J.T., Connelly, K.J., Moyer, D.N., 2022. A new virtual reality: benefits and barriers to providing pediatric gender-affirming health care through telehealth. Transgender Health 7 (2), 144–149. Available at https://doi.org/10.1089/trgh.2020.0159.

Hughes, L.D., Hughto, J.M.W., Wolfe, H.L., Kelly, P.J.A., Kidd, K.M., Restar, A.J., et al., 2026. Provider adaptations to legislative efforts to ban gender-affirming care for adolescents. Journal of Adolescent Health 78 (3), 504–512. Available at https://doi.org/10.1016/j.jadohealth.2025.11.005.

Inwards-Breland, D.J., Yeh, D., Marinkovic, M., Richardson, T.R., Marino-Kibbee, B., Bayley, A., et al., 2025. Facilitators and barriers to using telemedicine for gender-affirming care in gender-diverse youth: a qualitative study. Journal of Telemedicine and Telecare 31 (6), 873–881. Available at https://doi.org/10.1177/1357633X241231015.

Johnson, H.E., Logan, T.K., Hill, K.K., 2025. Pharmacy barriers for transgender and gender diverse young adults. Journal of the American Pharmacists Association 65 (1). Available at https://doi.org/10.1016/j.japh.2024.102251.

Kahn, N.F., Asante, P.G., Coker, T.R., Kidd, K.M., Christakis, D.A., Richardson, L.P., et al., 2024. Demographic differences in gender dysphoria diagnosis and access to gender-affirming care among adolescents. LGBT Health 11 (5), 348–358. Available at https://doi.org/10.1089/lgbt.2023.0273.

Kearns, S., O'Shea, D., Neff, K., 2024. Factors that help and hinder transgender and nonbinary youth accessing gender care in Ireland: a multistakeholder exploration. Journal of Nursing Scholarship 56 (1), 60–75. Available at https://doi.org/10.1111/jnu.12907.

Keuroghlian, A.S., Radix, A.E., 2024. Transgender health research needed. Science (New York, N.Y.) 385 (6711), 809. Available at https://doi.org/10.1126/science.ads3727.

Kidd, K.M., Sequeira, G.M., Katz-Wise, S.L., Fechter-Leggett, M., Gandy, M., Herring, N., et al., 2023. ''Difficult to find, stressful to navigate'': parents' experiences accessing affirming care for gender-diverse youth. LGBT Health 10 (7), 496–504. Available at https://doi.org/10.1089/lgbt.2021.0468.

Kidd, K.M., Sequeira, G.M., Fechter-Leggett, M., Kahn, N.F., Salyer, R., Chinn, E., et al., 2025. "From having regular care with a team of doctors to fleeing the state": parent experiences supporting gender diverse youth in states with care bans. Journal of Adolescent Health 77 (4), 620–627. Available at https://doi.org/10.1016/j.jadohealth.2025.06.001.

Lee, W.Y., Hobbs, J.N., Hobaica, S., DeChants, J.P., Price, M.N., Nath, R., 2024. State-level anti-transgender laws increase past-year suicide attempts among transgender and non-binary young people in the USA. Nature Human Behaviour 8 (11), 2096–2106. Available at https://doi.org/10.1038/s41562-024-01979-5.

Locke, E.R., Highland, K.B., Thornton, J.A., Sunderland, K.W., Funk, W., Pav, V., et al., 2024. Time to gender-affirming hormone therapy among US military-affiliated adolescents and young adults. JAMA Pediatrics 178 (10), 1049–1056. Available at https://doi.org/10.1001/jamapediatrics.2024.2835.

Mar, S.A., Earp, B.D., Indig, G., McNamara, M., Boskey, E.R., Dalke, K.B., et al., 2025. US laws restricting medical care for transgender minors statutory inconsistencies involving intersex and other individuals. JAMA Health Forum 6 (11). Available at https://doi.org/10.1001/jamahealthforum.2025.4157.

Nahata, L., Quinn, G.P., Caltabellotta, N.M., Tishelman, A.C., 2017. Mental health concerns and insurance denials among transgender adolescents. LGBT Health 4 (3), 188–193. Available at https://doi.org/10.1089/lgbt.2016.0151.

Nightingale, K.J., Jelinek, S.K., Jones, C., Bevington, L.M., Darien, K., Ding, A., et al., 2024. 19. Telehealth versus in-person instruction for adolescents and young adults initiating gender-affirming testosterone therapy. Journal of Adolescent Health 74 (3), S11. Available at https://doi.org/10.1016/j.jadohealth.2023.11.216.

Pastor Bravo, M.del M., Linander, I., Morgan, J., 2024. Access to healthcare among transgender and non-binary youth in Sweden and Spain: a qualitative analysis and comparison. PLoS One 19 (5), e0303339. Available at https://doi.org/10.1371/journal.pone.0303339.

Preston, S., Lopez, X., 2024. Legislation, medicine, and politics: care for gender diverse youth. Current Opinion in Endocrinology, Diabetes & Obesity 31 (1), 43–52. Available at https://doi.org/10.1097/med.0000000000000845.

Puckett, J.A., Cleary, P., Rossman, K., Mustanski, B., Newcomb, M.E., 2018. Barriers to gender-affirming care for transgender and gender nonconforming individuals. Sexuality Research and Social Policy 15 (1), 48–59. Available at https://doi.org/10.1007/s13178-017-0295-8.

Rider, G.N., Mcmorris, B.J., Gower, A.L., Coleman, E., Eisenberg, M.E., 2018. Health and care utilization of transgender and gender nonconforming youth: a population-based study. Pediatrics 141 (3). Available at https://doi.org/10.1542/peds.2017–1683.

Rider, G.N., McMorris, B.J., Gower, A.L., Coleman, E., Brown, C., Eisenberg, M.E., 2019. Perspectives from nurses and physicians on training needs and comfort working with transgender and gender-diverse youth. Journal of Pediatric Health Care 33 (4), 379–385. Available at https://doi.org/10.1016/j.pedhc.2018.11.003.

Rodriguez, M., O'Neil, J., 2025. The use of telemedicine by transgender and gender nonconforming youth. The Journal for Nurse Practitioners 21 (2), 105294. Available at https://doi.org/10.1016/j.nurpra.2024.105294.

Rusow, J.A., Hidalgo, M.A., Calvetti, S., Quint, M., Wu, S., Bray, B.C., et al., 2022. Health and service utilization among a sample of gender-diverse youth of color: the TRUTH study. BMC Public Health 22 (1). Available at https://doi.org/10.1186/s12889-022-14585-9.

Samargia, L.A., Saewyc, E.M., Elliott, B.A., 2006. Foregone Mental Health Care and Self-Reported Access Barriers Among Adolescents.

Sequeira, G.M., Asante, P.G., Bocek, K., Kahn, N.F., Sethness, J.L., Hodax, J.K., et al., 2024. Evaluating an electronic consultation platform to support pediatric primary care providers in caring for transgender and nonbinary adolescents. Telemedicine and e-Health 30 (2), 595–600. Available at https://doi.org/10.1089/tmj.2023.0266.

Sequeria, G.M., Guler, J., Reyes, V., Asante, P.G., Kahn, N., Anan, Y., et al., 2023. Adolescent and caregiver perspectives on receiving gender-affirming care in primary care. Pediatrics 152 (4). Available at https://doi.org/10.1542/peds.2023–062210.

Shi, Q., Castillo, F., Viswanathan, K., Kupferman, F., MacDermid, J.C., 2020. Facilitators and barriers to access to pediatric medical services in a community hospital. Journal of Primary Care and Community Health 11. Available at https://doi.org/10.1177/2150132720904518.

Silveira, J.C.P., Souza, D.Md, Cardoso, Cd.S., Oliveira, M.A.Fd, 2025. Barriers and facilitating strategies for healthcare access and reception for transgender children and adolescents. Revista Brasileira de Enfermagem 78. Available at https://doi.org/10.1590/0034–7167-2024-0266.

Taliaferro, L.A., Harder, B.M., Lampe, N.M., Carter, S.K., Rider, G.N., Eisenberg, M.E., 2019. Social connectedness factors that facilitate use of healthcare services: comparison of transgender and gender nonconforming and cisgender adolescents. Journal of Pediatrics 211, 172–178. Available at https://doi.org/10.1016/j.jpeds.2019.04.024.

Vance, S.R., Halpern-Felsher, B.L., Rosenthal, S.M., 2015. Health care providers' comfort with and barriers to care of transgender youth. Journal of Adolescent Health 56 (2), 251–253. Available at https://doi.org/10.1016/j.jadohealth.2014.11.002.

Williams, K.A., Chapman, M.V., 2011. Comparing health and mental health needs, service use, and barriers to services among sexual minority youths and their peers. Health & Social Work 36 (3), 197–206. Available at https://doi.org/10.1093/hsw/36.3.197.

Yeh, D., Inwards-Breland, D.J., Richardson, T.L., Marinkovic, M., Marino-Kibbee, B., Patel, A., et al., 2025. Designing a better health system: patient perspectives on gender inclusive care. Clinical Pediatrics 64 (9), 1218–1226. Available at https://doi.org/10.1177/00099228251325448.

Part II

Pediatric gender-affirming healthcare

Chapter 5

Holistic and interdisciplinary healthcare for transgender and gender diverse youth

T. Zachary Huit[1,2,3], Amanda F. Terry[1,2,3], Michaela Patton[2,4,5], Claire Coyne[1,2,3], and Diane Chen[1,2,3]

[1]*The Potocsnak Family Division of Adolescent and Young Adult Medicine, Ann & Robert H. Lurie Children's Hospital of Chicago, Chicago, IL, United States,* [2]*Pritzker Department of Psychiatry & Behavioral Health Ann & Robert H. Lurie Children's Hospital of Chicago, Chicago, IL, United States,* [3]*Department of Psychiatry and Behavioral Sciences, Northwestern University Feinberg School of Medicine, Chicago, IL, United States,* [4]*Division of Endocrinology & Diabetes, Seattle Children's Hospital, Seattle, WA, United States,* [5]*Embedded Behavioral Health, Division of Psychiatry and Behavioral Medicine, Seattle Children's Hospital, Seattle, WA, United States*

Learning objectives

- Define holistic, interdisciplinary approaches to gender-affirming healthcare for transgender and genderdiverse youth.
- Describe existing models of care and the current evidence base, highlighting areas of need for further research.
- Examine international differences in pediatric gender-affirming healthcare, discuss culturally inclusive healthcare practices, and identify barriers to equitable and inclusive care.
- Highlight interventions, approaches, and gaps in providing holistic, integrated, evidence-based gender-affirming care, making recommendations for research and practice within rapidly evolving healthcare and sociopolitical contexts.

Key terms and definitions

- **Gender-affirming care** refers to the range of individually tailored care that aims to support transgender and genderdiverse (TGD) youth, including individual and family psychosocial supports, medical and surgical interventions, and community building and social connection.
- **Holistic, interdisciplinary healthcare** refers to comprehensive and coordinated care that promotes overall physical, mental, and social health and well-being for TGD youth and families, which involves collaboration and education of providers and community partners to provide quality care and support.
- **Gender embodiment desires and goals** include any desired physical or social changes to better align with their gender identity or reduce gender dysphoria. Exploration of embodiment goals allows for more individualized healthcare planning than gender dysphoria or identity-based recommendations alone.

Transgender and genderdiverse (TGD) youth include children and adolescents for whom gender differs from the physical and/or social characteristics culturally aligned with the designated sex at birth. There has been a substantial increase in youth identifying as TGD (Herman et al., 2022) and increased diagnosis of gender dysphoria among youth internationally (Kaltiala et al., 2020), likely a result of increased global acceptance of gender diversity and LGBTQIA+ identities overall (Flores, 2019). Research shows that youth between ages 13 and 17 comprise about 18% of the 1.6

Advances in Research on Pediatric Gender Affirming Healthcare. DOI: https://doi.org/10.1016/B978-0-443-29019-0.00003-8
© 2026 Elsevier Inc. All rights are reserved, including those for text and data mining, AI training, and similar technologies.

million TGD population in the United States—a notable 8% increase since last reported in 2017 (Herman et al., 2022). As a result, there is an increased need for gender-affirming care for TGD youth and families.

The healthcare needs of TGD youth vary based on development stage, caregiver and social support, embodiment desires, and overall treatment goals. TGD youth and their families, particularly those of color (Goldenberg et al., 2021), encounter barriers to accessing quality healthcare, including stigma and discrimination in healthcare settings, lack of understanding of TGD identities by providers, and logistical challenges related to location, availability, and finances (Gridley et al., 2016; Kidd et al., 2023). For gender-affirming healthcare to be effective, it must consider individual, interpersonal, contextual, systemic, and sociopolitical experiences of TGD youth and their families (Katz-Wise et al., 2022). As such, integrated gender-affirming healthcare models are uniquely suited to meet the needs of TGD youth and their families, taking a holistic approach to health and well-being that incorporates both medical and nonmedical avenues for supporting TGD youth.

Interdisciplinary gender-affirming care

The World Professional Association of Transgender Health (WPATH) Standards of Care, Version 8 (SOC8) (Coleman et al., 2022) and the Endocrine Society Clinical Practice Guidelines (Hembree et al., 2017) provide evidence-based recommendations for multidisciplinary and interdisciplinary care for TGD individuals seeking medical gender affirmation. Multidisciplinary care refers to the need for collaboration among multiple disciplines to provide high-quality, holistic care. Not every TGD individual will need the services of every discipline at each visit, which is dependent upon patient and family goals and presenting concerns. Additionally, a lack of availability for one discipline should not preclude receiving services from other disciplines when needed (Coleman et al., 2022). Interdisciplinary care refers to a specific model of healthcare in which multiple disciplines are integrated, some of which are colocated and interface with youth and families during a set clinic visit, and others that operate separately but intentionally communicate and jointly coordinate care.

For youth, interdisciplinary care teams may include pediatric endocrinologists, adolescent medicine physicians, nurse practitioners, surgeons, clinical psychologists, social workers, and speech therapists, among others, to provide a range of developmentally appropriate treatment options. With a team of providers, healthcare can be individually tailored to address the emotional well-being of TGD youth and include strategies to support youth in expressing gender authentically, without prioritizing a particular outcome or pathway (Hidalgo et al., 2013; Huit et al., 2024). For prepubertal TGD youth, no medical interventions are provided, and care may include affirming social interventions (e.g., chosen name and pronouns) as well as psychoeducational and supportive family therapy (Huit et al., 2024).

For pubertal TGD youth, care may include medical services (e.g., pubertal suppression, menstrual suppression, gender-affirming hormones), mental health supports (e.g., psychotherapy and psychiatric care), and nonmedical gender-affirming services (e.g., voice therapy) (Huit et al., 2024). For those interested in medical interventions, both the SOC8 and Endocrine Society Guidelines recommend that TGD youth and their caregivers work with appropriately trained medical and behavioral health providers to receive information about the desired medical intervention and complete a comprehensive biopsychosocial assessment, most often with a mental health provider. According to the SOC8, any licensed professional (medical or mental health) with appropriate experience can conduct a biopsychosocial assessment, although mental health providers often have more time to build rapport and discuss sensitive topics (Coleman et al., 2022). Some youth do not want medical support for gender affirmation; however, they may benefit from mental health support or other resources.

Holistic, interdisciplinary care extends beyond the clinic setting and includes connecting TGD youth and families with community supports. Care teams often include social workers who, in addition to direct clinical service, provide care coordination to streamline services within gender specialty clinics and community spaces. Additionally, youth and families may need legal support to facilitate name or gender marker change, or to navigate consent and custody concerns. Care coordinators can also help navigate school supports (e.g., connecting youth with LGBTQIA+ affinity groups or Gender and Sexuality Alliances) (Inwards-Breland et al., 2019). Furthermore, TGD youth and families often seek connection to others with shared experiences, which can be found in youth and family support groups, in addition to community-led educational and social events (Huit et al., 2024). Care navigation is viewed by adolescents and parents as a particularly helpful aspect of interdisciplinary gender-affirming care teams (Inwards-Breland et al., 2019).

Well-trained providers with experience working with TGD communities also recognize the unique impact of gender-related marginalization and stigma on mental health and well-being (Chodzen et al., 2019). Using a developmentally adapted model of gender minority stress for TGD youth, providers can identify targets for intervention, which broadly include reducing exposure to nonaffirming environments, addressing gender dysphoria, and increasing gender identity pride and resilience.

Existing gender-affirming care models

Several approaches to gender-affirming care for TGD youth have been detailed in the literature. The first gender-affirming healthcare clinic dedicated specifically to youth, the Center for Expertise in Gender Dysphoria, opened in the Netherlands in 1987 (Cohen-Kettenis et al., 2011). The "Dutch model of care" (de Vries and Cohen-Kettenis, 2012), with its emphasis on assessment and mental health support, has since served as a template for programs established in other countries. The original Dutch approach to caring for adolescents encompassed three phases: (1) an extensive assessment and diagnostic phase, including psychological and psychiatric evaluation, psychoeducation, and multidisciplinary discussion and decision-making; (2) a "real life" phase that includes social transition to live in an affirmed gender role and gender-affirming medical treatments like pubertal suppression and gender-affirming hormones; and (3) gender-affirming surgery (Cohen-Kettenis et al., 2011).

Over time, models of gender-affirming healthcare have evolved to deemphasize protracted psychological assessment, although the inclusion of multiple disciplines, including mental health providers, is still recommended whenever possible (Coleman et al., 2022). Some programs intentionally have families meet with medical and mental health providers together early in care to facilitate multidisciplinary assessment and recommendations, to limit the need for youth and families to repeat sensitive information, and to emphasize the collaborative nature of care (Chen et al., 2016; Olezeski and Boulware, 2017).

Several interdisciplinary healthcare teams have published on their structure and process of providing gender-affirming services to TGD youth and their families (Allen et al., 2019; Chen et al., 2016; Edwards-Leeper and Spack, 2012; Cousino et al., 2014; Inwards-Breland et al., 2019; Olezeski and Boulware, 2017; Olson-Kennedy et al., 2019; Warwick and Shumer, 2023). Collectively, these publications detail the core components of modern interdisciplinary care and potential areas for continued improvement in holistic care models for TGD youth.

Initial assessment of care needs

Fairly consistently across programs, families establish care through initial intake calls with a social worker designed to gather information about psychosocial history, gender identity, and pubertal development to inform care options, share educational materials, and connect to community resources. During this call, staff can address urgent care needs and provide options for those with nonurgent needs. Social work generally remains a point of contact while families are waiting for services and after they establish care (Coyne et al., 2023a).

Psychosocial care and supportive interventions

As members of an interdisciplinary care team, mental health providers can offer a range of supports depending on the developmental and mental health needs of TGD youth. One role of mental health providers is to complete biopsychosocial assessments to support medical decision-making in the context of gender-affirming interventions. These assessments clarify a patient's gender history, gender-related healthcare goals, experience of gender dysphoria, and mental health concerns.

While interdisciplinary teams often include mental health providers, there are notable differences in how teams approach psychosocial assessment in terms of when and how mental health providers are involved (Coyne et al., 2023b). In some programs, mental health providers are the first point of clinical contact, with a comprehensive assessment required before referral for medical care. In these models, mental health providers primarily serve in an assessment role and refer patients to community providers for ongoing therapy as indicated. In other interdisciplinary care models, mental health providers may provide ongoing therapy for TGD youth.

More than psychosocial assessment, multidisciplinary and integrated mental healthcare is meant to serve the important role of offering support for youth across a range of cognitive, social, and emotional development to address the diverse needs of patients and families (Lee et al., 2024). The focus of integrated mental health support varies and may include work with TGD youth and their families to support gender exploration and identify affirming social changes (e.g., name, clothing, pronouns) that help youth to present as their authentic selves. Providers can also help youth navigate identity disclosure and social interactions within schools, healthcare, and other community settings. Family psychotherapy can increase effective communication between youth and their parents about gender-related issues and help caregivers understand their child's gender development.

Mental health providers are well-positioned to facilitate coping with gender-related marginalization. Individual and group therapy with TGD youth may focus on education about the impact of gender-related marginalization on mental health and developing strategies for responding to distress and stigma. Offering space for caregivers to meet separately from their children, either in groups or individual therapy, can help caregivers process their own reactions to gender identity disclosure and affirmation, and support informed decision-making in the context of their TGD child's healthcare.

Medical and surgical interventions

Medical providers offer education about gender-affirming medical interventions, including risks and benefits, reversible and irreversible effects, and the expected timeline of physical changes. In preparation for initiating medical interventions, medical providers may refer TGD youth and their caregivers to consult with fertility preservation experts to learn more about options for preserving sperm or ova. Medical providers may also present other options for gender affirmation, such as chest binding, packing, tucking, padding, laser hair removal, and/or speech therapy to better align voice with gender presentation.

Access to surgical interventions for older adolescents is rare, limited primarily to mastectomy for TGD youth assigned female at birth, and this access has become more limited amid recent changes in the sociopolitical climate. Adolescents may nonetheless benefit from support in planning for future surgery and navigating the process of referral to appropriate surgical services.

Adjunctive programming and collaboration

Interdisciplinary team members may facilitate inclusive education in schools, groups, and supportive programming. Support and affinity groups may provide social connection for TGD youth and families, who tend to be more isolated, to TGD peers and families with shared experiences. In addition, medical and mental health providers and other team members with specialized expertise (e.g., health educators) can educate trainees and providers in both healthcare and community contexts. Such training and education serve to increase capacity for providing gender-affirming support across service agencies and improve knowledge among community-based agencies to better incorporate TGD-inclusive practices.

While integrated interdisciplinary care clinics can address many of the needs of TGD youth, communication and collaboration between various providers and stakeholders are crucial to ensure wraparound services for families. This may include referring ongoing therapeutic services to community providers or other generalist providers within health systems. This may also include recommending or advocating for a neuropsychological assessment to support individual learning, academic, and/or social needs. Challenges in school may require additional intervention and coordination. Ideally, collaboration among multiple systems provides TGD youth and families with tailored, holistic care that best meets their individual needs.

Evidence base for holistic and interdisciplinary care

Gender-affirming, holistic, and interdisciplinary care is necessarily individualized. While this makes it challenging to evaluate outcomes specific to a particular model of care, some research has explored overall experiences of programs that generally reflect the components described above. One study surveyed TGD youth and their parents about their experiences of a multidisciplinary clinic in Australia and found high rates of satisfaction with services and providers (Tollit et al., 2019). Survey respondents largely rated their care as "good" or "excellent," described the process as "validating" and "normalizing," and indicated that distress had generally decreased since initiating care. Patients and families were least satisfied with the length of wait times and suggested improvements, including more frequent appointments, more parent support, and additional help with the transition from pediatric to adult care.

Another mixed methods study used questionnaires and qualitative interviews to assess the experiences of TGD youth and families who received care at a multidisciplinary clinic in the United States (Inwards-Breland et al., 2019). Questionnaire results similarly revealed high satisfaction rates and that most families had a good understanding of their next steps in the overall process. Qualitative results revealed that families found the clinical environment to be affirming and particularly appreciated the availability of care navigation to overcome systemic barriers and reduce stress. Families also highlighted access and scheduling barriers, including frustration with the assessment process, challenges related to distance, and the impact of lengthy wait times. More research has been conducted to evaluate the medical and psychosocial outcomes of specific gender-affirming social, psychological, and medical and surgical interventions.

Social interventions

Social interventions involve making intentional changes to gender presentation to live publicly in line with authentic gender versus the gender aligned with designated sex at birth. Changes can include name change, gender marker changes, and nonmedical changes in appearance (e.g., clothing, hair, makeup, chest binding, tucking, packing, padding). Research shows that prepubertal TGD children who make a binary social transition present with developmentally normative levels of depression and anxiety compared to cisgender counterparts (Durwood et al., 2017; Olson et al., 2016). Among TGD adolescents, use of an affirmed name in multiple environments (i.e., home, school, work, with friends) is protective against depression and suicidal ideation and behaviors (Russell et al., 2018).

Medical and surgical interventions

For some TGD youth who have begun pubertal changes and present with primary concerns related to body incongruence or gender dysphoria associated with primary and secondary sex characteristics, medical and/or surgical interventions may be helpful. For youth in the earlier stages of puberty, medications to suppress endogenous sex hormones (i.e., gonadotropin-releasing hormone analogues or GnRHa) are used to prevent the progression of a gender-incongruent puberty. While some studies suggest treatment with GnRHa improves aspects of mental health, others suggest mixed results. Longitudinal studies of GnRHa have reported improved psychosocial functioning (Costa et al., 2015; De Vries et al., 2011), while others have demonstrated no significant changes in overall emotional and behavioral challenges with treatment (Carmichael et al., 2021). Moreover, although improved psychosocial functioning has been reported, GnRHa does not always reduce gender dysphoria on its own (De Vries et al., 2011), likely because GnRHa medications do not result in desired physical changes.

For adolescents who are in the later stages of pubertal development and for whom social transition and nonmedical interventions are not effectively reducing gender dysphoria, access to gender-affirming hormones—testosterone and estrogen, which induce gender-congruent secondary sex characteristics—can be helpful. Gender-affirming hormone treatment is associated with improved quality of life and well-being, as well as decreases in body dissatisfaction, depression, suicidality, and anxiety after 1 year of treatment (Achille et al., 2020; Allen et al., 2019; Kuper et al., 2020). Gender-affirming hormones are also associated with increased appearance congruence and improvements in depression, anxiety, positive affect, and life satisfaction following 2 years of treatment (Chen et al., 2023).

Some older TGD adolescents may benefit from chest masculinization surgery, particularly when gender-affirming hormone treatment does not resolve chest dysphoria or for those who do not desire phenotypic changes associated with gender-affirming hormones but have specific concerns related to their chest. A multicenter, matched-control study of TGD youth designated female at birth found that chest masculinization surgery led to significant improvement in chest dysphoria, gender congruence, and body image satisfaction (Ascha et al., 2022).

Psychological assessments and interventions

There is no research to date that directly compares the relative quality and effectiveness of various mental health models in pediatric gender-affirming care. Some providers and researchers have questioned the utility of models that require psychological assessment as a prerequisite to medical care. Historically, psychological assessment has functioned to confirm the presence of gender dysphoria and to ensure an unspecified standard of mental health stability, in the service of identifying those who would most benefit from care (Mosier-Mills et al., 2024). Unfortunately, there is no evidence that these types of assessments prevent negative outcomes, and they may cause harm in some cases by incentivizing conformity to cisnormative stereotypes, conflating gender diversity with psychopathology, and encouraging adolescents to downplay mental health concerns in an effort to better ensure access to care (Mosier-Mills et al., 2024).

Recent qualitative research showed that young adults in Italy found these assessments to be a source of fear and anxiety and viewed the role of assessment as limiting access to care rather than providing support (Vincenzo et al., 2025). Participants described feeling as though they needed to adhere to a specific narrative that was aligned with the expectations of the evaluator. However, participants in this study also identified some positive aspects of the experience, such as facilitating self-understanding, as well as strengthening relationships and building healthy coping strategies. These results highlight the potential for psychological assessment and treatment to be beneficial when the function is to increase support and socioemotional well-being rather than to limit treatment access. Specifically, it has been recommended that the assessment in pediatric gender-affirming care focus more on the youth's psychosocial functioning and informed consent, including identifying embodiment goals, ensuring a strong understanding of desired interventions, and addressing needed supports (Mosier-Mills et al., 2024).

In recognition of these disparities related to psychosocial assessment and the lack of established standards for assessment, The Companionship Model for Youth (TCM-Y) has been proposed as a structured, patient- and family-centered assessment that centers support and access to resources rather than gatekeeping (Lee et al., 2024). The TCM-Y offers general guidance to mental health providers working with TGD youth, and it also outlines a step-by-step approach to conducting an assessment that supports families in navigating systemic and institutional barriers and inequities, assesses capacity for informed consent and assent, and facilitates access to important resources. While this approach centers on building support and resources that have been identified in the literature as helpful and avoids unnecessary barriers in care access that perpetuate stigma and preclude authenticity, further research on the quality and effectiveness of this model is needed.

Outside of the role of the psychosocial readiness assessment, many TGD youth face additional psychosocial challenges that necessitate evidence-based psychological interventions. TGD youth experience increased rates of depression, anxiety, suicidality, self-harm, and eating disorders in comparison with their cisgender peers (Connolly et al., 2016; Newcomb et al., 2020; Veale et al., 2017; Becerra-Culqui et al, 2018), likely as a result of increased TGD-specific distal (i.e., discrimination, rejection, victimization, and nonaffirmation) and proximal (i.e., internalized transphobia, negative future expectations, and stress related to identity concealment) marginalization factors (Chodzen et al., 2019; Hunter et al., 2021; Testa et al., 2015).

Due to systematic risk for poorer mental health functioning, there is a need for evidence-based psychological intervention that considers unique aspects of TGD experiences. While a growing body of literature regarding adaptations and frameworks exists (Austin et al., 2016; Budge, 2013; Chang and Singh, 2016; Lange, 2020; Matsuno and Israel, 2018), there remain methodological challenges and limitations in prior knowledge, such as a lack of TGD-specific interventions and a dearth of youth-focused interventions (Expósito-Campos et al., 2023).

For some youth, gender-related stressors and stigma are the primary sources of distress. Appropriate interventions for these youth primarily focus on adapting evidence-based psychological interventions to address specific gender minority stressors (Huit et al., 2024). AFFIRM, an adapted group-based cognitive-behavioral intervention for sexual and gender minority (SGM) youth, has demonstrated positive outcomes for decreasing depression among TGD youth (Austin et al., 2018), and another iteration, including caregivers (AFFIRM Caregiver; Austin et al., 2021), has demonstrated increased caregiver affirmation and support for SGM youth. Similar individual approaches using transgender affirmative psychotherapy (TA) combined with building awareness of minority-related stressors (BAMS) have resulted in decreases in nonaffirmation and internalized transphobia (Budge et al., 2021).

Another approach for addressing internalized stigma is highlighted through a brief online, self-led intervention, releasing internalized stigma for empowerment (RISE)(Israel et al., 2021). RISE demonstrated better outcomes in the intervention group for the reduction of self-stigma related to identity and marginalization factors. These interventions use evidence-based interventions, such as cognitive-behavioral therapy and values-based principles, to better allow youth to understand, process, and cope with experiences of gender-related stress, which may be effective for decreasing overall distress among this subset of youth.

TGD youth with complex psychosocial challenges that are not due to gender-related distress alone may require additional treatment planning to develop treatment goals and address additional psychosocial challenges (Coyne et al., 2020; Huit et al., 2024). These youth present with socioemotional and neurodevelopmental challenges that likely predate and are distinct from gender-related distress but are likely exacerbated by these stressors. Common mental health presentations in TGD clinical samples include youth with disordered eating, nonsuicidal self-injury, suicidal ideations and behaviors, and autism spectrum disorders (Wittlin et al., 2023). Interventions such as dialectical behavior therapy (DBT) are effective in addressing emotional dysregulation and coping with chronic invalidation (Rizvi et al., 2024) but may require tailoring to address overlap with genderminority stressors. Holistic, interdisciplinary care approaches simultaneously assess for and address ongoing concerns, while balancing overall gender and embodiment goals.

Caring for TGD young people with complex developmental and mental health concerns often requires ongoing close collaboration with a young person's ongoing mental health provider. Depending on the severity of presenting concerns, psychotropic medications may be a helpful adjunct to care to address mood and functional challenges, requiring collaboration with psychiatric providers for optimal care. For this subset of youth, it is especially critical for interdisciplinary collaboration that simultaneously addresses individual and interpersonal needs in supporting overall well-being, and planful intervention.

Regional and global distinctions and barriers to care

Access to gender health specialists and interdisciplinary teams varies both regionally and globally. The resources required to staff a holistic, interdisciplinary program are substantial, limiting access and availability. Lack of specialized providers across diverse geographical locations increases wait times for affirming providers. These challenges add to logistical and structural barriers in accessing integrated gender-affirming care for many youth and families (Gridley et al., 2016; Inwards-Breland et al., 2019; Pampati et al., 2021; Sequeira et al., 2022; Strauss et al., 2022; Tollit et al., 2019).

The United States is a large country with sociocultural and regional differences that impact healthcare. Despite a lack of attention in many large-scale studies, a majority of the country consists of rural areas far from dense cities that include tertiary care and large hospitals where gender clinics tend to be housed. Thus youth living in rural areas largely lack access to interdisciplinary gender care due to the fragmentation of pediatric specialty care (Manns et al., 2023). In addition, youth in rural settings experience barriers in social support, particularly as these areas have higher concentrations of religious and/or conservative populations that may hold negative views around TGD people (Poquiz et al., 2021). Moreover, TGD youth may experience increased harassment and victimization within their schools, which may be less likely to support affirming clubs or other LGBTQIA+-related resources. TGD individuals have described issues with gender-affirming care provided in rural areas, including intrusive and inappropriate questioning, unnecessary gatekeeping, socio-contextual stigmatization, and high instances of being "deadnamed" (intentionally called by a nonaffirming previously used name) and misgendered (Meyer et al., 2020).

These care issues are compounded by limited options for continuing education in evidence-based gender-affirming care available to rural healthcare providers (Holt et al., 2020). Training programs like Trans-Extension for Community Healthcare Outcomes (TransECHO), a national professional tele-education model, facilitate training for behavioral and medical care providers in primary care settings, including those in rural areas, to provide interdisciplinary gender-affirming care. This comprehensive training provides education, case discussion, and consultation, which empowers primary care physicians to provide gender-affirming care in underserved areas (Bruno et al., 2023).

Despite US-based research reporting improved psychosocial outcomes among youth receiving gender-affirming medical care (Allen et al., 2019; Chen et al., 2023), an unprecedented amount of legislation has been introduced in recent years aiming to restrict access to gender-affirming medical interventions (Redfield et al., 2024) or criminalize healthcare professionals who provide gender-affirming medical interventions to youth (Turban et al., 2021). The negative impact of these bills becoming law is far-reaching, including barring insurance providers from covering gender-affirming care and restricting the use of public funds and facilities to support this care.

It is estimated that 113,900 TGD youth live in states impacted by gender-affirming care restrictions, following over 500 bills filed with state legislatures in 2023 regarding the restriction of rights for LGBTQIA+ communities (Redfield et al., 2024), with the number of impacted youth likely increasing due to ongoing legislative and policy changes. Gender-affirming care bans exacerbate existing structural barriers, compounding stressors for a population already at disproportionate rates for worsening psychosocial outcomes and suicidal ideation (Kuper et al., 2022). These policies leave families to navigate care access outside of their home state and add to the already long waitlists, in addition to the increased hostility and discrimination to which TGD youth and their families are increasingly subjected.

Legislative efforts to ban gender-affirming healthcare for youth have emerged in other countries besides the United States. In the United Kingdom a ruling in December 2020 that was later overturned (Bell v. Tavistock) stated that youth under the age of 16 were required to obtain court approval before initiating gender-affirming medical treatment through the National Health Service, and youth were deemed unable to provide consent to start puberty-blocking medications (Kremen et al., 2021). Increasingly harmful political climates negatively impact access to appropriate, pediatric gender-affirming care in numerous ways: (1) family doctors refusing to make patient referrals to specialty care clinics; (2) long waitlists, which become particularly challenging in the context of accessing GnRH agonists in a timely manner to prevent pubertal progression; (3) protracted assessment procedures; (4) lack of awareness and education in mental health services, including youth being constantly questioned around the validity of their TGD identity; and (5) accessibility issues due to geographical constraints (Carlile et al., 2021).

Other UK-based studies have found that TGD youth and families commonly experience healthcare professionals' lack of knowledge regarding gender-affirming care, perpetuation of gender stereotypes among providers, overt experiences of discrimination within healthcare, and lack of shared decision-making and inclusion of parent and child perspectives in care practices (Carlile, 2020). Policies in the United Kingdom serve to highlight similar barriers and disparities in care and the inherent harm that TGD youth experience in nongender-affirming healthcare approaches.

In Canada, TGD youth accessing care from gender specialty clinics were generally healthy overall and reported symptoms of depression, anxiety, and suicidality (Bauer et al., 2021). A study examining pathways to accessing pediatric gender-affirming care in Canada found that youth usually interacted with multiple providers before being referred for gender-affirming medical care and, after referral, experienced significant wait times (an average of 269 days) for their first appointment (Lawson et al., 2024). The study also found that Indigenous heritage was related to even longer care initiation timelines. Regarding psychosocial health the majority of youth reported a history of self-harm (67.6%), while 58.1% had seriously considered suicide at one point, and 36.0% of youth had attempted suicide, with longer wait times associated with increased likelihood of suicidal ideation and self-harm. Overall, TGD youth and their families in Canada would benefit from more timely access to gender-affirming care.

Australian clinics approach gender dysphoria from a biopsychosocial perspective, utilizing interdisciplinary teams to assess both the medical and the psychological needs of youth (Kozlowska et al., 2021). Consistent with global trends, wait times for gender-affirming care for youth are increasing in Australia, with families waiting up to 2 years to access services (Allen et al., 2021). To address these long wait times for care an Australian clinic piloted a single-session intervention for waitlisted youth focused on support and triage to appropriate multidisciplinary care waitlists that resulted in increased rates of social transition and menstrual suppression, improvements in mental health, quality of life, and family functioning, and an overall increased sense of agency (Allen et al., 2021). Single-session interventions may be a flexible adaptation to interdisciplinary care to reduce wait times and increase access to services.

Cultural context, inclusivity, and equity in gender-affirming healthcare practices

A higher percentage of youth and young adults of color, specifically Latine, African American, American Indian, or Alaska Native, biracial, and multiracial groups, identify as TGD, compared to white youth and young adults (Herman et al., 2022). A robust body of literature documents the additive stressors and barriers that those with multiple marginalized identities experience due to interpersonal and structural factors, impacting the ability to access necessary and quality healthcare (Call et al., 2021; Johns et al., 2023; Kattari et al., 2015). Additionally, existing information on TGD youth who are accessing integrated gender-affirming care is predominantly based on experiences of white TGD youth, highlighting inclusivity and equity concerns (Vincent, 2018).

TGD youth of color are at increased risk of experiencing different forms of stigma and discrimination (e.g., rejection, isolation, violence, and harassment) due to their intersecting identities (Lett et al., 2020). The complex patterns of intersectionality and associated experiences systemically place these youth at higher risk for minority stress and related psychosocial concerns, such as depression, anxiety, and suicidality (Tankersley et al., 2021; Lefevor et al., 2019). Experiences of racial and ethnic minorities who also identify as TGD remain particularly underexplored, ultimately limiting the available resources designed to help improve psychosocial outcomes and quality of life. However, initial evidence suggests that among youth who access interdisciplinary gender-affirming care, there are comparable rates of depression, anxiety, and gender affirmation for Black and Latine TGD youth compared to white TGD youth. For these youth, irrespective of racial or ethnic identity, lower levels of gender affirmation were associated with poorer mental health functioning (Vance et al., 2023).

Accessing gender-affirming healthcare requires that youth feel comfortable discussing gender identity with healthcare providers. Voluntary disclosure of gender identity, particularly nonbinary gender identities, can be challenging for youth who may anticipate negative responses and rejection from healthcare providers (Sequeira et al., 2020). This lack of disclosure can postpone exploration and identity affirmation, and ultimately inhibit the ability to access gender-affirming interdisciplinary care.

Nonbinary and genderqueer youth may be concerned that providers will not understand gender expansive identities or how to align treatment goals that do not follow a linear transition between "one" gender to "the other" (Hastings et al., 2021). Providers should be familiar with nonmedical (e.g., binding, tucking, packing, voice therapy, hair removal) gender-affirming interventions for patients who may not desire medical interventions. Gender-inclusive clinical practices and procedures, such as front desk staff asking for and using chosen name and pronouns and offering forms that allow space for disclosing gender identity, are strategies that can increase comfort with disclosure in any clinic (Sequeira et al., 2020).

Recommendations for research, practice, and advocacy

Tailoring inclusive care approaches

Gender-affirming interdisciplinary healthcare practices would benefit from considering methods of supporting TGD children at various stages of gender exploration, disclosure, and family support. Providers can advocate within their own institutions and communities to increase the availability and quality of services for TGD youth and families. Within clinics and institutions, providers can create more welcoming and affirming environments through education and normalization of the use of pronouns and other gender-inclusive language, ensuring that intake forms and discussions allow expression of diverse gender identities, updating names and pronouns in electronic records systems to be consistent with affirmed gender (with the consent of patient and family), and providing diverse support and educational materials that consider gender and other intersections of identity. Program evaluation and quality improvement methods are needed to evaluate the benefit of these changes and to identify further areas of improvement in creating affirming environments.

Gender-affirming interdisciplinary care providers that serve TGD youth with multiple marginalized identities should utilize antioppressive practices that address racism in healthcare, cultural humility, community engagement, and competent care practices. Examples of these approaches have been detailed in the literature (Chang and Singh, 2016). Specifically, addressing biases, tailoring language use, utilizing inclusive and appropriate measures during evaluation, valuing patient cultural beliefs during care discussions, and addressing intersectional factors can aid in trust and interventions that meet the needs of racial and ethnic minority TGD youth. Community-based participatory research designs are recommended to collect high-quality data on the experiences of those with intersecting identities and their specific needs within healthcare programs.

Gender-affirming mental health providers can adapt the transgender resilience intervention model (TRIM) that builds on the minority stress model by identifying group and individual resilience factors (Matsuno and Israel, 2018). Programs can use the TRIM and knowledge on TGD-specific resilience factors to form the basis of psychological interventions and target resilience factors or marginalization stressors that racial and ethnic minority TGD youth face. Additionally, focusing care on ways to help patients flourish and experience gender euphoria can alleviate experiences that pathologize and stigmatize gender in healthcare settings. The experience of gender euphoria is particularly under-researched, and providers and researchers would both benefit from measurement tools that evaluate positive experiences of gender in addition to distress and dysphoria.

Providers in all healthcare settings can create gender-inclusive care practices through ongoing education about current professional guidelines, research, and best practices for working with TGD youth and their families. Providers should also be up to date on policy and legal changes that may affect TGD young people and how these may increase fear and experiences of internal and external marginalization processes. Providers may address gaps in care access through educating other healthcare professionals and trainees, providing training in educational contexts, building community capacity for supporting TGD youth and families, and disseminating information to educate broader audiences and dispel misinformation (Coyne et al., 2023b; Hope et al., 2022; Puckett et al., 2018).

Few training protocols on interdisciplinary pediatric gender-affirming care are widely available. A small number of studies have demonstrated the benefit of multidisciplinary training programs for youth (Hodax et al., 2024; Vance et al., 2020). More educators are encouraged to evaluate their training programs to better understand the effectiveness of building competencies in working with TGD youth and to consider publishing a curriculum to aid in the dissemination of competent, affirming care practices.

Building social supports and community connection

Given the often-challenging nature of navigating gender in individual, social, systemic, and cultural contexts, there is a need to consider how to rewrite narratives of despair and struggle into hope and flourishing. For sexual and gender minorities in particular, microlevel (i.e., significant others, friends, family) and macrolevel supports (i.e., community integration, participation, organization) are significantly associated with feelings of hope and dimensions of flourishing (Bragg et al., 2022).

Support groups for parents, as well as safe spaces for youth, foster community connection, build empowerment, and increase resilience. While these may be hosted by gender specialty programs or mental health programs, they may also be part of community organizations such as PFLAG (formerly Parents, Family, and Friends of Lesbians and Gays, however was updated to PFLAG as a name versus acronym to reflect broad inclusivity of LGBTQIA+ communities), Glisten (formerly GLSEN; Gay, Lesbian, and Straight Education Network, which was similarly updated for greater inclusivity of LGBTQIA+ youth), or other TGD-inclusive organizations. Future research is needed to demonstrate the benefits of these programs, especially for parents and caregivers who may be struggling with support and acceptance of their TGD child, and to identify barriers to participation and adherence among families who would likely benefit most.

Navigating healthcare and addressing care barriers

Development of protocols for the care of young TGD patients, including guidance with insurance coverage processes and support with identifying financial assistance programs, can address concerns surrounding financial burden. Establishing individualized clinical care roadmaps for families, as well as utilizing social workers or care coordinators as designated navigators in clinics for TGD patients and families, are ways to address the noted navigation barriers to accessing interdisciplinary gender-affirming clinics. Designated care navigators could aid in establishing trust and rapport between patients and the care team, as well as being a go-to for accessing needed supports and interventions. Existing research demonstrates the benefit of care navigation to patients and families (Inwards-Breland et al., 2019). Additional research that demonstrates the efficiency, cost-effectiveness, and patient retention benefits that result from appropriate care navigation is needed to advocate for this service within healthcare institutions.

As the number of TGD youth seeking care continues to increase, there is a need to expand access to gender-affirming interdisciplinary care. Existing interdisciplinary programs can explore adaptations that provide evidence-based care in flexible ways. More research is needed to understand which aspects of interdisciplinary care are most beneficial and how additional supportive services can augment positive youth and family outcomes. This includes evaluation of various approaches to the biopsychosocial assessment process to optimize the experience for youth, families, and providers alike and to reduce barriers in care access. Greater attention to methods of assessment, such as program evaluation, measurement of gender-related distress, and overall youth and family satisfaction, could improve care within integrated interdisciplinary gender clinics.

Adaptations to current processes could involve longer in-person appointments to allow for more care to be provided when families are able to travel to out-of-state clinics. Scheduling families to see multiple providers on the same day they are presenting to the clinic can be another way to increase access to care for these individuals and families. To address access barriers related to distance from hospitals and gender clinics, providers should offer virtual appointments when possible and appropriate (Poquiz et al., 2021), such as for appointments that do not require a physical work-up or in-person observation due to safety concerns. Programs and providers with gender-affirming care expertise should also consult with and educate community providers to increase availability and capacity for offering needed services. Given the sociopolitical efforts to limit access to gender-affirming care, healthcare providers need ways to collaborate with one another to work within the bounds of the law, identify organizational guardrails, and provide professional support and guidance on continuing supportive care to youth and families.

Conclusion

The holistic, interdisciplinary gender-affirming care model is focused on the well-being of TGD youth and offers tailored, comprehensive assessment and care planning to connect youth and their families with appropriate services. This approach to healthcare evaluates how intersecting social and cultural identities can impact care needs and considers both medical and nonmedical interventions to help TGD youth. With quality improvement and community-based participatory research designs, researchers and healthcare providers will continue to refine and improve holistic, interdisciplinary pediatric gender-affirming care. Legislative actions that ban TGD youth from accessing interdisciplinary care are stigmatizing and compound the existing health disparities between TGD youth and their cisgender peers. Providers and staff on interdisciplinary teams are not only clinicians but also advocates for the reduction of the social and structural factors that negatively impact the well-being of TGD youth and their families.

References

Achille, C., Taggart, T., Eaton, N.R., Osipoff, J., Tafuri, K., Lane, A., et al., 2020. Longitudinal impact of gender-affirming endocrine intervention on the mental health and well-being of transgender youths: preliminary results. International Journal of Pediatric Endocrinology 2020 (1). Available at https://doi.org/10.1186/s13633-020-00078-2.

Allen, L.R., Watson, L.B., Egan, A.M., Moser, C.N., 2019. Well-being and suicidality among transgender youth after gender-affirming hormones. Clinical Practice in Pediatric Psychology 7 (3), 302–311. Available at https://doi.org/10.1037/cpp0000288.

Allen, S.D., Tollit, M.A., McDougall, R., Eade, D., Hoq, M., Pang, K.C., 2021. A waitlist intervention for transgender young people and psychosocial outcomes. Pediatrics 148 (2). Available at https://doi.org/10.1542/peds.2020-042762.

Ascha, M., Sasson, D.C., Sood, R., Cornelius, J.W., Schauer, J.M., Runge, A., et al., 2022. Top surgery and chest dysphoria among transmasculine and nonbinary adolescents and young adults. JAMA Pediatrics 176 (11), 1115–1122. Available at https://doi.org/10.1001/jamapediatrics.2022.3424.

Austin, A., Craig, S.L., Alessi, E.J., Wagaman, M.A., Paceley, M.S., Dziengel, L., et al., 2016. Guidelines for transgender and gender nonconforming (TGNC) affirmative education: Enhancing the climate for TGNC students, staff and faculty in social work education. Council on Social Work Education, Alexandria, VA. Available at https://www.cswe.org/getmedia/e95d6dce-9e22-4c1f-a7af-374d398fd8de/Guidelines-for-Trasngender-and-Gender-Nonconforming-(TGNC)-Affirmative-Education-CSOGIE_WP2_TGNC_final_web.pdf.

Austin, A., Craig, S.L., D'Souza, S.A., 2018. An AFFIRMative cognitive behavioral intervention for transgender youth: preliminary effectiveness. Professional Psychology: Research and Practice 49 (1), 1–8. Available at https://doi.org/10.1037/pro0000154.

Austin, A., Craig, S.L., Matarese, M., Greeno, E.J., Weeks, A., Betsinger, S.A., 2021. Preliminary effectiveness of an LGBTQ+ affirmative parenting intervention with foster parents. Children and Youth Services Review 127. Available at https://doi.org/10.1016/j.childyouth.2021.106107.

Bauer, G.R., Pacaud, D., Couch, R., Metzger, D.L., Gale, L., Gotovac, S., et al., 2021. Transgender youth referred to clinics for gender-affirming medical care in canada. Pediatrics 148 (5). Available at https://doi.org/10.1542/peds.2020-047266.

Becerra-Culqui, T.A., Liu, Y., Nash, R., Cromwell, L., Flanders, W.D., Getahun, D., et al., 2018. Mental health of transgender and gender nonconforming youth compared with their peers. Pediatrics 141 (5). Available at https://doi.org/10.1542/peds.2017-3845.

Bragg, J.E., Brady, S.R., Howell, D., 2022. The role of positive social institutions in promoting hope and flourishing among sexual and gender minorities: a multi-group analysis. Youth 2 (4), 654–667. Available at https://doi.org/10.3390/youth2040046.

Bruno, J., Vedilago, V., Goldhammer, H., Campbell, J., Keuroghlian, A.S., 2023. TransECHO: a national tele-education program for expanding transgender and gender diverse healthcare. LGBT Health 10 (6), 456–462. Available at https://doi.org/10.1089/lgbt.2022.0355.

Budge, S.L., 2013. Interpersonal psychotherapy with transgender clients. Psychotherapy (Chicago, Ill.) 50 (3), 356–359. Available at https://doi.org/10.1037/a0032194.

Budge, S.L., Sinnard, M.T., Hoyt, W.T., 2021. Longitudinal effects of psychotherapy with transgender and nonbinary clients: a randomized controlled pilot trial. Psychotherapy 58 (1), 1–11. Available at https://doi.org/10.1037/pst0000310.

Call, D.C., Challa, M., Telingator, C.J., 2021. Providing affirmative care to transgender and gender diverse youth: disparities, interventions, and outcomes. Current Psychiatry Reports 23 (6). Available at https://doi.org/10.1007/s11920-021-01245-9.

Carlile, A., 2020. The experiences of transgender and non-binary children and young people and their parents in healthcare settings in England, UK: interviews with members of a family support group. International Journal of Transgender Health 21 (1), 16–32. Available at https://doi.org/10.1080/15532739.2019.1693472.

Carlile, A., Butteriss, E., Sansfaçon, A.P., 2021. "It's like my kid came back overnight": experiences of trans and non-binary young people and their families seeking, finding and engaging with clinical care in England. International Journal of Transgender Health 22 (4), 412–424. Available at https://doi.org/10.1080/26895269.2020.1870188.

Carmichael, P., Butler, G., Masic, U., Cole, T.J., de Stavola, B.L., Davidson, S., et al., 2021. Short-term outcomes of pubertal suppression in a selected cohort of 12 to 15 year old young people with persistent gender dysphoria in the UK. PLoS One 16 (2). Available at https://doi.org/10.1371/journal.pone.0243894.

Chang, S.C., Singh, A.A., 2016. Affirming psychological practice with transgender and gender nonconforming people of color. Psychology of Sexual Orientation and Gender Diversity 3 (2), 140–147. Available at https://doi.org/10.1037/sgd0000153.

Chen, D., Hidalgo, M.A., Leibowitz, S., Leininger, J., Simons, L., Finlayson, C., et al., 2016. Multidisciplinary care for gender-diverse youth: a narrative review and unique model of gender-affirming care. Transgender Health 1 (1), 117–123. Available at https://doi.org/10.1089/trgh.2016.0009.

Chen, D., Berona, J., Chan, Y.M., Ehrensaft, D., Garofalo, R., Hidalgo, M.A., et al., 2023. Psychosocial functioning in transgender youth after 2 years of hormones. New England Journal of Medicine 388 (3), 240–250. Available at https://doi.org/10.1056/NEJMoa2206297.

Chodzen, G., Hidalgo, M.A., Chen, D., Garofalo, R., 2019. Minority stress factors associated with depression and anxiety among transgender and gender-nonconforming youth. Journal of Adolescent Health 64 (4), 467–471. Available at https://doi.org/10.1016/j.jadohealth.2018.07.006.

Cohen-Kettenis, P.T., Steensma, T.D., de Vries, A.L.C., 2011. Treatment of adolescents with gender dysphoria in the Netherlands. Child and Adolescent Psychiatric Clinics of North America 20 (4), 689–700. Available at https://doi.org/10.1016/j.chc.2011.08.001.

Coleman, E., Radix, A.E., Bouman, W.P., Brown, G.R., de Vries, A.L.C., Deutsch, M.B., et al., 2022. Standards of care for the health of transgender and gender diverse people, version 8. International Journal of Transgender Health 23 (sup1), S1. Available at https://doi.org/10.1080/26895269.2022.2100644.

Connolly, M.D., Zervos, M.J., Barone, C.J., Johnson, C.C., Joseph, C.L.M., 2016. The mental health of transgender youth: advances in understanding. Journal of Adolescent Health 59 (5), 489–495. Available at https://doi.org/10.1016/j.jadohealth.2016.06.012.

Costa, R., Dunsford, M., Skagerberg, E., Holt, V., Carmichael, P., Colizzi, M., 2015. Psychological support, puberty suppression, and psychosocial functioning in adolescents with gender dysphoria. The Journal of Sexual Medicine 12 (11), 2206–2214. Available at https://doi.org/10.1111/jsm.13034.

Cousino, M.K., Davis, A., Ng, H., Stancin, T., 2014. An emerging opportunity for pediatric psychologists: our role in a multidisciplinary clinic for youth with gender dysphoria. Clinical Practice in Pediatric Psychology 2 (4), 400–411. Available at https://doi.org/10.1037/cpp0000077.

Coyne, C.A., Zachary Huit, T., Janssen, A., Chen, D., 2023a. Supporting the mental health of transgender and gender-diverse youth. Pediatric Annals 52 (12). Available at https://doi.org/10.3928/19382359-20231016-02.

Coyne, C.A., Yuodsnukis, B.T., Chen, D., 2023b. Gender dysphoria: optimizing healthcare for transgender and gender diverse youth with a multidisciplinary approach. Neuropsychiatric Disease and Treatment 19, 479–493. Available at https://doi.org/10.2147/NDT.S359979.

Coyne, C.A., Poquiz, J.L., Janssen, A., Chen, D., 2020. Evidence-based psychological practice for transgender and non-binary youth: defining the need, framework for treatment adaptation, and future directions. Evidence-Based Practice in Child and Adolescent Mental Health 1–14. Available at https://doi.org/10.1080/23794925.2020.1765433.

de Vries, A.L.C., Cohen-Kettenis, P.T., 2012. Clinical management of gender dysphoria in children and adolescents: the Dutch approach. Journal of Homosexuality 59 (3), 301–320. Available at https://doi.org/10.1080/00918369.2012.653300.

De Vries, A.L.C., Steensma, T.D., Doreleijers, T.A.H., Cohen-Kettenis, P.T., 2011. Puberty suppression in adolescents with gender identity disorder: a prospective follow-up study. Journal of Sexual Medicine 8 (8), 2276–2283. Available at https://doi.org/10.1111/j.1743-6109.2010.01943.x.

Durwood, L., McLaughlin, K.A., Olson, K.R., 2017. Mental health and self-worth in socially transitioned transgender youth. Journal of the American Academy of Child and Adolescent Psychiatry 56 (2), 116. Available at https://doi.org/10.1016/j.jaac.2016.10.016.

Edwards-Leeper, L., Spack, N.P., 2012. Psychological evaluation and medical treatment of transgender youth in an interdisciplinary "gender management service" (GeMS) in a major pediatric center. Journal of Homosexuality 59 (3), 321–336. Available at https://doi.org/10.1080/00918369.2012.653302.

Expósito-Campos, P., Pérez-Fernández, J.I., Salaberria, K., 2023. Empirically supported affirmative psychological interventions for transgender and non-binary youth and adults: a systematic review. Clinical Psychology Review 100. Available at https://doi.org/10.1016/j.cpr.2022.102229.

Flores, A.R., 2019. Social acceptance of LGBT people in 174 countries: 1981 to 2017.

Goldenberg, T., Gamarel, K.E., Reisner, S.L., Jadwin-Cakmak, L., Harper, G.W., 2021. Gender affirmation as a source of resilience for addressing stigmatizing healthcare experiences of transgender youth of color. Annals of Behavioral Medicine 55 (12), 1168–1183. Available at https://doi.org/10.1093/abm/kaab011.

Gridley, S.J., Crouch, J.M., Evans, Y., Eng, W., Antoon, E., Lyapustina, M., et al., 2016. Youth and caregiver perspectives on barriers to gender-affirming healthcare for transgender youth. Journal of Adolescent Health 59 (3), 254–261. Available at https://doi.org/10.1016/j.jadohealth.2016.03.017.

Hastings, J., Bobb, C., Wolfe, M., Jimenez, Z.A., Amand, C.S., 2021. Medical care for nonbinary youth: individualized gender care beyond a binary framework. Pediatric Annals 50 (9), e384. Available at https://doi.org/10.3928/19382359-20210818-03.

Hembree, W.C., Cohen-Kettenis, P.T., Gooren, L., Hannema, S.E., Meyer, W.J., Murad, M.H., et al., 2017. Endocrine treatment of gender-dysphoric/gender-incongruent persons: an endocrine society*clinical practice guideline. Journal of Clinical Endocrinology and Metabolism 102 (11), 3869–3903. Available at https://doi.org/10.1210/jc.2017-01658.

Herman, J.L., Flores, A.R. & O'Neill, K.K., 2022. How many adults and youth identify as transgender in the United States? https://williamsinstitute.law.ucla.edu/publications/trans-adults-united-states/

Hidalgo, M.A., Ehrensaft, D., Tishelman, A.C., Clark, L.F., Garofalo, R., Rosenthal, S.M., et al., 2013. The gender affirmative model: what we know and what we aim to learn. Human Development 56 (5), 285–290. Available at https://doi.org/10.1159/000355235.

Hodax, J.K., Kahn, N.F., Crouch, J.M., Sethness, J.L., Bocek, K.M., Sumerwell, C., et al., 2024. Online learning modules improve confidence in providing gender affirming care for youth. BMC Medical Education 24 (1). Available at https://doi.org/10.1186/s12909-024-06517-5.

Holt, N.R., Hope, D.A., Mocarski, R., Meyer, H., King, R., Woodruff, N., 2020. The provider perspective on behavioral healthcare for transgender and gender nonconforming individuals in the central great plains: a qualitative study of approaches and needs. American Journal of Orthopsychiatry 90 (1), 136–146. Available at https://doi.org/10.1037/ort0000406.

Hope, D.A., Holt, N.R., Woodruff, N., Mocarski, R., Meyer, H.M., Puckett, J.A., et al., 2022. Bridging the gap between practice guidelines and the therapy room: community-derived practice adaptations for psychological services with transgender and gender diverse adults in the central united states. Professional Psychology: Research and Practice 53 (4), 351–361. Available at https://doi.org/10.1037/pro0000448.

Huit, T.Z., Coyne, C., Chen, D., 2024. State of the science: gender-affirming care for transgender and gender diverse youth. Behavior Therapy 55 (6), 1335–1347. Available at https://doi.org/10.1016/j.beth.2024.02.010.

Hunter, J., Butler, C., Cooper, K., 2021. Gender minority stress in trans and gender diverse adolescents and young people. Clinical Child Psychology and Psychiatry 26 (4), 1182–1195. Available at https://doi.org/10.1177/13591045211033187.

Inwards-Breland, D.J., Divall, S., Salehi, P., Crouch, J.M., Negaard, M., Lu, A., et al., 2019. Youth and parent experiences in a multidisciplinary gender clinic. Transgender Health 4 (1), 100–106. Available at https://doi.org/10.1089/trgh.2018.0046.

Israel, T., Choi, A.Y., Lin, Y.J., Matsuno, E., Goodman, J.A., Kary, K.G., et al., 2021. Reducing internalized transnegativity: randomized controlled trial of an online intervention. Psychology of Sexual Orientation and Gender Diversity 8 (4), 429–439. Available at https://doi.org/10.1037/sgd0000447.

Johns, M.M., Gordon, A.R., Andrzejewski, J., Harper, C.R., Michaels, S., Hansen, C., et al., 2023. Differences in healthcare experiences among transgender and gender diverse youth by gender identity and race/ethnicity. Prevention Science 24 (6), 1128–1141. Available at https://doi.org/10.1007/s11121-023-01521-5.

Kaltiala, R., Heino, E., Työläjärvi, M., Suomalainen, L., 2020. Adolescent development and psychosocial functioning after starting cross-sex hormones for gender dysphoria. Nordic Journal of Psychiatry 74 (3), 213–219. Available at https://doi.org/10.1080/08039488.2019.1691260.

Kattari, S.K., Walls, N.E., Whitfield, D.L., Langenderfer-Magruder, L., 2015. Racial and ethnic differences in experiences of discrimination in accessing health services among transgender people in the United States. International Journal of Transgenderism 16 (2), 68–79. Available at https://doi.org/10.1080/15532739.2015.1064336.

Katz-Wise, S.L., Godwin, E.G., Parsa, N., Brown, C.A., Sansfaçon, A.P., Goldman, R., et al., 2022. Using family and ecological systems approaches to conceptualize family and community-based experiences of transgender and/or nonbinary youth from the trans teen and family narratives project. Psychology of Sexual Orientation and Gender Diversity 9 (1), 21–36. Available at https://doi.org/10.1037/sgd0000442.

Kidd, K.M., Sequeira, G.M., Katz-Wise, S.L., Fechter-Leggett, M., Gandy, M., Herring, N., et al., 2023. Difficult to find, stressful to navigate": parents' experiences accessing affirming care for gender-diverse youth. LGBT Health 10 (7), 496–504. Available at https://doi.org/10.1089/lgbt.2021.0468.

Kozlowska, K., McClure, G., Chudleigh, C., Maguire, A.M., Gessler, D., Scher, S., et al., 2021. Australian children and adolescents with gender dysphoria: clinical presentations and challenges experienced by a multidisciplinary team and gender service. Culture and Attachments 1 (1), 70–95. Available at https://doi.org/10.1177/26344041211010777.

Kremen, J., Williams, C., Barrera, E.P., Harris, R.M., McGregor, K., Millington, K., et al., 2021. Addressing legislation that restricts access to care for transgender youth. Pediatrics 147 (5). Available at https://doi.org/10.1542/peds.2021-049940.

Kuper, L.E., Stewart, S., Preston, S., Lau, M., Lopez, X., 2020. Body dissatisfaction and mental health outcomes of youth on gender-affirming hormone therapy. Pediatrics 145 (4). Available at https://doi.org/10.1542/peds.2019-3006.

Kuper, L.E., Cooper, M.B., Mooney, M.A., 2022. Supporting and advocating for transgender and gender diverse youth and their families within the sociopolitical context of widespread discriminatory legislation and policies. Clinical Practice in Pediatric Psychology 10 (3), 336–345. Available at https://doi.org/10.1037/cpp0000456.

Lange, T.M., 2020. Trans-affirmative narrative exposure therapy (TA-NET): a therapeutic approach for targeting minority stress, internalized stigma, and trauma reactions among gender diverse adults. Practice Innovations 5 (3), 230–245. Available at https://doi.org/10.1037/pri0000126.

Lawson, M.L., Gotovac, S., Couch, B., Gale, L., Vandermorris, A., Ghosh, S., et al., 2024. Pathways to care for adolescents attending a first hormone appointment at canadian gender affirming medical clinics: a cross-sectional analysis from the trans youth can! study. Journal of Adolescent Health 74 (1), 140–147. Available at https://doi.org/10.1016/j.jadohealth.2023.07.021.

Lee, J., Kim, H.H., Budge, S.L., 2024. Psychological assessment for gender-affirming care in transgender and nonbinary youth: the companionship model. Harvard Review of Psychiatry 32 (5), 183–192. Available at https://doi.org/10.1097/HRP.0000000000000404.

Lefevor, G.T., Boyd-Rogers, C.C., Sprague, B.M., Janis, R.A., 2019. Health disparities between genderqueer, transgender, and cisgender individuals: an extension of minority stress theory. Journal of Counseling Psychology 66 (4), 385–395. Available at https://doi.org/10.1037/cou0000339.

Lett, E., Dowshen, N.L., Baker, K.E., 2020. Intersectionality and health inequities for gender minority blacks in the U.S. American Journal of Preventive Medicine 59 (5), 639–647. Available at https://doi.org/10.1016/j.amepre.2020.04.013.

Manns, P.Z., Buckman, C., Tumin, D., Mathai, A., Dendy, C., 2023. Rural perspectives on health services for transgender and gender diverse youth. Health Promotion Practice 24 (4), 609–611. Available at https://doi.org/10.1177/15248399221146805.

Matsuno, E., Israel, T., 2018. Psychological interventions promoting resilience among transgender individuals: transgender resilience intervention model (TRIM. The Counseling Psychologist 46 (5), 632–655. Available at https://doi.org/10.1177/0011000018787261.

Meyer, H.M., Mocarski, R., Holt, N.R., Hope, D.A., King, R.E., Woodruff, N., 2020. Unmet expectations in healthcare settings: experiences of transgender and gender diverse adults in the central great plains. Qualitative Health Research 30 (3), 409–422. Available at https://doi.org/10.1177/1049732319860265.

Mosier-Mills, A., Kim, H.H., Keuroghlian, A.S., 2024. Removing barriers and honoring autonomy: rethinking mental health professional assessments in adolescent gender-affirming medical care. Harvard Review of Psychiatry 32 (3), 96–100. Available at https://doi.org/10.1097/HRP.0000000000000397.

Newcomb, M.E., Hill, R., Buehler, K., Ryan, D.T., Whitton, S.W., Mustanski, B., 2020. High burden of mental health problems, substance use, violence, and related psychosocial factors in transgender, non-binary, and gender diverse youth and young adults. Archives of Sexual Behavior 49 (2), 645–659. Available at https://doi.org/10.1007/s10508-019-01533-9.

Olezeski, C.L., Boulware, S.D., 2017. Development of the yale gender center: an early progress report. Connecticut Medicine 81 (1), 41–44.

Olson, K.R., Durwood, L., Demeules, M., McLaughlin, K.A., 2016. Mental health of transgender children who are supported in their identities. Pediatrics 137 (3). Available at https://doi.org/10.1542/peds.2015-3223.

Olson-Kennedy, J., Chan, Y.-M., Garofalo, R., Spack, N., Chen, D., Clark, L., et al., 2019. Impact of early medical treatment for transgender youth: protocol for the longitudinal, observational trans youth care study. JMIR Research Protocols 8 (7), e14434. Available at https://doi.org/10.2196/14434.

Pampati, S., Andrzejewski, J., Steiner, R.J., Rasberry, C.N., Adkins, S.H., Lesesne, C.A., et al., 2021. "We deserve care and we deserve competent care": qualitative perspectives on healthcare from transgender youth in the Southeast United States. Journal of Pediatric Nursing 56, 54–59. Available at https://doi.org/10.1016/j.pedn.2020.09.021.

Poquiz, J., Moser, C.N., Grimstad, F., Boman, K., Sonneville, S.A., Turpin, A., et al., 2021. Gender-affirming care in the midwest: reaching rural populations. Journal of Rural Mental Health 45 (2), 121–128. Available at https://doi.org/10.1037/rmh0000174.

Puckett, J.A., Cleary, P., Rossman, K., Mustanski, B., Newcomb, M.E., 2018. Barriers to gender-affirming care for transgender and gender nonconforming individuals. Sexuality Research and Social Policy 15 (1), 48–59. Available at https://doi.org/10.1007/s13178-017-0295-8.

Redfield, E., Conron, K.J., Mallory, C., 2024. Youth impacted by anti-transgender legislation in 2024. Williams Institute, UCLA School of Law. Available from: https://williamsinstitute.law.ucla.edu/wp-content/uploads/2024-Anti-Trans-Legislation-Apr-2024.pdf.

Rizvi, S.L., Bitran, A.M., Oshin, L.A., Yin, Q., Ruork, A.K., 2024. The state of the science: dialectical behavior therapy. Behavior Therapy 55 (6), 1233–1248. Available at https://doi.org/10.1016/j.beth.2024.02.006.

Russell, S.T., Pollitt, A.M., Li, G., Grossman, A.H., 2018. Chosen name use is linked to reduced depressive symptoms, suicidal ideation, and suicidal behavior among transgender youth. Journal of Adolescent Health 63 (4), 503–505. Available at https://doi.org/10.1016/j.jadohealth.2018.02.003.

Sequeira, G.M., Ray, K.N., Miller, E., Coulter, R.W.S., 2020. Transgender youth's disclosure of gender identity to providers outside of specialized gender centers. Journal of Adolescent Health 66 (6), 691–698. Available at https://doi.org/10.1016/j.jadohealth.2019.12.010.

Sequeira, G.M., Boyer, T., Coulter, R.W.S., Miller, E., Kahn, N.F., Ray, K.N., 2022. Healthcare experiences of gender diverse youth across clinical settings. Journal of Pediatrics 240, 251–255. Available at https://doi.org/10.1016/j.jpeds.2021.08.089.

Strauss, P., Winter, S., Waters, Z., Wright Toussaint, D., Watson, V., Lin, A., 2022. Perspectives of trans and gender diverse young people accessing primary care and gender-affirming medical services: findings from trans pathways. International Journal of Transgender Health 23 (3), 295–307. Available at https://doi.org/10.1080/26895269.2021.1884925.

Tankersley, A.P., Grafsky, E.L., Dike, J., Jones, R.T., 2021. Risk and resilience factors for mental health among transgender and gender nonconforming (TGNC) youth: a systematic review. Clinical Child and Family Psychology Review 24 (2), 183–206. Available at https://doi.org/10.1007/s10567-021-00344-6.

Testa, R.J., Habarth, J., Peta, J., Balsam, K., Bockting, W., 2015. Development of the gender minority stress and resilience measure. Psychology of Sexual Orientation and Gender Diversity 2 (1), 65–77. Available at https://doi.org/10.1037/sgd0000081.

Tollit, M.A., Feldman, D., McKie, G., Telfer, M.M., 2019. Transgender health. Patient and Parent Experiences of Care at a Pediatric Gender Service 3 (1), 251–256. Available at https://doi.org/10.1089/trgh.2018.0016.

Turban, J.L., Kraschel, K.L., Cohen, I.G., 2021. Legislation to criminalize gender-affirming medical care for transgender youth. JAMA - Journal of the American Medical Association 325 (22), 2251–2252. Available at https://doi.org/10.1001/jama.2021.7764.

Vance, S.R., Buckelew, S.M., Dentoni-Lasofsky, B., Ozer, E., Deutsch, M.B., Meyers, M., 2020. A pediatric transgender medicine curriculum for multidisciplinary trainees. MedEdPORTAL the Journal of Teaching and Learning Resources 16, 10896. Available at https://doi.org/10.15766/mep_2374-8265.10896.

Vance, S.R., Chen, D., Garofalo, R., Glidden, D.V., Ehrensaft, D., Hidalgo, M., et al., 2023. Mental health and gender affirmation of black and latine transgender/nonbinary youth compared to white peers prior to hormone initiation. Journal of Adolescent Health 73 (5), 880–886. Available at https://doi.org/10.1016/j.jadohealth.2023.06.022.

Veale, J.F., Watson, R.J., Peter, T., Saewyc, E.M., 2017. Mental health disparities among canadian transgender youth. Journal of Adolescent Health 60 (1), 44–49. Available at https://doi.org/10.1016/j.jadohealth.2016.09.014.

Vincent, B., 2018. Transgender Health: A Practitioner's Guide to Binary and Non-Binary Trans Patient Care. Jessica Kingsley Publishers, London.

Vincenzo, C.D., Garolla, A., Delinna, S., Pizzolato, L., Testoni, I., 2025. Experiences and narratives of mandatory psychological assessment in trans and nonbinary people: an Italian qualitative analysis. BMC Psychology 13 (1). Available at https://doi.org/10.1186/s40359-025-02675-3.

Warwick, R.M., Shumer, D.E., 2023. Gender-affirming multidisciplinary care for transgender and non-binary children and adolescents. Children's Healthcare 52 (1), 91–115. Available at https://doi.org/10.1080/02739615.2021.2004146.

Wittlin, N.M., Kuper, L.E., Olson, K.R., 2023. Mental health of transgender and gender diverse youth. Annual Review of Clinical Psychology 19, 207–232. Available at https://doi.org/10.1146/annurev-clinpsy-072220-020326.

Chapter 6

Pubertal suppression

Kelsey Eitel[1], Karine Khatchadourian[2], and Daniel Klink[3]

[1]*Department of Pediatrics, University of Washington, Seattle, WA, United States,* [2]*Department of Pediatrics, Children's Hospital of Eastern Ontario and Children's Hospital of Eastern Ontario Research Institute, University of Ottawa, Ottawa, ON, Canada,* [3]*ZAS Center for Gender Care, Antwerp, Belgium; Department of Paediatric Endocrinology, Ghent University Hospital, Ghent, Belgium*

Learning objectives

- Summarize the research on pubertal suppression via gonadotropin-releasing hormone agonists among transgender and gender diverse (TGD) youth.
- Review relevant clinical outcomes, including bone health, other physical health metrics, and mental health and psychosocial outcomes, including differences in outcomes based on timing of pubertal suppression.
- Highlight gaps in the literature, particularly related to long-term follow-up, given the recency of access to pubertal suppression among TGD youth.
- Provide recommendations for ongoing future research initiatives on pubertal suppression in TGD youth.

Key terms and definitions

- Pubertal suppression is used to refer to the use of gonadotropin-releasing hormone agonists (GnRHa) to delay or prevent the development of sex characteristics associated with endogenous puberty. Other types of hormonal suppression (e.g., antiandrogens) are outside of the scope of this chapter.
- Gonadal suppression is used to refer to the use of GnRHa to suppress the endogenous sexual function in adolescents who have completed pubertal development.
- Sexual maturity rating, also known as Tanner staging, refers to the sequence of stages through which children and adolescents progress through puberty, from prepubertal (stage 1) to adult (stage 5).
- Bone mineral density refers to the amount of calcium and other minerals in bone, measured as the density of bone mass relative to others of the same age. Adolescence is a critical time of bone mass accrual, which is influenced by genetic, biological, nutritional, and environmental factors.

Puberty is a period of time during which a child develops and attains secondary sexual characteristics and reaches reproductive capability (Patton and Viner, 2007). People with ovaries, typically assigned female at birth (AFAB), enter puberty between the ages of 8 and 13 years, with the first physical indication of puberty being breast development (Rosenfield et al., 2009). People with testicles, typically assigned male at birth (AMAB), enter puberty between the ages of 9 and 14 years, with the first physical indication being enlargement of the testes. Puberty follows in general a consistent sequence of stages ranging from prepubertal (stage 1) to adult (stage 5) (Tanner, 1962). This staging, known as Tanner staging or sexual maturity rating (SMR), was developed by examining white youth and is still used by clinicians, although the timing and progression of puberty can vary among people of different ethnicities. Puberty takes on average 3 years from beginning to completion (Marshall and Tanner, 1969, 1970).

Advances in Research on Pediatric Gender Affirming Healthcare. DOI: https://doi.org/10.1016/B978-0-443-29019-0.00015-4
© 2026 Elsevier Inc. All rights are reserved, including those for text and data mining, AI training, and similar technologies.

On a molecular level, puberty begins with a reactivation of a complex neuroendocrine network, which ultimately leads to a reemergence of pulsatile release of gonadotropin-releasing hormone (GnRH) from the hypothalamus (Herbison, 2018; Ojeda et al., 2006). This, in turn, leads to increased pulsatile secretion of luteinizing hormone (LH) and follicle-stimulating hormone (FSH) from the pituitary gland, which in turn stimulates the ovaries and testes to produce sex hormones. Studies show genetic factors, metabolic factors, and nutrition are important in this process.

GnRH agonists (GnRHa) have been used for over 40 years in the treatment of central precocious puberty (Crowley et al., 1981). There are a variety of formulations of the medication with effectiveness demonstrated for all despite their different routes of administration, dosing, and duration of action (Crowley et al., 1981). The depot preparations are injected intramuscularly every 1–6 months. There are newer formulations of the medication administered subcutaneously every 4–6 months, and the medication is also available as a subcutaneous implant in certain countries. The medication is a synthetic nonapeptide analog of naturally occurring GnRH and exhibits superagonist effects relative to natural GnRH (Lahlou et al., 2000). The GnRHa bind to the GnRH receptor, leading to the internalization of the analog-receptor complex and resulting in the GnRH receptor desensitization. With the first injection, there is an initial increase in LH secretion with LH and sex steroids then decreasing to prepubertal values within 15–20 days (Lahlou et al., 2000). Continuous administration results in ongoing suppression of LH. GnRHa are not indicated before the onset of central puberty. Youth must reach SMR stage 2 or later to be eligible for GnRHa. Most patients in studies of GnRHa used to treat precocious puberty are AFAB.

Pubertal suppression in gender-affirming care

Endocrine treatment for transgender and gender diverse (TGD) adolescents was first documented in the Netherlands when an increasing number of TGD adolescents began presenting for care, pursuing medical intervention. In 1996 the Center for Expertise in Gender Dysphoria in the Netherlands published the first article detailing the use of GnRHa for gender dysphoria in adolescents (Gooren and Delemarre-van de Waal, 1996). After careful deliberation, gender-affirming hormone (GAH) treatment was made available for thoroughly screened, well-functioning young people between the ages of 16 and 18 years (Cohen-Kettenis and Van Goozen, 1997). The group also published a case report of a 13-year-old adolescent who was started on GnRHa for gender dysphoria (Cohen-Kettenis and van Goozen, 1998). The treated individual later accessed GAH and surgery, and as an adult demonstrated improvement in depression, anxiety, and quality of life, and resolution of gender dysphoria.

The use of GnRHa for the treatment of gender dysphoria became more widespread in the early 2000s. The Endocrine Society included the use of GnRHa for pubertal suppression in the first version of their clinical practice guidelines in 2009, and this was also included in the World Professional Association for Transgender Health (WPATH) Standards of Care version 7, published in 2012 (Coleman et al., 2012; Hembree et al., 2009). The first long-term data were published in 2011 in a case report of an individual AFAB who started GnRHa at the age of 13.7 years, followed by GAH starting at the age of 18.5 years, with doses increased incrementally (Cohen-Kettenis et al., 2011). Gonadectomy was performed at the age of 20 years. A somatic evaluation at the age of 35 demonstrated anthropometrical measurements (height, weight, body mass index [BMI]) in the normal reference values of Dutch males and females. Blood pressure, lipid profile, and glucose homeostasis did not deviate from the normal reference values. Bone mineral density (BMD) measurements (Z- and t-scores of the total hip, lumbar spine, and total body) showed normal values for both sexes.

GnRHa pause further development of primary and secondary sex characteristics. In TGD individuals AFAB, this includes preventing further chest development, uterine enlargement, menarche, height acceleration, and regional fat distribution in the hips and buttocks (Krishna and Witchel, 2024; Hembree et al., 2017; Kelly et al., 2014; Schagen et al., 2016; Schulmeister et al., 2022; Willemsen et al., 2023). If GnRHa are started in early puberty in individuals AFAB, existing chest tissue may decrease and, in some cases, resolve completely (Coleman et al., 2022; Shumer et al., 2016). In TGD individuals AMAB, GnRHa prevent further testicular enlargement, penile and scrotal growth, voice deepening, laryngeal prominence (also known as Adam's apple) growth, height acceleration, increased muscle mass, and regional fat distribution in the abdomen (Krishna and Witchel, 2024; Boogers et al., 2022; Hembree et al., 2017; Kelly et al., 2014; Schagen et al., 2016; Schulmeister et al., 2022).

Prevention of these primary and secondary sex characteristics is reversible with endogenous puberty resuming within months to 1.5 years of GnRHa discontinuation, as has been demonstrated in cisgender youth with central precocious puberty (Krishna and Witchel, 2024; Krishna et al., 2019; Bertelloni et al., 2000; Hembree et al., 2017). The mean time to menarche in cisgender females with central precocious puberty following GnRHa discontinuation is 10–16 months (Krishna and Witchel, 2024; Pasquino et al., 2008). Adrenarche (pubic hair, axillary hair) continues to progress during GnRHa treatment in all individuals (Krishna and Witchel, 2024; Hembree et al., 2017).

Clinical outcomes

Systematic reviews provide the most robust data and conclusions regarding the research to date on GnRHa in TGD youth, although specific methodological limitations of these reviews are also relevant to interpretation. A 2024 systematic review identified 23 moderate- to high-quality studies across physical and psychosocial outcomes based on a modified version of the Newcastle–Ottawa Scale for cohort studies (Taylor et al., 2024). This review has been critiqued for deviations from the preregistered protocol, use of a search strategy and appraisal method that excluded relevant studies, and for offering conclusions that sometimes conflict with presented findings (McNamara et al., 2024). A 2025 systematic review identified 22 studies of moderate- to high-quality studies based on the Grading of Recommendations Assessment, Development, and Evaluation system (GRADE) (Tornese et al., 2025). The GRADE approach to evidence appraisal is meant to evaluate a body of research rather than individual studies, and this review has been critiqued for this misuse of the GRADE framework as well as for fragmenting the research question into five separate ones based on different outcomes. No moderate- or high-quality evidence has been found for fertility, sexual functioning, or cancer risk, each of which remains a priority for future research initiatives. Preliminary research with smaller sample sizes, missing data, lack of adjustment for important confounders, and shorter time frames supplements the research summarized in existing systematic reviews. More research using rigorous, prospective research designs, particularly in the area of mental health and psychosocial outcomes of pubertal suppression, is needed.

Efficacy

Research has consistently shown that GnRHa effectively suppresses gonadotropins and sex steroids in TGD youth, thereby preventing further development of primary and secondary sex characteristics (Fisher et al., 2024; Schagen et al., 2016; van de Grift et al., 2020). When initiated early in puberty, this can prevent the need for later gender-affirming surgeries, such as mastectomy in those AFAB and facial feminization surgery in those AMAB (van de Grift et al., 2020). Two recent prospective studies in the United States (histrelin implants) and Italy (triptorelin injections) demonstrated GnRHa efficacy (2–12 months after initiation) according to biochemically demonstrated suppression of gonadotropins and sex steroids (Fisher et al., 2024; Olson-Kennedy et al., 2020). One of the studies also demonstrated efficacy according to clinical examination (Fisher et al., 2024).

Research has generally demonstrated similar efficacy in TGD youth compared to children with central precocious puberty and when comparing different GnRHa formulations. A retrospective chart review from the United States found overall similar suppression among TGD youth and those with CPP, with one exception (Mejia-Otero et al., 2021). LH, FSH, and testosterone were similarly suppressed across youth AMAB in both groups. LH and FSH were also suppressed across groups in those AFAB, although estrogen was higher in TGD youth compared to those with CPP. Among TGD youth in the study, FSH was more suppressed in those with the histrelin implant compared to leuprolide. Another chart review study on leuprolide found biochemical suppression 1 hour postinjection for Eligard (90%) and Lupron (69%), and this difference was not statistically significant (Eitel et al., 2023). Pubertal suppression, as evidenced by clinical exam, occurred for all patients in the study.

Finally, a study of the efficacy of long-term use of the histrelin implant (17+ months) found that most of the implants in the study (42 out of 50) maintained clinical and biochemical efficacy at an average of 37.5 months (Pine-Twaddell et al., 2023). Of the eight remaining implants, most demonstrated biochemical changes only, with only three resulting in physical pubertal progression. For those in place for 3 years or longer, breakage upon removal occurred for only two. Compared to youth with CPP for whom strict biochemical suppression is necessary, clinical suppression may be sufficient for TGD youth. Therefore extended histrelin use is often efficacious, decreases risks associated with multiple procedures, and may be desirable for some patients. Risks of extended use include difficulty with removal, breakage during removal, and the potential for pubertal progression if not monitored.

Mental health and psychosocial outcomes

Although the use of GnRHa was originally intended as an extended diagnostic phase, research has investigated the mental health benefits of GnRHa before initiation of and along with GAH (Cohen-Kettenis and Klink, 2015; Fisher et al., 2024). In some studies, TGD adolescents on GnRHa alone have demonstrated lower anxiety, depression, and suicidal ideation compared to before initiating treatment (Fisher et al., 2024; van der Miesen et al., 2020) and compared to those unable to access pubertal suppression (McGregor et al., 2024). Among TGD young adults who reported retrospectively on their experiences as part of the 2015 US Transgender Survey, those who had access to pubertal suppression as adolescents similarly reported lower odds of lifetime suicidal ideation compared to those who desired but could not access this intervention

(Turban et al., 2020). Improvements have also been found in quality of life, well-being, and overall psychosocial functioning (Achille et al., 2020; Costa et al., 2015; Fisher et al., 2024). However, some studies have shown mixed or no changes in psychosocial measures on GnRHa alone, sometimes followed by improvements with the introduction of other affirming interventions (De Vries et al., 2014; Becker-Hebly et al., 2021; Carmichael et al., 2021; De Vries et al., 2011).

In the first cohort of 70 subjects described, behavioral and emotional problems and depressive symptoms decreased while general functioning improved significantly (De Vries et al., 2011). However, feelings of anxiety and anger, gender dysphoria, and body satisfaction did not change. Before GnRHa treatment, clinic-referred adolescents showed more internalizing problems and reported increased self-harm and suicidality and poorer peer relations compared with their age-equivalent peers (van der Miesen et al., 2020). However, while receiving pubertal suppression, TGD adolescents had similar or fewer problems than same-age cisgender peers on the Youth Self-Report domains. In addition, when compared to psychological support alone, pubertal suppression together with psychological support was reported to improve psychosocial functioning better after 12 months of GnRHa treatment (Costa et al., 2015).

Although some of these individual studies have found mental health benefits associated with pubertal suppression, systematic reviews have overall shown that this body of evidence is inconsistent and uncertain (Taylor et al., 2024). Conversely, no research to date has found worsening mental health outcomes associated with pubertal suppression. It may be that, rather than leading to improvements in mental health, pubertal suppression more often serves to prevent worsening mental health until other affirming treatments can be accessed. However, more research is needed to clarify the association between pubertal suppression and mental health, over time, and with and without the presence of affirming hormones.

Findings have also been mixed with regard to changes in gender dysphoria and body dissatisfaction, with some studies showing no change on GnRHa alone, followed by improvements with other affirming interventions (De Vries et al., 2014, 2011). However, one prospective study showed that improvements in mental health and psychosocial functioning were specifically linked to improvements in body dissatisfaction (Fisher et al., 2024). This was also after a period of worsening mental health and body dissatisfaction between referral and initiation of GnRHa, further suggesting that pubertal suppression may serve to prevent worsening distress. Qualitatively, many youth report feelings of relief while on pubertal suppression, specifically related to feeling more aligned with affirmed gender (Carmichael et al., 2021). More research is needed to clarify the impact of GnRHa on gender dysphoria and body satisfaction, and the role of these changes on other mental health and psychosocial outcomes.

Bone health

Bone mass accrual occurs from infancy to young adulthood, with peak bone mass density occurring at 22 years of age for individuals AFAB and 26 years of age for individuals AMAB (Lu et al., 2016). Puberty is a crucial time for bone mass accrual. Bone density is influenced by calcium intake, vitamin D, BMI, weight-bearing activity, and sex steroid hormones. BMD in pediatrics is clinically measured by dual-energy X-ray absorptiometry (DXA) and reported in age- and sex-matched Z-scores. In pediatrics, sites of focus include the lumbar spine and total body less head (Lee, 2023; Shuhart et al., 2019).

Clinical practice guidelines have differing recommendations for monitoring bone health in TGD youth. The Endocrine Society recommends DXA at baseline and every 1–2 years until peak bone mass is achieved, and beyond that in individuals with risk factors for bone loss or with nonadherence to GAH treatment (Hembree et al., 2017). The International Society for Clinical Densitometry recommends baseline DXA for individuals on GnRHa therapy, then follow-up DXA when the results are likely to influence patient management (Shuhart et al., 2019). WPATH recommends discussing bone health and ways to enhance bone health with TGD individuals, in addition to starting osteoporosis screening with DXA when indicated based on an individual's history and risk factors (Coleman et al., 2022).

Most research studies in TGD youth use DXA and report areal BMD (aBMD) Z-scores for sex assigned at birth (Lee, 2023). However, the International Society for Clinical Densitometry suggests using Z-scores based on gender identity and, if nonbinary or gender nonconforming, then using Z-scores based on sex assigned at birth (Shuhart et al., 2019). A smaller number of studies have used quantitative computed tomography (QCT) to assess volumetric BMD (vBMD), which has the advantage of not being dependent on bone size (Lee et al., 2020; Nasomyont et al., 2022). Structural bone geometry, which is measured as subperiosteal width and endocortical diameter, is largely influenced by the timing of puberty and sex steroid hormones (van der Loos et al., 2021). Bone turnover markers have also been examined in a limited number of studies (Lee, 2023). These include formation markers, such as procollagen type 1 N-terminal propeptide (P1NP), osteocalcin, amino terminal of type III procollagen peptide (P3NP), in addition to resorption markers, cross-linked telopeptide of type I collagen (ICTP), and serum C-terminal telopeptide (s-CTX). Studies have found that trends of bone markers seem to follow growth velocity patterns; however, further studies are needed to elucidate the utility of the measurement of these markers (Schagen et al., 2020).

Before any treatment with GnRHa, TGD youth have a higher incidence of low aBMD Z-scores and vBMD Z-scores than their cisgender peers, regardless of puberty status, with these results more pronounced in youth AMAB (Hodax et al., 2023; Klink et al., 2015; Lee et al., 2020; Navabi et al., 2021; Vlot et al., 2017). Pretreatment BMI has been positively correlated with lumbar spine and total body less head aBMD Z-scores, while lower activity levels, lower calcium intake, and vitamin D deficiency are negative predictors of aBMD Z-scores (Hodax et al., 2023; Lee et al., 2020; Marwa et al., 2022; Roy et al., 2024). Among a cohort of young TGD adolescents initiating GnRHa at SMR stage 2 or 3 in the prospective Trans Youth Care Study across four US institutions, baseline BMD Z-scores were below average compared to sex-matched reference standards, especially among those AMAB (Lee et al., 2020). Prospective methodology allowed for the identification of risk factors without postintervention confounding variables, and lower physical activity was identified as a likely contributing factor. Calcium and vitamin D were not associated with lower BMD in this study.

There are several potential explanations for lower baseline BMD among TGD youth compared to their cisgender peers, but more research is needed to examine these hypotheses and clarify risk factors prospectively. One potential explanation for baseline differences is that TGD youth may be less engaged in physical activity due to gender dysphoria, sports team restrictions, or a lack of social supports. Another hypothesis is that TGD youth may participate in physical activities more consistently with their affirmed gender rather than sex assigned at birth, meaning that youth AMAB get less physical activity than their sex-matched peers while also having less exposure to sex steroid hormones due to starting puberty at later ages (Lee et al., 2020).

During treatment with GnRHa monotherapy, sex steroid levels decline, leading to a predictable slowing of bone mass accrual. This results in lower aBMD Z-scores than the age- and assigned sex at birth-matched population who is progressing through typical endogenous puberty (Joseph et al., 2019; Roy et al., 2024; Van Der Loos et al., 2023; Vlot et al., 2017). A retrospective longitudinal study that followed TGD youth through age 28 years found that vitamin D level and duration of GnRHa monotherapy were not associated with long-term follow-up BMD Z-scores (Van Der Loos et al., 2023). In contrast, BMI Z-scores were positively associated with long-term follow-up BMD Z-scores at the total hip and femoral neck, but not the lumbar spine. One small prospective study found that compared to controls, TGD youth demonstrate lower BMD Z-scores and greater bone marrow adipose tissue following 12 months of GnRHa monotherapy, suggesting that early changes in bone marrow adipose tissue may contribute to BMD accrual during pubertal suppression (Nasomyont et al., 2022).

While BMD accrual slows on GnRHa alone, research among young adults has demonstrated a return to expected levels following the introduction of GAH, especially testosterone (Boogers et al., 2023; Klink et al., 2015; Van Der Loos et al., 2023; Vlot et al., 2017). In a small cohort of transgender girls (AMAB, n=15; median age at start of GnRHa 14.9 years; duration of GnRHa monotherapy 1.3 years), aBMD in the lumbar spine and hip increased by age 22 years following initiation of estrogen, but the Z-score for the lumbar spine continued to decrease (Klink et al., 2015). Three participants AMAB had a t-score < -2.5. In transgender boys (AFAB, n=19; median age start GnRHa 15.0 years; median duration GnRHa monotherapy 1.5 years) between the start of testosterone therapy and age 22, aBMD of the lumbar spine and hip increased significantly. The Z-score of the lumbar spine showed a significant increase, whereas the Z-score of the hip remained the same. At age 22 years the aBMD Z-score of the lumbar spine was lower than when GnRHa was started, but this was not significant. No participant AFAB had a t-score < -2.5. Thus, at age 22, the Z-score of the lumbar spine was under pretreatment level, implying a possible delay in or loss of peak bone mass, which was more pronounced in those AMAB.

In another retrospective cohort study of TGD young adults following at least 9 years of GAH, BMD Z-scores decreased following GnRHa initiation, but returned to expected levels in all measurements except lumbar spine among those AMAB (Van Der Loos et al., 2023). This exception was hypothesized to be related to low estradiol concentrations. The same research group examined differences in BMD among TGD youth AMAB following GnRHa plus one of three hormone regimens: (1) estradiol up to 2 mg, (2) estradiol up to 6 mg, and (3) ethinyl estradiol (Boogers et al., 2023). Results demonstrated greater increases in BMD Z-scores with 6 mg of estradiol or ethinyl estradiol, suggesting that typical doses of estradiol may be insufficient to optimize bone health. However, higher estradiol dosing must be weighed against other increased risks (Boogers et al., 2022).

In conclusion, BMD accrual slows during GnRHa monotherapy and subsequently increases to pretreatment levels after long-term use of GAH (Van Der Loos et al., 2023; Vlot et al., 2017). Those AMAB in particular often present with lower BMD at baseline, and this can persist for the first few years on GAH, particularly at the lumbar spine (Tornese et al., 2025). The cause of the lack of this catch-up is most likely to be multifactorial, but estrogen levels and dosage may play a role (Boogers et al., 2023). Bone turnover markers also appear to decrease on GnRHa (Schagen et al., 2020; Vlot et al., 2017). However, the added value of evaluating these markers may be limited (Tornese et al., 2025).

Research methods used to study BMD have included retrospective and prospective observational studies. Strengths include baseline DXA before GnRHa initiation and separation of participants by pubertal status. The follow-up period of BMD on GnRHa monotherapy is an average of 2 years (Klink et al., 2015; Joseph et al., 2019), and up to 36 months before GAH are started (Carmichael et al., 2021; Schagen et al., 2020). One study has followed TGD youth treated with GnRHa and GAH through 28 years, presumably when peak bone mass has been obtained (Van Der Loos et al., 2023). To date, there are no long-term studies following BMD from early pubertal GnRHa initiation through later adulthood when age-related bone mass loss occurs and the risk for osteoporosis increases. This is a direct result of only recent access to these medications, and long-term follow-up studies will be possible as initial cohorts of TGD youth reach later adulthood. While lower than average BMD has been demonstrated in TGD youth before initiation of affirming treatments, there has been no correlation to increased fracture risk or risk of pediatric osteoporosis, which cannot be diagnosed with low BMD in isolation (Shuhart et al., 2019).

Prolonged delay of BMD accrual may decrease the potential for attaining peak bone mass. The use of pediatric bone-accrual velocity calculations can be explored much like pediatric endocrinologists use growth velocity scores for linear growth as a novel technique for bone health surveillance (Kelly et al., 2019). One study showed TGD youth had below-average aBMD velocity Z-scores while on GnRHa monotherapy (Roberge et al., 2024). Future studies should include assessment of aBMD velocity Z-scores. Additional studies are needed to assess the impacts of GnRHa monotherapy on bone marker turnover, determine long-term fracture risk in TGD individuals, and provide guidance on what duration of GnRHa monotherapy can be safely prescribed without problematic long-term impact on bone health.

Height and skeletal structure

Puberty is a key time for linear growth acceleration, with height velocity increasing up to 10 cm/year for individuals AFAB and 12 cm/year for those AMAB (Kelly et al., 2014). GnRHa slow pubertal growth spurts, with height velocity maintained at prepubertal rates (ranging 5–8 cm/year) (Krishna and Witchel, 2024; Boogers et al., 2022; Kelly et al., 2014). If a GnRHa is initiated early in puberty and subsequently discontinued, growth velocity increases to pubertal rates as endogenous puberty resumes (Krishna and Witchel, 2024).

Several retrospective and prospective studies have consistently shown a decrease in height velocity as a result of GnRHa monotherapy (Boogers et al., 2022; Fisher et al., 2024; Schagen et al., 2016; Schulmeister et al., 2022; Willemsen et al., 2023). This effect may be more pronounced for those AMAB (Fisher et al., 2024) and those who initiate pubertal suppression at later pubertal stages (Schulmeister et al., 2022). Data from the prospective US Trans Youth Care study demonstrated that height velocity for those who initiated puberty blockers in earlier stages of puberty was equivalent to prepubertal youth after 1 year on GnRHa, while youth who began GnRHa at SMS stage 4 or later demonstrated even lower height velocity of 1.5–2.9 cm/year (Schulmeister et al., 2022). This finding may be due to prior hormonal exposure, either as a compensatory process or reflecting an inability to grow without hormone exposure, but more research is needed to replicate and clarify these results.

Like BMD, research has also consistently shown that height velocity returns to expected levels with the introduction of GAH (Boogers et al., 2022; Ciancia et al., 2023; Willemsen et al., 2023). However, studies that have examined final height in relation to predicted adult height (according to baseline bone age) and target height (according to parental height) have had mixed results. Some studies have shown that pubertal suppression may result in decreased final height for those AMAB (Boogers et al., 2022) and increased final height for those AFAB (Willemsen et al., 2023). Among those AMAB with a history of pubertal suppression in one study, only those on higher estradiol doses saw reduced final height, which would need to be weighed against potential risks (Boogers et al., 2022). Among those AFAB with a history of pubertal suppression in another study, final height was higher than predicted, which was more pronounced among those with earlier initiation of GnRHa (Willemsen et al., 2023). In both studies, differences were more pronounced when final height was compared to predicted adult height rather than target height.

Other studies have shown no differences in final height related to GnRHa use. One study measured bone age at initiation of pubertal suppression and later GAH and compared final height to predicted adult height and target height based on both sex assigned at birth and gender identity (Ciancia et al., 2023). In this study, target height based on sex assigned at birth was the best predictor of final height. Another study compared final height across individuals with a history of early GnRHa initiation, late GnRHa initiation, and no history of pubertal suppression (van de Grift et al., 2020). No differences in final height were found across any group, and this was true for both AMAB and AFAB.

In contrast to height, bone geometry may be impacted by pubertal suppression. Individuals who start GnRHa therapy in early puberty, defined as testes ≤9 mL or chest SMR stage 2, have bone geometry that follows the reference curve of cisgender peers of the same affirmed gender. Those who start GnRHa therapy in later puberty demonstrate bone geometry similar to cisgender peers with the same sex assigned at birth (van der Loos et al., 2021). A retrospective cohort study examined DXA results among young adults in one of four groups: (1) early pubertal suppression and GAH, (2) late pubertal suppression and GAH, (3) GAH only, and (4) no medical interventions (Boogers et al., 2025). Results demonstrated pelvis and shoulder width similar to affirmed gender only among those with early pubertal suppression, with the exception of greater pelvis width among those AMAB with late pubertal suppression. No participant groups without pubertal suppression demonstrated bone geometry similar to cisgender peers of the same affirmed gender. These results suggest that skeletal dimensions may be altered by affirming hormones, but only for those who have less exposure to endogenous pubertal hormones by using pubertal suppression.

Together, these results suggest that early pubertal suppression may result in skeletal development more consistent with gender identity than sex assigned at birth, but this effect is more consistent and pronounced for bone geometry than final adult height. These differences are likely due to a lack of exposure to endogenous hormones, followed by affirming hormone exposure during the critical window of puberty prior to the completion of skeletal structure development. While height velocity slows significantly when GnRHa is initiated in later stages of puberty, final height is most consistent with the target height for sex assigned at birth following GAH. More research is needed to prospectively evaluate differences in height and skeletal structure based on the precise timing of GnRHa initiation. There is also a need to evaluate whether the timing of GAH initiation impacts skeletal development in any way.

Body composition and cardiometabolic health

Results of body composition in TGD adolescents on GnRHa monotherapy have been mixed, with more changes found in absolute values of body fat and lean mass compared with Z-scores. According to Z-scores across two retrospective studies, youth AMAB have demonstrated increased total body fat and decreased lean body mass, while no significant changes have been found in youth AFAB on GnRHa monotherapy (Boogers et al., 2024; Navabi et al., 2021). In one of these studies, only absolute increases in total body fat and lean body mass were seen in youth AFAB (Navabi et al., 2021). The second study found only small changes in body composition for youth AFAB over the first year of treatment, and no changes after the first year (Boogers et al., 2024). In contrast, a prospective study found an increase in absolute body fat and a decrease in absolute lean body mass in the first year of GnRHa monotherapy for participants both AFAB and AMAB (Schagen et al., 2016). A cross-sectional study that matched TGD adolescents on GnRHa monotherapy to cisgender controls based on age, BMI, and sex assigned at birth found that TGD AFAB participants had lower body fat percentage, and TGD AMAB participants had higher body fat percentage and lower lean mass compared to matched controls (Nokoff et al., 2021).

Following the introduction of GAH, body composition appears to shift toward being more consistent with cisgender peers of the same gender. The change in body shape from the start of treatment with GnRHa until age 22 years was studied in 192 TGD individuals (Klaver et al., 2018). At a mean age of 14.5 years, 71 transgender girls started GnRHa, followed by the addition of 17-beta estradiol (mean age: 16.4 years). During treatment the waist and hip circumference increased by 8 and 17 cm, respectively. Waist-to-hip ratio (WHR) decreased from 0.81 to 0.77. In 121 transgender men who started GnRHa at a mean age of 15.3 years, followed by the addition of testosterone (mean age: 16.9 years), the waist and hip circumference increased by 6 and 5 cm, respectively, by the age of 22 years. WHR increased from 0.77 to 0.80. When compared at 22 years of age with age-matched peers, young adult transgender women showed a greater similarity to cisgender women than to cisgender men for WHR SDS (+0.3, 95% CI=0.0–0.5). In transgender men, SDS for WHR was within reference values for cisgender women (+0.6, 95% CI=0.4–0.8) and men (-0.5, 95% CI=-0.8 to -0.3). Initiation of GnRHa treatment earlier in puberty appeared to be associated with a closer resemblance of body shape to cisgender peers of the same affirmed gender.

Long-term data on cardiovascular risk associated with GnRHa are still accumulating as the earliest cohorts of TGD youth to access pubertal suppression reach older ages. However, the availability of these treatments in adulthood is highly limited in many countries, and thus long-term use in later adulthood may be less relevant for understanding outcomes for TGD youth. One study measured BMI, blood pressure, glucose homeostasis, lipid values, the prevalence of obesity, and dyslipidemia in a cohort of 71 transgender women and 121 transgender men at the start of GnRHa, the start of GAH, and at the age of 22 years compared to cisgender controls (Klaver et al., 2020). This long-term study showed that changes in cardiometabolic risk factors on GnRHa monotherapy and after the addition of GAH were not significantly different than changes occurring among cisgender adolescents of the same affirmed gender. There were also no significant differences in any risk factor between those who pursued gonadectomy versus continued use of pubertal suppression.

Most longitudinal studies have shown no change in BMI Z-scores during GnRHa treatment before initiation of GAH (Boogers et al., 2024; Klink et al., 2015; Navabi et al., 2021). One study of 116 TGD adolescents found a significant increase in BMI Z-scores for AFAB but not AMAB individuals in the first year on GnRHa monotherapy (Schagen et al., 2016). One study found an increase in absolute BMI, although the increase was similar to BMI changes in the cisgender population (Klaver et al., 2020). However, a higher prevalence of obesity in TGD young adults was found compared to their cisgender peers, suggesting that factors other than medical interventions, such as minority stress, may contribute to health risks such as obesity in adulthood. A meta-analysis of cisgender children with central precocious puberty found a 0.14 increase in BMI Z-score at the end of GnRHa therapy that returned to baseline after reaching final adult height (Hou et al., 2024). Finally, a retrospective study found no differences in BMI across TGD adolescents who initiated GnRHa in early puberty, late puberty, or not at all (van de Grift et al., 2020).

Research demonstrates no significant changes in systolic and diastolic blood pressures during GnRHa treatment (Taylor et al., 2024). A retrospective study of 112 cisgender girls on GnRHa for central precocious puberty found no changes in mean systolic or diastolic blood pressure percentiles (Fisch-Shvalb et al., 2023). A longitudinal study of 192 TGD adolescents found that systolic blood pressure increased by +1 mmHg in those AMAB and +2 mmHg in those AFAB, but these changes did not differ significantly from cisgender adolescents of the same affirmed gender (Klaver et al., 2020). However, it is plausible that inducing a hypogonadal state in adolescence may impact the cardiovascular system. Indeed, hypertension is a very rare reversible complication of GnRHa monotherapy in those AFAB (Klink et al., 2015)

Compared to matched cisgender controls, one study found TGD adolescents to have lower estimated insulin sensitivity and higher glycemic markers (Nokoff et al., 2021). However, other research has found no differences in insulin sensitivity compared to cisgender controls (Klaver et al., 2020). This study also examined cholesterol levels after 2 years on GnRHa monotherapy and found small but significant changes in total cholesterol, LDL, and HDL, but these changes were consistent with cisgender, age-matched controls. Long-term data from this study similarly showed that ongoing changes with the addition of GAH do not significantly differ from those of the general population.

Future research should include larger longitudinal studies that utilize BMI Z-scores and blood pressure percentiles and account for baseline BMI Z-score, recent BMI rate of change, activity levels, nutrition assessment, and other socioeconomic confounding factors such as family income, race, ethnicity, minority stress, and food insecurity. Because cardiovascular risk is multifactorial and GnRHa monotherapy lasts only a few years, a large powered study will be needed to determine the degree of BMI changes from GnRHa therapy and its effect on long-term cardiovascular risk.

Brain development and cognition

Sex steroids have a role in adolescent brain organization and maturation (Sisk and Zehr, 2005). One cross-sectional study examined executive functioning in TGD youth who were treated with GnRHa compared to TGD youth not receiving GnRHa and age-matched cisgender controls (Staphorsius et al., 2015). The study found no significant differences in executive functioning in TGD youth receiving GnRHa compared to TGD youth not on any medical treatment, although youth AMAB treated with GnRHa showed worse executive functioning when compared to cisgender controls.

Another study of 124 TGD youth examined the contribution of autism diagnosis, internalizing symptoms (depression and anxiety), and gender-affirming medications (GnRHa and/or GAH) to executive functioning (Strang et al., 2022). The Behavior Rating Inventory of Executive Function (BRIEF), a well-validated scale of executive functioning, was used as the primary metric in this study. The study showed that poorer executive functioning was associated with autism diagnosis, anxiety symptoms, and GnRHa monotherapy for >1 year. TGD youth receiving GAH had stronger executive functioning than those not receiving GAH. This study did not account for pubertal timing when GnRHa was initiated, which could have an impact on executive functioning while treated with GnRHa. Finally, one study compared IQ scores in TGD adolescents before any gender-affirming treatments to later educational attainment following GnRHa and GAH and found a strong correlation, suggesting no overall impact on cognitive functioning as a result of pubertal suppression (Arnoldussen et al., 2022).

Future research should examine the potential neurocognitive effects of GnRHa monotherapy. This is an especially important consideration for individuals who may desire ongoing pubertal suppression with GnRHa but are unsure about initiating GAH (Notini et al., 2020). Brain development is complex and multifactorial. Studies should examine both structural changes as well as different measures of neurocognitive function and should consider confounding factors such as educational level, family education, and minority stress. A consensus outlining research methodologies for evaluating neurodevelopmental effects of GnRHa therapy includes the following recommendations: (1) minimum of three time point measurements, (2) pubertal staging at baseline, and (3) use of multiple comparison groups made up of untreated TGD youth with matched pubertal staging, cisgender youth with matched pubertal staging, and/or a large-scale youth development database (Chen et al., 2020).

Fertility and fertility preservation

GnRHa pause both the development of secondary sex characteristics and reproductive maturation. Depending on when GnRHa are initiated in pubertal development, this can include the onset of spermatogenesis (also known as spermarche) in youth AMAB and uterine growth, oocyte maturation, and menarche in youth AFAB (Krishna and Witchel, 2024). GnRHa monotherapy results in a reversible pausing of reproductive maturation. Return of fertility potential has been repeatedly demonstrated in cisgender youth treated with GnRHa for central precocious puberty (Krishna and Witchel, 2024; Lazar et al., 2014; Pasquino et al., 2008). In cisgender boys, spermarche occurred in 0.7–3 years after GnRHa discontinuation, and there were no abnormalities in semen analysis. In cisgender girls the mean time to menarche was 10–16 months after GnRHa discontinuation (Krishna and Witchel, 2024; Pasquino et al., 2008). A longitudinal study following cisgender girls with central precocious puberty treated with GnRHa into their third to fifth decades of life found that, compared to age-matched controls, there was no difference in spontaneous pregnancies, use of assisted reproductive technologies, or number of children (Lazar et al., 2014). There is no data on the long-term reproductive function of TGD youth who discontinue GnRHa monotherapy because this population is small, but theoretically similar outcomes would be expected.

TGD youth who receive GnRHa in early puberty and go on to start GAH do not undergo maturation of their endogenous reproductive system and are unlikely to be able to have biological children without discontinuing GnRHa and GAH, and in some cases, fertility can be permanently impacted by GAH (Hembree et al., 2017). Medical providers discuss potential impacts on long-term fertility potential and options for fertility preservation, if any, before initiating GnRHa treatment in all TGD youth, regardless of whether they intend to initiate GAH in the future.

Fertility preservation rates in general are low in TGD youth in North America, with higher rates seen in the Netherlands and Australia (Chiniara et al., 2019; Lai et al., 2020). Lack of access to affirming fertility specialists, as well as cost and lack of insurance coverage, are frequently reported barriers to fertility preservation. Reproductive health counseling should be encouraged at follow-up visits, as fertility-related perspectives may evolve over time (Nahata et al., 2020). Although GnRHa therapy does not permanently impact fertility, many TGD youth who started on puberty blockers will pursue GAHs as a future step and will not discontinue GnRHa to pursue fertility preservation (Lai et al., 2020). In TGD youth, GnRHa have commonly been used as part of a treatment pathway that progresses from pubertal suppression to the initiation of GAHs. Clinically, all potential pathways are discussed with patients, including the option to fully discontinue GnRHa before the initiation of GAHs to revisit relevant clinical topics such as fertility.

The techniques for fertility preservation depend on an individual's pubertal development before initiation of GnRHa. In those AMAB, sperm cryopreservation can be performed once spermarche is achieved (Coleman et al., 2022; Glodowski et al., 2024). A sperm sample may be obtained by ejaculation or testicular sperm extraction (Glodowski et al., 2024). An Australian study in youth AMAB indicated that a testicular volume >10 mL had 92% sensitivity and 71% specificity in predicting successful sperm retrieval in 25 subjects who received testicular biopsies (Peri et al., 2021). Before spermarche, options are limited. Testicular tissue cryopreservation is experimental, with no live births reported; however, a number of programs have been emerging across the world (Glodowski et al., 2024; McCauley et al., 2025; Moussaoui et al., 2022). Youth who were AMAB and their parents and guardians must balance the decision between sperm cryopreservation and initiating GnRHa to alleviate the psychological distress of unwanted permanent changes, such as voice deepening and hair growth, which often occur by the time of spermarche.

In youth AFAB, oocyte retrieval and cryopreservation can occur in individuals who have reached the later stages of puberty, usually postmenarche (Coleman et al., 2022; Glodowski et al., 2024; Hembree et al., 2017). Oocyte retrieval requires hormone-stimulating medications and transvaginal procedures and, therefore, may be perceived as invasive or cause worsening dysphoria in TGD youth. Ovarian tissue cryopreservation is approved for all ages and is the only option for prepubertal or early pubertal TGD youth (Coleman et al., 2022; Glodowski et al., 2024). It requires auto transplantation into the pelvis, with the ovarian tissue regaining function within 60–240 days. Over 100 live births with ovarian tissue cryopreservation have been reported in cisgender females (Glodowski et al., 2024).

The options for fertility preservation in prepubertal or early pubertal TGD youth are limited, especially for AMAB youth. Fertility preservation may be perceived as invasive, worsening gender dysphoria, and/or expensive (Lai et al., 2020). Future research in assisted reproductive endocrinology should focus on the development of testicular tissue cryopreservation, as well as in vitro maturation of oocytes and sperm from ovarian and testicular tissues, respectively.

Surgical considerations

Starting GnRHa therapy in early puberty (SMR stage 2 or 3) prevents secondary sex characteristics and therefore decreases the need for some gender-affirming surgeries (Coleman et al., 2022; Hembree et al., 2017; Shumer et al., 2016). In those AMAB, this includes facial feminization surgeries, tracheal shave (to reduce the size of the laryngeal tissue, or

Adam's apple), vocal cord surgery (shortening of the vocal cords to create higher frequency and voice pitch), and procedures for facial hair removal (Panagiotakopoulos et al., 2020; Shumer et al., 2016). In those AFAB, early GnRHa therapy prevents glandular chest tissue growth and decreases the need for gender-affirming mastectomy (Panagiotakopoulos et al., 2020; Shumer et al., 2016).

Extra considerations are needed for individuals AMAB who received GnRHa treatment in early puberty and are considering gender-affirming genital surgery, as GnRHa early in puberty leads to penoscrotal hypoplasia (Bene et al., 2024; Safa et al., 2019). The degree of penoscrotal hypoplasia may impact the type of gender-affirming genital surgery an individual AMAB is eligible for (Bene et al., 2024). Historically, the most common technique for creating a neovagina has been the penile inversion procedure. However, if there is an insufficient amount of penile tissue, other techniques must be used for the creation of the neovagina. Scrotal skin grafts, scrotal flaps, or other skin grafts are options that can be considered (Bene et al., 2024; Safa et al., 2019). More recently the robotic-assisted peritoneal flap vaginoplasty technique is being used more commonly for individuals with penoscrotal hypoplasia due to pubertal suppression (Peters et al., 2022).

Two studies have evaluated sexual satisfaction following gender-affirming surgery among TGD young adults with a history of pubertal suppression and found no differences in sexual satisfaction, sexual dysfunction, or ability to orgasm between those who initiated GnRHa in early versus late puberty (Van Der Meulen et al., 2025a, b). This research is limited by the small number of people with a history of pubertal suppression at SMR stage 2 (5 out of 70 participants). Additionally, over half of the participants still experienced at least one type of sexual dysfunction regardless of the timing of initiation of GnRHa. No other large prospective studies have examined the need for or outcomes of gender-affirming surgeries based on pubertal stage at GnRHa initiation. Future research should examine the association between pubertal stage at the time of GnRHa initiation and acute and long-term gender-affirming surgical outcomes.

Access, decision-making, and continuation of treatment

Despite the effectiveness, potential benefits, and relatively low risk profile, GnRHa access remains limited, especially in countries in which cost and insurance coverage are relevant. A large retrospective study of TGD adults in the United States found that only 2.5% of those who desired pubertal suppression as adolescents were able to access this intervention (Turban et al., 2020). A qualitative study with TGD youth and their parents in the United Kingdom identified three major themes related to pubertal suppression availability and access (Horton, 2024). First, TGD youth and their parents experience increasing anxiety as the onset of puberty approaches, and learning about options for pubertal suppression provides reassurance. Second, participants described difficulty with access, including delays in obtaining a referral and an extensive assessment period for eligibility. Finally, a mix of emotions on pubertal suppression was described. Specifically, many youth and parents experience relief knowing that pubertal progression is no longer advancing, but this is sometimes accompanied by frustration related to inflexible protocols for pursuing GAH in addition to pubertal suppression.

Research has also examined perspectives on youth decision-making and capacity to consent to pubertal suppression. A qualitative study was conducted among 30 parents of TGD youth in the United Kingdom whose children socially transitioned at a young age, around the time of puberty (Horton, 2023). Some of these youth were not yet in puberty, some were in early stages of puberty but not yet on pubertal suppression, and others had initiated GnRHa. Major themes included the importance of pubertal suppression for promoting mental health and well-being, aspects of the informed consent process that were perceived as either appropriate or inappropriate, and the difficulty of making decisions without certainty. Aspects of consent that were perceived as appropriate to parents included discussion of the impacts, risks, and benefits, and answering all questions by the youth and parents. Aspects perceived as inappropriate included discussions about other interventions, especially later surgery, describing pubertal suppression as a "huge" decision, and making the consenting process feel more like a test of knowledge and gender identity.

Another qualitative study in the Netherlands interviewed 14 TGD youth and 12 parents, and conducted a focus group with 10 clinicians who provide gender-affirming care about medical decision-making competence related to pubertal suppression (Vrouenraets et al., 2023). Major themes included challenges with assessing competence and additional considerations. Challenges with assessing competence included difficulty defining what constitutes medical decision-making competence, uncertainty about whether young people can understand and appreciate potential long-term effects, and the role that parents play in responsibility for the decision. Additional considerations included the reversibility of pubertal suppression and the role of age, intelligence, and mental health. One clinician questioned the relevance of assessing competence. Using a standardized measurement tool, the MacArthur Competence Assessment Tool for Treatment, this research group showed that approximately 90% of youth aged 10–18 years were competent to consent to medical treatment (Vrouenraets et al., 2021). Despite some youth being as young as 10 years, age was not significantly related to decision-making capacity, nor were emotional or behavioral challenges, although IQ was significantly associated with scores.

The emphasis on medical decision-making to initiate GnRHa appears to be related to concerns about whether gender incongruence and thus the wish for endocrine treatment would sustain into adulthood. In a single-center study, 720 TGD young adults were included who initiated GnRHa before the age of 18 years and used GnRHa for a minimum duration of 3 months before addition of GAH (van der Loos et al., 2022). This cohort was linked to a nationwide prescription registry to verify the prescription of GAH at follow-up. Of these 720 TGD young adults, 97.8% (n=704) who had started pubertal suppression treatment in adolescence continued to use GAHs at follow-up. Of the 16 (2.2%) individuals without an ongoing prescription, nine were AMAB and seven AFAB. Discontinuation of gender-affirming medical treatments was not associated with age or pubertal stage at start of GnRHa, age at start of GAH, or having pursued gender-affirming surgery.

Summary and future research directions

Multiple forms of GnRHa, including leuprolide, triptorelin, and histrelin, have been shown to effectively suppress puberty, including, in many cases, long-term use of the subcutaneous histrelin implant. Future research should continue to amass long-term mental health and psychosocial outcomes among TGD individuals who access GnRHa as adolescents. Additional areas of needed research include the direct effects of pubertal suppression on body dissatisfaction and gender dysphoria, and the consequences of lack of access and specific barriers to accessing pubertal suppression for those who desire it. Longitudinal research across multiple institutions is needed to clarify differences in mental health outcomes based on timing of GnRHa initiation, length of time on monotherapy before initiating GAH, and other risk and protective factors such as parental support, minority stress, and access. Clinically, mental health should be monitored during initiation of GnRHa and beyond, and additional mental health support should be provided to address ongoing gender dysphoria that may not be relieved by GnRHa alone, residual anxiety, and any other mental health concerns unrelated to gender (Tornese et al., 2025).

TGD youth have a higher incidence of low aBMD Z-scores than their cisgender peers before GnRHa treatment, regardless of puberty status, which is more pronounced in AMAB youth. During GnRHa monotherapy, there is an expected slowing of bone mass accrual, resulting in lower aBMD Z-scores, but with long-term use of GAH through age 28, Z-scores generally return to pretreatment levels regardless of duration of GnRHa monotherapy (van der Loos et al., 2021). It is unknown if there are differences in bone mass accrual based on pubertal stage when GnRHa treatment is started. However, bone geometry does follow the reference curve of the experienced gender for youth who start GnRHa in early puberty, whereas bone geometry follows the reference curve for sex assigned at birth in individuals starting GnRHa in later puberty. Final height is most likely to align with target height based on parental height, regardless of treatment with GnRHa, although some research suggests pubertal suppression initiated early in puberty may minimally alter outcomes.

Standardized, long-term follow-up studies are needed to better understand skeletal outcomes in later adulthood. Specifically, research is needed to clarify the duration of GnRHa monotherapy that can be safely prescribed without a problematic long-term impact on bone health. Large patient cohorts will likely be needed to better understand what effects, if any, the timing of GnRHa has on final adult height. Clinically, ongoing monitoring of bone health and BMD optimization through physical activity, calcium intake, and vitamin D supplementation is recommended (Tornese et al., 2025).

AMAB individuals have increases in total body fat Z-scores and decreases in lean body mass Z-scores on GnRHa treatment (Boogers et al., 2024; Navabi et al., 2021; Nokoff et al., 2021). Studies on body composition Z-score changes in AFAB individuals on GnRHa monotherapy are conflicting (Boogers et al., 2024; Navabi et al., 2021), but a cross-sectional study with cisgender controls matched based on age, BMI, and sex assigned at birth found that TGD AFAB participants had lower body fat percentage (Nokoff et al., 2021). BMI Z-scores do not appear to be impacted by GnRHa. While TGD individuals are at higher risk of obesity regardless of gender-affirming medical interventions, changes in BMI during GnRHa use are consistent with cisgender individuals of the same age and pubertal stage (Klaver et al., 2020). Research on body composition has been complicated by variation in treatment protocols, duration of GnRHa monotherapy, and age of GnRHa initiation, as well as an inability to control for other factors that likely contribute to body composition outside of gender-affirming care. Future research on body composition should account for additional contributing factors, such as activity level before and after treatment, weight-related risk factors, and timing of GnRHa initiation.

There are several areas of research that are more limited and remain priorities for future initiatives, including neurocognitive development and fertility (Tornese et al., 2025). While pubertal suppression has not been found to impact long-term cognitive ability in a small number of preliminary studies, additional studies are needed to clarify the short- and long-term effects of GnRHa with and without GAH on neurocognitive development (Chen et al., 2020). While the effect of GnRHa on fertility development is reversible, early initiation followed by later GAH prevents the progression of fertility development and limits fertility preservation options. Future research should prioritize options for fertility preservation in early pubertal development. Different pathways to fertility preservation, such as discontinuing GnRHa before initiation of gender-affirming hormones to allow fertility preservation while closely monitoring an adolescent's pubertal progression, should be explored. Qualitative research on the perspectives of youth and families on decision-making related to pubertal suppression and fertility is also needed.

While pubertal suppression with GnRHa is an effective and beneficial treatment for TGD youth, access to this intervention remains limited. Barriers to access exist on multiple levels, including cost and insurance, regional availability, institutional policies, clinician protocols, and ideas about consent, parental concerns, and sometimes individual fears related to injections or procedures. As research on the risks, benefits, and alternatives to pubertal suppression continues to amass, results can support efforts to reduce barriers and improve health equity for TGD youth.

References

Achille, C., Taggart, T., Eaton, N.R., Osipoff, J., Tafuri, K., Lane, A., et al., 2020. Longitudinal impact of gender-affirming endocrine intervention on the mental health and well-being of transgender youths: preliminary results. International Journal of Pediatric Endocrinology 2020 (1). Available at https://doi.org/10.1186/s13633-020-00078-2.

Arnoldussen, M., Hooijman, E.C., Kreukels, B.P.C., de Vries, A.L.C., 2022. Association between pre-treatment IQ and educational achievement after gender-affirming treatment including puberty suppression in transgender adolescents. Clinical Child Psychology and Psychiatry 27 (4), 1069–1076. Available at https://doi.org/10.1177/13591045221091652.

Bangalore Krishna, K., Fuqua, J.S., Rogol, A.D., Klein, K.O., Popovic, J., Houk, C.P., et al., 2019. Use of gonadotropin-releasing hormone analogs in children: update by an international consortium. Hormone Research in Paediatrics 91 (6), 357–372. Available at https://doi.org/10.1159/000501336.

Becker-Hebly, I., Fahrenkrug, S., Campion, F., Richter-Appelt, H., Schulte-Markwort, M., Barkmann, C., 2021. Psychosocial health in adolescents and young adults with gender dysphoria before and after gender-affirming medical interventions: a descriptive study from the Hamburg Gender Identity Service. European Child & Adolescent Psychiatry 30 (11), 1755–1767. Available at https://doi.org/10.1007/s00787-020-01640-2.

Bene, N.C., Ferrin, P.C., Xu, J., Dy, G.W., Dugi, D., Peters, B.R., 2024. Tissue options for construction of the neovaginal canal in gender-affirming vaginoplasty. Journal of Clinical Medicine 13 (10). Available at https://doi.org/10.3390/jcm13102760.

Bertelloni, S., Baroncelli, G.I., Ferdeghini, M., Menchini-Fabris, F., Saggese, G., 2000. Final height, gonadal function and bone mineral density of adolescent males with central precocious puberty after therapy with gonadotropin- releasing hormone analogues. European Journal of Pediatrics 159 (5), 369–374. Available at https://doi.org/10.1007/s004310051289 Available from: link.springer.de/link/service/journals/00431/index.htm.

Boogers, L.S., Wiepjes, C.M., Klink, D.T., Hellinga, I., Van Trotsenburg, A.S.P., Den Heijer, M., et al., 2022. Transgender girls grow tall: adult height is unaffected by GnRH analogue and estradiol treatment. Journal of Clinical Endocrinology and Metabolism 107 (9), e3805–e3815. Available at https://doi.org/10.1210/clinem/dgac349.

Boogers, L.S., Van Der Loos, M.A.T.C., Wiepjes, C.M., Van Trotsenburg, A.S.P., Den Heijer, M., Hannema, S.E., 2023. The dose-dependent effect of estrogen on bone mineral density in trans girls. European Journal of Endocrinology 189 (2), 290–296. Available at https://doi.org/10.1093/ejendo/lvad116.

Boogers, L.S., Reijtenbagh, S.J.P., Wiepjes, C.M., van Trotsenburg, A.S.P., den Heijer, M., Hannema, S.E., 2024. Time course of body composition changes in transgender adolescents during puberty suppression and sex hormone treatment. Journal of Clinical Endocrinology and Metabolism 109 (8), e1593. Available at https://doi.org/10.1210/clinem/dgad750.

Boogers, L.S., Sikma, B.T., Bouman, M.B., van Trotsenburg, A.S.P., den Heijer, M., Wiepjes, C.M., et al., 2025. Shaping the skeleton: impact of gnrh analogue and sex hormone therapy on skeletal dimensions in transgender individuals. Journal of Clinical Endocrinology and Metabolism 110 (5), e1411. Available at https://doi.org/10.1210/clinem/dgae574.

Carmichael, P., Butler, G., Masic, U., Cole, T.J., de Stavola, B.L., Davidson, S., et al., 2021. Short-term outcomes of pubertal suppression in a selected cohort of 12 to 15 year old young people with persistent gender dysphoria in the UK. PLoS One 16 (2). Available at https://doi.org/10.1371/journal.pone.0243894.

Chen, D., Strang, J.F., Kolbuck, V.D., Rosenthal, S.M., Wallen, K., Waber, D.P., et al., 2020. Consensus parameter: research methodologies to evaluate neurodevelopmental effects of pubertal suppression in transgender youth. Transgender Health 5 (4), 246–257. Available at https://doi.org/10.1089/trgh.2020.0006.

Chiniara, L.N., Viner, C., Palmert, M., Bonifacio, H., 2019. Perspectives on fertility preservation and parenthood among transgender youth and their parents. Archives of Disease in Childhood 104 (8), 739–744. Available at https://doi.org/10.1136/archdischild-2018-316080.

Ciancia, S., Klink, D., Craen, M., Cools, M., 2023. Early puberty suppression and gender-affirming hormones do not alter final height in transgender adolescents. European Journal of Endocrinology 189 (3), 396–401. Available at https://doi.org/10.1093/ejendo/lvad125.

Cohen-Kettenis, P.T., Van Goozen, S.H.M., 1997. Sex reassignment of adolescent transsexuals: a follow-up study. Journal of the American Academy of Child and Adolescent Psychiatry 36 (2), 263–271. Available at https://doi.org/10.1097/00004583-199702000-00017.

Cohen-Kettenis, P.T., van Goozen, S.H.M., 1998. Pubertal delay as an aid in diagnosis and treatment of a transsexual adolescent. European Child & Adolescent Psychiatry 7 (4), 246–248. Available at https://doi.org/10.1007/s007870050073.

Cohen-Kettenis, P.T., Klink, D., 2015. Adolescents with gender dysphoria. Best Practice and Research: Clinical Endocrinology and Metabolism 29 (3), 485–495. Available at https://doi.org/10.1016/j.beem.2015.01.004.

Cohen-Kettenis, P.T., Schagen, S.E.E., Steensma, T.D., De Vries, A.L.C., Delemarre-Van De Waal, H.A., 2011. Puberty suppression in a gender-dysphoric adolescent: a 22-year follow-up. Archives of Sexual Behavior 40 (4), 843–847. Available at https://doi.org/10.1007/s10508-011-9758-9.

Coleman, E., Bockting, W., Botzer, M., Cohen-Kettenis, P., DeCuypere, G., Feldman, J., et al., 2012. Standards of care for the health of transsexual, transgender, and gender-nonconforming people, version 7. International Journal of Transgenderism 13 (4), 165–232. Available at https://doi.org/10.1080/15532739.2011.700873.

Coleman, E., Radix, A.E., Bouman, W.P., Brown, G.R., de Vries, A.L.C., Deutsch, M.B., et al., 2022. Standards of care for the health of transgender and gender diverse people, version 8. International Journal of Transgender Health 23 (1), S1. Available at https://doi.org/10.1080/26895269.2022.2100644.

Costa, R., Dunsford, M., Skagerberg, E., Holt, V., Carmichael, P., Colizzi, M., 2015. Psychological support, puberty suppression, and psychosocial functioning in adolescents with gender dysphoria. The Journal of Sexual Medicine 12 (11), 2206–2214. Available at https://doi.org/10.1111/jsm.13034.

Crowley, W.F., Comitf, F., Vale, W., Rivier, J., Loriaux, D.L., Cutler, G.B., 1981. Therapeutic use of pituitary oesensitization with a long-acting lhrm agonist: a potential new treatment for idiopathic precocious puberty* The Journal of Clinical Endocrinology & Metabolism 52 (2), 370–372. Available at https://doi.org/10.1210/jcem-52-2-370.

De Vries, A.L.C., Steensma, T.D., Doreleijers, T.A.H., Cohen-Kettenis, P.T., 2011. Puberty suppression in adolescents with gender identity disorder: a prospective follow-up study. Journal of Sexual Medicine 8 (8), 2276–2283. Available at https://doi.org/10.1111/j.1743-6109.2010.01943.x.

De Vries, A.L.C., McGuire, J.K., Steensma, T.D., Wagenaar, E.C.F., Doreleijers, T.A.H., Cohen-Kettenis, P.T., 2014. Young adult psychological outcome after puberty suppression and gender reassignment. Pediatrics 134 (4), 696–704. Available at https://doi.org/10.1542/peds.2013-2958.

Eitel, K.B., Hodax, J.K., DiVall, S., Kidd, K.M., Salehi, P., Sequeira, G.M., 2023. Leuprolide acetate for puberty suppression in transgender and gender diverse youth: a comparison of subcutaneous eligard versus intramuscular lupron. Journal of Adolescent Health 72 (2), 307–311. Available at https://doi.org/10.1016/j.jadohealth.2022.09.017.

Fisch-Shvalb, N., Alfandary-Harani, H., Lazar, L., Davidovits, M., Shvalb, N., Demol-Eliaz, S., et al., 2023. Blood pressure in girls with central precocious puberty receiving GnRH analogue therapy. Journal of Pediatric Endocrinology and Metabolism 36 (8), 726–731. Available at https://doi.org/10.1515/jpem-2023-0118.

Fisher, A.D., Ristori, J., Romani, A., Cassioli, E., Mazzoli, F., Cocchetti, C., et al., 2024. Back to the future: Is GnRHa treatment in transgender and gender diverse adolescents only an extended evaluation phase? Journal of Clinical Endocrinology and Metabolism 109 (6), 1565–1579. Available at https://doi.org/10.1210/clinem/dgad729.

Glodowski, M.B., Parra, C.M., Xin, M.K., Fino, M.E., 2024. Reproductive considerations in the transgender and gender diverse population: a review. Endocrinology and Metabolism Clinics of North America 53 (3), 471–482. Available at https://doi.org/10.1016/j.ecl.2024.05.008.

Gooren, L., Delemarre-van de Waal, H., 1996. The feasibility of endocrine interventions in juvenile transsexuals. Journal of Psychology & Human Sexuality 8 (4), 69–74. Available at https://doi.org/10.1300/j056v08n04_05.

Hembree, W.C., Cohen-Kettenis, P., Delemarre-Van De Waal, H.A., Gooren, L.J., Meyer, W.J., Spack, N.P., et al., 2009. Endocrine treatment of transsexual persons: an endocrine society clinical practice guideline. Journal of Clinical Endocrinology and Metabolism 94 (9), 3132–3154. Available at https://doi.org/10.1210/jc.2009-0345.

Hembree, W.C., Cohen-Kettenis, P.T., Gooren, L., Hannema, S.E., Meyer, W.J., Murad, M.H., et al., 2017. Endocrine treatment of gender-dysphoric/gender-incongruent persons: an endocrine society*clinical practice guideline. Journal of Clinical Endocrinology and Metabolism 102 (11), 3869–3903. Available at https://doi.org/10.1210/jc.2017-01658.

Herbison, A.E., 2018. The gonadotropin-releasing hormone pulse generator. Endocrinology 159 (11), 3723–3736. Available at https://doi.org/10.1210/en.2018-00653.

Hodax, J.K., Brady, C., DiVall, S., Ahrens, K.R., Carlin, K., Khalatbari, H., et al., 2023. Low pretreatment bone mineral density in gender diverse youth. Transgender Health 8 (5), 467–471. Available at https://doi.org/10.1089/trgh.2021.0183.

Horton, C., 2023. I didn't want him to disappear" parental decision-making on access to puberty blockers for trans early adolescents. The Journal of Early Adolescence 43 (4), 490–515. Available at https://doi.org/10.1177/02724316221107076.

Horton, C., 2024. Experiences of puberty and puberty blockers: insights from trans children, trans adolescents, and their parents. Journal of Adolescent Research 39 (1), 77–103. Available at https://doi.org/10.1177/07435584221100591.

Hou, L., Ying, Y., Wu, W., Ye, F., Zhang, C., Luo, X., 2024. The effect of GnRHa treatment on body mass index in central precocious puberty: a systematic review and meta-analysis. Hormone Research in Paediatrics 97 (5), 419–432. Available at https://doi.org/10.1159/000535132.

Joseph, T., Ting, J., Butler, G., 2019. The effect of GnRH analogue treatment on bone mineral density in young adolescents with gender dysphoria: findings from a large national cohort. Journal of Pediatric Endocrinology and Metabolism 32 (10), 1077–1081. Available at https://doi.org/10.1515/jpem-2019-0046.

Kelly, A., Winer, K.K., Kalkwarf, H., Oberfield, S.E., Lappe, J., Gilsanz, V., et al., 2014. Age-based reference ranges for annual height velocity in US children. Journal of Clinical Endocrinology and Metabolism 99 (6), 2104–2112. Available at https://doi.org/10.1210/jc.2013-4455.

Kelly, A., Shults, J., Mostoufi-Moab, S., McCormack, S.E., Stallings, V.A., Schall, J.I., et al., 2019. Pediatric bone mineral accrual Z-score calculation equations and their application in childhood disease. Journal of Bone and Mineral Research 34 (1), 195–203. Available at https://doi.org/10.1002/jbmr.3589.

Klaver, M., de Mutsert, R., Wiepjes, C.M., Twisk, J.W.R., den Heijer, M., Rotteveel, J., et al., 2018. Early hormonal treatment affects body composition and body shape in young transgender adolescents. Journal of Sexual Medicine 15 (2), 251–260. Available at https://doi.org/10.1016/j.jsxm.2017.12.009.

Klaver, M., De Mutsert, R., Van Der Loos, M.A.T.C., Wiepjes, C.M., Twisk, J.W.R., Den Heijer, M., et al., 2020. Hormonal treatment and cardiovascular risk profile in transgender adolescents. Pediatrics 145 (3). Available at https://doi.org/10.1542/peds.2019-0741.

Klink, D., Caris, M., Heijboer, A., van Trotsenburg, M., Rotteveel, J., 2015. Bone mass in young adulthood following gonadotropin-releasing hormone analog treatment and cross-sex hormone treatment in adolescents with gender dysphoria. The Journal of Clinical Endocrinology & Metabolism 100 (2), E270. Available at https://doi.org/10.1210/jc.2014-2439.

Krishna, K.B., Witchel, S.F., 2024. Normal and abnormal puberty. In: Feingold, K.R., Adler, R.A., Ahmed, S.F. (Eds.), Endotext [Internet]. MDText.com, Inc., South Dartmouth (MA). Available at https://www.ncbi.nlm.nih.gov/books/NBK279024/.

Lahlou, N., Carel, J.-C., Chaussain, J.-L., Roger, M., 2000. Pharmacokinetics and pharmacodynamics of GnRH agonists: clinical implications in pediatrics. Journal of Pediatric Endocrinology and Metabolism 13 (Supplement). Available at https://doi.org/10.1515/jpem.2000.13.s1.723.

Lai, T.C., McDougall, R., Feldman, D., Elder, C.V., Pang, K.C., 2020. Fertility counseling for transgender adolescents: a review. Journal of Adolescent Health 66 (6), 658–665. Available at https://doi.org/10.1016/j.jadohealth.2020.01.007.

Lazar, L., Meyerovitch, J., de Vries, L., Phillip, M., Lebenthal, Y., 2014. Treated and untreated women with idiopathic precocious puberty: long-term follow-up and reproductive outcome between the third and fifth decades. Clinical Endocrinology 80 (4), 570–576. Available at https://doi.org/10.1111/cen.12319.

Lee, J.Y., 2023. Bone health in the transgender and gender diverse youth population. Current Osteoporosis Reports 21 (4), 459–471. Available at https://doi.org/10.1007/s11914-023-00799-2.

Lee, J.Y., Finlayson, C., Olson-Kennedy, J., Garofalo, R., Chan, Y.M., Glidden, D.V., et al., 2020. Low bone mineral density in early pubertal transgender/gender diverse youth: findings from the trans youth care study. Journal of the Endocrine Society 4 (9). Available at https://doi.org/10.1210/jendso/bvaa065.

Lu, J., Shin, Y., Yen, M.S., Sun, S.S., 2016. Peak bone mass and patterns of change in total bone mineral density and bone mineral contents from childhood into young adulthood. Journal of Clinical Densitometry 19 (2), 180–191. Available at https://doi.org/10.1016/j.jocd.2014.08.001.

Marshall, W.A., Tanner, J.M., 1969. Variations in pattern of pubertal changes in girls. Archives of Disease in Childhood 44 (235), 291–303. Available at https://doi.org/10.1136/adc.44.235.291.

Marshall, W.A., Tanner, J.M., 1970. Variations in the pattern of pubertal changes in boys. Arch Dis Child 45 (239), 13–23. Available at 10.1136/adc.45.239.13.

Marwa, A., Misra, M., Lopez, X., 2022. Determinants of bone mineral density in transgender youth. Transgender Health 7 (3), 213–218. Available at https://doi.org/10.1089/trgh.2020.0111.

McCauley, C.J., McElhinney, K.L., Lockart, B.A., Schmitt, L., Zielen, A., Orwig, K.E., et al., 2025. A report on testicular tissue cryopreservation for fertility preservation in children: confirmation of germ cell preservation and few surgical complications. Journal of Pediatric Surgery 60 (9). Available at https://doi.org/10.1016/j.jpedsurg.2025.162390.

McGregor, K., McKenna, J.L., Williams, C.R., Barrera, E.P., Boskey, E.R., 2024. Association of pubertal blockade at tanner 2/3 with psychosocial benefits in transgender and gender diverse youth at hormone readiness assessment. Journal of Adolescent Health 74 (4), 801–807. Available at https://doi.org/10.1016/j.jadohealth.2023.10.028.

McNamara, M., McLamore, Q., Meade, N., Olgun, M., Robinson, H., Alstott, A., 2024. A thematic analysis of disinformation in gender-affirming healthcare bans in the United States. Soc Sci Med 351, 116943. Available at 10.1016/j.socscimed.2024.116943.

Mejia-Otero, J.D., White, P., Lopez, X., 2021. Effectiveness of puberty suppression with gonadotropin-releasing hormone agonists in transgender youth. Transgender Health 6 (1), 31–35. Available at https://doi.org/10.1089/trgh.2020.0007.

Moussaoui, D., Surbone, A., Adam, C., Diesch-Furlanetto, T., Girardin, C., Bénard, J., et al., 2022. Testicular tissue cryopreservation for fertility preservation in prepubertal and adolescent boys: a 6 year experience from a Swiss multi-center network. Frontiers in Pediatrics 10. Available at https://doi.org/10.3389/fped.2022.909000.

Nahata, L., Chen, D., Quinn, G.P., Travis, M., Grannis, C., Nelson, E., et al., 2020. Reproductive attitudes and behaviors among transgender/nonbinary adolescents. Journal of Adolescent Health 66 (3), 372–374. Available at https://doi.org/10.1016/j.jadohealth.2019.09.008.

Nasomyont, N., Meisman, A.R., Ecklund, K., Vajapeyam, S., Cecil, K.M., Tkach, J.A., et al., 2022. Changes in bone marrow adipose tissue in transgender and gender non-conforming youth undergoing pubertal suppression: a pilot study. Journal of Clinical Densitometry 25 (4), 485–489. Available at https://doi.org/10.1016/j.jocd.2022.06.006.

Navabi, B., Tang, K., Khatchadourian, K., Lawson, M.L., 2021. Pubertal suppression, bone mass, and body composition in youth with gender dysphoria. Pediatrics 148 (4). Available at https://doi.org/10.1542/peds.2020-039339.

Nokoff, N.J., Scarbro, S.L., Moreau, K.L., Zeitler, P., Nadeau, K.J., Reirden, D., et al., 2021. Body composition and markers of cardiometabolic health in transgender youth on gonadotropin-releasing hormone agonists. Transgender Health 6 (2), 111–119. Available at https://doi.org/10.1089/trgh.2020.0029.

Notini, L., Earp, B.D., Gillam, L., McDougall, R.J., Savulescu, J., Telfer, M., et al., 2020. Forever young? The ethics of ongoing puberty suppression for non-binary adults. Journal of Medical Ethics 46 (11), 743–752. Available at https://doi.org/10.1136/medethics-2019-106012.

Ojeda, S.R., Lomniczi, A., Mastronardi, C., Heger, S., Roth, C., Parent, A.S., et al., 2006. Minireview: the neuroendocrine regulation of puberty: Is the time ripe for a systems biology approach? Endocrinology 147 (3), 1166–1174. Available at https://doi.org/10.1210/en.2005-1136.

Olson-Kennedy, J., Streeter, L.H., Garofalo, R., Chan, Y.M., Rosenthal, S.M., 2021. Histrelin implants for suppression of puberty in youth with gender dysphoria: A comparison of 50 mcg/day (Vantas) and 65 mcg/Day (SupprelinLA). Transgend Health 6 (1), 36–42. Available at 10.1089/trgh.2020.0055.

Panagiotakopoulos, L., Chulani, V., Koyama, A., Childress, K., Forcier, M., Grimsby, G., et al., 2020. The effect of early puberty suppression on treatment options and outcomes in transgender patients. Nature Reviews Urology 17 (11), 626–636. Available at https://doi.org/10.1038/s41585-020-0372-2.

Pasquino, A.M., Pucarelli, I., Accardo, F., Demiraj, V., Segni, M., Di Nardo, R., 2008. Long-term observation of 87 girls with idiopathic central precocious puberty treated with gonadotropin-releasing hormone analogs: impact on adult height, body mass index, bone mineral content, and reproductive function. Journal of Clinical Endocrinology and Metabolism 93 (1), 190–195. Available at https://doi.org/10.1210/jc.2007-1216.

Patton, G.C., Viner, R., 2007. Pubertal transitions in health. Lancet 369 (9567), 1130–1139. Available at https://doi.org/10.1016/S0140-6736(07)60366-3.

Peri, A., Ahler, A., Gook, D., O'Connell, M.A., Bourne, H., Nightingale, M., et al., 2021. Predicting successful sperm retrieval in transfeminine adolescents after testicular biopsy. Journal of Assisted Reproduction and Genetics 38 (10), 2735–2743. Available at https://doi.org/10.1007/s10815-021-02293-z.

Peters, B.R., Martin, L.H., Butler, C., Dugi, D., Dy, G.W., 2022. Robotic peritoneal flap vs. perineal penile inversion techniques for gender-affirming vaginoplasty. Current Urology Reports 23 (10), 211–218. Available at https://doi.org/10.1007/s11934-022-01106-9.

Pine-Twaddell, E., Newfield, R.S., Marinkovic, M., 2023. Extended use of histrelin implant in pediatric patients. Transgender Health 8 (3), 264–272. Available at https://doi.org/10.1089/trgh.2021.0130.

Roberge, S., Roberge, T., Corathers, S., Nasomyont, N., 2024. Determinants of bone mass accrual in transgender and gender diverse youth undergoing pubertal suppression therapy. Journal of Clinical Densitometry 27 (3), 101505. Available at https://doi.org/10.1016/j.jocd.2024.101505.

Rosenfield, R.L., Lipton, R.B., Drum, M.L., 2009. Thelarche, pubarche, and menarche attainment in children with normal and elevated body mass index. Pediatrics 123 (1), 84–88. Available at https://doi.org/10.1542/peds.2008-0146.

Roy, M.K., Bothwell, S., Kelsey, M.M., Ma, N.S., Moreau, K.L., Nadeau, K.J., et al., 2024. Bone density in transgender youth on gender-affirming hormone therapy. Journal of the Endocrine Society 8 (5). Available at https://doi.org/10.1210/jendso/bvae045.

Safa, B., Lin, W.C., Salim, A.M., Deschamps-Braly, J.C., Poh, M.M., 2019. Current concepts in feminizing gender surgery. Plastic and Reconstructive Surgery 143 (5), 1081–1091. Available at https://doi.org/10.1097/PRS.0000000000005595.

Schagen, S.E.E., Cohen-Kettenis, P.T., Delemarre-van de Waal, H.A., Hannema, S.E., 2016. Efficacy and safety of gonadotropin-releasing hormone agonist treatment to suppress puberty in gender dysphoric adolescents. Journal of Sexual Medicine 13 (7), 1125–1132. Available at https://doi.org/10.1016/j.jsxm.2016.05.004.

Schagen, S.E.E., Wouters, F.M., Cohen-Kettenis, P.T., Gooren, L.J., Hannema, S.E., 2020. Bone development in transgender adolescents treated with gnrh analogues and subsequent gender-affirming hormones. Journal of Clinical Endocrinology and Metabolism 105 (12). Available at https://doi.org/10.1210/clinem/dgaa604.

Schulmeister, C., Millington, K., Kaufman, M., Finlayson, C., Kennedy, J.O., Garofalo, R., et al., 2022. Growth in transgender/gender-diverse youth in the first year of treatment with gonadotropin-releasing hormone agonists. Journal of Adolescent Health 70 (1), 108–113. Available at https://doi.org/10.1016/j.jadohealth.2021.06.022.

Shuhart, C.R., Yeap, S.S., Anderson, P.A., Jankowski, L.G., Lewiecki, E.M., Morse, L.R., et al., 2019. Executive summary of the 2019 ISCD position development conference on monitoring treatment, DXA cross-calibration and least significant change, spinal cord injury, peri-prosthetic and orthopedic bone health, transgender medicine, and pediatrics. Journal of Clinical Densitometry 22 (4), 453–471. Available at https://doi.org/10.1016/j.jocd.2019.07.001.

Shumer, D.E., Nokoff, N.J., Spack, N.P., 2016. Advances in the care of transgender children and adolescents. Advances in Pediatrics 63 (1), 79–102. Available at https://doi.org/10.1016/j.yapd.2016.04.018.

Sisk, C.L., Zehr, J.L., 2005. Pubertal hormones organize the adolescent brain and behavior. Frontiers in Neuroendocrinology 26 (3-4), 163–174. Available at https://doi.org/10.1016/j.yfrne.2005.10.003.

Staphorsius, A.S., Kreukels, B.P.C., Cohen-Kettenis, P.T., Veltman, D.J., Burke, S.M., Schagen, S.E.E., et al., 2015. Puberty suppression and executive functioning: An fMRI-study in adolescents with gender dysphoria. Psychoneuroendocrinology 56, 190–199. Available at https://doi.org/10.1016/j.psyneuen.2015.03.007.

Strang, J.F., Chen, D., Nelson, E., Leibowitz, S.F., Nahata, L., Anthony, L.G., et al., 2022. Transgender youth executive functioning: relationships with anxiety symptoms, autism spectrum disorder, and gender-affirming medical treatment status. Child Psychiatry and Human Development 53 (6), 1252–1265. Available at https://doi.org/10.1007/s10578-021-01195-6.

Tanner, J.M., 1962. Growth at Adolescence, second ed. Thomas, Springfield, Ill.

Taylor, J., Mitchell, A., Hall, R., Heathcote, C., Langton, T., Fraser, L., et al., 2024. Interventions to suppress puberty in adolescents experiencing gender dysphoria or incongruence: a systematic review. Archives of Disease in Childhood 109. Available at https://doi.org/10.1136/archdischild-2023-326669 s33.

Tornese, G., Di Mase, R., Munarin, J., Ciancia, S., Santamaria, F., Fava, D., et al., 2025. Use of gonadotropin-releasing hormone agonists in transgender and gender diverse youth: a systematic review. Frontiers in Endocrinology 16. Available at https://doi.org/10.3389/fendo.2025.1555186.

Turban, J.L., King, D., Carswell, J.M., Keuroghlian, A.S., 2020. Pubertal suppression for transgender youth and risk of suicidal ideation. Pediatrics 145 (2). Available at https://doi.org/10.1542/peds.2019-1725.

van de Grift, T.C., van Gelder, Z.J., Mullender, M.G., Steensma, T.D., de Vries, A.L.C., Bouman, M.B., 2020. Timing of puberty suppression and surgical options for transgender youth. Pediatrics 146 (5). Available at https://doi.org/10.1542/peds.2019-3653.

van der Loos, M.A.T.C., Hannema, S.E., Klink, D.T., den Heijer, M., Wiepjes, C.M., 2022. Continuation of gender-affirming hormones in transgender people starting puberty suppression in adolescence: a cohort study in the Netherlands. The Lancet Child and Adolescent Health 6 (12), 869–875. Available at https://doi.org/10.1016/S2352-4642(22)00254-1.

van der Loos, M.A.T.C., Hellinga, I., Vlot, M.C., Klink, D.T., den Heijer, M., Wiepjes, C.M., 2021. Development of hip bone geometry during gender-affirming hormone therapy in transgender adolescents resembles that of the experienced gender when pubertal suspension is started in early puberty. Journal of Bone and Mineral Research 36 (5), 931–941. Available at https://doi.org/10.1002/jbmr.4262.

Van Der Loos, M.A.T.C., Vlot, M.C., Klink, D.T., Hannema, S.E., Den Heijer, M., Wiepjes, C.M., 2023. Bone mineral density in transgender adolescents treated with puberty suppression and subsequent gender-affirming hormones. JAMA Pediatrics 177 (12), 1332–1341. Available at https://doi.org/10.1001/jamapediatrics.2023.4588.

Van Der Meulen, I.S., Arnoldussen, M., Van Der Miesen, A.I.R., Hannema, S.E., Steensma, T.D., De Vries, A.L.C., et al., 2025a. Sexual satisfaction and dysfunction in transgender adults following puberty suppression treatment during adolescence. Journal of Sexual Medicine 22 (8), 1493–1503. Available at https://doi.org/10.1093/jsxmed/qdaf095.

Van Der Meulen, I.S., Bungener, S.L., Van Der Miesen, A.I.R., Hannema, S.E., Kreukels, B.P.C., Steensma, T.D., et al., 2025b. Timing of puberty suppression in transgender adolescents and sexual functioning after vaginoplasty. Journal of Sexual Medicine 22 (1), 196–204. Available at https://doi.org/10.1093/jsxmed/qdae152.

van der Miesen, A.I.R., Steensma, T.D., de Vries, A.L.C., Bos, H., Popma, A., 2020. Psychological functioning in transgender adolescents before and after gender-affirmative care compared with cisgender general population peers. Journal of Adolescent Health 66 (6), 699–704. Available at https://doi.org/10.1016/j.jadohealth.2019.12.018.

Vlot, M.C., Klink, D.T., den Heijer, M., Blankenstein, M.A., Rotteveel, J., Heijboer, A.C., 2017. Effect of pubertal suppression and cross-sex hormone therapy on bone turnover markers and bone mineral apparent density (BMAD) in transgender adolescents. Bone 95, 11–19. Available at https://doi.org/10.1016/j.bone.2016.11.008.

Vrouenraets, L.J.J.J., de Vries, A.L.C., de Vries, M.C., van der Miesen, A.I.R., Hein, I.M., 2021. Assessing medical decision-making competence in transgender youth. Pediatrics 148 (6). Available at https://doi.org/10.1542/peds.2020-049643, https://doi.org/10.1542/peds.2020-049643.

Vrouenraets, L.J.J.J., de Vries, A.L.C., Arnoldussen, M., Hannema, S.E., Lindauer, R.J.L., de Vries, M.C., et al., 2023. Medical decision-making competence regarding puberty suppression: perceptions of transgender adolescents, their parents and clinicians. European Child and Adolescent Psychiatry 32 (11), 2343–2361. Available at https://doi.org/10.1007/s00787-022-02076-6.

Willemsen, L.A., Boogers, L.S., Wiepjes, C.M., Klink, D.T., Van Trotsenburg, A.S.P., Den Heijer, M., et al., 2023. Just as tall on testosterone; a neutral to positive effect on adult height of gnrha and testosterone in trans boys. Journal of Clinical Endocrinology and Metabolism 108 (2), 414–421. Available at https://doi.org/10.1210/clinem/dgac571.

Chapter 7

Testosterone and estradiol

Kara J. Connelly and Danielle N. Moyer
Department of Pediatrics, Oregon Health & Science University, Portland, OR, United States

Learning objectives

- Describe the use of gender-affirming hormones—specifically testosterone and estradiol—in pediatric gender-affirming healthcare, including risks and benefits.
- Summarize the research on testosterone and estradiol treatment among transgender and gender diverse youth receiving gender-affirming healthcare.
- Recommend future directions for research, including prospective studies on the benefits and risks of affirming hormones based on timing of initiation, standardized measurement outcomes, and research that explores different pathways of care, such as short-term use or consistent low-dose use of gender-affirming hormones.

Key terms and definitions

- **Gender-affirming hormones** refer here to sex steroid hormones that are used to cause the development of secondary sex characteristics that more closely align with a person's gender identity—specifically testosterone and estradiol. Other hormone treatments used within gender-affirming care, such as progesterone for menstrual suppression, are outside of the scope of this chapter.
- **Gender embodiment goals** include desired changes to the physical body or social interactions to better align with gender identity or reduce gender dysphoria.
- **Appearance congruence** is a measure of how closely an individual perceives their external appearance to align with their gender identity.

Transgender and gender diverse (TGD) individuals, including adolescents, who are interested in developing secondary sex characteristics more aligned with their gender identity may be prescribed gender-affirming testosterone or estradiol. Robust research demonstrates the benefits of these gender-affirming hormones (GAHs) on mental health and well-being, including among youth. As access to GAH among adolescents grows, more long-term research on physical health and risks is also beginning to emerge.

The earliest research on GAH did not differentiate adolescents from adults, as there was no established theoretical reason that the health risks and benefits would differ between the two populations. Furthermore, the establishment of dedicated pediatric gender clinics is a more recent development, and adolescents previously sought care in the same clinics as adults, making up the clinical samples of most research in this area. Therefore, while the research on the risks and benefits of GAH for adolescents may appear more recent, this is actually not the case. In fact, many decades of research, including youth as young as 15 years, have demonstrated lowrisk profiles and long-term positive improvements to mental health and well-being as a result of GAH (Wilson et al., 2021; Doyle et al., 2023; van Leerdam et al., 2023; Nguyen et al., 2018; D'hoore and T'Sjoen, 2022).

Advances in Research on Pediatric Gender Affirming Healthcare. DOI: https://doi.org/10.1016/B978-0-443-29019-0.00007-5
© 2026 Elsevier Inc. All rights are reserved, including those for text and data mining, AI training, and similar technologies.

Overview of testosterone and estradiol

Published clinical practice guidelines provide recommended doses for hormone initiation and titration schedules over time (Coleman et al., 2022; Hembree et al., 2017). Adolescents on gonadotropin-releasing hormone agonists (GnRHa) for pubertal suppression generally start at lower initiation doses than older adolescents who have completed more endogenous puberty. The goal for many TGD individuals prescribed estradiol or testosterone is to ultimately achieve circulating sex steroid hormone levels in the physiologic reference range for cisgender individuals, in the interest of achieving gender embodiment goals. More recently, clinical practice has evolved to be more individualized, and some individuals choose to remain on lower doses or to only take GAH for a period of time until desired permanent changes to secondary sex characteristics have been achieved to satisfaction. Research is needed to clarify differences in physical and mental health outcomes based on this more individualized approach to GAH use.

Gender-affirming testosterone is prescribed to TGD individuals assigned female at birth (AFAB) to help achieve secondary sex characteristics more aligned with gender identity, which may be male, nonbinary, or another specified identity. Some physical changes are reversible: body fat redistribution, suppression/cessation of menstrual cycles, increased muscle mass, and acne. Other physical changes are permanent: voice deepening, growth of the clitoris, and facial and body hair.

Gender-affirming estradiol is prescribed to TGD individuals assigned male at birth (AMAB) to help achieve secondary sex characteristics aligned with gender identity, which may be female, nonbinary, or another specified identity. Some of the physical changes are reversible: redistribution of facial and body fat and reduction of body hair. Other physical changes, such as breast development, are permanent. For some physical changes, namely decreased sperm count, there is limited available evidence of potential reversibility, and clinicians therefore discuss this change as likely permanent.

Dosing methods

Testosterone therapy for TGD youth is most commonly prescribed as intramuscular (IM) or subcutaneous (SC) injections, or as a gel or cream that is applied topically. One observational study investigating the efficacy of testosterone in transgender adults, measured by total testosterone levels and achievement of amenorrhea, showed no significant differences between SC injections, IM injections, or gel (Pelusi et al., 2014). One study randomized adolescent patients to either SC or IM injections and found that testosterone levels and perceived physical effects were similar across the two methods (Baines and Connelly, 2023). Some studies suggest that SC injections are preferred over IM injections as they are less painful and easier to self-administer; SC injection teaching can be done via telehealth (Nightingale et al., 2025). Testosterone transdermal patches are no longer widely available and were discontinued in the United States and Canada in 2022; allergic contact dermatitis was commonly reported (Ta et al., 2014). Testosterone is also available as an oral tablet (testosterone undecanoate); however, at the time of this publication, its use has only been studied in adult cisgender men (Miner et al., 2025).

The 17-β-estradiol form is the most commonly prescribed class of estradiol for inducing secondary sex characteristics in both cisgender and TGD individuals because it is bioidentical to estradiol produced by a human ovary. Ethinyl estradiol is a synthetic form of estradiol that is used in oral contraceptive pills and has a higher risk profile for venous thromboembolic events than 17-β-estradiol. Estradiol can be taken as an oral tablet, transdermal patch, or injection of a conjugated ester (estradiol valerate or estradiol cypionate). Most studies of differences between dosing modalities have been conducted in adult cisgender and transgender populations (Carlson et al., 2025; Misakian et al., 2025). Measurement of efficacy is complicated by differences in estradiol assays and inconsistent dose–response relationships with estradiol levels and clinical effects or testosterone suppression (Slack et al., 2025). Oral estradiol can either be swallowed or dissolved sublingually. A study investigating the pharmacokinetics of sublingual (SL) versus oral estradiol in adult transgender women found that, compared with oral estradiol, SL estradiol produced significantly higher peak serum estradiol levels at earlier times after administration, followed by a fall in estradiol levels, suggesting that multidaily dosing of estradiol may be needed to sufficiently suppress endogenous testosterone production (Doll et al., 2022). A randomized controlled trial comparing SL estradiol taken once or twice daily to transdermal estradiol in adult transgender females found that transdermal estradiol suppressed endogenous testosterone production more effectively compared to SL dosing regimens; all dosing regimens resulted in similar estradiol levels (Cortez et al., 2024).

Studies of TGD adolescents and young adults are more limited. One retrospective study that compared different routes of estradiol administration in TGD adolescents found that transdermal estradiol resulted in lower testosterone levels compared to oral and subcutaneously injected estradiol (Sumerwell et al., 2024). Another small retrospective study in TGD adolescents and young adults receiving IM and SC estradiol injections found that the majority had therapeutic estradiol levels and suppressed testosterone levels, though nearly all were also taking antiandrogen medications (Aaron and Phulwani, 2025). No new adverse effects were reported. Transdermal estradiol is associated with a lower risk of venous thromboembolic events in adults compared to oral pills or injections, whether intramuscular or subcutaneous (King et al., 2024). Studies have also shown greater efficacy of transdermal estradiol in improving bone mineral density in cisgender female adolescent athletes compared to oral contraceptive pills containing ethinyl estradiol (Ackerman et al., 2020; Singhal et al., 2019). However, studies comparing the route of administration to bone health in TGD youth are still needed.

Physical health, body composition, and risk profiles

There are known sex differences in body composition, muscle mass, lipids, and glucose homeostasis that become apparent at the onset of puberty, primarily under the influence of sex steroid hormones (Goossens et al., 2021). Body mass index, body composition, blood pressure, and lipid measurements change somewhat for all adolescents, regardless of sex assigned at birth, as they progress through puberty. Many studies have demonstrated that, over time, laboratory and anthropometric parameters for TGD youth become more closely approximated to those of age-matched cisgender individuals of the same gender. Study conclusions vary by duration of time on GAH, whether patients had received pubertal suppression with GnRHa, and whether outcomes in TGD individuals were compared with age-matched cisgender individuals or a change in the TGD cohort over time.

Anthropometric measurement and body mass index

Research studies are mixed on the incidence of overweight and obesity in TGD youth before the initiation of GAH. One large cohort study of 296 TGD youth aged 12–20 years matched with data from the 2015–16 National Health and Nutrition Examination Survey (NHANES) showed no significant differences in baseline body mass index (BMI) and prevalence of overweight and obesity compared to cisgender youth of the same age and sex assigned at birth (Millington et al., 2020). However, a cross-sectional study of 153 TGD youth aged 8–18 years from the Israeli Children and Adolescents Gender Clinic did show baseline differences in BMI (Borger et al., 2024). Among those AFAB, TGD youth had a higher incidence of overweight and obesity compared to their cisgender peers, whereas among those AMAB, TGD youth had higher variability in BMI and a higher prevalence of underweight compared to their cisgender peers (Borger et al., 2024). A retrospective review of 269 TGD youth similarly showed a higher prevalence of overweight and obesity, especially among those AFAB (Knaus et al., 2024). Differences that occur before the initiation of hormones are most likely due to structural discrimination and other minority stress factors rather than inherent to gender diversity.

In most studies, BMI and BMI standard deviation scores do not change significantly before and after initiating GAH among TGD youth (Knaus et al., 2024; Nokoff et al., 2024; Stoffers et al., 2019). However, some studies have shown small changes, particularly for those AFAB treated with testosterone (Jarin et al., 2017). Some research suggests that initiation of pubertal suppression and GAH at earlier ages may result in body shape and composition that more closely resembles that of cisgender individuals of the same affirmed gender. One study found that, approximately 6 years after initiation of GAH, body shape and composition of young adults AMAB were closely aligned with those of cisgender women, and body shape and composition of young adults AFAB fell between the reference ranges for cisgender men and women (Klaver et al., 2018).

In other words, while some studies show an increase in BMI over time in adolescents AMAB treated with estradiol (Millington et al., 2022), this may reflect changes that are consistent with cisgender adolescent girls. Similarly, a longitudinal study of adolescents AFAB after 12 months of testosterone treatment showed no significant change in weight, but there were increases in fat-free mass and decreases in fat mass, similar to cisgender adolescent boys (Nokoff et al., 2024). This study also showed more substantial changes among youth who had received GnRHa, further suggesting that adolescents who primarily undergo puberty most closely aligned with their gender identity may develop similarly to their cisgender peers of the same gender.

Glucose metabolism

Some studies suggest that there are sex-specific differences in insulin sensitivity and hormones on pancreatic beta cell function (Xu et al., 2019). Among TGD youth, a longitudinal observational study of adolescents showed that after 12 months of testosterone therapy, there were no changes in parameters of insulin sensitivity, regardless of whether or not they were receiving concurrent GnRHa therapy for pubertal suppression (Nokoff et al., 2024). Other studies have produced similar findings, with no change in homeostatic model assessment for insulin resistance (HOMA-IR) or hemoglobin A1c in TGD adolescents treated with testosterone, with some data collected after 8 years of treatment (Stoffers et al., 2019; Klaver et al., 2020; Millington et al., 2024).

Similarly, among TGD youth AMAB, one retrospective study of 27 adolescents treated with cypoterone acetate, an antiandrogenic progestin, and estradiol showed no change in parameters of glucose metabolism or insulin sensitivity after a mean treatment duration of 16 months (Tack et al., 2017). Another retrospective study showed a greater increase in HOMA-IR in transgender women at age 22 years, approximately 8 years after initiating treatment with estradiol, compared to age-matched cisgender women (Klaver et al., 2020).

Blood pressure

Hypertension is more common in TGD youth before initiation of GAH compared to their cisgender peers, suggesting that other factors such as minority stress, mental health, and medication use are more likely causes of differences in blood pressure than gender diversity itself (Winer et al., 2025). Similar to findings in body composition, research shows that blood pressure values either show no significant differences following GAH use or more closely approximate typical values for cisgender individuals of the same affirmed gender (Stoffers et al., 2019; Jarin et al., 2017; Klaver et al., 2020; Hannema et al., 2017; Perl et al., 2020).

Lipids

Most studies looking at changes in lipids in individuals AFAB have found that high density lipoprotein (HDL) levels decrease after initiation of testosterone therapy (Stoffers et al., 2019; Valentine et al., 2021). A later retrospective study showed that at the age of 22, approximately 8 years after initiation of testosterone therapy, individuals AFAB had lipid levels similar to those of cisgender men at the same age, with the exception of slightly higher HDL levels (Klaver et al., 2020). One study examined changes in lipoprotein subtypes in adolescents AFAB after treatment with testosterone for a mean duration of 1.2 years and found that TGD adolescents treated with testosterone had higher concentrations of smaller low density lipoprotein (LDL) particles and lower concentrations of HDL particles compared to age-matched cisgender females. This lipoprotein profile may suggest a more atherogenic cardiovascular profile, more closely aligned with that of cisgender men than cisgender women.

In contrast to testosterone, estradiol treatment appears to have a somewhat favorable impact on lipid levels in individuals AMAB, leading to lower total cholesterol and LDL levels, and higher HDL compared to age-matched cisgender women (Klaver et al., 2020). In this study, estradiol did result in an increase in triglycerides over time; however, this was not significantly different from cisgender women. Together, this research shows that GAH use among TGD youth follows a similar pattern of some negative lipid changes with testosterone but not estrogen, but long-term research will help to clarify if there are any differences based on age of initiation or duration of use (Winer et al., 2025).

Hemoglobin and hematocrit

A retrospective analysis of 25 adolescents treated with lynestrenol (an androgenic progestin) and testosterone for a mean duration of 11.4 months showed that testosterone increases hemoglobin and hematocrit, shifting the values into the reference range for cisgender men but not exceeding it (Tack et al., 2016). A larger study of 200 adolescents AFAB treated with testosterone showed a progressive increase in hemoglobin and hematocrit over the 24-month observation period, with higher serum testosterone levels associated with higher hematocrit (Millington et al., 2024). Hemoglobin and hematocrit levels exceeding the typical reference range for cisgender males are atypical, with a less than 7% incidence of hematocrit >50% (Jarin et al., 2017; Millington et al., 2024).

Studies have consistently shown a decrease in hemoglobin and hematocrit in adolescents AMAB treated with estradiol, with values trending toward typical measurements for cisgender females. This change is presumed to be due to lower testosterone levels, as testosterone is known to stimulate production of erythropoietin and hemoglobin (Millington et al., 2024; Olson-Kennedy et al., 2018).

Renal function and serum creatinine

Serum creatinine is often used as a surrogate marker of glomerular filtration rate (GFR) and renal function. It is important in determining the degree of kidney disease and injury, dosing of medications, assessing need for dialysis, and, in some cases, can play a role in determining eligibility for kidney transplants. Due to challenges in direct measurement of GFR, equations using sex or gender-based coefficients to calculate estimated GFR (eGFR) are often used instead. Factors influencing GFR and eGFR are not completely understood. Serum creatinine can be influenced by muscle mass and other hormone-related factors. TGD adolescents AFAB treated with testosterone have higher lean muscle mass than cisgender female adolescents (Nokoff et al., 2021). However, research is lacking on the direct role of sex steroids and body composition on GFR, creatinine, and eGFR.

A study of 194 adolescents AFAB showed an increase in serum creatinine over the first 12 months of treatment (Millington et al., 2022). At that time point, serum creatinine was more similar to that of cisgender males than cisgender females. The study also found a positive association between increases in serum creatinine and increases in serum testosterone levels. In contrast, a study of 92 individuals AMAB treated with estradiol showed a slight decrease in serum creatinine over the first 6 months of treatment, with no further changes following the 6-month visit (Millington et al., 2022). After 12 months of estradiol the serum creatinine of individuals AMAB on estradiol more closely approximated the mean for cisgender females than cisgender males.

Across multiple studies, liver enzymes alanine transaminase (ALT), aspartate aminotransferase (AST), and gamma-glutamyl transferase (GGT) showed no clinically significant difference after treatment with testosterone or estradiol (Jarin et al., 2017; Millington et al., 2024; Olson-Kennedy et al., 2018). Some studies found slight changes that mimicked changes seen in cisgender individuals of the same affirmed gender (Stoffers et al., 2019; Tack et al., 2016).

Hyperprolactinemia

Clinical practice guidelines recommend monitoring prolactin levels in individuals AMAB receiving estradiol treatment based on early studies in adults suggesting that estradiol leads to hyperprolactinemia and occasionally prolactinomas (Hembree et al., 2017). More recently, studies have demonstrated that the higher prolactin levels seen in transgender women in earlier studies were more likely due to concurrent treatment with cyproterone acetate, an antiandrogenic progestin (Nota et al., 2017; Defreyne et al., 2017). Cyproterone acetate is no longer part of most dosing protocols in current clinical practice guidelines. More recent studies in adolescents AMAB treated with estradiol have found either no change in prolactin (Jarin et al., 2017) or statistical increases in prolactin not considered to be clinically increased (Olson-Kennedy et al., 2018). A study of 110 adolescents AMAB treated with estradiol found that only three individuals had prolactin levels ≥20 ng/mL at any point in the study period, and all three were on additional medications known to cause elevation in prolactin (Millington et al., 2024). Two of the three had follow-up prolactin levels <20 ng/mL at the 24-month follow-up.

Long-term health risks

Long-term risks associated with GAH treatment in TGD individuals are largely derived from evidence in TGD adults. Research on long-term health outcomes in TGD populations should always be interpreted within the context of minority stress and healthcare discrimination. Research has shown that TGD individuals, including adolescents, often avoid or lack access to preventive medical care, which can lead to progression or worsening of certain health problems (Rider et al., 2018). One of the greatest risks of long-term negative health outcomes is avoidance of preventative screening, whether due to healthcare discrimination, lack of access, or avoidance of procedures that may worsen gender dysphoria due to the specific focus on sex-specific aspects of the body.

Cancer

Some early studies suggested an increase in certain cancers in TGD adults treated with GAH. However, subsequent studies demonstrate that increased risk can be attribquted to additional medications, such as cyproterone acetate, or other hormone formulations (i.e., ethinyl estradiol). In addition, TGD individuals may avoid or decline screening for body parts that do not align with their gender identity, which could lead to delayed diagnoses and treatment of oncologic conditions.

A large retrospective cohort study with long-term follow-up of TGD individuals in the Netherlands treated with GAH demonstrated that transgender men treated with testosterone had a lower breast cancer risk compared with cisgender women (De Blok et al., 2019). Transgender women treated with estradiol, and in some cases, antiandrogens, including cyproterone acetate or spironolactone, had increased breast cancer risk compared to cisgender men, but lower risk compared to cisgender women. A systematic review assessing breast cancer risk in individuals AFAB treated with testosterone found that the incidence of breast cancer is significantly lower compared to cisgender women, but that the age of diagnosis was younger in TGD individuals (Gurrala et al., 2023). In fact, many breast cancer diagnoses occurred incidentally at the time of gender-affirming mastectomy, potentially indicating a protective factor based on early identification.

A retrospective cohort study of nearly 2000 individuals AFAB undergoing hysterectomy and/or oophorectomy at a median age of 24 years and a median duration of testosterone use of 1.7 years showed no cases of endometrial, ovarian, vulvar, or vaginal cancer (Vestering et al., 2025). A systematic review of 17 studies with an average age of 28.5 years for participants and average duration of testosterone use of 30.2 months showed that while patients AFAB on testosterone have equal rates of atrophic and proliferative endometrial lining, they do not have an increased risk of precancerous or cancerous endometrial lesions compared to cisgender populations (Toland et al., 2024). This review also demonstrated that the incidence of precancerous ovarian lesions is low.

Studies have generally shown a lower incidence of prostate cancer for TGD individuals on gender-affirming estradiol therapy compared with cisgender men, and an incidence of testicular cancer comparable to that of cisgender men (De Blok et al., 2019).

Venous thromboembolism

Studies in adult peri- and postmenopausal cisgender women have shown increased overall risk of venous thromboembolism (VTE) in individuals taking estrogen, with some studies showing higher risk associated with increased estrogen dose or when estrogen treatment was combined with a progestin (Nelson et al., 2002). The body of literature on the risk of VTE in TGD individuals on estradiol is growing; however, it is still mainly limited to studies in adults. Current research in TGD adults demonstrates a lower risk of VTE with transdermal estradiol compared to oral and injectable, and a higher risk of VTE in individuals with BMI >35 mg/m^2 or nicotine use (King et al., 2024).

Emerging research among TGD youth suggests a very low risk for VTE associated with GAH use, at least into young adulthood. One large retrospective chart review of TGD young adults who were prescribed GAH in adolescence found no difference in VTE compared to the young adults who had never been prescribed GAH (Baskaran et al., 2024). Only 3 of 1860 patients across both groups had any history of VTE, and all three were within the context of known VTE risk factors, and only one of the three occurred anytime after initiation of GAH. As time progresses and these patients become older, long-term outcomes will further clarify whether there are any differences in VTE risk based on GAH use in adolescence.

Mental health, body satisfaction, and well-being

Decades of research in adult TGD populations that included adolescents as young as 15 years have consistently found that gender-affirming testosterone and estradiol lead to improvements in anxiety, depression, suicidal ideation, gender dysphoria, body satisfaction, social distress, self-esteem, quality of life, and life satisfaction (Wilson et al., 2021; Doyle et al., 2023; van Leerdam et al., 2023; Nguyen et al., 2018; D'hoore and T'Sjoen, 2022). While results have varied with regard to the degree of improvement, decreasing depression and overall distress has had the strongest evidence (Doyle et al., 2023). The results of longitudinal cohort studies have demonstrated more consistent results than cross-sectional studies (van Leerdam et al., 2023).

Over time and as international standards of care have begun to more explicitly outline treatment recommendations for adolescents (Coleman et al., 2012, 2022), the number of research studies specific to young people has increased substantially (Chew et al., 2018; Taylor et al., 2024; Connolly et al., 2016; Mahfouda et al., 2019). The earliest research studies provided preliminary, qualified support for the mental health benefits of hormones for youth with an eye toward exciting large-scale research initiatives that were beginning to emerge (Chew et al., 2018; Connolly et al., 2016; Mahfouda et al., 2019), leading to the surge of research that has been conducted and disseminated in the last several years. Some critics have attacked this body of research specifically on the basis of a lack of randomized controlled trials (RCTs), despite finding moderate and high-quality studies within the larger body of scientific evidence (Taylor et al., 2024). Unfortunately, these critiques are not scientifically based, but rather are motivated by political opposition to gender-affirming care (McNamara et al., 2024). While limitations to the body of evidence on gender-affirming hormone use in adolescents do exist and will be discussed, neither the absence of RCTs is a limitation, nor are RCTs recommended for future research due to the fact that they are methodologically and ethically inappropriate (Ashley et al., 2024).

As with research that has included a mix of adolescents and adults, research on the benefits of GAH in youth is largely made up of longitudinal cohort studies, with or without a comparison group, and cross-sectional studies with comparison groups. The most common mental health outcomes that have been studied are anxiety, depression, suicidality, gender dysphoria, and body dissatisfaction, with some studies examining other outcomes such as well-being, quality of life, self-esteem, and life satisfaction. Most studies examine outcomes such as depression and anxiety individually, although some have used composite outcomes, such as internalizing symptoms, severe distress, or psychosocial functioning. Nearly every study has found positive improvements in mental health, body satisfaction, and well-being, with few studies having either mixed results (e.g., significant improvement in self-reported depression but not clinician-reported symptoms) or improvements that do not reach the level of statistical significance. No studies have demonstrated significant negative outcomes in these areas.

Longitudinal research methods

The most robust research comes from the longitudinal cohort studies. At the time of this writing, and with the knowledge that there are many research studies currently underway and likely to be published in the near future, 20 peer-reviewed, longitudinal cohort studies on the mental health, body satisfaction, and quality-of-life benefits of gender-affirming hormone use in youth are included in this summary. See Table 7.1 for a list of these studies with the results related to the most commonly studied mental health benefits.

The earliest known research studies specifically on youth were conducted by a group in the Netherlands, first at the University Medical Center Utrecht and later at VU University Medical Center in Amsterdam, which has continued to conduct this research and is now known as the Center of Expertise on Gender Dysphoria (CEGD) (Cohen-Kettenis and Van Goozen, 1997; Smith et al., 2001; De Vries et al., 2014; Arnoldussen et al., 2022a,b). These studies demonstrate the benefits of gender-affirming treatment overall, including hormones, and typically at least 1 year following surgery. By evaluating mental health and well-being at this point in time, these studies show long-term outcomes (up to 7 years after

TABLE 7.1 Longitudinal cohort studies on the mental health benefits of gender-affirming hormones.

First author	Year	Country	Sample	DEP	SI	ANX	GD	BD
Cohen-Kettenis and Van Goozen (1997)	1997	The Netherlands	Clinic				↓	↓
Smith et al. (2001)	2001	The Netherlands	Clinic	↓		↓	↓	↓
De Vries et al. (2014)	2014	The Netherlands	Clinic	~		~	↓	↓
Arnoldussen et al. (2022b)	2022	The Netherlands	Clinic					↓
Arnoldussen et al. (2022a)	2022	The Netherlands	Clinic					↓
Allen et al. (2026, 2019)[a]	2019/2025	United States	Clinic		↓			
Cantu et al. (2020)	2020	United States	Clinic	~		~		
Achille et al. (2020)	2020	United States	Clinic	↓	↓			
de Lara et al. (2020)	2020	Spain	Clinic	↓		↓	↓	
Kaltiala et al. (2020)	2020	Finland	Clinic	↓	↓	↓		
Kuper et al. (2020)	2020	United States	Clinic	~		↓		↓
Chelliah et al. (2024)	2024	United States	Clinic	↓		↓		↓
Becker-Hebly et al. (2021)	2021	Germany	Clinic					
Tordoff et al. (2022)	2022	United States	Clinic	↓	↓	~		
Hisle-Gorman et al. (2021)	2021	United States	Database					
Chen et al. (2023)	2023	United States	Clinic	↓		↓		↓
Chen et al. (2024)	2024	United States	Clinic					
Olson-Kennedy et al. (2025)	2025	United States	Clinic					↓
Wittlin et al. (2025)	2025	United States	Community	~		~		
Lavender et al. (2023)	2023	United Kingdom	Clinic		↓		↓	~

Note: Study outcomes are only reported when distinct measures were used, and some studies that have shown improvements to these outcomes will not be reflected in this table due to the use of composite outcomes. *ANX*, Anxiety; *BD*, body dissatisfaction; *DEP*, depression; *GD*, gender dysphoria; *SI*, suicidal ideation; ↓, decrease/improvement in this area; ~, either mixed results (e.g., improved self-report but not clinician-report of symptoms) or positive improvement that did not reach statistical significance.
[a] *2025 study represents a follow-up to the 2019 study and includes a substantially larger sample size over a longer observation period*

initiating hormones), though using this time point creates difficulty in discerning the specific impact of hormones distinct from other gender-affirming interventions. In their landmark study, this research group assessed gender dysphoria and psychological functioning after access to pubertal suppression, GAH, and surgery (De Vries et al., 2014). Over the course of the study, overall psychological functioning steadily improved, and at the completion of the study, quality of life and subjective happiness were similar to, or better than, those of age-matched peers. Gender dysphoria and body image difficulties, while not initially improving with pubertal suppression alone, significantly improved following GAH and surgery.

Since that landmark study, many more research groups have disseminated outcomes, largely across the United States and Europe. There are several large-scale, longitudinal research studies on TGD youth overall that have published findings specific to the mental health benefits of affirming hormone use. The Trans Youth Care—United States study is one of the largest prospective studies of TGD youth seeking care at one of four major US pediatric gender clinics, which is now disseminating the results from up to 4 years after initiating GAH (Chen et al., 2023, 2024; Olson-Kennedy et al., 2025). Other large-scale research initiatives include the Trans Youth Longitudinal study (Kuper et al., 2020; Chelliah et al., 2024) and the TransYouth Project (Wittlin et al., 2025). Notably, the Trans Youth CAN! study (Bauer et al., 2021a) is a prospective study underway across pediatric clinics in Canada, although mental health outcomes related to the use of GAH remain forthcoming. These research initiatives are expected to continue to provide new insights over time as the length of time that participant cohorts have been on hormones continues to expand.

The majority of longitudinal studies have used clinic-based samples, although one study is based on robust community recruitment (Wittlin et al., 2025), and another utilized a large database of youth seen within military hospitals (Hisle-Gorman et al., 2021). The Trans Youth Care–United States study and the research out of the Netherlands have presented results of hormone use up to 4 and 7 years, respectively, while most other studies have presented the mental health benefits of hormones after approximately 1–2 years. One study that was published in 2019 on improved suicidality in a small cohort of 47 TGD youth recently published an updated follow-up study, demonstrating similar results in a substantially larger sample and over a longer observational period, showing the robust nature of these results over time (Allen et al., 2019, 2026). The results of these 20 studies consistently show improved anxiety, depression, suicidality, gender dysphoria, and body dissatisfaction.

In addition to following patients over time, some studies have used comparison groups to further clarify the mental health benefits of GAH, including TGD youth not prescribed hormones (Achille et al., 2020; Becker-Hebly et al., 2021; Tordoff et al., 2022), or cisgender controls (de Lara et al., 2020; Wittlin et al., 2025). For example, the TransYouth Project did not see improvements in anxiety and depression that reached significance, but found that while the TGD youth demonstrated consistent depression and anxiety over time, cisgender girls demonstrated worsening mental health and cisgender boys demonstrated improvements in mental health (Wittlin et al., 2025). This particular study cohort is known to have a high level of support, early access to gender affirmation, and, therefore, lower baseline mental health concerns. Together, this information suggests that the lack of significant improvement may actually be a positive effect for some (i.e., preventing worsening of distress) and is also likely influenced by better baseline mental health compared to some other groups of TGD youth. Other longitudinal studies have been able to show that mental health not only improves over time but can reach a level consistent with age-matched peers following the use of GAH (De Vries et al., 2014; de Lara et al., 2020; Becker-Hebly et al., 2021).

Researchers are also beginning to look at mediating and moderating factors within longitudinal cohorts to clarify risk and protective factors and mechanisms of change, in addition to overall mental health and quality-of-life outcomes. For example, appearance incongruence, or how closely an individual perceives their external appearance as aligning with their gender identity, has emerged as a likely mechanism of change toward improved mental health and well-being (Chen et al., 2023; Olson-Kennedy et al., 2025). Similar outcomes were found in two cross-sectional studies that used functional MRI data to show that these relationships may be associated with neural connectivity and down-regulation of emotions (Grannis et al., 2021, 2023). In other words, improvements in appearance congruence and body satisfaction, such that an individual's body more closely aligns with their gender identity, may lead to improved mental health by decreasing the executive functioning load and supporting emotion regulation.

Parental support is another important contextual factor that has been found to impact the degree to which GAH leads to positive mental health improvements (Chen et al., 2024). Specifically, adolescents with poor mental health at baseline but with strong parental support are those most likely to see the positive mental health benefits over the course of taking hormones, while those without parental support tend to improve less quickly or not at all. In other words, in a supportive home environment, adolescents are able to gain the most benefits from hormones. Unsupportive environments, on the contrary, appear to be harmful enough to temper the potential mental health benefits of GAH.

While most studies have examined testosterone and estradiol together, some studies have attempted to explore differences in outcomes between the two hormones. For example, one study showed a reduction in depression and suicidal ideation over time after initiating either hormone, but a greater impact among transfeminine patients treated with both pubertal suppression and estradiol (Achille et al., 2020). Another study found that while most outcomes improved on either treatment, conduct problems specifically improved in transmasculine patients after testosterone, whereas there was no significant change among transfeminine patients after estrogen (Arnoldussen et al., 2022b). These differences may be due to the different hormonal effects, or they may be due to differences in those assigned female versus male at birth at baseline (Grannis et al., 2023).

Cross-sectional research methods

The primary limitations of longitudinal research, such as that reviewed above, include barriers to isolating the effects of hormones and the length of time on hormones that can be assessed. Longitudinal studies, by design, take time before long-term outcomes can be evaluated. Furthermore, GAH use often occurs along with, or in sequence with, other affirming interventions, making it difficult to clarify the mental health benefits of hormones alone, distinct from other treatments. Cross-sectional studies with comparison groups provide helpful supplemental information to complement longitudinal research. Ten such studies are reviewed here. See Table 7.2 for a list of cross-sectional studies with results related to the most commonly studied mental health benefits.

Cross-sectional studies address some of the limitations of longitudinal designs because they can employ more diverse sampling strategies, examine long-term outcomes without following people over time, and assess outcomes specific to a particular intervention. For example, by comparing youth who are not taking affirming hormones to those who are, cross-sectional studies have been able to demonstrate lower levels of anxiety, depression, and suicidality in the latter group, distinguishing the impact of hormones from that of surgery or other interventions (Grannis et al., 2021, 2023; Fontanari et al., 2020; Nunes-Moreno et al., 2025; Olsavsky et al., 2023). Community recruitment is also more common in cross-sectional studies (Fontanari et al., 2020; Green et al., 2022; Turban et al., 2022; Lee et al., 2024).

While these cross-sectional studies largely examined similar outcomes, such as anxiety, depression, and gender dysphoria, one study found better executive functioning among adolescents prescribed affirming hormones compared to those who desired hormones but lacked access (Strang et al., 2022). This is a novel outcome that will benefit from replication in ongoing longitudinal cohort studies. Only one longitudinal study has studied the potential impact of GAH on cognitive functioning, which showed that hormonal treatment did not change the correlation between intelligence (pretreatment) and educational attainment (posttreatment) (Arnoldussen et al., 2022a). This study also found that while

TABLE 7.2 Cross-sectional studies on the mental health benefits of gender-affirming hormones.

First author	Year	Country	Sample	DEP	SI	ANX	GD	BD
Arcelus et al. (2016)	2016	United Kingdom	Clinic					
Fontanari et al. (2020)	2020	Brazil	Community	↓		↓	~	
Green et al. (2022)	2022	United States	Community	↓	↓			
Nunes-Moreno et al. (2025)	2025	United States	Database		↓			
Olsavsky et al. (2023)	2023	United States	Clinic	↓	~	↓		
Grannis et al. (2021)	2021	United States	Clinic	↓	~	↓		↓
Grannis et al. (2023)	2023	United States	Clinic	↓	↓	~		↓
Turban et al. (2022)	2022	United States	Community		↓			
Lee et al. (2024)	2024	United States	Community		↓			
Strang et al. (2022)	2022	United States	Clinic					

Note: Study outcomes are only reported when distinct measures were used, and some studies that have shown improvements to these outcomes will not be reflected in this table due to the use of composite outcomes. *ANX*, Anxiety; *BD*, body dissatisfaction; *DEP*, depression; *GD*, gender dysphoria; *SI*, suicidal ideation; ↓, decrease/improvement in this area; ~, either mixed results (e.g., improved self-report but not clinician-report of symptoms) or positive improvement that did not reach statistical significance intelligence was comparable to the Dutch national norm, educational attainment was actually higher than the national norm, suggesting the potential for a positive effect from hormones. However, given that the pretreatment and posttreatment measures were different, and intelligence in particular would be unlikely to change as a result of intervention, more research is needed to clarify the specific cognitive effects of GAH use.

Cross-sectional studies using large community samples have also been able to clarify that the particular timing of hormone initiation is relevant to long-term mental health outcomes. Specifically, the 2015 US Trans Survey asked about the timing of hormone initiation and current mental health and received over 27,000 responses. Two studies analyzed the results in slightly different ways, and both found that access to hormones in adolescence was related to significantly lower lifetime levels of severe distress and suicidality compared to access in adulthood or no access at all (Turban et al., 2022; Lee et al., 2024). In fact, those who accessed hormones at the age of 14 or 15 reported significantly better current mental health even than those who accessed hormones at the age of 16 or 17 (Turban et al., 2022). While cross-sectional studies are limited in their ability to demonstrate causation, and it is likely that other factors also contributed to this result (e.g., youth accessing hormones at earlier ages likely also have more parent support), large sample sizes and diverse methodology provide further support for the causal outcomes established in longitudinal designs.

Measurement tools

Across both longitudinal cohort and cross-sectional studies a variety of self-report measurements have been used to evaluate outcomes. For anxiety and depression, some studies have used parallel measures of the two, such as the Patient Health Questionnaire for depression (PHQ-9) (Kroenke et al., 2001) and anxiety (GAD-7) (Spitzer et al., 2006) or the Patient Reported Outcomes Measurement Information System (PROMIS) measures of depression and anxiety (Irwin et al., 2010). Other individual measures of depression include the Beck Depression Inventory (BDI-II) (Beck et al., 1996), the Children's Depression Inventory (CDI) (Kovacs, 1985), and the Center for Epidemiological Studies Depression Scale (CESD-R) (Eaton et al., 2004; Haroz et al., 2014). Other individual measures of anxiety include the Screen for Anxiety Related Emotional Disorders (SCARED) (Birmaher et al., 1999), the State-Trait Anxiety Inventory (STAI) (Spielberger et al., 1983), and the Revised Children's Manifest Anxiety Scale (RCMAS-2) (Reynolds and Richmond, 2008). Individual measures of suicidal ideation and behaviors that have been used include the Ask Suicide Screening Questionnaire (ASQ) (Horowitz et al., 2012), the Suicidal Behavior Questionnaire-Revised (SBQ-R) (Osman et al., 2001), the Columbia-Suicide Severity Rating Scale (C-SSRS) (Posner et al., 2011), and the Self-Injury Questionnaire-Treatment Related (SIQ-TR) (Claes and Vandereycken, 2007).

Instead of individual measures of anxiety and depression, some studies have used larger composite measures to evaluate psychological distress and other outcomes. The most common composite measures used across all studies were those from the Achenbach System of Empirically Based Assessment (ASEBA), including the Youth Self-Report (YSR), and the parent-report Child Behavior Checklist (CBCL) (Achenbach, 2009). These studies often grouped distress as internalizing and externalizing symptoms, consistent with the results of these measures. While most studies have evaluated anxiety, depression, suicidality, and other aspects of distress, fewer studies have examined changes to positive aspect of mental health such as quality of life (Achille et al., 2020; Becker-Hebly et al., 2021), self-esteem and self-worth (Cohen-Kettenis and Van Goozen, 1997; Arnoldussen et al., 2022a,b), well-being (Allen et al., 2019), life satisfaction (Chen et al., 2023), and overall psychosocial functioning (De Vries et al., 2014; Chelliah et al., 2024; Becker-Hebly et al., 2021).

The most commonly used measure of body-related satisfaction and dissatisfaction is the Body Image Scale (BIS) (Lindgren and Pauly, 1975). The original measure created in 1975 has two versions based on either sex or gender (depending on the transition steps taken). Both versions ask respondents to rate their satisfaction with 30 body parts, which were categorized as primary sex characteristics, secondary sex characteristics, and so-called "neutral" characteristics. The accuracy and psychometric validity of that factor structure have been disputed; however, the measure has also been criticized for its binary distinction (Strauss et al., 2024). Some researchers have attempted to update the measure in a variety of ways, including one research group that combined the versions into one 33-item measure meant to be more inclusive and useful across time, calling it the Body Image Scale—Gender Spectrum (BIS-GS) (McGuire et al., 2016).

One of the earliest and most commonly used measures of gender dysphoria is the Utrecht Gender Dysphoria Scale (UGDS) (Cohen-Kettenis and Van Goozen, 1997). This 12-item measure has also been criticized for creating two binary versions based on sex, in addition to having asymmetrical questions across versions (Strauss et al., 2024). The same research group that revised the BIS also has a revised version of this measure, the Utrecht Gender Dysphoria Scale—Gender Spectrum (UGDS-GS), which has demonstrated strong psychometric properties and may be a better choice for researchers in the future (McGuire et al., 2020). Another promising measure that has been used to evaluate outcomes is the Transgender Congruence Scale (TCS), which asks 12-questions across two subscales measuring appearance congruence and gender-identity acceptance (Kozee et al., 2012). The appearance congruence subscale, in particular, has been used to measure the degree to which youth perceive their appearance to be congruent or incongruent with their gender identity.

Finally, the Trans Youth CAN! research group has developed two measures: the Gender Distress Scale (TYC-GDS) (Bauer et al., 2021c) and the Gender Positivity Scale (TYC-GPS) (Bauer et al., 2021b). The TYC-GDS asks 14-items across two subscales measuring social distress and body-related distress. It has two versions based on affirmed gender, which has been seen as an improvement over measures based on sex assigned at birth, but can nonetheless be exclusionary of nonbinary identities. The TYC-GPS only has one version for all genders, and it asks 11 questions across two subscales measuring positive social experiences and body-specific positivity. Unlike previous measures described above, which were developed among adults and subsequently adapted and validated among youth, the TYC-GDS and TYC-GPS were developed explicitly among youth and benefited greatly from having TGD youth included in the measurement development process.

Summary and future research directions

In summary, research as far back as 1975 has demonstrated that GAHs result in improved mental health and quality of life for TGD adolescents, and research on youth specifically has increased substantially in the last decade with similarly positive results. Longitudinal cohort studies and cross-sectional designs consistently demonstrate that access to affirming hormones is associated with improvements in anxiety, depression, suicidality, gender dysphoria, body dissatisfaction, quality of life, and life satisfaction. No studies have ever demonstrated worsening mental health as a result of hormone use among TGD youth, but there are important contextual factors that can facilitate or impede positive change. Most notably, youth who are lacking parent support for their identity and embodiment goals may not benefit from hormones to the same magnitude as youth who do have parental support. Furthermore, improvements to mental health and well-being appear to be mediated by improvements in gender appearance congruence and body satisfaction, which may facilitate better executive functioning and emotion regulation.

With this well-established and ongoing body of research consistently demonstrating mental health benefits, researchers can now turn toward these more nuanced research questions about the mediators and moderators of change and the specific neuropsychological processes that facilitate these relationships. Cross-institutional collaborations and standardized methodology, particularly with regard to measurement tools, will be helpful in comparing more nuanced results across studies. Future research should also consider any differences in the mental health improvements between testosterone and estradiol. While RCTs are ethically and methodologically inappropriate for studying GAH in adolescents, the well-designed studies summarized here have produced robust and consistent results. This knowledge base will continue to benefit from the ongoing results of large-scale, longitudinal cohort studies.

Research further shows low risk of GAH impact on physical health among TGD adolescents, although long-term outcomes with large sample sizes are still forthcoming as access to GAH at younger ages continues to improve. Baseline differences in physical health suggest that factors other than testosterone or estradiol, such as minority stress and healthcare discrimination, are more likely causes of differences in physical health outcomes seen in early research among adult TGD populations. Where physical health differences are seen among TGD youth following GAH, changes appear more consistent with cisgender individuals of the same affirmed gender rather than increasing over and above what is expected in the general population.

As access to GAH for TGD adolescents continues to grow, long-term physical health outcomes will become available, allowing for clarification of any differences compared to TGD individuals who initiated GAH in adulthood. Prospective research studies that specifically compare physical and mental health outcomes based on timing of GAH initiation and duration of use are also needed. There is no theoretical reason why physical health outcomes would be more unfavorable for TGD individuals initiating GAH in adolescence compared to adulthood. In fact, long-term outcomes may prove more favorable for those who initiated GAH in adolescence, which inherently requires greater parental support and fewer barriers to accessing gender-affirming healthcare. Retrospective research has shown more favorable mental health outcomes among those initiating GAH sooner, likely for these reasons. Prospective studies are needed to replicate this finding and potentially control for other factors, such as parental support and resources, to clarify the causal relationship to GAH specifically.

Research on physical health outcomes should incorporate measures of minority stress and barriers to healthcare access into methodology, and results should be interpreted within a minority stress framework. More research is needed to clarify whether there are differences in physical health or mental health outcomes when GAH is prescribed in more individualized ways, including consistent low-dose or short-term use to achieve select physical changes most consistent with gender embodiment goals. As with all aspects of pediatric gender-affirming healthcare and the experiences of TGD youth, community-based research methods that take into consideration the priorities and outcomes of interest of these youth and families are considered optimal for ethical, informative, and methodologically robust research outcomes on gender-affirming testosterone and estradiol for youth.

References

Aaron, M.S., Phulwani, P., 2025. Single-center retrospective analysis of safety and efficacy of subcutaneous estradiol use in transgender and nonbinary adolescents and young adults. Transgender Health 10 (4), 316–324. Available at https://doi.org/10.1089/trgh.2024.0264.

Achenbach, T.M., 2009. The Achenbach System of Empirically Based Assessment (ASEBA): Development, Findngs, Theory, and Applications. University of Vermont, Research Center for Children, Youth, and Families, Burlington.

Achille, C., Taggart, T., Eaton, N.R., Osipoff, J., Tafuri, K., Lane, A., et al., 2020. Longitudinal impact of gender-affirming endocrine intervention on the mental health and well-being of transgender youths: preliminary results. International Journal of Pediatric Endocrinology 2020, 8. Available at https://doi.org/10.1186/s13633-020-00078-2.

Ackerman, K.E., Singhal, V., Slattery, M., Eddy, K.T., Bouxsein, M.L., Lee, H., et al., 2020. Effects of estrogen replacement on bone geometry and microarchitecture in adolescent and young adult oligoamenorrheic Athletes: a randomized trial. Journal of Bone and Mineral Research 35 (2), 248–260. Available at https://doi.org/10.1002/jbmr.3887.

Allen, L.R., Watson, L.B., Egan, A.M., Moser, C.N., 2019. Well-being and suicidality among transgender youth after gender-affirming hormones. Clinical Practice in Pediatric Psychology 7 (3), 302–311. Available at https://doi.org/10.1037/cpp0000288.

Allen, L.R., Dodd, C.G., Moser, C.N., Knoll, M.M., 2026. Changes in suicidality among transgender adolescents following hormone therapy: an extended study. Journal of Pediatrics 289. Available at https://doi.org/10.1016/j.jpeds.2025.114883.

Arcelus, J., Claes, L., Witcomb, G.L., Marshall, E., Bouman, W.P., 2016. Risk factors for non-suicidal self-injury among trans youth. Journal of Sexual Medicine 13 (3), 402–412. Available at https://doi.org/10.1016/j.jsxm.2016.01.003.

Arnoldussen, M., Hooijman, E.C., Kreukels, B.P.C., de Vries, A.L.C., 2022a. Association between pre-treatment IQ and educational achievement after gender-affirming treatment including puberty suppression in transgender adolescents. Clinical Child Psychology and Psychiatry 27 (4), 1069–1076. Available at https://doi.org/10.1177/13591045221091652.

Arnoldussen, M., Van Der Miesen, A.I.R., Elzinga, W.S., Alberse, A.M.E., Popma, A., Steensma, T.D., et al., 2022b. Self-perception of transgender adolescents after gender-affirming treatment: a follow-up study into young adulthood. LGBT Health 9 (4), 238–246. Available at https://doi.org/10.1089/lgbt.2020.0494.

Ashley, F., Tordoff, D.M., Olson-Kennedy, J., Restar, A.J., 2024. Randomized-controlled trials are methodologically inappropriate in adolescent transgender healthcare. International Journal of Transgender Health 25 (3), 407–418. Available at https://doi.org/10.1080/26895269.2023.2218357.

Baines, H.K., Connelly, K.J., 2023. A prospective comparison study of subcutaneous and intramuscular testosterone injections in transgender male adolescents. Journal of Pediatric Endocrinology and Metabolism 36 (11), 1028–1036. Available at https://doi.org/10.1515/jpem-2023-0237.

Baskaran, C., Roberts, S.A., Barrera, E., Pilcher, S., Kumar, R., 2024. Venous thromboembolism in transgender and gender non-binary youth is rare and occurs in the setting of secondary risk factors: a retrospective cohort study. Pediatric Blood & Cancer 71 (11). Available at https://doi.org/10.1002/pbc.31284.

Bauer, G.R., Pacaud, D., Couch, R., Metzger, D.L., Gale, L., Gotovac, S., et al., 2021a. Transgender youth referred to clinics for gender-affirming medical care in canada. Pediatrics 148 (5). Available at https://doi.org/10.1542/peds.2020-047266.

Bauer, G., Churchill, S., Ducharme, J., Feder, S., Gillis, L., Gotovac, S., et al., 2021. Trans Youth CAN! Gender Positivity Scale (TYC-GPS). London, Canada, Trans Youth CAN! Research Team.

Bauer, G., Churchill, S., Ducharme, J., Feder, S., Gillis, L., Gotovac, S., et al., 2021. Trans Youth CAN! Gender Distress Scale (TYC-GDS). Trans Youth CAN! Research Team, London, Canada.

Beck, A.T., Steer, R.A., Brown, G., 1996. Beck Depression Inventory–II (BDI-II) [Database record]. APA PsycTests. Available at https://doi.org/10.1037/t00742-000.

Becker-Hebly, I., Fahrenkrug, S., Campion, F., Richter-Appelt, H., Schulte-Markwort, M., Barkmann, C., 2021. Psychosocial health in adolescents and young adults with gender dysphoria before and after gender-affirming medical interventions: a descriptive study from the Hamburg Gender Identity Service. European Child and Adolescent Psychiatry 30 (11), 1755–1767. Available at https://doi.org/10.1007/s00787-020-01640-2.

Birmaher, B., Brent, D.A., Chiappetta, L., Bridge, J., Monga, S., Baugher, M., 1999. Psychometric properties of the screen for child anxiety related emotional disorders (SCARED): a replication study. Journal of the American Academy of Child and Adolescent Psychiatry 38 (10), 1230–1236. Available at https://doi.org/10.1097/00004583-199910000-00011.

Borger, O., Segev-Becker, A., Perl, L., Ben Simon, A., Yackobovitch-Gavan, M., Sheppes, T., et al., 2024. Body composition, metabolic syndrome, and lifestyle in treatment-naïve gender-diverse youth in Israel. Pediatric Obesity 19 (10). Available at https://doi.org/10.1111/ijpo.13159.

Cantu, A.L., Moyer, D.N., Connelly, K.J., Holley, A.L., 2020. Changes in anxiety and depression from intake to first follow-up among transgender youth in a pediatric endocrinology clinic. Transgender Health 5 (3), 196–200. Available at https://doi.org/10.1089/trgh.2019.0077.

Carlson, S.M., Dominguez, C., Jeevananthan, A., Crowley, M.J., 2025. Follow-up estradiol levels based on regimen formulation with guideline-concordant gender-affirming hormone therapy. Journal of the Endocrine Society 9 (3). Available at https://doi.org/10.1210/jendso/bvae205.

Chelliah, P., Lau, M., Kuper, L.E., 2024. Changes in gender dysphoria, interpersonal minority stress, and mental health among transgender youth after one year of hormone therapy. Journal of Adolescent Health 74 (6), 1106–1111. Available at https://doi.org/10.1016/j.jadohealth.2023.12.024.

Chen, D., Berona, J., Chan, Y.M., Ehrensaft, D., Garofalo, R., Hidalgo, M.A., et al., 2023. Psychosocial functioning in transgender youth after 2 years of hormones. New England Journal of Medicine 388 (3), 240–250. Available at https://doi.org/10.1056/NEJMoa2206297.

Chen, D., Berona, J., Coyne, C.A., Chan, Y.-M., Ehrensaft, D., Hidalgo, M.A., et al., 2024. 9. Predictors of diverse mental health trajectories following 4 years of treatment with gender-affirming hormones. Journal of Adolescent Health 74 (3), S5. Available at https://doi.org/10.1016/j.jadohealth.2023.11.028.

Chew, D., Anderson, J., Williams, K., May, T., Pang, K., 2018. Hormonal treatment in young people with gender dysphoria: a systematic review. Pediatrics 141 (4). Available at https://doi.org/10.1542/peds.2017-3742.

Claes, L., Vandereycken, W., 2007. The self-injury questionnaire–treatment related (SIQ-TR): construction, reliability, and validity in a sample of female eating disorder patients. In: Psychological tests and testing: research trends

Cohen-Kettenis, P.T., Van Goozen, S.H.M., 1997. Sex reassignment of adolescent transsexuals: a follow-up study. Journal of the American Academy of Child and Adolescent Psychiatry 36 (2), 263–271. Available at https://doi.org/10.1097/00004583-199702000-00017.

Coleman, E., Bockting, W., Botzer, M., Cohen-Kettenis, P., DeCuypere, G., Feldman, J., et al., 2012. Standards of care for the health of transsexual, transgender, and gender-nonconforming people, version 7. International Journal of Transgenderism 13 (4), 165–232. Available at https://doi.org/10.1080/15532739.2011.700873.

Coleman, E., Radix, A.E., Bouman, W.P., Brown, G.R., de Vries, A.L.C., Deutsch, M.B., et al., 2022. Standards of care for the health of transgender and gender diverse people, version 8. International Journal of Transgender Health 23 (1), S1. Available at https://doi.org/10.1080/26895269.2022.2100644.

Connolly, M.D., Zervos, M.J., Barone, C.J., Johnson, C.C., Joseph, C.L.M., 2016. The mental health of transgender youth: advances in understanding. Journal of Adolescent Health 59 (5), 489–495. Available at https://doi.org/10.1016/j.jadohealth.2016.06.012.

Cortez, S., Moog, D., Lewis, C., Williams, K., Herrick, C.J., Fields, M.E., et al., 2024. Effectiveness and safety of different estradiol regimens in transgender females: a randomized controlled trial. Journal of the Endocrine Society 8 (8). Available at https://doi.org/10.1210/jendso/bvae108.

De Blok, C.J.M., Wiepjes, C.M., Nota, N.M., Van Engelen, K., Adank, M.A., Dreijerink, K.M.A., et al., 2019. Breast cancer risk in transgender people receiving hormone treatment: nationwide cohort study in the Netherlands. The BMJ 365. Available at https://doi.org/10.1136/bmj.l1652.

Defreyne, J., Nota, N., Pereira, C., Schreiner, T., Fisher, A.D., Den Heijer, M., et al., 2017. Transient elevated serum prolactin in trans women is caused by cyproterone acetate treatment. LGBT Health 4 (5), 328–336. Available at https://doi.org/10.1089/lgbt.2016.0190.

De Vries, A.L.C., McGuire, J.K., Steensma, T.D., Wagenaar, E.C.F., Doreleijers, T.A.H., Cohen-Kettenis, P.T., 2014. Young adult psychological outcome after puberty suppression and gender reassignment. Pediatrics 134 (4), 696–704. Available at https://doi.org/10.1542/peds.2013-2958.

D'hoore, L., T'Sjoen, G., 2022. Gender-affirming hormone therapy: an updated literature review with an eye on the future. Journal of Internal Medicine 291 (5), 574–592. Available at https://doi.org/10.1111/joim.13441.

Doll, E., Gunsolus, I., Thorgerson, A., Tangpricha, V., Lamberton, N., Sarvaideo, J.L., 2022. Pharmacokinetics of sublingual versus oral estradiol in transgender women. Endocrine Practice: Official Journal of the American College of Endocrinology and the American Association of Clinical Endocrinologists 28 (3), 237–242.

Doyle, D.M., Lewis, T.O.G., Barreto, M., 2023. A systematic review of psychosocial functioning changes after gender-affirming hormone therapy among transgender people. Nature Human Behaviour 7 (8), 1320–1331. Available at https://doi.org/10.1038/s41562-023-01605-w.

Eaton, W.W., Smith, C., Ybarra, M., Muntaner, C., Tien, A., 2004. Center for epidemiologic studies depression scale: review and revision (CESD and CESD-R). In: The use of psychological testing for treatment planning and outcomes measurement.

Fontanari, A.M.V., Vilanova, F., Schneider, M.A., Chinazzo, I., Soll, B.M.H., Schwarz, K., et al., 2020. Gender affirmation is associated with transgender and gender nonbinary youth mental health improvement. LGBT Health 7 (5), 237–247. Available at https://doi.org/10.1089/lgbt.2019.0046.

Goossens, G.H., Jocken, J.W.E., Blaak, E.E., 2021. Sexual dimorphism in cardiometabolic health: the role of adipose tissue, muscle and liver. Nature Reviews Endocrinology 17 (1), 47–66. Available at https://doi.org/10.1038/s41574-020-00431-8.

Grannis, C., Leibowitz, S.F., Gahn, S., Nahata, L., Morningstar, M., Mattson, W.I., et al., 2021. Testosterone treatment, internalizing symptoms, and body image dissatisfaction in transgender boys. Psychoneuroendocrinology 132, 105358. Available at https://doi.org/10.1016/j.psyneuen.2021.105358.

Grannis, C., Mattson, W.I., Leibowitz, S.F., Nahata, L., Chen, D., Strang, J.F., et al., 2023. Expanding upon the relationship between gender-affirming hormone therapy, neural connectivity, mental health, and body image dissatisfaction. Psychoneuroendocrinology 156. Available at https://doi.org/10.1016/j.psyneuen.2023.106319.

Green, A.E., DeChants, J.P., Price, M.N., Davis, C.K., 2022. Association of gender-affirming hormone therapy with depression, thoughts of suicide, and attempted suicide among transgender and nonbinary youth. Journal of Adolescent Health 70 (4), 643–649. Available at https://doi.org/10.1016/j.jadohealth.2021.10.036.

Gurrala, R.R., Kumar, T., Yoo, A., Mundinger, G.S., Womac, D.J., Lau, F.H., 2023. The impact of exogenous testosterone on breast cancer risk in transmasculine individuals. Annals of Plastic Surgery 90 (1), 96–105. Available at https://doi.org/10.1097/SAP.0000000000003321.

Hannema, S.E., Schagen, S.E.E., Cohen-Kettenis, P.T., Delemarre-Van De Waal, H.A., 2017. Efficacy and safety of pubertal induction using 17β-estradiol in transgirls. Journal of Clinical Endocrinology and Metabolism 102 (7), 2356–2363. Available at https://doi.org/10.1210/jc.2017-00373.

Haroz, E.E., Ybarra, M.L., Eaton, W.W., 2014. Psychometric evaluation of a self-report scale to measure adolescent depression: the CESDR-10 in two national adolescent samples in the United States. Journal of Affective Disorders 158, 154–160. Available at https://doi.org/10.1016/j.jad.2014.02.009.

Hembree, W.C., Cohen-Kettenis, P.T., Gooren, L., Hannema, S.E., Meyer, W.J., Murad, M.H., et al., 2017. Endocrine treatment of gender-dysphoric/gender-incongruent persons: an endocrine society*clinical practice guideline. Journal of Clinical Endocrinology and Metabolism 102 (11), 3869–3903. Available at https://doi.org/10.1210/jc.2017-01658.

Hisle-Gorman, E., Schvey, N.A., Adirim, T.A., Rayne, A.K., Susi, A., Roberts, T.A., et al., 2021. Mental healthcare utilization of transgender youth before and after affirming treatment. Journal of Sexual Medicine 18 (8), 1444–1454. Available at https://doi.org/10.1016/j.jsxm.2021.05.014.

Horowitz, L.M., Bridge, J.A., Teach, S.J., Ballard, E., Klima, J., Rosenstein, D.L., et al., 2012. Ask suicide-screening questions (ASQ): a brief instrument for the pediatric emergency department. Archives of Pediatrics and Adolescent Medicine 166 (12), 1170–1176. Available at https://doi.org/10.1001/archpediatrics.2012.1276.

Irwin, D.E., Stucky, B., Langer, M.M., Thissen, D., DeWitt, E.M., Lai, J.S., et al., 2010. An item response analysis of the pediatric PROMIS anxiety and depressive symptoms scales. Quality of Life Research 19 (4), 595–607. Available at https://doi.org/10.1007/s11136-010-9619-3.

Jarin, J., Pine-Twaddell, E., Trotman, G., Stevens, J., Conard, L.A., Tefera, E., et al., 2017. Cross-sex hormones and metabolic parameters in adolescents with gender dysphoria. Pediatrics 139 (5). Available at https://doi.org/10.1542/peds.2016-3173.

Kaltiala, R., Heino, E., Työläjärvi, M., Suomalainen, L., 2020. Adolescent development and psychosocial functioning after starting cross-sex hormones for gender dysphoria. Nordic Journal of Psychiatry 74 (3), 213–219. Available at https://doi.org/10.1080/08039488.2019.1691260.

King, H., Kelley, T.P., Shatzel, J.J., 2024. Gender-affirming hormone therapy in the transgender patient: influence on thrombotic risk. Hematology (United States) 2024 (1), 652–663. Available at https://doi.org/10.1182/hematology.2024000592.

Klaver, M., de Mutsert, R., Wiepjes, C.M., Twisk, J.W.R., den Heijer, M., Rotteveel, J., et al., 2018. Early hormonal treatment affects body composition and body shape in young transgender adolescents. Journal of Sexual Medicine 15 (2), 251–260. Available at https://doi.org/10.1016/j.jsxm.2017.12.009.

Klaver, M., De Mutsert, R., Van Der Loos, M.A.T.C., Wiepjes, C.M., Twisk, J.W.R., Den Heijer, M., et al., 2020. Hormonal treatment and cardiovascular risk profile in transgender adolescents. Pediatrics 145 (3). Available at https://doi.org/10.1542/peds.2019-0741.

Knaus, S., Steininger, J., Klinger, D., Riedl, S., 2024. Body mass index distributions and obesity prevalence in a transgender youth cohort – a retrospective analysis. Journal of Adolescent Health 75 (1), 127–132. Available at https://doi.org/10.1016/j.jadohealth.2024.03.005.

Kovacs, M., 1985. The children's depression inventory (CDI). Psychopharmacology Bulletin 21 (4), 995 995.

Kozee, H.B., Tylka, T.L., Bauerband, L.A., 2012. Measuring transgender individuals' comfort with gender identity and appearance. Psychology of Women Quarterly 36 (2), 179–196. Available at https://doi.org/10.1177/0361684312442161.

Kroenke, K., Spitzer, R.L., Williams, J.B.W., 2001. Validity of a brief depression severity measure. Journal of General Internal Medicine 16.

Kuper, L.E., Stewart, S., Preston, S., Lau, M., Lopez, X., 2020. Body dissatisfaction and mental health outcomes of youth on gender-affirming hormone therapy. Pediatrics 145 (4). Available at https://doi.org/10.1542/peds.2019-3006.

Lavender, R., Shaw, S., Maninger, J.K., Butler, G., Carruthers, P., Carmichael, P., et al., 2023. Impact of hormone treatment on psychosocial functioning in gender-diverse young people. LGBT Health 10 (5), 382–390. Available at https://doi.org/10.1089/lgbt.2022.0201.

Lee, M.K., Yih, Y., Willis, D.R., Fogel, J.M., Fortenberry, J.D., 2024. The impact of gender affirming medical care during adolescence on adult health outcomes among transgender and gender diverse individuals in the United States: the role of state-level policy stigma. LGBT Health 11 (2), 111–121. Available at https://doi.org/10.1089/lgbt.2022.0334.

Lindgren, T.W., Pauly, I.B., 1975. A body image scale for evaluating transsexuals. Archives of Sexual Behavior 4 (6), 639–656. Available at https://doi.org/10.1007/BF01544272.

López de Lara, D., Pérez Rodríguez, O., Cuellar Flores, I., Pedreira Masa, J.L., Campos-Muñoz, L., Cuesta Hernández, M., et al., 2020. Psychosocial assessment in transgender adolescents. Anales de Pediatría (English Edition) 93 (1), 41–48. Available at https://doi.org/10.1016/j.anpede.2020.01.004.

Mahfouda, S., Moore, J.K., Siafarikas, A., Hewitt, T., Ganti, U., Lin, A., et al., 2019. Gender-affirming hormones and surgery in transgender children and adolescents. The Lancet Diabetes and Endocrinology 7 (6), 484–498. Available at https://doi.org/10.1016/S2213-8587(18)30305-X.

McGuire, J.K., Spencer, K., Rider, G.N., Berg, D.R., 2016. Body image scale – gender spectrum. University of Minnesota.

McGuire, J.K., Berg, D., Catalpa, J.M., Morrow, Q.J., Fish, J.N., Nic Rider, G., et al., 2020. Utrecht gender dysphoria scale - gender spectrum (UGDS-GS): construct validity among transgender, nonbinary, and LGBQ samples. International Journal of Transgender Health 21 (2), 194–208. Available at https://doi.org/10.1080/26895269.2020.1723460.

McNamara, M., McLamore, Q., Meade, N., Olgun, M., Robinson, H., Alstott, A., 2024. A thematic analysis of disinformation in gender-affirming healthcare bans in the United States. Social Science & Medicine 351, 116943. Available at https://doi.org/10.1016/j.socscimed.2024.116943.

Millington, K., Schulmeister, C., Finlayson, C., Grabert, R., Olson-Kennedy, J., Garofalo, R., et al., 2020. Physiological and metabolic characteristics of a cohort of transgender and gender-diverse youth in the United States. Journal of Adolescent Health 67 (3), 376–383. Available at https://doi.org/10.1016/j.jadohealth.2020.03.028.

Millington, K., Barrera, E., Daga, A., Mann, N., Olson-Kennedy, J., Garofalo, R., et al., 2022. The effect of gender-affirming hormone treatment on serum creatinine in transgender and gender-diverse youth: implications for estimating GFR. Pediatric Nephrology 37 (9), 2141–2150. Available at https://doi.org/10.1007/s00467-022-05445-0.

Millington, K., Lee, J.Y., Kennedy, J.O., Garofalo, R., Rosenthal, S.M., Chan, Y.M., 2024. Laboratory changes during gender-Affirming hormone therapy in transgender adolescents. Pediatrics 153 (5). Available at https://doi.org/10.1542/peds.2023-064380.

Miner, M., Wang, C., Kaminetsky, J., Khera, M., Goldstein, I., Carson, C., et al., 2025. Safety, efficacy, and pharmacokinetics of oral testosterone undecanoate in males with hypogonadism. Andrology 13 (4), 882–893. Available at https://doi.org/10.1111/andr.13747.

Misakian, A., Ariel, D., Sullivan, E., Singh, G., Loeb, D., Strickland, T., et al., 2025. Injectable estradiol monotherapy effectively suppresses testosterone in gender-affirming hormone therapy. Endocrine Practice: Official Journal of the American College of Endocrinology and the American Association of Clinical Endocrinologists.

Nelson, H.D., Humphrey, L.L., Nygren, P., Teutsch, S.M., Allan, J.D., 2002. Postmenopausal hormone replacement therapy: scientific review. JAMA: the Journal of the American Medical Association 288 (7), 872–881. Available at https://doi.org/10.1001/jama.288.7.872.

Nguyen, H.B., Chavez, A.M., Lipner, E., Hantsoo, L., Kornfield, S.L., Davies, R.D., et al., 2018. Gender-affirming hormone use in transgender individuals: impact on behavioral health and cognition. Current Psychiatry Reports 20 (12). Available at https://doi.org/10.1007/s11920-018-0973-0.

Nightingale, K.J., Jelinek, S.K., Jones, C., Bevington, L.M., Darien, K., Ding, A., et al., 2025. Telehealth versus in-person injection instruction for adolescents and young adults initiating gender-affirming testosterone therapy. Transgender health 10 (4), 325–333. Available at https://doi.org/10.1089/trgh.2024.0130.

Nokoff, N.J., Scarbro, S.L., Moreau, K.L., Zeitler, P., Nadeau, K.J., Reirden, D., et al., 2021. Body composition and markers of cardiometabolic health in transgender youth on gonadotropin-releasing hormone agonists. Transgender Health 6 (2), 111–119. Available at https://doi.org/10.1089/trgh.2020.0029.

Nokoff, N.J., Bothwell, S., Rice, J.D., Cree, M.G., Kelsey, M.M., Moreau, K.L., et al., 2024. Insulin sensitivity, body composition and bone mineral density after testosterone treatment in transgender youth with and without prior GnRH agonist therapy. Journal of Clinical & Translational Endocrinology 36, 100356. Available at https://doi.org/10.1016/j.jcte.2024.100356.

Nota, N.M., Dekker, M.J.H.J., Klaver, M., Wiepjes, C.M., van Trotsenburg, M.A., Heijboer, A.C., et al., 2017. Prolactin levels during short- and long-term cross-sex hormone treatment: an observational study in transgender persons. Andrologia 49 (6). Available at https://doi.org/10.1111/and.12666.

Nunes-Moreno, M., Furniss, A., Cortez, S., Davis, S.M., Dowshen, N., Kazak, A.E., et al., 2025. Mental health diagnoses and suicidality among transgender youth in hospital settings. LGBT Health 12 (1), 20–28. Available at https://doi.org/10.1089/lgbt.2023.0394.

Olsavsky, A.L., Grannis, C., Bricker, J., Chelvakumar, G., Indyk, J.A., Leibowitz, S.F., et al., 2023. Associations among gender-affirming hormonal interventions, social support, and transgender adolescents' mental health. Journal of Adolescent Health 72 (6), 860–868. Available at https://doi.org/10.1016/j.jadohealth.2023.01.031.

Olson-Kennedy, J., Okonta, V., Clark, L.F., Belzer, M., 2018. Physiologic response to gender-affirming hormones among transgender youth. Journal of Adolescent Health 62 (4), 397–401. Available at https://doi.org/10.1016/j.jadohealth.2017.08.005.

Olson-Kennedy, J., Wang, L., Wong, C.F., Chen, D., Ehrensaft, D., Hidalgo, M.A., et al., 2025. Emotional health of transgender youth 24 months after initiating gender-affirming hormone therapy. Journal of Adolescent Health 77 (1), 41–50. Available at https://doi.org/10.1016/j.jadohealth.2024.11.014.

Osman, A., Bagge, C.L., Gutierrez, P.M., Konick, L.C., Kopper, B.A., Barrios, F.X., 2001. The suicidal behaviors questionnaire-revised (SBQ-R): validation with clinical and nonclinical samples. Assessment 8 (4), 443–454. Available at https://doi.org/10.1177/107319110100800409.

Pelusi, C., Costantino, A., Martelli, V., Lambertini, M., Bazzocchi, A., Ponti, F., et al., 2014. Effects of three different testosterone formulations in female-to-male transsexual persons. The Journal of Sexual Medicine 11 (12), 3002–3011. Available at https://doi.org/10.1111/jsm.12698.

Perl, L., Segev-Becker, A., Israeli, G., Elkon-Tamir, E., Oren, A., 2020. Blood pressure dynamics after pubertal suppression with gonadotropin-releasing hormone analogs followed by testosterone treatment in transgender male adolescents: a pilot study. LGBT Health 7 (6), 340–344. Available at https://doi.org/10.1089/lgbt.2020.0026.

Posner, K., Brown, G.K., Stanley, B., Brent, D.A., Yershova, K.V., Oquendo, M.A., et al., 2011. The columbia-suicide severity rating scale: initial validity and internal consistency findings from three multisite studies with adolescents and adults. American Journal of Psychiatry 168 (12), 1266–1277. Available at https://doi.org/10.1176/appi.ajp.2011.10111704.

Reynolds, C.R., Richmond, B.O., 2008a. Revised Children's Manifest Anxiety Scale–Second Edition (RCMAS-2). Western Psychological Services, Los Angeles, CA.

Rider, G.N., Mcmorris, B.J., Gower, A.L., Coleman, E., Eisenberg, M.E., 2018. Health and care utilization of transgender and gender nonconforming youth: a population-based study. Pediatrics 141 (3). Available at https://doi.org/10.1542/peds.2017-1683.

Singhal, V., Ackerman, K.E., Bose, A., Flores, L.P.T., Lee, H., Misra, M., 2019. Impact of route of estrogen administration on bone turnover markers in oligoamenorrheic athletes and its mediators. Journal of Clinical Endocrinology and Metabolism 104 (5), 1449–1458. Available at https://doi.org/10.1210/jc.2018-02143.

Slack, D., Krishnamurthy, N., Garrity, M., Chen, D., Kyweluk, M., Trakhtenberg, E., et al., 2025. Choice of hormone assay to monitor feminizing gender-affirming hormone therapy. Endocrine Practice: Official Journal of the American College of Endocrinology and the American Association of Clinical Endocrinologists.

Smith, Y.L.S., van Goozen, S.H.M., Cohen-Kettenis, P.T., 2001. Accepted or rejected for sex reassignement surgery: a prospective follow-up study. Journal of the American Academy of Child and Adolescent Psychiatry 40 (4).

Spielberger, C.D., Gorsuch, R.L., Lushene, R., Vagg, P.R., Jacobs, G.A., 1983. Manual for the State Trait Anxiety Inventory. (Form Y). Consulting Psychologists Press, Palo Alto.

Spitzer, R.L., Kroenke, K., Williams, J.B.W., Löwe, B., 2006. A brief measure for assessing generalized anxiety disorder: the GAD-7. Archives of Internal Medicine 166 (10), 1092–1097. Available at https://doi.org/10.1001/archinte.166.10.1092.

Stoffers, I.E., de Vries, M.C., Hannema, S.E., 2019. Physical changes, laboratory parameters, and bone mineral density during testosterone treatment in adolescents with gender dysphoria. Journal of Sexual Medicine 16 (9), 1459–1468. Available at https://doi.org/10.1016/j.jsxm.2019.06.014.

Strang, J.F., Chen, D., Nelson, E., Leibowitz, S.F., Nahata, L., Anthony, L.G., et al., 2022. Transgender youth executive functioning: relationships with anxiety symptoms, autism spectrum disorder, and gender-affirming medical treatment status. Child Psychiatry and Human Development 53 (6), 1252–1265. Available at https://doi.org/10.1007/s10578-021-01195-6, www.wkap.nl/journalhome.htm/0009-398X.

Strauss, P., Ball, J., Bonney, S., Costanza, M., Cavve, B.S., Hird, K., et al., 2024. A critical discussion of pediatric gender measures to clarify the utility and purpose of "measuring" gender. International Journal of Transgender Health 1–25. Available at https://doi.org/10.1080/26895269.2024.2375409.

Sumerwell, C., Carlin, K., Walsh, E., Hodax, J.K., 2024. Serum hormone concentrations in transgender youth receiving estradiol. Endocrine practice: official journal of the American College of Endocrinology and the American Association of Clinical Endocrinologists 30 (2), 155–159.

Ta, V., Chin, W.K., White, A.A., 2014. Allergic contact dermatitis to testosterone and estrogen in transdermal therapeutic systems. Dermatitis: Contact, Atopic, Occupational, Drug 25 (5). Available at https://doi.org/10.1097/DER.0000000000000062.

Tack, L.J.W., Craen, M., Dhondt, K., Vanden Bossche, H., Laridaen, J., Cools, M., 2016. Consecutive lynestrenol and cross-sex hormone treatment in biological female adolescents with gender dysphoria: a retrospective analysis. Biology of Sex Differences 7 (1). Available at https://doi.org/10.1186/s13293-016-0067-9.

Tack, L.J.W., Heyse, R., Craen, M., Dhondt, K., Bossche, H.V., Laridaen, J., et al., 2017. Consecutive cyproterone acetate and estradiol treatment in late-pubertal transgender female adolescents. Journal of Sexual Medicine 14 (5), 747–757. Available at https://doi.org/10.1016/j.jsxm.2017.03.251.

Taylor, J., Mitchell, A., Hall, R., Langton, T., Fraser, L., Hewitt, C.E., 2024. Masculinising and feminising hormone interventions for adolescents experiencing gender dysphoria or incongruence: a systematic review. Archives of Disease in Childhood 109 (2), s48. Available at https://doi.org/10.1136/archdischild-2023-326670.

Toland, M.K., Bonasia, K., Bentz, J., DelBaugh, R.M., Vitale, E.J., Scudder, P.N., et al., 2024. Uterine and ovarian histopathology after testosterone for gender affirmation: a systematic review. Transgender Health 9 (4), 288–297. Available at https://doi.org/10.1089/trgh.2022.0207.

Tordoff, D.M., Wanta, J.W., Collin, A., Stepney, C., Inwards-Breland, D.J., Ahrens, K., 2022. Mental health outcomes in transgender and nonbinary youths receiving gender-affirming care. JAMA Network Open 5 (2), e220978. Available at https://doi.org/10.1001/jamanetworkopen.2022.0978.

Turban, J.L., King, D., Kobe, J., Reisner, S.L., Keuroghlian, A.S., Radix, A.E., 2022. Access to gender-affirming hormones during adolescence and mental health outcomes among transgender adults. PLoS One 17 (1), e0261039. Available at https://doi.org/10.1371/journal.pone.0261039.

Valentine, A., Nokoff, N., Bonny, A., Chelvakumar, G., Indyk, J., Leibowitz, S., et al., 2021. Cardiometabolic parameters among transgender adolescent males on testosterone therapy and body mass index-matched cisgender females. Transgender Health 6 (6), 369–373. Available at https://doi.org/10.1089/trgh.2020.0052.

van Leerdam, T.R., Zajac, J.D., Cheung, A.S., 2023. The effect of gender-affirming hormones on gender dysphoria, quality of life, and psychological functioning in transgender individuals: a systematic review. Transgender Health 8 (1), 6–21. Available at https://doi.org/10.1089/trgh.2020.0094.

Vestering, A., van Vugt, W.L.J., Berner, A.M., Snijders, M.L.H., Heijer, Md, Groenman, F.A., et al., 2025. Incidence of gynaecological (pre-) malignancies and endometrial activity in transmasculine and gender diverse individuals using testosterone: a retrospective, single-centre cohort study. eClinicalMedicine 84. Available at https://doi.org/10.1016/j.eclinm.2025.103248.

Wilson, L.M., Sharma, R., Dukhanin, V., McArthur, K., Robinson, K.A., Baker, K.E., 2021. Hormone therapy, mental health, and quality of life among transgender people: a systematic review. Journal of the Endocrine Society 5 (4). Available at https://doi.org/10.1210/jendso/bvab011.

Winer, J., McKee, M., Lemelman, M.B., 2025. Metabolic effects of gender-affirming hormonal therapy in youth. Pediatric Annals 54 (10), e355. Available at https://doi.org/10.3928/19382359-20250828-05.

Wittlin, N.M., Gallagher, N.M., Atwood, S., Olson, K.R., 2025. Mental health during medical transition in a US and Canadian sample of early socially transitioned transgender youth. Journal of Adolescent Health 76 (2), 228–237. Available at https://doi.org/10.1016/j.jadohealth.2024.10.023.

Xu, W., Morford, J., Mauvais-Jarvis, F., 2019. Emerging role of testosterone in pancreatic β cell function and insulin secretion. Journal of Endocrinology 240 (3), R97. Available at https://doi.org/10.1530/JOE-18-0573.

Chapter 8

Other medical and nonmedical treatments for gender dysphoria

Matthew Goldenberg
Department of Adolescent Medicine, Seattle Children's Hospital, Seattle, WA, United States

Learning objectives

- Summarize existing research on gender-affirming medical interventions other than pubertal suppression, testosterone, or estradiol, such as menstrual cycle suppression, antiandrogens, and acne medication.
- Summarize existing research on nonmedical gender-affirming interventions, such as social and legal transition steps, concealing unwanted sex characteristics, and voice training.
- Highlight the importance of individualized gender-affirming care and treatment planning for all transgender and gender diverse youth, and especially those whose gender identity does not align with binary notions of gender.

Key terms and definitions

- **Embodiment** refers to the process of making external what is felt internally. This term has been proposed as an alternative to transition that is less binary and more inclusive. Exploring embodiment goals can help to inform individualized gender-affirming healthcare.
- **Transition** is also sometimes used, particularly when discussing social and legal steps toward gender affirmation. This general term is nonetheless meant to refer to taking steps in the direction of individual gender identity, regardless of whether identity is binary or not.
- **Nonbinary** is an umbrella term for gender identities that do not align with the binary notions of male and female. Some individuals use nonbinary as a more specific term for their own gender, but it is also used to capture a broader spectrum of identity.
- **Gender dysphoria** refers to the distress someone feels when their physical body or social perception does not align with their felt gender identity. Not everyone experiences gender dysphoria, and specific experiences with gender dysphoria vary greatly across individuals.
- **Nonmedical interventions** refer to any gender-affirming treatments or behaviors that do not require a prescription or oversight from a medical provider.

Gender dysphoria has been mischaracterized as an overall feeling of "trapped in the wrong body," a phrase that overstates an inherent wrongness in a person's entire body, as well as the assumption that a total change in physical body is wanted (Bettcher, 2014). Certainly, there are individuals who desire comprehensive physical changes to bring alignment between gender identity and how others perceive them through a gendered lens. While some individuals may feel that the phrase "trapped in the wrong body" is a close enough paraphrase of their experiences of self, the flaws in that terminology do potentially overemphasize a biological basis of gender and imply that a medically supported transition is most necessary. Furthermore, this phrase implies high distress in a continual state, which is not reflective of all transgender and gender diverse (TGD) individuals, and the phrase suggests that individuals with gender dysphoria experience distress in all aspects of the physical self. Research shows that individuals with gender

Advances in Research on Pediatric Gender Affirming Healthcare. DOI: https://doi.org/10.1016/B978-0-443-29019-0.00005-1
© 2026 Elsevier Inc. All rights are reserved, including those for text and data mining, AI training, and similar technologies.

dysphoria may feel distress regarding a specific location of their body, such as their chest, but have neutral or positive feelings in other aspects of the body (Kausar et al., 2024).

Similarly, some individuals may experience distress related to the sound of their voice, the curvature of their hips, the size of their hands, or other parts of the body, which may not be soothed with mainstream medical gender-affirming treatment such as hormone therapy. Additionally, some effects from hormone therapy may not be desired by the individual. As the scientific basis that supports gender-affirming medical care as a primary treatment for gender dysphoria grows, there must also be an expansion of available treatments that are suited to the individual's subjective sense of dysphoria. Researchers should also aim to include individuals who experience gender euphoria in the study of TGD individuals, as focusing exclusively on negative emotions can increase stigma that transgender individuals suffer with inherent mental health challenges (Ashley, 2021).

Gender euphoria refers to the positive emotions that result from affirmation of gender identity or expression, including reaching particular gender development and affirmation milestones (Reisner et al., 2023). Resetting gender dysphoria assessment goals from assuming and confirming distress to including curiosity about which aspects of an individual's gender expression lead to pride, satisfaction, gender euphoria, or other positive emotions is key to the affirming approach. Another reason for providers to include assessment questions about what a TGD person appreciates about their physical self is that patients may be more authentic in their request for healthcare services, as they may feel less challenged to present as more distressed than they really are, as a pathway to receive gender-affirming care. In a study of TGD patients receiving nursing care, participants reported that it is important to them to be treated as a whole person, and not just a focus on gender identity. To quote the conclusion of the study, "They [study participants] ask us to take accountability for our ongoing professional development, honoring each individual's expression of their identity, affirmed name, and pronouns. Treating our transgender patients as a person first can improve health outcomes at the individual and population level. We owe our patients nothing less." (p. 20) (Mikovits, 2022)

The TGD community is not a monolith, and there is not one single desired outcome in the provision of gender-affirming medical care. Additionally, there is documentation of TGD individuals throughout generations and across cultures (Feinberg, 1997). Given that there is wide variance in the availability of gender-affirming medical care, it should be understood that TGD individuals cannot and will not necessarily depend on gender-affirming medical care becoming available before enhancing their lived experiences. This can be done through "do it yourself" methods of gender transition, which are inclusive of any way a person changes their outer appearance for the purpose of gender congruency without the oversight of a credentialed medical provider (Rotondi et al., 2013).

There is also increasing awareness of the unique needs of individuals whose gender identity does not align with the binary notions of man or woman, and for whom more nuanced affirmation options may be beneficial. Research has shown that nonbinary youth experience more health disparities than their binary transgender peers (Williams et al., 2024). At the same time, they are less likely to be living as their identified gender or to receive desired affirming medical care (Todd et al., 2019) and less likely to be supported in social and legal affirmation than some of their binary transgender peers (Amos et al., 2025). In response to healthcare that is biased toward binary identities and individuals, nonbinary young people will often change their narrative to include a more binary identity or expression, alter the healthcare they are provided, or forego medical treatment altogether despite their affirmation goals (Lykens et al., 2018). Nonbinary youth may or may not desire medical and nonmedical affirmation options similar to binary transgender youth, and their individual experiences with gender dysphoria and embodiment goals should be evaluated to provide the most appropriate clinical recommendations. Embodiment refers to the process of making external what is felt internally. This term has been proposed as an alternative to gender transition that is less binary and more inclusive (Hastings et al., 2021).

In addition to decreasing gender dysphoria and realizing goals for embodiment, individuals may choose to physically transition as a means of increasing safety in public spaces. TGD individuals may be prone to anxiety that they will be targeted or harassed when accessing public spaces such as bathrooms (Weinhardt et al., 2017). The 2022 US Transgender Survey indicated that 4% of respondents were denied access to a restroom in a public place, at work, or at school in the last 12 months, while 6% of respondents had been verbally harassed, physically attacked, or experienced unwanted sexual contact when accessing or using a restroom (James et al., 2024).

Research suggests that reported hate crimes against TGD people are increasing, though there is also strong evidence that hate crimes against TGD people are currently underreported (Gauthier et al., 2021). In a 2021 study of TGD people who did experience a hate crime, nearly half did not report these crimes to the police due to doubt in any positive results and fear of being arrested or victimized by the police (Gauthier et al., 2021). The growing discrimination against TGD people, which can elevate to victimization from a hate crime, is a valid factor in why TGD individuals may be in search of safety. For some individuals, gender transition can represent an opportunity to increase a sense of safety in public. Understanding the dynamic reasons that motivate gender transition for the individual is key to developing an affirming and comprehensive treatment plan.

In other cases, TGD patients may look toward treatments that are better suited to their embodiment goals or are within their financial or geographic means. Treatment availability is dependent on several factors, including but not limited to provider knowledge, pharmaceutical supply, cost of medications or cost of healthcare services, and legislative limitations on gender-affirming care. Further research examining the medical and psychological risks and benefits of these additional interventions is needed to supplement existing literature.

Social and legal transition steps

The positive effects of some social transition steps require the involvement of others. For example, two separate surveys of TGD adults found that misgendering (referred to as gender misclassification in the study) led to increased negative affect and less strength and coherence in identity (McLemore, 2014), consistent with observations of TGD youth collected through years of clinical practice. A more recent survey of TGD adults similarly found that both nonaffirmation and antitransgender bias are potentially traumatic to the person and can increase risk of developing posttraumatic stress disorder (Barr et al., 2021).

Research among youth has similarly demonstrated the benefits of social and legal affirmation and the potential harm of not being supported in these ways by others. Children who have supportive parents and are allowed to socially transition at a young age demonstrate developmentally normative levels of distress, similar to their cisgender peers (Olson et al., 2016). Being allowed to legally change identity documents is similarly associated with improved mental health (De Chants et al., 2022).

Because social and legal affirmation inherently includes the perceptions and behaviors of others, the mental health benefits of these affirmation steps may also depend on the reactions from others. A large national survey in Australia found that young people who were supported in their desire for social and legal affirmation experienced fewer symptoms of anxiety, distress, and suicidal ideation compared to those who desired these steps but lacked support to enact them (Amos et al., 2025). Similarly, a longitudinal panel study found no mental health differences between youth who had or had not chosen an affirmed name; however, there was a significant difference among those who had chosen an affirmed name based on whether others were actually using the affirmed name (Pollitt et al., 2021). When a chosen name is used across multiple contexts, there are significant reductions in depression and suicidal ideation (Russell et al., 2018). These studies suggest that the positive impact of social or legal transitions is not just additive to overall health, but indeed can be understood to prevent negative health outcomes.

Management of the appearance of sex characteristics

In addition to or separate from social or legal transition, some adolescents may pursue medical and/or nonmedical interventions aimed at managing the appearance of or experience with primary or secondary sex characteristics.

Chest binding

Chest binding is a common practice within TGD communities and can be done safely using a variety of techniques to lessen the appearance of breast tissue. Chest binding is primarily done to provide relief for individuals assigned female at birth who experience chest gender dysphoria, and it is becoming more common among adolescents and young adults (Diana et al., 2024). Young people report that when they are not able to use a chest binder, it is harder to engage with new people or participate in social activities due to both dysphoria and challenges with gender expression (Tüzün et al., 2025). To determine which adolescents are more likely to use binding, one study examined the results of the 2022 Lesbian, Gay, Bisexual, Transgender, Queer National Teen Survey (LGBTQ) National Teen Survey, which had a large number of TGD respondents (Bishop et al., 2024). According to the results, most TGD adolescents assigned female at birth (63.8%) had a history of chest binding, and many (24.7%) used a chest binder 7 days per week. Adolescents who reported chest binding were more likely to be older, to have a binary male identity, and to report more masculine gender expression, more outness of identity, and younger ages of realization and disclosure.

Regarding potential risks the most common negative effects of binding are pain and skin problems, with more serious impacts being rarer (Akgul et al., 2025; Diana et al., 2024; Peitzmeier et al., 2021). In a large cohort of young adults who participated in the Binding Health Project, shoulder and breast pain were the most common out of 27 potential outcomes, and pain was also more likely to emerge early in binding, whereas more serious symptoms (rib fractures, rib or spine changes, or respiratory infections) tended to emerge after long-term frequent binding (Peitzmeier et al., 2021). There is limited data with mixed findings about the impact of chest binding on skin quality that may impact the outcome of affirming top surgery (Diana et al., 2024; Monstrey et al., 2008). Finally, there may be risks associated with participating in physical exercise while wearing a chest binder (Akgul et al., 2025).

Despite potential negative effects, it appears that individuals may choose to bind anyway, prioritizing the benefits of chest binding (Diana et al., 2024; Julian et al., 2021). One study used an anonymous online survey to assess patterns and trends in chest binding among adolescents and young adults (Julian et al., 2021). The study used the Chest Dysphoria Scale, a subscale of the Gender Congruence and Life Satisfaction Scale, and several additional questions related to chest binding. Greater chest dysphoria was associated with lower life satisfaction, and respondents reported continuing to use a chest binder despite negative effects due to feeling more comfortable in public spaces, thinking it was important to have a masculine chest, and feeling safer in public. For most participants the source of learning about how to bind was the internet, with a few receiving guidance from medical providers. This suggests that medical providers could increase their ability to provide scientifically sound guidance to patients who are binding or are considering using a binder.

The Chest Dysphoria Scale is a 17-item measure of chest dysphoria and its impact on intimate relationships, hygiene, exercise, and other important areas of functioning (Olson-Kennedy et al., 2018). Respondents are asked to rate each item on a scale from 0 (never) to 3 (all the time), with higher scores indicating more severe chest dysphoria. A full list of scale items is presented in the validation study (Olson-Kennedy et al., 2018). The Gender Congruence and Life Satisfaction Scale is a 38-item measure of gender (in)congruence and related well-being and life satisfaction, which can be given to anyone regardless of sex assigned at birth (Jones et al., 2019). The measure asks how often over the past 6 months each item occurred as a result of gender dysphoria on a scale from never (5) to always (1), with high scores indicating more congruence and satisfaction. This measure is not specific to chest dysphoria but can be used to assess the functional impact of dysphoria more broadly. The full measure can be found in the Appendix of the validation study (Jones et al., 2019).

To explore the potential risks of chest binding while enjoying physical activity, two studies had adolescents and young adults complete pulmonary function tests and cardiopulmonary tests with and without a chest binder. One small sample study tested adolescents with and without a binder at one or two time points and found that chest binding did not greatly reduce capacity for physical activity (Voss et al., 2024). Another study tested adolescents and young adults with and without a binder as well as a control group (Akgul et al., 2025). Results revealed that some pulmonary functions were lower with a binder than without a binder, but there were no significant differences between the two groups. In other words, binding may result in acute restrictive impairment of airflow, but there is no evidence to suggest long-lasting effects of binding on pulmonary function over time.

Some youth may utilize exercise as a method to attempt to reduce chest dysphoria, both because exercise is positively associated with reduced psychological distress and because individuals may have awareness of specific physical activities that they believe will create a more congruent appearance (Linsenmeyer et al., 2022). Thus it is important for providers who offer guidance to balance the many positive benefits of physical activity with the potential concerns of using a binder with high-intensity physical activity.

Gender-affirming care providers can support their patients by providing education and resources on safe chest binding practices and locating appropriate chest binders. Recommendations for safe binding practices include the following: (1) ensuring appropriate sizing; (2) avoiding elastic bandages, plastic wrap, or the use of multiple binders; (3) removing the binder while sleeping and exercising; (4) wearing for a maximum of 8 hours per day and taking at least two breaks per week; and (5) practicing good skin hygiene (Diana et al., 2024). More research is needed with adolescents specifically, and this body of research would benefit from more prospective designs with a priori hypotheses about chest binding behaviors, benefits, and adverse outcomes.

Packing, tucking, and the use of breast and body forms

A comprehensive review of the literature and available clinical resources found no research to date on other methods of manipulating the appearance of primary sex characteristics among TGD adolescents (Diana et al., 2024). The review provides the following summary and recommendations based on what is known through clinical resources. Packing involves the use of an object or undergarment that creates the appearance of having a penis, typically used by TGD individuals assigned female at birth. This can also include devices that allow an individual to stand for urination. The only known risk of genital packing is the rare potential for skin irritation, depending on the material of the packer being used.

TGD individuals assigned male at birth may use genital tucking strategies and/or breast and body forms to appear more feminine. Genital tucking involves repositioning the penis and testicles to lessen the appearance of the genitals. Either the penis and testicles are tucked between the legs and secured with underwear or a gaff, or the testicles are tucked inside the inguinal canal while the penis is held between the legs. Recommendations for safe genital tucking include the following: (1) maintaining hygiene and considering shaving to prevent irritation, (2) avoiding duct tape and only using medical tape if tape is needed, (3) using the restroom before tucking to prevent urinary retention, and (4) untucking any time pain is experienced. Some young people assigned male at birth benefit from using undergarments to enhance the appearance of breasts and hips, typically in the form of breast and body forms. There are no known risks associated with use other than the rare potential for skin irritation depending on the material used (Diana et al., 2024).

Voice training

Voice is a cause of dysphoria for some TGD individuals, including adolescents. Some TGD people feel that voice is more important than other outward changes (Smith et al., 2024). There is also a risk of voice strain among those attempting to force their voice to a higher or lower pitch on their own without support, for which adolescents might be particularly vulnerable (Russell and Abrams, 2019). The relationship between voice and gender is often discussed in terms of pitch, with higher pitch being associated with more feminine voices and lower pitch being associated with more masculine voices (Chang and Yung, 2021). Fundamental frequency (f_0) refers to the base pitch of someone's voice, or the lowest note they can produce, and it is measured in Hertz (Hz). Cisgender men typically demonstrate f_0 between 100 and 120 Hz, while cisgender women typically demonstrate f_0 between 190 and 220 Hz (Chang and Yung, 2021).

Vocal quality and perception are made up of more than simple pitch and can include resonance, articulation patterns, inflection, speech rate and volume, and intonation (Chang and Yung, 2021; Davies et al., 2015). Speech-language pathologists have the necessary skills to engage in voice training with TGD people (Smith et al., 2024). Voice training can help TGD individuals feel more affirmed while speaking by modifying both vocal features and communication behaviors (Davies et al., 2015). Vocal therapy may be a better initial strategy than surgery, given that it is noninvasive and has resulted in high satisfaction for many people (Nolan et al., 2019). One of the biggest challenges to voice training compared to more invasive methods is the importance of adherence (Rapoport et al., 2023), for which some researchers are exploring using new technology such as virtual reality to assist (Smith et al., 2024).

One prospective study of TGD individuals seeking vocal feminization summarized available measurement tools and clarified some differences that emerge with different measures (Quinn et al., 2022). Specifically, there are three general categories of outcome measures: (1) objective acoustic outcomes such as f_0 and voice quality, (2) auditory-perceptual outcomes such as ratings from the self and others, and (3) self-report questionnaires of functioning and satisfaction. Auditory-perceptual ratings may be made by the individual themselves, by clinically-experienced listeners (e.g., speech-language pathologists), or by naïve listeners who do not have a specific frame of reference for the person or voice metrics in general. This study discussed the most common strategy for assessing perceptual ratings, visual analog scales, and compared two wording options (very female to very male vs very feminine to very masculine) across multiple listeners and found a very high correlation between the two wording choices (Quinn et al., 2022). Common self-report measures include the Transgender Self-Evaluation Questionnaire (TSEQ), the Transsexual Voice Questionnaire (TVQ^{MtF}), and the Trans Woman Voice Questionnaire (TWVQ), each of which has been modifications of the prior (Chadwick et al., 2022; Dacakis et al., 2013). Other self-report measures might ask about satisfaction with current voice, with overall communication, or with aspects of a voice training program (Quinn et al., 2022).

Vocal masculinization is typically thought to occur as a result of the effects of testosterone on voice in individuals assigned female at birth (Chang and Yung, 2021; Davies et al., 2015). Pitch lowering typically begins between 0 and 6 months of initiating testosterone, and many individuals achieve some lowering of f_0 in the first year. Even with testosterone, however, results are variable and not always to each person's satisfaction (Myers and Bell, 2020), and some people can develop vocal instability or strain, or lose vocal endurance over time with testosterone (Chang and Yung, 2021). Testosterone does not impact other aspects of voice and communication, such as resonance, volume, intonation, gestures, or posture (Myers and Bell, 2020). Studies investigating the impacts of vocal training on voice masculinization are scarce. One retrospective study of 10 adult transgender men demonstrated high levels of satisfaction with a group voice training program (Mills et al., 2019).

In contrast, there have been more published studies of voice feminization training (Schwarz et al., 2023). Both voice training and vocal surgery may be good options for individuals assigned male at birth, depending on financial means, desired vocal changes, and tolerance for potential risk (Nolan et al., 2019). A longitudinal study found that perceptual ratings of voice (from self and others) and satisfaction with voice all increased immediately after 8–12 sessions, which was maintained at 3 months after completing the program (Quinn et al., 2022). One retrospective chart review of adult transgender women who received voice training alone found successful pitch increases after an average of six sessions, which was sustained at a delayed follow-up (Rapoport et al., 2023). Research using visual analog scales to assess perceptual ratings has found that self-report does not always correlate with outside listener ratings, but self-report correlates much more strongly with satisfaction than outside listener ratings (Quinn et al., 2022).

There has been almost no published information on nonbinary and genderfluid individuals, despite the flexibility of voice training to help develop multiple speech patterns (Davies et al., 2015).

Published research pertaining to voice training outcomes in adolescents is limited to case reports, despite the fact that adolescents may find voice training particularly helpful given their developmental stage and voice health (Davies et al., 2015). One case report presented the results of weekly assessments of a transgender female adolescent who completed 15 sessions over the course of 7 months, which found positive changes in acoustic, perceptual, and self-report outcomes (Hancock and Helenius, 2012). The researchers emphasized that evaluation and treatment of adolescents likely need to be more nuanced, given varying levels of experienced endogenous pubertal-driven changes to voice, as well as varying degrees of exposure to exogenous hormones. The second published work is a series of three case studies, including one transgender female, one transgender male, and one nonbinary adolescent assigned female at birth (Russell and Abrams, 2019). The results of this case series suggest that adolescents have a variety of needs and goals when it comes to seeking voice training, including self-perception, reducing strain, and safety.

Future research is needed to better understand the unique experiences of and outcomes related to adolescents, including those who identify as binary male, binary female, and a myriad of nonbinary identities. Research is also needed to help clarify what length and time of training is likely to result in positive change while remaining efficient and cost-effective (Rapoport et al., 2023).

Hair removal

Some TGD individuals may benefit from hair removal, whether that be managed with professional support, such as with electrolysis or laser removal, or utilizing more do-it-yourself methods such as shaving, waxing, plucking, or using at-home hair removal lotions. In some cases, presurgical patients may need to complete hair removal to accommodate surgical site preparation. Hair removal includes a spectrum of methods that TGD people may use to reach their embodiment goals. A survey of adults found that an overwhelming majority of individuals assigned male at birth (over 95% of survey respondents) sought facial hair removal (Ginsberg et al., 2016).

Although more research on the necessity and outcomes of hair removal for TGD individuals is needed, preliminary research among adults suggests that hair removal has significant positive benefits. One survey of nearly 300 adults assigned male at birth found that hair removal was associated with reduced anxiety, depression, and negative affective states (Bradford et al., 2021). Unfortunately, cost is consistently identified as a barrier to hair removal services, made worse by a pervasive lack of coverage by insurance companies (Thoreson et al., 2020). Research is also needed to explore the use and outcomes of hair removal among young people specifically.

Facilitating facial hair

Minoxidil is an accessible, over-the-counter medication that has been used to treat hair loss for many decades (Pang et al., 2021). Minoxidil is widely available as a topical foam, and it is also available as an oral prescription (Gao et al., 2023). Use of minoxidil for beard enhancement is less common, but some TGD young people are interested in this potential use. Minoxidil has been shown to enhance beard growth in cisgender men wanting fuller beards (Ingprasert et al., 2016), but very little research has reported on this use in TGD individuals.

In a study of 16 TGD adults on testosterone, topical minoxidil 2% lotion 2 mL was applied to the face daily, and Ferriman Gallwey (FGm) scores were measured at baseline, 3 months, and 6 months (Marinelli et al., 2024). Results found significant changes in beard growth at 3 months, with less change between 3 and 6 months. Only one participant reported skin irritation and discontinued use. Three participants discontinued between 3 and 6 months due to being satisfied with hair growth. In both the participants and the control group, hirsutism before testosterone initiation was a significant predictor of ultimate hair growth outcome, regardless of minoxidil use.

One case report describes minoxidil use in a 17-year-old who was not able to access testosterone (Pang et al., 2021). Obvious hair pigmentation was observed after only 3 months of using minoxidil, and the only reported negative effect was skin dryness. Surprisingly, body hair pigmentation was also observed, suggesting there may be a more systemic absorption and effect. This limited research suggests that minoxidil may be a safe and effective intervention for enhancing beard growth. Patients should be made aware of the potential for skin irritation or dryness. More research among adolescents using minoxidil over time and compared to control groups is greatly needed to increase confidence in these results. Additionally, there may be distinct risks and benefits to using oral minoxidil compared to topical minoxidil, and practitioner education is necessary to ensure fully informed consent for the use of each method (Onejeme et al., 2025).

Hormone management

Some young people are interested in managing their endogenous hormones or using gender-affirming exogenous hormones in ways that vary from traditional protocols. For those assigned female at birth, this might include methods of menstrual cycle suppression, low-dose testosterone, or short-term testosterone. For those assigned male at birth, this might include antiandrogens, progesterone, low-dose estrogen, or short-term estrogen.

Menstrual cycle suppression

Many TGD youth assigned female at birth experience dysphoria related to menstrual cycles and desire options to suppress cycles. Even for those interested in testosterone, it may be an extended period of time before cycles are effectively suppressed on testosterone, and other options may be used to achieve amenorrhea sooner. Some adolescents may want to stay on a lower dose of testosterone, which may not lead to amenorrhea, and others may have achieved amenorrhea using testosterone but also desire an effective contraceptive. Parents are also relatively supportive of cycle suppression methods (Schwartz et al., 2022), and these methods can therefore provide relief from dysphoria among youth navigating lower parental support for other aspects of gender-affirming care. The most common methods of menstrual cycle suppression, effectiveness rates, and additional considerations for choosing a method are summarized in Table 8.1 (Berrahou et al., 2022; Schwartz et al., 2023a).

The majority of research on menstrual cycle suppression among TGD adolescents has utilized retrospective chart review, with fewer studies using prospective recruitment in combination with chart review. For example, one prospective study recruited adolescents to complete a survey after counseling on cycle suppression but before initiation of a particular method, and data were supplemented with a chart review (Stewart et al., 2025). Other studies have included a broader sample, including youth already prescribed cycle suppression from an outside provider. The majority of these studies included youth prescribed testosterone but excluded those using gonadotropin-releasing hormone agonists (GnRHa) (Kanj et al., 2019; Schwartz et al., 2023a,b). A few retrospective studies have also examined multiple time points or changes over time (Alaniz et al., 2023; Kanj et al., 2019; Schwartz et al., 2023a).

Studies have consistently shown high rates (over 92%) of dysphoria and distress associated with menstrual cycles (Moussaoui et al., 2023; Schwartz et al., 2022; Stewart et al., 2025). In one study, participants who identified as male reported more distress related to menstrual cycles compared to those who identified as female, nonbinary, or unsure (Moussaoui et al., 2023).

Studies of youth starting cycle suppression in specialty gender clinics show that progesterone-only pills are the most often selected method, particularly norethindrone acetate, followed by intrauterine device (IUD) and depot

TABLE 8.1 Menstrual cycle suppression options.

		Amenorhea	Contains estrogen	Effective form of contraception	Considerations
Combined hormonal contraception	Oral pill, transdermal patch, vaginal ring	50%–88%	+	+	Used continuously
Norethindrone ("mini pill")	Oral pill	10%–15.3%		+	
Norethindrone acetate	Oral pill	76%			Can be titrated
Medroxyprogesterone	Oral pill				
Depot medroxyprogesterone acetate	Intramuscular injection	52%–71%		+	Bone density caution
Levonorgestrel	Intrauterine device	12%–60%		+	Pelvic exam needed
Etonogestrel	Subcutaneous implant	20%–30%		+	

medroxyprogesterone acetate, with relatively few to none choosing combined hormonal methods or the etonogestrel implant (Schwartz et al., 2023a; Stewart et al., 2025). In contrast, research reporting on youth accessing cycle suppression medications in nonspecialty clinics found that the combined hormonal contraceptive was most commonly prescribed, followed by norethindrone acetate, depot medroxyprogesterone acetate, the etonogestrel implant, and the levonorgestrel IUD (Schwartz et al., 2023b).

Some studies have looked at reasons for starting cycle suppression, reasons for choosing particular methods, and reasons for discontinuation. The most common reasons for pursuing cycle suppression are desire for amenorrhea and decreasing dysphoria, followed by improving symptoms such as dysmenorrhea, heavy bleeding, and mood, with relatively fewer opting specifically for contraception (Moussaoui et al., 2023; Schwartz et al., 2022). Common reasons for choosing one method over the other include not wanting estrogen, ease of use, discomfort with the internal device, and not wanting a pelvic exam (Berrahou et al., 2022; Stewart et al., 2025). Cost of method, provider recommendation, parent preference, and friend recommendation are not commonly endorsed as reasons (Stewart et al., 2025). In one study the most common reasons for changing methods were breakthrough bleeding and no longer needing contraception, and the most common reasons for stopping the method altogether were amenorrhea due to testosterone, followed by irregular bleeding (Kanj et al., 2019).

Some centers offer IUD placement along with a pelvic exam to be done under sedation, which may make the IUD a more desirable method than when in-office placement is the only option. Studies conducted within clinics that offer sedation have found that the majority of those who chose IUD opted to have it placed under sedation (Kanj et al., 2019; Schwartz et al., 2023b). In the studies reviewed here the percentage of participants who were sexually active and had the potential to become pregnant varied from under 4% (Schwartz et al., 2022) to more than one-third (Kanj et al., 2019). Similarly, some studies have found that up to 20% of patients desire contraception (Stewart et al., 2025), while others have found a fairly low interest in contraception and instead a high rate of explicitly indicating that contraception was not a reason for medications to suppress cycles (Moussaoui et al., 2023; Schwartz et al., 2022). One study found that some sexually active patients discontinued their contraception method after achieving amenorrhea (Kanj et al., 2019), highlighting the importance of counseling about testosterone not being an effective form of contraception.

Studies investigating the effectiveness of achieving amenorrhea and satisfaction with various methods have varied in results (Moussaoui et al., 2023; Schwartz et al., 2023a). In one study, amenorrhea was achieved for 80% of participants at approximately 3 months and for 93% of participants at 1 year; similarly, 78% were satisfied at 3 months, and 97% were satisfied at 1 year (Schwartz et al., 2023a). This study also found that those taking norethindrone acetate were more likely than those with an IUD to report amenorrhea at 3 months, but there was no difference at 1 year (Schwartz et al., 2023a).

Another study found an overall median time to achieving amenorrhea of 182 days, meaning that fewer than half of participants achieved amenorrhea within 6 months on any method (Alaniz et al., 2023). The fastest single method (other than GnRHa) was norethindrone acetate (78 days). For those prescribed testosterone and norethindrone acetate simultaneously the time to amenorrhea was 188 days. The time to amenorrhea was the same (168 days) for those starting testosterone and depot medroxyprogesterone acetate together. A final study found high rates of effectiveness at achieving amenorrhea with little to no breakthrough bleeding and no difference in effectiveness between methods (Moussaoui et al., 2023).

There is limited data concerning the mental health benefits of menstrual suppression in TGD individuals. Most youth expect improvements in quality of life from menstrual suppression (Stewart et al., 2025), and they appear to be relatively satisfied with the methods they choose (Moussaoui et al., 2023). However, there is limited longitudinal data to track changes in mental health over time. One study showed that despite a high level of satisfaction, patients using menstrual cycle suppression at the time of the visit did not show any differences in mental health compared to those not prescribed menstrual suppression (Moussaoui et al., 2023). The study posited that the use of menstrual suppression, though often effective in achieving a reduction or complete pause of a menstrual cycle, may not be effective in treating gender dysphoria because having to take a daily pill, for example, potentially reminds individuals of their dysphoria. Additionally, dysphoria often impacts an individual in more than one way. Therefore using menstrual suppression may not address all contributors to gender dysphoria.

In summary, research has consistently shown that TGD youth assigned female at birth experience distress and dysphoria related to their menstrual cycles, many desire methods to suppress cycles, and these patients often have individual priorities that guide their choice of methods. Norethindrone acetate appears to be associated with a high rate of satisfaction and effectiveness in achieving amenorrhea and may be the preferred choice of patients who do not require contraception. Pelvic exams are a barrier to IUD placement, and the option to have the exam and placement conducted under sedation will increase access to this method. More research is needed in this area, particularly longitudinal research on mental health outcomes as well as studies with community samples of TGD youth who may represent different degrees of distress and dysphoria, risk and protective factors, and parental support.

Progesterone for individuals assigned male at birth

Some TGD individuals assigned male at birth may also find benefit from progesterone. Research in this area is limited but growing. No research studies have looked at the effects of the form of progesterone use in adolescents. Studies in adult transgender women have looked at the potential benefits of progesterone on breast development, gynoid body fat changes, and overall satisfaction and well-being. However, published studies acknowledge the challenge of separating the physical changes of progesterone only, as most patients prescribed progesterone are also taking estradiol. A 2022 systematic review was unable to draw conclusions about the positive or negative impact of progesterone on breast development, thrombotic or cardiovascular risk, bone health, or quality of life due to the inconsistency of study design as well as type and dose of progesterone used (Patel et al., 2022). However, more recent studies have demonstrated a positive impact of progesterone on patient satisfaction with breast development and mental health, with low risk of complications (Bahr et al., 2024; Chang et al., 2025). The first randomized controlled trial investigating the clinical effects and side effects of various doses of progesterone in TGD adults taking estradiol is currently underway (Dijkman et al., 2023).

Antiandrogens

Individuals born with testes may benefit from medications that aim specifically to decrease the effect of testosterone on the body, which is why an antiandrogen may be used. Currently, it is unclear which antiandrogen medication is most effective at producing desired clinical effects for patients, due in part to patients desiring different effects as well as a lack of population-specific research (Angus et al., 2021). It is more common for antiandrogens to be used in combination with estrogen therapy than as a standalone medication, and the only published randomized controlled trials comparing antiandrogens are confounded by patients concurrently prescribed estradiol (Smith et al., 2014). In addition, there are no studies investigating antiandrogen use in adolescents and young adults alone. More research is needed on the use and potential mental health benefits of antiandrogen medications in youth (Angus et al., 2025; Burinkul et al., 2021).

Spironolactone is a commonly prescribed antiandrogen for TGD adolescents assigned male at birth who are not treated with GnRHa. Spironolactone works both as an androgen receptor antagonist and by reducing testosterone production through decreasing the activity of 17α-hydroxylase in the testis (Fagart et al., 2010; Kossor et al., 1991). Spironolactone is also a weak agonist at the estrogen receptor (Levy et al., 1980). The combination of these effects leads to a reduction in testosterone's effects as well as a minimal estrogenic effect. Because of the antagonistic effects on the androgen receptor, spironolactone use often does not result in lower testosterone levels, despite its clinical effects. Spironolactone has been shown to stimulate the development of breast tissue in both cisgender men and TGD individuals (Rose et al., 1977). However, studies of breast development in TGD populations all involve concurrent estradiol use (Angus et al., 2025).

Spironolactone can reduce facial and body hair growth. Most of the evidence on hair growth is from spironolactone used as a treatment for androgen-driven acne and hirsutism in cisgender women (Ameri et al., 2026; Moghetti et al., 2000). Spironolactone is also a mineralocorticoid receptor antagonist and is considered a potassium-sparing diuretic. Adults with heart failure treated with spironolactone have an increased risk of hyperkalemia (Juurlink et al., 2004). However, a retrospective chart review of 85 TGD youth with a mean ± SD age of 16.6 ± 1.7 years treated with spironolactone as part of their gender care demonstrated that hyperkalemia in this population is rare and, if it does occur, it is transient and asymptomatic (Millington et al., 2019).

Bicalutamide is an androgen receptor antagonist that is traditionally used in the treatment of prostate cancer. Testosterone levels rise with this treatment due to the loss of negative feedback of androgens on the hypothalamic-pituitary-gonadal axis. The higher testosterone levels, unable to act at the androgen receptor, are aromatized to estradiol, which can cause an often desired side effect of breast tissue development in TGD individuals assigned male at birth. Bicalutamide has not been widely used in TGD individuals due to a rare but serious side effect of hepatotoxicity, which has been reported in studies of bicalutamide use in the treatment of prostate cancer in cisgender men (Hussain et al., 2014). Doses used for prostate cancer treatment are higher than those prescribed for TGD individuals. In addition, individuals receiving the medication for treatment of prostate cancer are typically older and may have more overall risk factors for hepatotoxicity compared to TGD adolescents and young adults.

There has been one published case report of bicalutamide-associated liver toxicity in a 17-year-old transgender female following 3 months of treatment (Wilde et al., 2024). One retrospective review of 40 adolescents assigned male at birth, aged 15.5 ± 1.55 years, treated with either bicalutamide alone (53%) or bicalutamide and estradiol (47%) reported no episodes of hepatotoxicity after a treatment of 2.94 ± 18.4 months (Fuqua et al., 2024). Of those on bicalutamide alone, 90.4% reported breast development. The largest study investigating bicalutamide use in TGD youth was a retrospective analysis of 84 individuals (median age: 18 years) taking bicalutamide compared to 69 individuals not taking it (median age: 19 years) (Burgener et al., 2026). After adjusting for age, race, BMI, baseline alanine aminotransferase (ALT) or aspartate aminotransferase (AST), and alcohol use, there were no clinically significant differences in the change of AST or ALT levels in the bicalutamide-treated group and the comparison groups over a mean treatment time of 113 days.

Managing aspects of other interventions

In addition to medical and nonmedical interventions that facilitate gender embodiment, young people may seek professional support related to adherence to or management of unwanted aspects of those same interventions. For example, adolescents who are prescribed testosterone may experience worsening acne, which may be especially distressing during a life period in which there is already a high risk of acne. Other young people may need support with adherence to treatments, especially if they struggle with the methods of administration due to needle phobia or aversion to pill swallowing. Adherence is a well-documented difficulty during the adolescent years, specifically across a range of healthcare disciplines (DiMatteo, 2004). Although TGD youth may appear more motivated to adhere to gender-affirming treatments than other types of interventions, research shows that adherence is nonetheless a struggle for some (Crabtree et al., 2024). Ongoing research efforts should aim to identify appropriate interventions for managing these peripheral aspects of care.

Managing dermatologic effects

In some individuals, there may be unwanted dermatological effects correlated with the use of antiandrogens or hormone replacement therapy. For instance, estrogen and progesterone can cause melasma, or uneven skin pigmentation, especially on the face. The risk of melasma may be more prominent or bothersome to individuals with darker skin pigmentation, although the use of sunblock or other dermatological interventions may help.

Individuals who use testosterone may be susceptible to acne or postinflammatory hyperpigmentation (Erickson-Schroth, 2022). Available research on acne, primarily conducted among TGD adults, has shown that individuals assigned female at birth are at higher risk of acne than those assigned male at birth, due in large part to testosterone use and occasionally worsened by chest binding, especially when binders are worn excessively or without good hygiene practices (Boos et al., 2022). One study specifically of adolescents used a retrospective chart review to track acne before and after testosterone initiation (Chu et al., 2023). The study found that approximately 23% of adolescents struggled with acne before initiation, and that acne worsened after initiation of testosterone for all individuals. Approximately half of those without baseline acne developed acne once testosterone was initiated. No individual factors were associated with the development of acne except for concurrent progestin use.

Treatment of acne with topical agents, oral antibiotics, and isotretinoin would generally be similar to treatment for cisgender youth (Hollingshead et al., 2022). In a retrospective chart review study of TGD adolescents and adults on testosterone, isotretinoin showed similar results as prior studies with cisgender individuals in terms of improvement and risk of recurrence (Choe et al., 2024). The only difference this study found from prior research was a slightly lower rate of complete clearance of acne, suggesting a longer course of treatment may be needed for those prescribed testosterone. For those who experience acne following initiation of testosterone, it tends to occur in the first 6 months of use, suggesting that early dermatological intervention may be helpful (Boos et al., 2022). In summary, TGD youth with acne benefit from similar treatment as their cisgender peers, but risk factors may be slightly different. Clinicians can also offer general recommendations for hygiene to improve overall skincare, especially in the context of testosterone use or chest binding.

Needle phobia and pill swallowing aversion

Research has shown that TGD youth consider difficulty with intervention methods, such as fear of needles, when making decisions about gender-affirming care (Daley et al., 2019; Stewart et al., 2025) and that adherence to intervention methods can be a barrier to continuous care (Crabtree et al., 2024). Needle phobia is a clinical diagnosis found within both the international classification of disease, 10th edition (ICD-10) and the diagnostic and statistical manual, 5th edition (DSM-5), and it is characterized by marked fear or anxiety regarding needles lasting more than 6 months in duration and typically related to anxiety (American Psychiatric Association, 2014). While more commonly experienced in children and adolescents, adults can also develop needle phobia. Even when symptoms are not severe enough to meet a diagnostic threshold, needle worry and needle phobia can be significant enough that it impedes a person's ability to receive gender-affirming healthcare. The results of one study suggest that 63% of youth aged 6–17 years reported a fear of needles, suggesting that the prevalence of needle worries may be high enough that it could be advantageous to discuss comfort with needles with all younger patients (Taddio et al., 2012).

If needle worries are present for a patient, the recommendation may be to offer behavioral counseling, including exposure-based treatment using a cognitive-behavioral framework. Another evidence-based treatment is distraction, in which a patient is directed to divert attention to another stimulus that is more emotionally neutral than the needle procedure. There is also research to support the use of hypnosis as well as medications such as topical analgesics, nitrous oxide gas, or benzodiazepines (Orenius et al., 2018).

An aversion to swallowing capsules or pills can also be a deterrent or promote delay in the delivery of gender-affirming medical treatments. The typical age of onset for aversion to swallowing pills varies; thus the condition can develop even in adulthood (American Psychiatric Association, 2014). There is considerable evidence that it is possible to treat pill swallowing aversion within a single session, which utilizes exposure-based treatment and follows the patient as they create and overcome increasingly difficult activities using a hierarchical framework (Davis et al., 2013). Indeed, given that many gender-affirming medical treatments require either injection, retrieval of blood before or over the course of treatment, or swallowing pills, careful planning of how a patient will endure the treatment is a critical factor in successful treatment. In larger institutions, interventions or resources may be available through child life or social work departments. More research, particularly using qualitative methodology, would help to clarify the experiences of TGD young people specifically and to identify population-specific interventions or strategies for reducing barriers related to intervention methods.

Conclusion

There is growing recognition that binary transition options may not adequately meet the gender embodiment goals of all TGD youth, especially for those who identify as nonbinary or otherwise genderqueer (Hastings et al., 2021). For TGD youth across the spectrum of gender, supplementary medical and nonmedical options may further reduce dysphoria and support affirmation. In contrast to the large body of research literature documenting the risks, benefits, and experiences with gender-affirming hormones and pubertal suppression, less research has been conducted on other medical and nonmedical options to support gender affirmation and embodiment goals. Some interventions, such as chest binding, menstrual cycle suppression, and voice training, have emerging bodies of research, while other areas, such as the use of shapewear and packers or progesterone as a supplement to estrogen, have almost no research to date. Both quantitative and qualitative research methods are needed to clarify the safety, potential benefits, and overall perception of these other medical and nonmedical gender-affirming interventions.

References

Akgul, S., Tuzun, Z., Pehlivanturk Kizilkan, M., Alboga, D., Yalcin, E., Ozcelik, U., et al., 2025. The impact of chest binding on pulmonary functions of trans and gender diverse youth. Journal of Adolescent Health 76 (1), 37–43. Available at https://doi.org/10.1016/j.jadohealth.2024.07.031.

Alaniz, V.I., Sheeder, J.L., Whitmore, G.T., Wilde, M.D., Hutchens, K.J., Nokoff, N.J., et al., 2023. Menstrual suppression in adolescent and young adult transgender males. Journal of Pediatric and Adolescent Gynecology 36 (2), 116–121. Available at https://doi.org/10.1016/j.jpag.2022.10.007.

Ameri, K., Ghafoor, A.A.A., Maazi, M., Rehmus, W., Gregory, A., 2026. Efficacy and safety of spironolactone for adolescent acne vulgaris. International Journal of Dermatology. Available at https://doi.org/10.1111/ijd.70337.

American Psychiatric Association, 2014. Diagnostic and Statistical Manual of Mental Disorders. American Psychiatric Association. Available at http://doi.org/10.1176/appi.books.9780890425596.

Amos, N., Grant, R., Lin, A., Hill, A.O., Pang, K.C., Skinner, S.R., et al., 2025. Mental health and wellbeing outcomes among trans young people in australia who are supported to affirm their gender. Journal of Adolescent Health 77 (1), 51–58. Available at https://doi.org/10.1016/j.jadohealth.2025.03.011.

Angus, L.M., Nolan, B.J., Zajac, J.D., Cheung, A.S., 2021. A systematic review of antiandrogens and feminization in transgender women. Clinical Endocrinology 94 (5), 743–752. Available at https://doi.org/10.1111/cen.14329.

Angus, L.M., Leemaqz, S.Y., Kasielska-Trojan, A.K., Mikołajczyk, M., Doery, J.C.G., Zajac, J.D., et al., 2025. Effect of spironolactone and cyproterone acetate on breast growth in transgender people: a randomized clinical trial. Journal of Clinical Endocrinology and Metabolism 110 (6), e1874. Available at https://doi.org/10.1210/clinem/dgae650.

Ashley, F., 2021. The misuse of gender dysphoria: toward greater conceptual clarity in transgender health. Perspectives on Psychological Science 16 (6), 1159–1164. Available at https://doi.org/10.1177/1745691619872987.

Bahr, C., Ewald, J., Dragovich, R., Gothard, M.D., 2024. Effects of progesterone on gender affirmation outcomes as part of feminizing hormone therapy. Journal of the American Pharmacists Association 64 (1), 268–272. Available at https://doi.org/10.1016/j.japh.2023.08.001.

Barr, S.M., Snyder, K.E., Adelson, J.L., Budge, S.L., 2021. Posttraumatic stress in the trans community: the roles of anti-transgender bias, non-affirmation, and internalized transphobia. Psychology of Sexual Orientation and Gender Diversity 9 (4), 410–421. Available at https://doi.org/10.1037/sgd0000500.

Berrahou, I.K., Grimes, A., Autry, A.M., Hawkins, M., 2022. Management of menstruation in transgender and gender nonbinary adolescents. Clinical Obstetrics and Gynecology 65 (4), 753–767. Available at https://doi.org/10.1097/GRF.0000000000000710.

Bettcher, T.M., 2014. Trapped in the wrong theory: rethinking trans oppression and resistance. Signs 39 (2), 383–406. Available at https://doi.org/10.1086/673088.

Bishop, M.D., Caba, A.E., Watson, R.J., Fish, J.N., 2024. Chest binding: sociodemographic characteristics among a national sample of transgender and gender diverse adolescents. Journal of Adolescent Health 74 (6), 1256–1259. Available at https://doi.org/10.1016/j.jadohealth.2024.01.013.

Boos, M.D., Hollingshead, N., Hodax, J.K., 2022. Management of acne in transgender and gender diverse youth part 1: gender affirming care and risk factors for the development of acne. Pediatric Dermatology 39 (6), 866–869. Available at https://doi.org/10.1111/pde.15113.

Bradford, N.J., Rider, G.N., Spencer, K.G., 2021. Hair removal and psychological well-being in transfeminine adults: associations with gender dysphoria and gender euphoria. Journal of Dermatological Treatment 32 (6), 635–642. Available at https://doi.org/10.1080/09546634.2019.1687823.

Burgener, K., DeBosch, B., Wang, J., Lewis, C., Herrick, C., 2026. Bicalutamide does not raise transaminases clinically significantly compared to alternative antiandrogen regimens among transfeminine adolescents and young adults: a retrospective cohort study. International Journal of Transgender Health 27 (1), 503–512. Available at https://doi.org/10.1080/26895269.2025.2452184.

Burinkul, S., Panyakhamlerd, K., Suwan, A., Tuntiviriyapun, P., Wainipitapong, S., 2021. Anti-androgenic effects comparison between cyproterone acetate and spironolactone in transgender women: a randomized controlled trial. The Journal of Sexual Medicine 18 (7), 1299–1307. Available at https://doi.org/10.1016/j.jsxm.2021.05.003.

Chadwick, K.A., Coleman, R., Andreadis, K., Pitti, M., Rameau, A., 2022. Outcomes of gender-affirming voice and communication modification for transgender individuals. Laryngoscope 132 (8), 1615–1621. Available at https://doi.org/10.1002/lary.29946.

Chang, J., Yung, K., 2021. Gender affirming voice care: a literature review. International Journal of Head and Neck Surgery 12 (3), 93–97. Available at https://doi.org/10.5005/jp-journals-10001-1448.

Chang, J.J., Tran, N.K., Flentje, A., Lubensky, M.E., Obedin-Maliver, J., Lunn, M.R., et al., 2025. Progestogen experience among transgender women and gender diverse adults assigned male at birth in the United States. Endocrine Practice 31 (11), 1449–1461. Available at https://doi.org/10.1016/j.eprac.2025.06.028.

Choe, J., Shields, A., Ferreira, A., Gold, S., Gotschall, J.W., Kamal, K., et al., 2024. Isotretinoin for acne in transgender and gender-diverse individuals receiving masculinizing hormone therapy. JAMA Dermatology 160 (7), 741–745. Available at https://doi.org/10.1001/jamadermatol.2024.1420.

Chu, L., Gold, S., Harris, C., Lawley, L., Gupta, P., Tangpricha, V., et al., 2023. Incidence and factors associated with acne in transgender adolescents on testosterone: a retrospective cohort study. Endocrine Practice 29 (5), 353–355. Available at https://doi.org/10.1016/j.eprac.2023.02.002.

Crabtree, L., Connelly, K.J., Guerriero, J.T., Battison, E.A.J., Tiller-Ormord, J., Sutherland, S.M., et al., 2024. A more nuanced story: pediatric gender-affirming healthcare is associated with satisfaction and confidence. Journal of Adolescent Health 75 (5), 772–779. Available at https://doi.org/10.1016/j.jadohealth.2024.06.016.

Dacakis, G., Davies, S., Oates, J.M., Douglas, J.M., Johnston, J.R., 2013. Development and preliminary evaluation of the transsexual voice questionnaire for male-to-female transsexuals. Journal of Voice 27 (3), 312–320. Available at https://doi.org/10.1016/j.jvoice.2012.11.005.

Daley, T., Grossoehme, D., McGuire, J.K., Corathers, S., Conard, L.A., Lipstein, E.A., 2019. "I couldn't see a downside": decision-making about gender-affirming hormone therapy. Journal of Adolescent Health 65 (2), 274–279. Available at https://doi.org/10.1016/j.jadohealth.2019.02.018.

Davies, S., Papp, V.G., Antoni, C., 2015. Voice and communication change for gender nonconforming individuals: giving voice to the person inside. International Journal of Transgenderism 16 (3), 117–159. Available at https://doi.org/10.1080/15532739.2015.1075931.

Davis, T.E., Reuther, E.T., Rudy, B.M., 2013. One-session treatment of a specific phobia of swallowing pills. Clinical Case Studies 12 (6), 399–410. Available at https://doi.org/10.1177/1534650113497533.

De Chants, J.P., Price, M.N., Green, A.E., Davis, C.K., Pick, C.J., 2022. Association of updating identification documents with suicidal ideation and attempts among transgender and nonbinary youth. International Journal of Environmental Research and Public Health 19 (9). Available at https://doi.org/10.3390/ijerph19095016.

Diana, P., Belluzzi, B., Corona, F., Barbi, E., Tornese, G., 2024. Nonmedical gender-affirming practices in transgender and gender diverse adolescents: a narrative review. Transgender Health. Available at https://doi.org/10.1089/trgh.2023.0194.

Dijkman, B.A.M., Helder, D., Boogers, L.S., Gieles, N.C., van Heesewijk, J.O., Slaa, St, et al., 2023. Addition of progesterone to feminizing gender-affirming hormone therapy in transgender individuals for breast development: a randomized controlled trial. BMC Pharmacology and Toxicology 24 (1). Available at https://doi.org/10.1186/s40360-023-00724-4.

DiMatteo, M.R., 2004. Variations in patients' adherence to medical recommendations: a quantitative review of 50 years of research. Medical Care 42 (3), 200–209. Available at https://doi.org/10.1097/01.mlr.0000114908.90348.f9.

Erickson-Schroth, L., 2022. Trans Bodies, Trans Selves: A Resource by and for Transgender Communities. Oxford University Press vol. 2.

Fagart, J., Hillisch, A., Huyet, J., Bärfacker, L., Fay, M., Pleiss, U., et al., 2010. A new mode of mineralocorticoid receptor antagonism by a potent and selective nonsteroidal molecule. Journal of Biological Chemistry 285 (39), 29932–29940. Available at https://doi.org/10.1074/jbc.m110.131342.

Feinberg, L., 1997. Transgender Warriors: Making History from Joan of Arc to Dennis Rodman. Beacon Press.

Fuqua, J.S., Shi, E., Eugster, E.A., 2024. A retrospective review of the use of bicalutamide in transfeminine youth: a single center experience. International Journal of Transgender Health 25 (3), 533–537. Available at https://doi.org/10.1080/26895269.2023.2294321.

Gao, J.L., Streed, C.G., Thompson, J., Dommasch, E.D., Peebles, J.K., 2023. Androgenetic alopecia in transgender and gender diverse populations: a review of therapeutics. Journal of the American Academy of Dermatology 89 (4), 774–783. Available at https://doi.org/10.1016/j.jaad.2021.08.067.

Gauthier, J., Medina, K., Dierkhising, C., 2021. Analysis of hate crimes in transgender communities. Journal of Hate Studies 17 (2). Available at https://doi.org/10.33972/jhs.158.

Ginsberg, B.A., Calderon, M., Seminara, N.M., Day, D., 2016. A potential role for the dermatologist in the physical transformation of transgender people: a survey of attitudes and practices within the transgender community. Journal of the American Academy of Dermatology 74 (2), 303–308. Available at https://doi.org/10.1016/j.jaad.2015.10.013.

Hancock, A., Helenius, L., 2012. Adolescent male-to-female transgender voice and communication therapy. Journal of Communication Disorders 45 (5), 313–324. Available at https://doi.org/10.1016/j.jcomdis.2012.06.008.

Hastings, J., Bobb, C., Wolfe, M., Jimenez, Z.A., Amand, C.S., 2021. Medical care for nonbinary youth: individualized gender care beyond a binary framework. Pediatric Annals 50 (9), e384. Available at https://doi.org/10.3928/19382359-20210818-03.

Hollingshead, N., Hodax, J.K., Boos, M.D., 2022. Management of acne in transgender and gender diverse youth part 2: unique considerations and strategies in medical treatment. Pediatric Dermatology 39 (6), 870–875. Available at https://doi.org/10.1111/pde.15114.

Hussain, S., Haidar, A., Bloom, R.E., Zayouna, N., Piper, M.H., Jafri, S.M.R., 2014. Bicalutamide-induced hepatotoxicity: a rare adverse effect. American Journal of Case Reports 15, 266–270. Available at https://doi.org/10.12659/AJCR.890679.

Ingprasert, S., Tanglertsampan, C., Tangphianphan, N., Reanmanee, C., 2016. Efficacy and safety of minoxidil 3% lotion for beard enhancement: a randomized, double-masked, placebo-controlled study. The Journal of Dermatology 43 (8), 968–969. Available at https://doi.org/10.1111/1346-8138.13312.

James, S.E., Herman, J.L., Durso, L.E., 2024. Early Insights: A Report of the 2022. U.S. Transgender Survey.

Jones, B.A., Bouman, W.P., Haycraft, E., Arcelus, J., 2019. The gender congruence and life satisfaction scale (GCLS): development and validation of a scale to measure outcomes from transgender health services. International Journal of Transgenderism 20 (1), 63–80. Available at https://doi.org/10.1080/15532739.2018.1453425.

Julian, J.M., Salvetti, B., Held, J.I., Murray, P.M., Lara-Rojas, L., Olson-Kennedy, J., 2021. The impact of chest binding in transgender and gender diverse youth and young adults. Journal of Adolescent Health 68 (6), 1129–1134. Available at https://doi.org/10.1016/j.jadohealth.2020.09.029.

Juurlink, D.N., Mamdani, M.M., Lee, D.S., Kopp, A., Austin, P.C., Laupacis, A., et al., 2004. Rates of hyperkalemia after publication of the Randomized Aldactone Evaluation Study. Journal of Medicine 351 (6), 543–551. Available at https://doi.org/10.1056/NEJMoa040135.

Kanj, R.V., Conard, L.A.E., Corathers, S.D., Trotman, G.E., 2019. Hormonal contraceptive choices in a clinic-based series of transgender adolescents and young adults. International Journal of Transgenderism 20 (4), 413–420. Available at https://doi.org/10.1080/15532739.2019.1631929.

Kausar, N., Mumtaz, S., Shoaib, M., 2024. Lived experiences of individuals with gender dysphoria: a qualitative analysis. Sexuality & Culture 28 (2), 692–709. Available at https://doi.org/10.1007/s12119-023-10140-9.

Kossor, D.C., Kominami, S., Takemori, S., Colby, H.D., 1991. Role of the steroid 17 alpha-hydroxylase in spironolactone-mediated destruction of adrenal cytochrome P-450. Molecular Pharmacology 40 (2), 321–325. Available at https://doi.org/10.1016/s0026-895x(25)12955-6.

Levy, J., Burshell, A., Marbach, M., Afllalo, L., Glick, S.M., 1980. Interaction of spironolactone with oestradiol receptors in cytosol. Journal of Endocrinology 84 (3), 371–379. Available at https://doi.org/10.1677/joe.0.0840371.

Linsenmeyer, W., Heiden-Rootes, K., Drallmeier, T., Thomure, M., Nye, E., Armstrong, E., 2022. Nutrition and exercise as a 'source of empowerment': a narrative inquiry of transgender men. SSM - Qualitative Research in Health 2, 100128. Available at https://doi.org/10.1016/j.ssmqr.2022.100128.

Lykens, J.E., Leblanc, A.J., Bockting, W.O., 2018. Healthcare experiences among young adults who identify as genderqueer or nonbinary. LGBT Health 5 (3), 191–196. Available at https://doi.org/10.1089/lgbt.2017.0215.

Marinelli, L., Bichiri, A., Cagnina, S., Castella, L., Ghigo, E., Motta, G., 2024. Efficacy of topical minoxidil in enhancing beard growth in a group of transgender assigned female at birth individuals on gender affirming hormone therapy. Journal of Endocrinological Investigation 47 (11), 2843–2850. Available at https://doi.org/10.1007/s40618-024-02373-8.

McLemore, K.A., 2014. Experiences with misgendering: identity misclassification of transgender spectrum individuals. Self and Identity 14 (1), 51–74. Available at https://doi.org/10.1080/15298868.2014.950691.

Mikovits, J.C., 2022. "I don't feel like I'm a person": nursing knowledge of transgender care through the lens of transgender people. Journal of Advanced Nursing 78 (9), 3012–3024. Available at https://doi.org/10.1111/jan.15308.

Millington, K., Liu, E., Chan, Y.M., 2019. The utility of potassium monitoring in gender-diverse adolescents taking spironolactone. Journal of the Endocrine Society 3 (5), 1031–1038. Available at https://doi.org/10.1210/js.2019-00030.

Mills, M., Stoneham, G., Davies, S., 2019. Toward a protocol for transmasculine voice: a service evaluation of the voice and communication therapy group program, including long-term follow-up for trans men at the London Gender Identity Clinic. Transgender Health 4 (1), 143–151. Available at https://doi.org/10.1089/trgh.2019.0011.

Moghetti, P., Tosi, F., Tosti, A., Negri, C., Misciali, C., Perrone, F., et al., 2000. Comparison of spironolactone, flutamide, and finasteride efficacy in the treatment of hirsutism: a randomized, double blind, placebo-controlled trial 1. The Journal of Clinical Endocrinology & Metabolism 85 (1), 89–94. Available at https://doi.org/10.1210/jcem.85.1.6245.

Monstrey, S., Selvaggi, G., Ceulemans, P., Van Landuyt, K., Bowman, C., Blondeel, P., et al., 2008. Chest-wall contouring surgery in female-to-male transsexuals: a new algorithm. Plastic and Reconstructive Surgery 121 (3), 849–859. Available at https://doi.org/10.1097/01.prs.0000299921.15447.b2.

Moussaoui, D., O'Connell, M.A., Elder, C.V., Grover, S.R., Pang, K.C., 2023. Characteristics of menstrual suppression and its association with mental health in transgender adolescents. Obstetrics and Gynecology 142 (5), 1096–1104. Available at https://doi.org/10.1097/AOG.0000000000005317.

Myers, B., Bell, T., 2020. Adapting vocal function exercises for voice masculinization. Perspectives of the ASHA Special Interest Groups 5 (4), 861–866. Available at https://doi.org/10.1044/2020_persp-20-00076.

Nolan, I.T., Morrison, S.D., Arowojolu, O., Crowe, C.S., Massie, J.P., Adler, R.K., et al., 2019. The role of voice therapy and phonosurgery in transgender vocal feminization. Journal of Craniofacial Surgery 30 (5), 1368–1375. Available at https://doi.org/10.1097/SCS.0000000000005132.

Olson, K.R., Durwood, L., Demeules, M., McLaughlin, K.A., 2016. Mental health of transgender children who are supported in their identities. Pediatrics 137 (3). Available at https://doi.org/10.1542/peds.2015-3223.

Olson-Kennedy, J., Warus, J., Okonta, V., Belzer, M., Clark, L.F., 2018. Chest reconstruction and chest dysphoria in transmasculine minors and young adults comparisons of nonsurgical and postsurgical cohorts. JAMA Pediatrics 172 (5), 431–436. Available at https://doi.org/10.1001/jamapediatrics.2017.5440.

Onejeme, C., Fitzgibbon, M., Jimenez, A., Ross, L., 2025. Clinical pearls addressing hair growth and loss in transgender patients on hormone replacement therapy. Dermatology Online Journal 31 (1). Available at https://doi.org/10.5070/d331164961.

Orenius, T., LicPsych, Säilä, H., Mikola, K., Ristolainen, L., 2018. Fear of injections and needle phobia among children and adolescents: an overview of psychological, behavioral, and contextual factors. SAGE Open Nursing 4. Available at https://doi.org/10.1177/2377960818759442.

Pang, K.C., Nguyen, T.P., Upreti, R., 2021. Case report: successful use of minoxidil to promote facial hair growth in an adolescent transgender male. Frontiers in Endocrinology 12. Available at https://doi.org/10.3389/fendo.2021.725269.

Patel, K.T., Adeel, S., Rodrigues Miragaya, J., Tangpricha, V., 2022. Progestogen use in gender-affirming hormone therapy: a systematic review. Endocrine Practice 28 (12), 1244–1252. Available at https://doi.org/10.1016/j.eprac.2022.08.012.

Peitzmeier, S.M., Silberholz, J., Gardner, I.H., Weinand, J., Acevedo, K., 2021. Time to first onset of chest binding-related symptoms in transgender youth. Pediatrics 147 (3). Available at https://doi.org/10.1542/PEDS.2020-0728.

Pollitt, A.M., Ioverno, S., Russell, S.T., Li, G., Grossman, A.H., 2021. Predictors and mental health benefits of chosen name use among transgender youth. Youth and Society 53 (2), 320–341. Available at https://doi.org/10.1177/0044118X19855898.

Quinn, S., Oates, J., Dacakis, G., 2022. Perceived gender and client satisfaction in transgender voice work: comparing self and listener rating scales across a training program. Folia Phoniatrica et Logopaedica 74 (5), 364–379. Available at https://doi.org/10.1159/000521226.

Rapoport, S.K., Varelas, E.A., Park, C., Brown, S.K., Goldberg, L., Courey, M.S., 2023. Patient satisfaction and acoustic changes in trans women after gender affirming voice training. Laryngoscope 133 (9), 2340–2345. Available at https://doi.org/10.1002/lary.30543.

Reisner, S.L., Pletta, D.R., Harris, A., Campbell, J., Asquith, A., Pardee, D.J., et al., 2023. Exploring gender euphoria in a sample of transgender and gender diverse patients at two U.S. urban community health centers. Psychiatry Research 329. Available at https://doi.org/10.1016/j.psychres.2023.115541.

Rose, L.I., Underwood, R.H., Newmark, S.R., Kisch, E.S., Williams, G.H., 1977. Pathophysiology of spironolactone induced gynecomastia. Annals of Internal Medicine 87 (4), 398–403. Available at https://doi.org/10.7326/0003-4819-87-4-398.

Rotondi, N.K., Bauer, G.R., Scanlon, K., Kaay, M., Travers, R., Travers, A., 2013. Nonprescribed hormone use and self-performed surgeries: "do-it-yourself" transitions in transgender communities in Ontario, Canada. American Journal of Public Health 103 (10), 1830–1836. Available at https://doi.org/10.2105/AJPH.2013.301348.

Russell, M.R., Abrams, M., 2019. Transgender and nonbinary adolescents: the role of voice and communication therapy. Perspectives of the ASHA Special Interest Groups 4 (6), 1298–1305. Available at https://doi.org/10.1044/2019_persp-19-00034.

Russell, S.T., Pollitt, A.M., Li, G., Grossman, A.H., 2018. Chosen name use is linked to reduced depressive symptoms, suicidal ideation, and suicidal behavior among transgender youth. Journal of Adolescent Health 63 (4), 503–505. Available at https://doi.org/10.1016/j.jadohealth.2018.02.003.

Schwartz, B.I., Effron, A., Bear, B., Short, V.L., Eisenberg, J., Felleman, S., et al., 2022. Experiences with menses in transgender and gender nonbinary adolescents. Journal of Pediatric and Adolescent Gynecology 35 (4), 450–456. Available at https://doi.org/10.1016/j.jpag.2022.01.015.

Schwartz, B.I., Bear, B., Short, V.L., Kazak, A.E., 2023a. Outcomes of menstrual management use in transgender and gender-diverse adolescents. Obstetrics and Gynecology 141 (4), 748–755. Available at https://doi.org/10.1097/AOG.0000000000005123.

Schwartz, B.I., Bear, B., Kazak, A.E., 2023b. Menstrual management choices in transgender and gender diverse adolescents. Journal of Adolescent Health 72 (2), 207–213. Available at https://doi.org/10.1016/j.jadohealth.2022.09.023.

Schwarz, K., Cielo, C.A., Spritzer, P.M., Villas-Boas, A.P., Costa, A.B., Fontanari, A.M.V., et al., 2023. A speech therapy for transgender women: an updated systematic review and meta-analysis. Systematic Reviews 12 (1). Available at https://doi.org/10.1186/s13643-023-02267-5.

Smith, K.P., Madison, C.M., Milne, N.M., 2014. Gonadal suppressive and crosssex hormone therapy for gender dysphoria in adolescents and adults. Pharmacotherapy 34 (12), 1282–1297. Available at https://doi.org/10.1002/phar.1487.

Smith, C., Gregory, C., Bryant, L., 2024. Utilizing virtual reality for gender-affirming voice training: surveying the attitudes and perspectives of potential consumers. International Journal of Language & Communication Disorders 59 (3), 976–990. Available at https://doi.org/10.1111/1460-6984.12968.

Stewart, H.L., Manos, B.E., Chelvakumar, G., Bonny, A.E., 2025. Menstrual suppression in gender-diverse youth: what's most important to patients? Journal of Pediatric and Adolescent Gynecology 38 (2), 195–198. Available at https://doi.org/10.1016/j.jpag.2024.11.004.

Taddio, A., Ipp, M., Thivakaran, S., Jamal, A., Parikh, C., Smart, S., et al., 2012. Survey of the prevalence of immunization noncompliance due to needle fears in children and adults. Vaccine 30 (32), 4807–4812. Available at https://doi.org/10.1016/j.vaccine.2012.05.011.

Thoreson, N., Marks, D.H., Peebles, J.K., King, D.S., Dommasch, E., 2020. Health insurance coverage of permanent hair removal in transgender and gender-minority patients. JAMA Dermatology 156 (5), 561–565. Available at https://doi.org/10.1001/jamadermatol.2020.0480.

Todd, K., Peitzmeier, S.M., Kattari, S.K., Miller-Perusse, M., Sharma, A., Stephenson, R., 2019. Demographic and behavioral profiles of nonbinary and binary transgender youth. Transgender Health 4 (1), 254–261. Available at https://doi.org/10.1089/trgh.2018.0068.

Tüzün, Z., Akgül, S., Pehlivantürk Kızılkan, M., Başar, K., 2025. Psychosocial correlates of chest binding in transgender and gender-diverse youth. Transgender Health 10 (4), 355–363. Available at https://doi.org/10.1089/trgh.2024.0240.

Voss, R.V., Carl, R., Liem, R., Kuhns, L., Rybacki, D., Wolf, S.F., et al., 2024. 12. Impact of chest binding on exercise capacity in transgender and gender-diverse youth assigned female at birth. Journal of Adolescent Health 74 (3), S7. Available at https://doi.org/10.1016/j.jadohealth.2023.11.031.

Weinhardt, L.S., Stevens, P., Xie, H., Wesp, L.M., John, S.A., Apchemengich, I., et al., 2017. Transgender and gender nonconforming youths' public facilities use and psychological well-being: a mixed-method study. Transgender Health 2 (1), 141–150. Available at https://doi.org/10.1089/trgh.2017.0020.

Wilde, B., Diamond, J.B., Laborda, T.J., Frank, L., O'Gorman, M.A., Kocolas, I., 2024. Bicalutamide-induced hepatotoxicity in a transgender male-to-female adolescent. Journal of Adolescent Health 74 (1), 202–204. Available at https://doi.org/10.1016/j.jadohealth.2023.08.024.

Williams, C.R., McGregor, K., Feld, A., Boskey, E.R., 2024. Understanding their experiences: psychosocial functioning of nonbinary and binary youth at the time of hormone readiness assessment. LGBT Health 11 (2), 164–169. Available at https://doi.org/10.1089/lgbt.2023.0056.

Chapter 9

Surgical interventions in adolescence and young adulthood

Sarah K. Fadich[1], Rebecca Rada[2], and Jessica A. Lee[3]

[1]*David Geffen School of Medicine, University of California, Los Angeles, CA, United States,* [2]*UCLA Gender Health Program, University of California, Los Angeles, CA, United States,* [3]*Children's Hospital Los Angeles, University of Southern California, Los Angeles, CA, United States*

Learning objective

- Describe existing gender-affirming surgeries and summarize the available research on surgical options and outcomes for transgender and gender diverse adolescents and young adults.
- Discuss age considerations, shared decision-making in adolescence, and the benefits of preparing for future surgeries even in the early stages of decision-making.
- Highlight gaps in the research, such as postoperative pain and recovery experiences among youth, risks and benefits of surgery at different ages, and surgical implications for young people who received early pubertal suppression.
- Provide recommendations for community-centered research design and answering newly emerging research questions arising from increased access to gender-affirming surgeries at younger ages.

Key terms and definitions

- **Top surgery** is an umbrella term for gender-affirming surgeries to the chest area, including mastectomy with chest reconstruction, chest reduction surgery, and breast augmentation.
- **Bottom surgery** is an umbrella term for gender-affirming surgeries that involve reproductive organs, such as orchiectomy and hysterectomy, or genitals, such as vulvoplasty, vaginoplasty, metoidioplasty, and phalloplasty.
- **Masculinizing surgery** is used to categorize surgical options for transgender men and nonbinary people assigned female at birth. **Transmasculine** is also sometimes used as an umbrella term to more clearly distinguish the individuals who may be seeking these surgeries, with the recognition that some gender identities do not align with this terminology.
- **Feminizing surgery** is used to categorize surgical options for transgender women and nonbinary people assigned male at birth. **Transfeminine** is also sometimes used as an umbrella term to more clearly distinguish the individuals who may be seeking these surgeries, with the recognition that some gender identities do not align with this terminology.
- **Passing** refers to the ability for a transgender person to navigate their daily life without their transgender identity being involuntarily revealed or inferred from external cues.
- **Early pubertal suppression** refers to the use of gonadotropin-releasing hormone agonists beginning in sexual maturity rating stages 2 or 3, as it relates to the prevention of later-stage pubertal changes and implications for surgery.

Gender-affirming surgery (GAS) is a broad category of surgical interventions that function to align a person's physical body with their gender identity. Colloquially, these are further categorized as top, bottom, body, vocal, or facial surgery. Top surgery can refer to masculinizing chest surgery, such as mastectomy or breast reduction, or feminizing chest surgery, such as breast augmentation. Bottom surgery is most often used to describe vulvoplasty, vaginoplasty, phalloplasty, or

Advances in Research on Pediatric Gender Affirming Healthcare. DOI: https://doi.org/10.1016/B978-0-443-29019-0.00011-7
© 2026 Elsevier Inc. All rights are reserved, including those for text and data mining, AI training, and similar technologies.

metoidioplasty. While these typically include gonadectomy and/or hysterectomy, they can also be pursued independently. Body masculinization or feminization surgery aims to alter body fat distribution to contour and enhance certain areas of the body. Facial GAS is a collection of procedures that aim to align both bony structure and soft tissues within the sexually dimorphic range that most closely aligns with a person's identified gender identity and expression.

The medical indications for all gender-affirming treatments, whether medical or surgical, are to treat gender dysphoria and its psychosocial sequelae by increasing gender congruence and improving quality of life and functioning across all domains of life. Aligning a person's primary and/or secondary sex characteristics is highly individualized, and thus, treatment interventions are likewise tailored to the needs of each individual.

When providing transition-related treatment recommendations, it is beneficial to consider a patient's priorities in the context of long-term outcomes and to understand how the sequence of each intervention might impact subsequent interventions. Access to gonadotropin-releasing hormone agonists (GnRHa), for example, reduces the need for some surgeries, compared to those who begin medical treatment after puberty has already resulted in permanent, unaffirming changes. An adolescent or adult who has completed testosterone-driven puberty and has severe dysphoria related to their chest may find that starting estrogen gender-affirming hormone (GAH) provides adequate breast growth to alleviate the need for breast augmentation. Because sex hormones and puberty modulators have an impact on surgical planning, clinicians and researchers should understand how medical and surgical modalities impact one another and inform an overall treatment plan.

Community acknowledgment and considerations

The authors would like to acknowledge the beauty and diversity within the transgender and gender diverse (TGD) community. Every person's gender identity and journey toward congruence is valid, unique, and individual, regardless of how much or how little treatment is pursued to affirm gender identity. There is no correct or incorrect way to transition, and clinicians and researchers must understand how deeply individual each person's journey is. What may be a source of severe dysphoria for one may be neutral or even cherished for another. The focus of GAS research and clinical practice includes the various anatomical features and sex characteristics, how these might align with identity or expression, or cause dysphoria, and how surgery may help achieve alignment, provide psychological relief, reduce dysphoria, and allow individuals to thrive in their affirmed gender.

The TGD community is particularly vulnerable to the harms of dehumanizing rhetoric, prohibitions on participation in public life (i.e., sports bans and bathroom restrictions), and discriminatory health policies that are increasingly prominent worldwide. For many, gender-affirming healthcare is medically necessary and lifesaving, and the medical model focuses on the impact of gender incongruence and how it contributes to the harm and suffering of an individual. However, the validity of someone's gender identity does not rely on approval or affirmation from a medical professional.

Drawing the delineation between gender theory and medical management is an important nuance to ensure that harm is not perpetuated by "medicalizing" TGD identities or allowing political rhetoric to impact medically critical care. This distinction is intuitively understood when treating cisgender people experiencing distress from secondary sex characteristics that are incongruent with their gender, such as a cisgender man with gynecomastia undergoing surgery to remove breast tissue or receiving testosterone replacement therapy for hypogonadism. The gender identity of a cisgender man in this example is not even central to or impacted by the decision to treat or not treat. Gender-affirming medical and surgical interventions aim to alleviate psychological distress regardless of whether a person is cisgender, transgender, or nonbinary.

Some TGD individuals pursuing surgical intervention may use the term "passing" to describe their goals of care. The term "passing" refers to the ability to navigate daily life without a transgender identity being involuntarily revealed or inferred from external cues. For many, being able to pass also has safety implications. TGD individuals experience a significantly higher rate of physical violence, especially transgender women of color (James et al., 2016). Passing can therefore be lifesaving, and it can also impact how individuals are able to navigate the world—in education, employment, public settings, and intimate relationships.

While passing may be identified by some patients as a pressing objective of treatment, the validity of someone's gender identity does not rely on adherence to this arbitrary standard. Furthermore, the term "passing" can also be a source of tremendous pain and invalidation, and it perpetuates an unhelpful, binary conceptualization of gender. Healthcare providers are encouraged to solicit individualized goals without imposing rigid stereotypes or assumptions about desired outcomes, and the term passing should only be used when mirroring a patient's own language. Researchers are similarly encouraged to take into consideration the diversity of experiences and perspectives throughout the TGD community when formulating research questions, designing methodology, and interpreting research results.

Overview of gender-affirming surgery

With growing acceptance and access to gender-affirming care the number of TGD adolescents and young adults seeking GAS is increasing (Robinson et al., 2023). Research clearly demonstrates that masculinizing top surgery results in improved mental health and well-being in adolescents, but research on other surgeries among young people remains limited (Mahfouda et al., 2019). This discrepancy in research results from the fact that masculinizing top surgery is the only GAS regularly accessed by adolescents, for several reasons. First, living with breast tissue as a teenage boy or nonbinary adolescent not only exacerbates gender dysphoria, but also causes substantial social isolation at a time when peer acceptance is particularly important. Unlike feminizing top surgery, which benefits from multiple years of GAH, mastectomy does not require prior GAH use, and unlike bottom surgeries, which require substantial preparation and recovery, masculinizing top surgery involves a relatively shorter recovery time and has fewer potential complications.

While other GASs are rarely accessed in adolescents, an increasing number of young adults are pursuing these surgeries. More research is therefore needed to better understand the unique considerations, outcome differences, and specific needs of TGD people seeking surgery during the adolescent and emerging adult stages of life. There are likely benefits to accessing GAS at an earlier age, but there are also developmental considerations to take into account. As with GAS for adults an individualized approach to clinical care that takes into consideration life stage, embodiment goals, and protective factors is most likely to facilitate optimal outcomes, but more research is needed to clarify risks, benefits, and developmental differences in this population.

Surgical trends among adolescents and young adults

Many TGD adolescents and young adults desire GAS. A cross-sectional survey of TGD 18–21-year-olds found that approximately 17% of respondents had already completed at least one surgery, most often masculinizing top surgery, and approximately 69% were planning to pursue surgery in the future (Abern et al., 2023). For those planning to have surgery the most common barrier was cost, followed by an inability to find a surgeon and the risk of losing family support. Approximately 7% of respondents had no plan to have surgery in the future. The results of this survey demonstrate that while the desire for GAS is quite common, access remains limited to a much smaller proportion of TGD young people.

In a large pediatric health system database of adolescents 10–18 years old, GAS accounted for only 0.1% of all pediatric surgeries (Uffman et al., 2023). Most adolescents who underwent surgery for a diagnosis of gender dysphoria were assigned female at birth and underwent top surgery. Over half of the adolescents identified had commercial insurance, and there was an underrepresentation of racial and ethnic minorities, further highlighting systemic barriers to surgical access. Another large database study found that among gender-affirming top surgery specifically, only 1.4% of adolescents who had surgery were assigned male at birth (Das et al., 2023). Researchers have used similar large health databases and medical claims data to explore trends in chest surgery, specifically among both cisgender and TGD adolescents, finding that these surgeries are accessed at a higher rate by cisgender males (Dai et al., 2024) and cisgender females (Hassan et al., 2024) compared to TGD adolescents. Cisgender females also appear to access breast reduction surgery at younger ages than TGD youth (Hassan et al., 2024).

Retrospective chart review of surgical patients within specific clinics has also been used to identify trends in patient characteristics. In one study of 158 adolescents and young adults who underwent masculinizing top surgery, 82% were prescribed testosterone GAH, but several patients had no interest in affirming hormones (Boskey et al., 2020). Male surgical patients were more likely to be prescribed testosterone than nonbinary patients, but not exclusively. In another study of 209 adolescents, 13–17 years old, nearly 90% were prescribed testosterone, and racial and ethnic minorities were underrepresented, consistent with larger-scale research models (Tang et al., 2022).

The body of research on surgical trends among TGD young people has several limitations, most notably that it is largely US-based and in some cases even regionally specific. More international research is needed to look at global trends and to identify regional differences that either facilitate or hinder access. Research is also limited with regard to types of surgeries, and particularly, the surgical goals and experiences of adolescents and young adults assigned male at birth.

Considerations for age and timing of interventions

GAS in adolescence remains a subject of ongoing research, as well as evolving clinical practice and medical guidelines. While most professional organizations recommend deferring genital surgeries until the age of majority, certain interventions may be considered earlier in specific and select cases. Chest surgeries may be an option for postpubertal adolescents with significant distress and functional impairment who are well-prepared and have appropriate support.

The medical necessity for such interventions is evaluated on a case-by-case basis, taking into account factors such as parental consent and the presence of a supportive family environment, the severity of dysphoria and its impact on mental health and daily functioning, the adolescent's capacity for informed decision-making, and, in some cases, the duration of hormone therapy. Surgical decisions are made through a comprehensive, multidisciplinary shared decision-making approach, involving mental health professionals, hormone specialists (endocrinologists, primary care physicians, and adolescent medicine physicians), surgeons, parents and caregivers, and the adolescent. As the field of transgender healthcare continues to evolve, medical providers must stay up to date on the latest research and guidelines to ensure that interventions are both medically indicated and in the best interest of the adolescent patient.

Most older adolescents and young adults have the capacity to make informed decisions and provide informed consent, including decisions related to gender-affirming healthcare (Vrouenraets et al., 2021; Hein et al., 2015). This is especially true within the context of a multidisciplinary, shared decision-making approach to care. However, research examining decision-making specifically among adolescents and young adults seeking GAS remains limited (Lambert et al., 2023). A clinical case report of a 15-year-old transgender male who underwent top surgery was published in 2019 to illustrate the developmental considerations for youth making surgical decisions and preparing for recovery (Bernacki and Weimer, 2019). The case demonstrated that while adolescents are typically able to engage in informed consent with support, they benefit from detailed discussion about surgical expectations, results, and multiple stages of recovery to prepare them for potential outcomes that younger people may not necessarily anticipate on their own. Specifically, ensuring that adolescents and young adults have realistic expectations for surgical outcomes, the recovery process, and potential complications can facilitate optimal physical and emotional recovery.

There are some benefits to undergoing GAS at younger ages, as well as potential challenges. Adolescents and young adults often have built-in time without obligations during breaks from school and inherent support for recovery if they are still living with family, minimizing challenges to recovery such as returning to work and living independently (Robinson et al., 2023). Some research also suggests that GAS facilitates romantic and sexual development by decreasing dysphoria that can arise in intimate relationships, which is an important developmental activity during young adulthood (Bungener et al., 2020). Earlier GAS may also reduce bullying and harassment by helping young people more closely resemble their affirmed gender (Robinson et al., 2023).

There may also be challenges to undergoing surgery during the period of emerging adulthood, a stage of life characterized by important life decisions, increasing independence sometimes without complete stability, and adherence difficulties compared to both younger children and older adults. Activity restrictions may also be more difficult to follow for younger TGD people if they are more physically active than older adults (Battison et al., 2025). Potential risks and benefits should be weighed for each individual, taking into consideration the impact of gender dysphoria on functioning, support and resources, and individual embodiment goals. Importantly, there is no evidence to support delaying surgery for TGD adolescents and young adults based on age alone (Ascha et al., 2022).

Timing of bottom surgery is also an individualized decision, but there are additional considerations with regard to younger TGD individuals. In addition to a comprehensive multidisciplinary evaluation of each patient's own needs and dysphoria, another major consideration is emotional maturity and support resources required during the recovery period and beyond. This can be a challenging balance, since genital dysphoria can be quite severe for many. However, there are some important procedure-specific considerations that can impact both short- and long-term outcomes.

The success of these procedures relies on each patient's psychological readiness, regardless of age. The World Professional Association for Transgender Health (WPATH) Standards of Care, Version 8 (SOC8), generally recommends that patients be of the age of majority in their country or region before undergoing irreversible genital surgeries, although exceptions may be made on a case-by-case basis (Coleman et al., 2022). Patients who have used GnRHa may need additional counseling about fertility preservation options, as these procedures can impact reproductive capabilities. Long-term follow-up care is essential for all patients, but particularly for younger individuals who may experience changes in their surgical outcomes as they continue to grow and develop (Olson-Kennedy et al., 2019).

The surgical options for TGD adolescents and young adults are similar to those available in adulthood, with some exceptions. One distinction depends on the degree of secondary sex development from exposure to endogenous sex hormones. Specifically, the potential surgical interventions for those who undergo pubertal suppression with GnRHa are substantially less extensive than for those who complete endogenous puberty. Broadly speaking, GAS addresses either primary sex characteristics, including genitals and reproductive organs, or secondary sex characteristics, which include breast development, muscle mass, fat distribution, facial bone structure, vocal cord thickening and vocal pitch, and facial and body hair distribution. In all cases the impact of pubertal suppression depends on the sexual maturity rating (SMR) stage in which it is initiated. Pausing endogenous puberty at SMR stage 2 or 3 can prevent some or all development of secondary sex characteristics, thereby reducing or eliminating the need for future surgeries. The use of GnRHa to suppress puberty at SMR stage 2 or 3 is referred to as early pubertal suppression.

In contrast, those who complete endogenous puberty through SMR stage 5 may desire surgery to modify secondary sex characteristics that have developed as a result. For example, the surgical options for someone assigned male at birth who completed endogenous puberty may include vocal feminization surgery, tracheal shave, facial feminization surgery (FFS), breast augmentation, body contouring, and bottom surgery. If this same individual had access to early pubertal suppression, they may not need face, voice, chest, or body contouring surgery (Robinson et al., 2023), thus reducing cost, exposure to anesthesia, potential complications, recovery time, and ultimately allowing for a proactive, rather than reactive, approach to gender affirmation.

Individualized surgical planning

Nonbinary individuals identify outside of the traditional binary categories of male and female. Their gender identity, expression, and embodiment goals may encompass elements of masculinity, femininity, androgyny, or a rejection of gendered expressions altogether. Each person's experience of gender dysphoria is unique, and those requiring surgery will require a highly individualized, patient-centered approach that begins with an understanding of what will provide optimal relief. Some may desire all available surgical options, while others may only need very specific or minimal intervention. Others still may require a variant of a traditional procedure.

The guiding principle is to begin by understanding the needs of each patient and avoiding assumptions or prescriptive surgical recommendations. This principle of individualized surgical planning, which is an absolute requisite for nonbinary patients, represents a unifying approach to care that will ultimately benefit all patients regardless of their gender identity. There is increasing evidence that transgender men and women who identify squarely within the gender binary also have varied needs and embodiment goals. Many are satisfied with certain anatomical areas, but may be debilitated by dysphoria from other areas. Taking a patient-centered approach ensures optimal treatment success and patient satisfaction.

Psychosocial outcomes after gender-affirming surgery

GAS has demonstrated significant positive impacts across multiple psychosocial domains (Murad et al., 2010; Wernick et al., 2019). A 2022 systematic review of 53 studies found that GAS leads to measurable improvements in quality of life, happiness, and life satisfaction (Swan et al., 2023). The review also documented reductions in suicide attempts, anxiety, depression, and levels of gender dysphoria. The durability of these improvements has been further supported through medium- and long-term studies, suggesting that the benefits of surgical intervention persist over time (Park et al., 2022; van de Grift et al., 2018).

Population-level studies have provided compelling evidence for the broader impact of GAS on mental health outcomes. A secondary analysis of data from 27,715 TGD adult respondents to the 2015 US Transgender Survey further demonstrated that those with a history of GAS reported significantly lower odds of past-month psychological distress and past-year rates of smoking and suicidal ideation compared to those who have had no GAS (James et al., 2016). A total population study in Sweden from 2005 to 2015 demonstrated that GAS led to a reduction in mental health treatment utilization, with another study demonstrating decreased psychotropic prescriptions and reduced postoperative anxiety, depression, suicidal ideation, suicidal attempts, and alcohol and drug misuse for those who underwent GAS (Bränström and Pachankis, 2020; Kilmer et al., 2024).

Recent prospective research has provided more granular insights into specific psychosocial improvements. A landmark study utilized 11 validated outcome measures from the Patient-Reported Outcomes Measurement Information System (PROMIS) to evaluate quality of life outcomes (Caprini et al., 2023). When compared to those awaiting surgery, those who had completed facial gender-affirming surgery reported better scores across multiple domains, including anger, anxiety, depression, as well as improved positive affect, decreased social isolation, enhanced global mental health, and a stronger sense of meaning and purpose. These improvements persisted after controlling for variables such as previous surgeries, GAH duration, mental health diagnoses, socioeconomic factors, and social relationships. Preliminary analysis of the 2022 US Transgender Survey of 92,329 TGD people, including adolescents aged 16 and 17 years old, demonstrates that nearly all respondents (97%) who had undergone at least one GAS reported improved satisfaction with life, with less than 2% reporting decreased satisfaction with life (James et al., 2024).

While fewer research studies focus specifically on adolescents and young adults due to the historical lack of access for younger TGD people and the relatively lower frequency even today, emerging research on masculinizing top surgery for youth suggests similarly positive outcomes. One prospective, longitudinal study found significant improvements in chest dysphoria and gender congruence at 1 month and 6 months postsurgery compared to baseline (Boskey et al., 2023). A cross-sectional study of youth who had surgery as adolescents and were approximately 3 years postsurgery similarly found high levels of gender congruence and low rates of anxiety and depression (Weixel et al., 2025). Finally, a multicenter, prospective, matched cohort study found significant improvements in chest dysphoria, gender congruence, and body image at 3 months postsurgery, regardless of baseline dysphoria, while no change occurred in the control group (Ascha et al., 2022). In addition to general psychosocial measurement tools, these studies all used the Chest Dysphoria Measure, a 17-item measure that asks respondents to rate how often chest dysphoria interferes with a variety of activities on a scale from 0 (never) to 3 (all the time) (Olson-Kennedy et al., 2018).

These positive outcomes underscore the role of GAS as a medically necessary intervention for the treatment of gender dysphoria, including among adolescents and young adults. The comprehensive improvement observed across multiple psychosocial domains, combined with the durability of these benefits, provides strong evidence supporting the integration of surgical care into gender-affirming treatment protocols. As research methodologies continue to evolve and longitudinal data accumulates, our understanding of these outcomes continues to strengthen, further supporting the vital role of GAS in the care of TGD patients, while delay or denial of care results in higher rates of suicidality and depression when compared to those who receive prompt treatment (Tordoff et al., 2022; Evans et al., 2020; Tan et al., 2023).

Satisfaction and risk of regret

Surgical satisfaction and regret are important considerations regarding the risks and benefits of any invasive procedure. Research has generally demonstrated a high level of satisfaction with GAS and low rates of regret. A 2024 systematic review evaluating the prevalence of regret after GAS among 3662 patients across four continents reported an aggregate prevalence of surgical regret of 1.94% (Ren et al., 2024). The durability of these low regret rates has been demonstrated in longitudinal studies, including a Dutch national health system cohort of 6793 people from 1972 to 2015, where only 0.6% of transwomen and 0.3% of transmen who underwent gonadectomy were identified as experiencing regret (Wiepjes et al., 2018). Another cross-sectional study of patients who underwent gender-affirming mastectomy between 1990 and 2020 demonstrated exceptionally high rates of satisfaction and low levels of decisional regret using validated instruments (Bruce et al., 2023). Another 2024 systematic review found that GAS had among the lowest regret rates across surgical specialties (Thornton et al., 2024).

Regarding adolescents specifically a retrospective chart review of 209 surgical patients who underwent masculinizing top surgery over a 7-year period found that only two patients (<1%) expressed any regret more than 2 years following surgery (Tang et al., 2022). Neither patient had surgical complications, nor did they request or pursue revision or reversal surgery. Although patients in this study were as young as 13 years old at the time of referral, both patients who expressed regret were 16 years old at the time of surgery. This study showed that despite a 13-fold increase in surgeries over the course of 7 years the rate of regret remained stable and low. In another study that surveyed TGD adolescents and young adults about their gender-affirming healthcare history, only one survey respondent (1.4% of those who ever underwent surgery) endorsed regret of having ever had surgery (Crabtree et al., 2024). This respondent was 18 years old at the time of surgery, further suggesting that younger age is not a risk factor for regret.

While the measurement of regret in studies is far from standardized, a common theme is that it tends to fit into categories such as gender-related regret and desire for detransition (the rarest type), social regret stemming from a lack of acceptance or support, and medical regret related to functional limitations or complications (Ren et al., 2024; Narayan et al., 2021). Additionally, transient regret in the immediate postoperative period is not uncommon, including among adolescents (Bernacki and Weimer, 2019), but this can be mitigated through preoperative counseling and expectation setting, and incorporating tools such as photos or descriptions of recovery at various time points. A multidisciplinary approach involving mental health professionals, primary care providers, physical therapists, speech therapists, and surgeons allows for a thorough evaluation of surgical readiness while ensuring realistic expectations for aesthetic outcome and potential complications, surgical appropriateness for embodiment goals, and appropriate support systems to increase the likelihood of satisfaction and decrease potential for regret.

Preoperative considerations

There are a variety of preoperative considerations for TGD adolescents and young adults, many of which are similar to adults seeking GAS, including a biopsychosocial evaluation to ensure the diagnostic criteria for gender dysphoria are met, obtaining informed consent from the individual and parent(s), optimization for surgery, and recovery planning. As many adolescents and young adults undergoing preoperative evaluation have fewer chronic comorbidities compared to their adult counterparts, planning for recovery primarily involves facilitating any needed mental health resources, identifying support for physical recovery after surgery, and ensuring that young people understand and have the ability to adhere to postoperative instructions.

When assessing for surgical readiness, optimization of mental health can play a significant role in recovery. While anxiety and depressive symptoms tend to be associated with gender dysphoria, it is also important to consider body dysmorphia and eating disorders in TGD youth. Pubertal changes and secondary sex characteristics that worsen gender dysphoria can, in turn, increase the risk of developing an eating disorder. Diet and exercise may be the first intervention a TGD youth employs to reduce dysphoria and body dissatisfaction (Griffiths and Yager, 2019). Of note, body dissatisfaction is one of the strongest predictors of developing an eating disorder (Stice and Shaw, 2002). One study found that approximately 18% of TGD individuals reported an eating disorder diagnosis in the past year compared to 1.8% of cisgender female youth and 0.2% of cisgender male youth (Duffy et al., 2019).

The nuances of discerning between body dysmorphia and gender dysphoria are complex, can be intertwined, and are multifactorial. In general, body dysmorphia is the perception that there is a major flaw or problem with a part of the body that impacts multiple areas of life. Therefore body dysmorphia cannot be treated without taking into account gender dysphoria in TGD individuals. An experienced mental health professional can assist in these conversations to better evaluate and advise whether manifestations are related to body dysmorphia, gender dysphoria, or a combination.

Clinical optimization for surgery includes cessation from nicotine-containing products, optimizing any chronic health conditions, and recovery planning. Assessment for stable housing, support person(s) after surgery, adequate nutrition, transportation, and mental health support are essential components to ensuring patients have a successful recovery after surgery. While practices vary between surgeons, there is no evidence to support the recommendation to stop GAH preoperatively. While there have been no large-scale prospective trials specifically evaluating the risk of venous thromboembolism (VTE) among those undergoing GAS between groups who stop and those who continue GAH, current evidence does not demonstrate an increased risk in VTE when continuing GAH (Boskey et al., 2019).

In one review of 138 transgender women undergoing gender-affirming genital surgery, only one experienced VTE, an incidence rate of 0.5% (Herndon et al., 2024). In another retrospective review of those who experienced VTE after vaginoplasty a history of previous VTE was the most common risk factor and present in 47.1% of those who experienced it after vaginoplasty (Snyder et al., 2023). A multimodal analysis of 11 studies looking at VTE incidence rate between 602 patients who stopped GAH preoperatively, of whom three experienced VTE (incidence of 0.49%), is compared to just one incident of VTE among 925 who continued GAH throughout the perioperative period (Laspro et al., 2025). Another large single-center study looked at a 38-year history of conducting FFS and 953 patients who continued some level of GAH (714 continuing full-dose, 197 continuing reduced dose) and incidence rate of VTE with other common orthognathic procedures of similar length, finding no increased risk (Li et al., 2024).

Postoperative care and support

Postoperative recovery requires both physical and emotional care and support. Physical care includes adherence to discharge instructions, adequate pain management, and addressing any potential complications. Emotional care includes individual coping strategies, social support, and psychological counseling when needed. In the general surgical population, both young people and their family members report challenges after surgery, and they often feel unprepared to manage physical recovery, pain, or the emotional experiences that come up as a result (Rabbitts et al., 2017). Among TGD individuals, research related to surgical recovery has primarily focused on adults with less known about adolescent and young adult experiences (Diaddigo et al., 2024).

Research on masculinizing top surgery has shown low rates of complications among TGD adolescents and young adults. A retrospective chart review found the rate of complication to be approximately 7% and the rate of revision to be approximately 11% (Tang et al., 2022). The most common complications were hematoma (3.6%), followed by wound infection (2.9%), hypertrophic scars (2.9%), seroma (0.7%), and suture granuloma (0.7%). A multicenter prospective study found similarly low rates of complications, including two cases of seroma, one of hematoma, and one of nipple loss (Ascha et al., 2018). In a large database study of masculinizing top surgery and cisgender breast reduction, complication rates were equally low across both groups (Hassan et al., 2024). The most common complication among breast reduction patients was surgical site infection, which occurred significantly more often in cisgender adolescents compared to TGD adolescents, while the most common complication for masculinizing top surgery patients was unplanned revision.

The results of another large database study suggest little difference in complication rates between adolescents and young adults, although there were more complications among Black and African American young people in the study compared to other races (Diaddigo et al., 2024). Low complication rates among TGD adolescents and young adults are expected, given the relatively fewer chronic comorbidities and health risks compared to older adults. More research is nonetheless needed to explore whether there are unique risk factors for complications in this age group, especially among young adults undergoing bottom surgeries, which have more potential risks than top surgeries. For example, younger surgical patients may have more difficulty with adherence to activity restrictions, given that emerging adulthood is characterized by increasing independence and self-reliance. Dilation may also be a more challenging activity for young people who often have less experience and exposure to their own genitals as a result of dysphoria and earlier options for medical interventions.

Emerging research is beginning to explore the pain experiences and pain management of adolescents and young adults following GAS. As part of a larger prospective longitudinal study of pediatric pain broadly a secondary analysis of participants recruited following GAS specifically was conducted to explore the postoperative pain experiences of TGD adolescents and young adults (Battison et al., 2025). Among 64 participants aged 14–25 years the most common surgery was masculinizing top surgery, as expected in this population (87.5%). Approximately 69% of participants reported notable pain interference at one month postsurgery, and 62% continued to report pain interference at 6 months. One unexpected finding was that young adults reported significantly more pain interference than adolescents. This may be a reflection of the increased responsibilities and independence among young adults, who may feel more limited in their functioning compared to adolescents who often have fewer responsibilities and more flexibility for recovery. Despite a lower rate of complications and physical risk factors, this study highlights the importance of incorporating pain management and coping strategies for functional limitations into surgical recovery planning.

While mental health and psychosocial outcomes have been shown to improve following GAS, there are sometimes increased emotional challenges in the immediate postoperative period. Despite growing anecdotal evidence that some TGD individuals experience postoperative depression following GAS, similar to other major surgeries, there has been little research on this phenomenon. In addition to general risk factors for postoperative depression, such as physical limitations, pain management, and excess physical stress, GAS has the potential to worsen gender dysphoria in the immediate postoperative period by drawing significant attention to an area of the body historically known to cause distress.

Thorough preparation, patient education, expectation setting, and psychosocial support are key to a smooth postoperative recovery, both emotionally and physically (Ross et al., 2024). Adequate preparation involves not only physical readiness but also mental and logistical planning, such as arranging for postoperative care, preparing the living space, and gathering necessary supplies. Setting realistic expectations about the recovery process, including potential challenges, timeline for healing, and ultimate outcomes, helps patients navigate the postoperative period with greater resilience and reduced anxiety.

A strong support system plays an equally significant role, potentially influencing both physical and emotional recovery outcomes. This network may include healthcare providers, mental health professionals, family members, friends, and peer support groups. Such comprehensive support can facilitate the management of postoperative stress, promote adherence to recovery protocols, and provide essential practical assistance during the healing process. By prioritizing these aspects in conjunction with medical care, patients can potentially enhance their recovery experience and improve overall outcomes.

Ongoing psychological support serves as an integral component in the recovery process following GAS. Professional counseling provides a confidential environment for individuals to process their emotions, address emerging challenges or anxieties, and develop effective coping strategies throughout the recovery journey. Therapists with expertise working with TGD people can offer valuable insights and techniques to manage postoperative stress, body image concerns, and the adjustment to physical changes.

Peer support can be an invaluable resource at every stage of the perioperative period. Nothing can provide the same level of empathy and insight as people who have recovered from a similar procedure. In-person support groups are increasingly common, with many organizations offering video-conference sessions weekly, facilitating connections between individuals who have undergone similar experiences, and enabling the exchange of coping strategies and reducing feelings of isolation during recovery. The combination of professional counseling and peer support can significantly enhance emotional resilience, foster a positive outlook, and contribute to overall well-being during the recovery period and beyond. It is important to recognize that the need for psychological support may evolve over time, and individuals should feel empowered to seek assistance at any stage of their journey. More research is needed to quantify and improve the benefits of these auxiliary supports.

The transition process extends beyond the individual patient, encompassing their entire support network. Integrating a patient's family and close friends into preoperative education, especially for TGD adolescents and young adults, represents a pivotal aspect of comprehensive care. This inclusion allows the support system to gain a deeper understanding of the surgical procedures, potential challenges, and the patient's needs during recovery. Care teams should be flexible in allowing a patient to define who "family" is for themselves and must recognize the impact that marginalization and social isolation may have on the ability to access support systems that may be more readily available to cisgender patients.

Family members and friends may experience their own emotional responses to a patient's transition, which can range from enthusiastic support to uncertainty or anxiety. Consequently, offering resources, guidance, and support tailored to the needs of the patient's support network becomes imperative. This may include educational materials, family counseling sessions, or referrals to support groups specifically designed for families of TGD individuals undergoing surgery. By addressing the needs of both the patient and their support system, healthcare providers can foster a more robust and informed care environment, potentially leading to improved outcomes and a smoother transition process for all involved parties.

Surgical interventions for transgender women and nonbinary people assigned male at birth

Transfeminine is an umbrella term used here for transgender women and nonbinary people assigned male at birth, with the recognition and acknowledgment that some people do not find this terminology to align with their gender identity. Feminizing surgery is also sometimes used to refer to the surgical options for this group of TGD individuals.

Top surgery: breast augmentation

Top surgery for transfeminine patients differs fundamentally from that of transmasculine patients. In transfeminine patients, top surgery affirms target secondary sex development rather than removing the incongruent sex characteristics. This is an important distinction when considering surgical timing in conjunction with hormone therapy. In general, after initiating GAH with estrogen, patients will experience rapid breast development in the first year with peak development after 2–3 years, although some may continue to experience breast growth for 4–5 years (De Blok et al., 2018; Fisher et al., 2016; Reisman et al., 2019; Wierckx et al., 2014). Transfeminine patients are recommended to be on estrogen for at least 2 years before undergoing breast augmentation to ensure breast growth from estrogen and/or progesterone has occurred to completion before any surgical intervention. For these reasons, it is rare to receive breast augmentation before the age of 18 years.

Anecdotally, patients who undergo pubertal suppression before starting GAH appear to be more likely to achieve SMR stage 5 breast development. However, to date, no longitudinal or large-scale studies have investigated this comparison. From a physiological standpoint, breast development would be expected given the hormonal milieu of puberty that promotes sexual development, including breast tissue maturation. This includes elevated levels of growth hormone, insulin-like growth factor 1 (IGF-1), prolactin, leptin, and sex hormone (Kleinberg and Ruan, 2008; Veldhuis et al., 2005). It stands to reason, therefore, that GnRHa started early in puberty is likely to reduce future indication for top surgery in transfeminine patients.

TGD youth who undergo endogenous testosterone puberty through SMR stage 3 or beyond are more likely to have body dimensions that merit referral to a surgeon with extensive experience with gender-affirming breast augmentation. Implant or fat graft placement requires special expertise for optimal outcomes, given the likelihood of broader shoulders, wide-set nipples, and a wider chest circumference. Transfeminine patients who undergo early pubertal suppression are less likely to require these unique considerations, given reduced exposure to testosterone during pubertal development. See Table 9.1 for a summary of outcomes and complications associated with breast augmentation surgery.

Bottom surgery: orchiectomy, vulvoplasty, vaginoplasty

Orchiectomy, or the removal of testicles, is typically done in conjunction with vaginoplasty or vulvoplasty. However, in some cases, it can be done as a standalone procedure with minimal impact on future bottom surgery, if desired. Such cases are typically pursued due to intolerance of testosterone blocking medications such as spironolactone (cypionate in

TABLE 9.1 Outcomes and complications following breast augmentation surgery.

Surgery type	Long-term outcomes and follow-up care	Common complications	Revision rates
Top surgery (breast augmentation)	Satisfaction rates vary (80%–95%) Regular breast exams and imaging as per guidelines (Schoffer et al., 2022)	Capsular contracture (3.8%) Implant malposition (5%–10%)	9% (Oles et al., 2022)

Europe), bicalutamide, or inadequate testosterone suppression despite maximal medical therapy. Other times, orchiectomy might be pursued in isolation to provide for less bulk and a more easily achieved "flat front" appearance via tucking.

Some TGD people may be satisfied with orchiectomy itself. For patients pursuing orchiectomy before vaginoplasty or vulvoplasty the surgeon performing the vaginoplasty or vulvoplasty should be consulted to solicit their preference for the surgical approach for orchiectomy. In general the midline scrotal approach is the least invasive and should not impact future tissue grafting, as vascular supply is not interrupted and scarring is minimal. The inguinal or transabdominal approach is more invasive and rarely indicated, except in circumstances where donor tissue is severely limited.

Vulvoplasty involves the removal of the penis and testicles and feminizing of the genital area, including the construction of a vulva, labia, and neoclitoris, which uses the glans of the penis for maintenance of sexual sensation, pleasure, and orgasm. Orgasm can be achieved by most people following surgery, regardless of whether early pubertal suppression was used or whether orgasm has ever been achieved before surgery (Van Der Meulen et al., 2025). Vaginoplasty is a procedure for TGD patients who desire the creation of a neovagina in addition to the outcomes of vulvoplasty. This combination of procedures is often accomplished in either one or two stages, although sometimes additional revision is needed.

The most common procedures for the construction of a neovagina use penile and scrotal tissue. For patients who undergo early pubertal suppression, there is often limited development of scrotal and penile tissue, potentially limiting donor tissue available for the neovagina and vulva. In these cases, creation of the vaginal canal may require grafting from the small or large bowel, the inner groin area, or the peritoneal lining. There is early evidence that both sigmoid vaginoplasty and robotic peritoneal flap vaginoplasty can have equivalent vaginal depth among those with and without genital hypoplasia (Blasdel et al., 2023; Bouman et al., 2016). As gender diverse youth continue to have access to GnRHa, surgical techniques and intervention will continue to evolve and innovate. Some less common alternative techniques include entero- or colo-vaginoplasty or partial peritoneal pull-through vaginoplasty.

The postoperative recovery in vaginoplasty can be an arduous and lengthy. One of the most critical aspects of both short- and long-term recovery is maintenance of vaginal canal patency by means of dilating. While specific regimens differ per surgeon, commonly patients must place increasingly large vaginal dilators for 20–30 minutes three times daily with eventual decrease in frequency. However, in all cases, periodic vaginal dilation is a lifetime commitment. Failure to adequately dilate, especially during the immediate postoperative period, can lead to loss of vaginal depth or stenosis.

The discipline and daily time commitment for dilating must be thoroughly taken into consideration and included in patient education and surgical planning. Anecdotally, young adults sometimes underestimate the importance of and time commitment required for an adequate dilation schedule. While this is a point for shared decision-making, it may be appropriate to encourage patients to delay vaginoplasty, or consider a two-stage approach with zero-depth vulvoplasty in the first stage, followed by vaginoplasty after some maturation or establishment of adequate support. There is a growing preference for vulvoplasty alone, without the creation of a neovagina, due to there being fewer risks, not wanting to dilate, and/or having no interest in vaginal penetrative sexual intercourse (Aaen et al., 2024). More research is needed to clarify the outcomes of single- versus two-stage vaginoplasty, especially for young adults. See Table 9.2 for a summary of outcomes and complications associated with vaginoplasty.

Facial gender-affirming surgery: facial feminization surgery

FFS encompasses a range of procedures designed to modify facial features to align more closely with feminine characteristics. Common procedures include forehead contouring, rhinoplasty, and jaw and chin contouring. Forehead

TABLE 9.2 Outcomes and complications following vaginoplasty.

Surgery type	Long-term outcomes and follow-up care	Common complications	Revision rates
Vaginoplasty	High satisfaction rates (>80%) (Papadopulos et al., 2017) Regular dilation required, decreasing in frequency over time Annual pelvic exams are recommended (Bizic et al., 2014)	Neovaginal stenosis (6%) (De Rosa et al., 2024) Rectovaginal fistula (1%–2%) Urinary issues (4%–13%) (Jarolím et al., 2009)	21%–31% (Frey et al., 2016)

contouring typically involves reducing the bony prominence of the supraorbital ridge and reshaping the frontal bone. This procedure often includes hairline advancement or lowering to create a more oval face shape. Rhinoplasty aims to refine the nose, often reducing its size, narrowing the bridge, and creating a more upturned tip. Jaw and chin contouring involves reducing the width and height of the mandible and tapering the chin to create a more V-shaped lower face (Altman, 2012).

Other common FFS procedures include chondrolaryngoplasty (tracheal shave) to reduce the prominence of the Adam's apple, lip lift to shorten the upper lip and increase tooth show, and blepharoplasty to feminize the eye area. These procedures, while generally less extensive than bony remodeling, still play crucial roles in creating an affirming facial appearance. For patients who accessed early pubertal suppression the need for these soft tissue procedures may be reduced, as the impacts of testosterone may be less prominent.

Any significant bony remodeling should occur after facial growth is complete, typically around age 18–20 years (Morrison et al., 2016a). Patients who accessed GnRHa may have less incongruent facial features, potentially requiring less extensive surgeries. This can lead to more subtle, natural-looking results and potentially reduce surgical risks. However, it may also mean that the changes achieved through FFS are less dramatic, which should be discussed with patients when setting expectations (De Vries et al., 2014). While these procedures can significantly align facial appearance with gender identity, they should be approached holistically as part of a comprehensive gender-affirming care plan. The decision to undergo FFS should be made in consultation with a multidisciplinary team, including mental health professionals, to ensure psychological readiness. For adolescents and young adults, especially those who have used GnRHa, careful consideration must be given to the timing of these procedures in relation to overall development and other aspects of transition.

In addition to the facial surgical procedures, there are also a variety of nonsurgical interventions that can be utilized. The use of dermal fillers can create contour, enhance certain areas of the face, and be a means to improve dysphoria without the downtime, recovery, potential complications, and cost of surgery. There are a variety of fillers on the market today, many of which last for a year or some up to 5 years, and may be easier to access compared to finding providers who specialize in gender-affirming surgeries. See Table 9.3 for a summary of outcomes and complications associated with FFS.

Gender-affirming vocal surgery: glottoplasty

Voice feminization surgery is a collection of specialized procedures designed to elevate the pitch and quality of an individual's voice to better align with their gender identity. The most common modern approach is the modified Wendler Glottoplasty, which is a procedure that shortens the vibrating portion of the vocal folds, resulting in a higher pitch (Rogalska et al., 2024). Another technique is laser-assisted voice adjustment (LAVA), where a laser is used to scar and shorten the vocal folds. These techniques have largely replaced cricothyroid approximation (CTA), which involves suturing the cricoid and thyroid cartilages closer together to increase vocal fold tension, thereby raising pitch. Early pubertal suppression can have a profound impact on voice development and thereby reduce the need for gender-affirming vocal surgery in the future. By suppressing testosterone production early in puberty, growth and thickening of the vocal cords are prevented, significantly reducing the need for speech therapy or vocal surgery.

For those who have undergone endogenous testosterone puberty, voice feminization surgery can be life-changing. However, speech therapy continues to play an essential part in pre- and postoperative care. Before surgery, speech therapy helps patients develop proper vocal techniques and establish baseline vocal patterns that will facilitate postsurgical adaptation to a new vocal instrument. The initial weeks after surgery require strict voice rest, followed by a carefully structured program of vocal rehabilitation guided by a speech therapist who specializes in gender-affirming care. This rehabilitation typically involves gradually reintroducing vocal use, learning new resonance patterns, and developing muscle memory for the modified vocal structures. Speech therapy also helps patients address nonpitch aspects of

TABLE 9.3 Outcomes and complications following facial feminization surgery.

Surgery type	Long-term outcomes and follow-up care	Common complications	Revision rates
Facial feminization surgery	High satisfaction rates (>80%) Potential need for touch-ups as aging occurs (Oles et al., 2022)	Temporary numbness (common) Asymmetry (varies) Hairline irregularities (forehead procedures) (Oles et al., 2022)	4% (Oles et al., 2022)

stereotypically feminine speech patterns, such as intonation, articulation, and resonance, which surgery cannot modify. Without proper speech therapy guidance, patients may develop compensatory behaviors that could strain or damage their surgical results, potentially leading to suboptimal outcomes or even the need for revision surgery. See Table 9.4 for a summary of outcomes and complications associated with voice feminization surgery.

Gender-affirming body contouring

Some TGD patients may find that certain aspects of their body shape continue to cause gender dysphoria or do not align with their desired gender expression, even after the use of GAH. In these cases, surgical interventions can be considered to further modify body contours. These procedures typically aim to either reduce features associated with the sex assigned at birth or enhance characteristics aligned with the individual's gender identity.

Common interventions include targeted liposuction to areas such as the abdomen, flanks, or hips to create a more masculine or feminine silhouette. Conversely, fat grafting techniques can be employed to augment areas like the breasts, hips, buttocks, cheeks, or lips, enhancing feminine features or creating a more curvaceous appearance (Aristizábal et al., 2024; Grimstad and Lee, 2024). The specific approach to body contouring in TGD individuals is highly personalized, taking into account unique anatomy, transition goals, and overall health status. For transfeminine patients, procedures often focus on creating a more feminine figure by redistributing fat to create curves, potentially including breast augmentation with fat transfer, hip and buttock enhancement, and facial feminization through fat grafting to the cheeks and lips.

It is recommended that individuals undergo body contouring only after GAH has been established for several years and weight has stabilized (Moghadam et al., 2024). This allows for the full effect of hormonal changes on fat distribution and muscle mass to manifest, providing a stable foundation for surgical intervention (Aristizábal et al., 2024). Additionally, patients should be in good overall health and at a stable weight to optimize surgical outcomes and minimize complications. The psychological impact of these procedures can be significant, often resulting in improved body satisfaction and reduced gender dysphoria. Risks and potential complications may include asymmetry, contour irregularities, changes in skin sensation, and risks associated with anesthesia.

Surgical interventions for transgender men and nonbinary people assigned female at birth

Transmasculine is an umbrella term used here for transgender men and nonbinary people assigned female at birth, with the recognition and acknowledgment that some people do not find this terminology to align with their gender identity. Masculinizing surgery is also sometimes used to refer to the surgical options for this group of TGD individuals.

Top surgery: mastectomy with chest reconstruction, chest reduction

There are a few primary approaches to masculinizing top surgery that are employed based on the unique anatomy of a patient, primarily breast size and body habitus. Keyhole mastectomy involves the removal of glandular breast tissue via a periareolar incision. Aesthetically, this approach provides the least visible scarring, but can only be offered to those with minimal breast tissue with good skin elasticity and no ptosis, which often includes those who underwent early pubertal suppression. The majority of patients, regardless of age, require a double incision with or without a free nipple graft. This

TABLE 9.4 Outcomes and complications following voice feminization surgery.

Surgery type	Long-term outcomes and follow-up care	Common complications	Revision rates
Voice feminization surgery	Variable long-term pitch elevation Ongoing voice therapy is often recommended (Van Damme et al., 2017)	Voice instability (10%–15%) Dysphagia (5%–10%) (Van Damme et al., 2017)	10%–20% (Casado et al., 2017)

approach to masculinizing the chest differs significantly from oncological mastectomy, so it is critical that patients are referred to surgeons who specialize in gender-affirming chest surgery.

The surgical needs of nonbinary patients who were assigned female at birth are often more varied than those of binary transgender masculine people. Because their medical transition is likewise variable, some may present without having GAH. One study compared postoperative psychosocial outcomes between binary and nonbinary transmasculine patients undergoing mastectomy and found that despite the increased variability of surgical outcomes desired the nonbinary group outcomes were similarly positive (Roblee et al., 2024). Some TGD individuals may prefer a chest reduction surgery to a full mastectomy, which may be more aligned with a nonbinary gender identity or may be desired for some other reason, such as preserving nipple sensation or achieving a specific aesthetic outcome. See Table 9.5 for a summary of outcomes and complications associated with mastectomy with chest reconstruction.

Bottom surgery: metoidioplasty, phalloplasty, hysterectomy, salpingo-oophorectomy

Hysterectomy is a surgical option that may be pursued as a prerequisite for metoidioplasty or phalloplasty, or it may be pursued as a standalone procedure, regardless of desire for genital surgery in the future. Many transmasculine or nonbinary individuals have dysphoria related to menstrual cycles, some can have persistent pelvic pain and cramping, and many have no desire to carry a pregnancy. Several techniques are available for hysterectomy, including laparoscopic, robotic, vaginal, or open. Total hysterectomy involves the removal of the uterus, uterine tubes, ovaries, and cervix. Bilateral oophorectomy is not a requirement for hysterectomy and may become an issue if access to testosterone is interrupted or no longer desired (Barrera et al., 2023). Some individuals, therefore, may want to keep their ovaries or decide to only remove one, which will reduce the risk of ovarian cancer, but also will continue to supply estrogen if there is any pause or cessation of gender-affirming hormones. Opting to have the cervix removed along with the uterus prevents the need for cervical cancer screenings, which are often a source of dysphoria for patients, but some individuals may not want to have the cervix removed.

Metoidioplasty involves the creation of a neophallus by releasing the suspensory ligament around the hypertrophied clitoris, which has been enlarged by testosterone therapy. Metoidioplasty is dependent upon total GAH testosterone exposure. This procedure often preserves sexual sensation but results in a smaller phallus compared to phalloplasty. Metoidioplasty does not result in penetrative function of the neophallus, but it may have tumescent capabilities and erectile function. There are also surgical options for urethral lengthening of the neophallus for a stand-to-pee function as well as scrotoplasty, all of which can be completed in one surgery. Metoidioplasty has a lower rate of complications compared to phalloplasty, high satisfaction rates, and shorter operating times, making it an appealing option for patients if it aligns with their goals (Bordas et al., 2021).

Phalloplasty, on the contrary, involves the creation of a neophallus using tissue from another part of the body, most commonly the forearm (free flap) or thigh (pedicle or free flap). Phalloplasty can result in a larger phallus compared to metoidioplasty, which can have penetrative function with the placement of a semifirm penile implant or a pump that allows for sufficient tumescence to allow for penetrative sex. However, this requires multiple surgeries and may result in less sexual sensation compared to metoidioplasty (Frey et al., 2016). Most phalloplasty approaches require three surgical phases, each with its own recovery. Given the extremely lengthy and arduous recovery, consideration of future phalloplasty merits special attention to patient education and shared decision-making about potential risks and

TABLE 9.5 Outcomes and complications following mastectomy with chest reconstruction.

Surgery type	Long-term outcomes and follow-up care	Common complications	Revision rates
Top surgery (mastectomy)	Improved psychosocial and sexual well-being, satisfaction, and body image-related quality of life Reduced anxiety and depression (Bertrand et al., 2024) High satisfaction rates (>90%) Annual chest exams are recommended (Oles et al., 2022)	Hypertrophic scar (34.2%) (Bertrand et al., 2024) Hematoma (5%–15%) Nipple necrosis (partial/ complete: 5%–10%) Contour irregularities (10%–15%) (Cregten-Escobar et al., 2012)	6.2% (Oles et al., 2022)

complications. Early pubertal suppression does not significantly impact phalloplasty procedures, as the tissue is harvested from other parts of the body (De Vries et al., 2014).

Vaginectomy, or the removal or closure of the vaginal canal, is another procedure that is often done in conjunction with other surgeries such as hysterectomy, metoidioplasty, or phalloplasty. There are multiple variations available that can be tailored to the needs of the patient. Total vaginectomy involves the complete removal of the vaginal mucosa and obliteration of the vaginal canal. This is most commonly performed as part of phalloplasty or metoidioplasty, as it reduces the risk of complications such as fistula formation. Another option that some may choose includes colpocleisis, which involves closing off the vaginal canal without complete removal of tissue, using techniques such as the Le Forte procedure or total colpocleisis. See Table 9.6 for a summary of outcomes and complications associated with masculinizing bottom surgeries.

Facial gender-affirming surgery: facial masculinization surgery

Facial masculinization surgery encompasses a range of surgical interventions aimed at enhancing masculine facial features. These procedures may include augmentation of the brow ridge using dermal fillers or autologous fat transfer, rhinoplasty to create a more prominent nasal bridge, mandibular augmentation to enhance jaw definition, genioplasty to increase chin projection, and tracheal augmentation using rib cartilage grafting to create a more prominent Adam's apple (Grimstad and Lee, 2024). The selection and approach to these procedures are highly individualized, taking into account the patient's existing facial anatomy, desired outcome, and overall transition goals. The extent of intervention required can vary significantly among patients, with some individuals requiring minimal alterations, while others may benefit from more comprehensive surgical planning. As with all gender-affirming procedures a thorough preoperative assessment and discussion are essential to align the surgical approach with a patient's expectations and to ensure optimal outcomes.

In addition to the surgical techniques and procedures used, there are also a variety of nonsurgical interventions that can be utilized. The use of dermal fillers can create contour, enhance certain areas of the face, and be a means to improve dysphoria without the downtime, recovery, potential complications, and cost of surgery. There are a variety of fillers on the market today, many of which last for a year, and some up to 5 years, and may be easier to access compared to finding providers who specialize in GAS.

TABLE 9.6 Outcomes and complications following masculinizing bottom surgeries.

Surgery type	Long-term outcomes and follow-up care	Common complications	Revision rates
Phalloplasty	Satisfaction rates vary (75%–100%) (Morrison et al., 2016b) Ongoing physical therapy and scar management (Frey et al., 2016; Ascha et al., 2018)	Urethral complications (25%–58%) Partial flap loss (7%–15%) Donor site morbidity (varies) (Frey et al., 2016; Ascha et al., 2018)	50%–70% (Frey et al., 2016; Ascha et al., 2018)
Metoidioplasty	High satisfaction rates and functional outcomes (>85%) Regular follow-up for urinary function (Frey et al., 2016)	Urethral complications (17%–39%) Postoperative hematoma (6%–13%) (Frey et al., 2016) Testicular implant rejection or displacement	15%–35%
Hysterectomy, salpingectomy, oophorectomy		Transfusion (2.4%) Dehiscence (0.4%) Pulmonary embolism (0.4%) Surgical infection (0.2%) Urinary tract infection (0.2%) (Bretschneider et al., 2018)	0.3% (Bretschneider et al., 2018)

Gender-affirming vocal surgery: vocal fold augmentation, thyroplasty

Vocal pitch in the cisgender male range is achieved by 80% of TGD individuals following 1 year of testosterone GAH (Ziegler et al., 2018). Of the 20% who do not achieve adequate masculinization, many can achieve a satisfactory pitch with gender-affirming speech therapy. Patients who remain severely dysphoric about their voice following testosterone GAH and speech therapy may consider one of the emerging options, including surgical or injectable options. However, these interventions remain very specialized and have limited evidence and accessibility.

Vocal fold augmentation with autologous fat grafting with or without filler injections is an existing technique used for other pathologies, such as vocal fold paralysis. This technique has been adapted to bolster the vocal folds, allowing for a decrease in pitch. In one case report where both autologous fat transfer and temporary fillers were used, the patient was able to achieve a reduction in fundamental pitch of about 30 Hz. However, this required multiple injections and was anticipated to require annual maintenance injections (Webb et al., 2022). Vocal fold testosterone injection is an experimental technique that, in one case report, caused a reduction in pitch that would, in theory, be permanent. Both vocal fold augmentation and vocal fold testosterone injections show promise; however, at the time of publishing, they remain experimental (Vahabzadeh-Hagh et al., 2023).

Voice masculinization surgery, first described in 1974 by Isshiki, also called a type III thyroplasty, is a technique that lengthens and relaxes the vocal folds to achieve a lower pitch and decreased maximum pitch. However, this is an emerging technique with limited empirical evidence. Across four case series reports a total of 24 patients experienced an average decrease in fundamental pitch of around 50 Hz (Bultynck et al., 2021; Catani et al., 2024; Haben, 2022; Slavit et al., 1990). These surgeries carry an unknown level of risk, including potential alterations in voice quality or breathing difficulties, and outcomes can be variable. Therefore careful patient selection, thorough preoperative counseling, and integration with comprehensive voice therapy are essential components of care.

Summary and research recommendations

Research on patient characteristics and surgical trends among TGD adolescents and young adults suggests that many barriers to access continue to exist, especially for racially and ethnically diverse communities. However, this body of research is largely based in the United States. International collaboration in research is needed to clarify global trends, regional differences in access and surgical options, and unique trends in surgical desires among young TGD people.

TGD adolescents and young adults may have a lower complication rate due to having relatively fewer chronic comorbidities and health risks compared to older adults. Research is nonetheless needed to explore unique risks and benefits within this age group, particularly for young adults undergoing bottom surgeries, which have more potential for complications compared to top surgeries.

There are other areas of potential clinical interest that warrant more empirical evaluation, such as postoperative depression, pain management during recovery, and surgical implications for young people who underwent early pubertal suppression. GAS research would benefit from community-centered research designs to ensure research priorities are aligned with the emerging needs of the population, especially as young adults are increasingly accessing GAS.

Conclusion

GAS encompasses a range of interventions meant to help align primary and/or secondary sex characteristics with gender identity and expression and has been found to result in significant positive impacts across multiple psychosocial domains. TGD adolescents and young adults are increasingly accessing GAS, and many of these young people have a history of early gender-affirming medical care, presenting new implications and considerations for surgery. Masculinizing top surgery is the only GAS regularly accessed in adolescence, and research supports the psychosocial benefits of this intervention for some younger individuals. Other surgeries are rarely performed during adolescence but are nonetheless occurring at a higher frequency among young adults, who may benefit from beginning to prepare for future surgeries even during adolescence. More research is therefore needed to clarify the unique considerations, risks, and benefits of GAS during this developmental period. GAS is highly individualized, and clinical recommendations should be tailored to the needs and priorities of each patient.

References

Aaen, E.K., Kesmodel, U.S., Pop, M.L., Højgaard, A.D., 2024. Requests for vulvoplasty as gender-affirming surgery: a cross-sectional study in Denmark. Journal of Sexual Medicine 21 (3), 262–269. Available at https://doi.org/10.1093/jsxmed/qdae003.

Abern, L., Diego, D., Krempasky, C., Cook, J., Maguire, K., 2023. Desire for gender-affirming surgery in a sample of transgender and gender diverse individuals in late adolescence assigned female at birth. Journal of Pediatric and Adolescent Gynecology 36 (5), 488–490. Available at https://doi.org/10.1016/j.jpag.2023.07.001.

Altman, K., 2012. Facial feminization surgery: current state of the art. International Journal of Oral and Maxillofacial Surgery 41 (8), 885–894. Available at https://doi.org/10.1016/j.ijom.2012.04.024.

Aristizábal, A., Ríos-Sánchez, M., Escandón, J.M., DeRoberts, D., Armenta, E., Del Corral, G., et al., 2024. Body contouring as gender-affirming surgery in transgender patients: a systematic review of the current literature. Journal of Clinical Medicine 13 (12), 3523. Available at https://doi.org/10.3390/jcm13123523.

Ascha, M., Massie, J.P., Morrison, S.D., Crane, C.N., Chen, M.L., 2018. Outcomes of single stage phalloplasty by pedicled anterolateral thigh flap versus radial forearm free flap in gender confirming surgery. Journal of Urology 199 (1), 206–214. Available at https://doi.org/10.1016/j.juro.2017.07.084.

Ascha, M., Sasson, D.C., Sood, R., Cornelius, J.W., Schauer, J.M., Runge, A., et al., 2022. Top surgery and chest dysphoria among transmasculine and nonbinary adolescents and young adults. JAMA Pediatrics 176 (11), 1115–1122. Available at https://doi.org/10.1001/jamapediatrics.2022.3424.

Barrera, E.P., Grimstad, F.W., Boskey, E.R., 2023. Young adult patients with testosterone management concerns after gender-affirming hysterectomy and bilateral oophorectomy: a case series. Journal of Pediatric and Adolescent Gynecology 36 (1), 89–91. Available at https://doi.org/10.1016/j.jpag.2022.07.010.

Battison, E.A.J., Kenyon, E.A., Murphy, C.M., Lenne, E.L., Moyer, D.N., Feldstein Ewing, S.W., et al., 2025. Postsurgical pain, psychosocial functioning, and cannabis use among adolescents and young adults undergoing gender-affirming surgery. Transgender Health 10 (4), 334–344. Available at https://doi.org/10.1089/trgh.2024.0173.

Bernacki, J.M., Weimer, A.K., 2019. Role of development on youth decision-making and recovery from gender-affirming surgery. Clinical Practice in Pediatric Psychology 7 (3), 312–321. Available at https://doi.org/10.1037/cpp0000294.

Bertrand, A.A., DeLong, M.R., McCleary, S.P., Nahabet, E.H., Slack, G.C., DaLio, A.L., et al., 2024. Gender-affirming mastectomy: psychosocial and surgical outcomes in transgender adults. Journal of the American College of Surgeons 238 (5), 890–899. Available at https://doi.org/10.1097/XCS.0000000000000940.

Bizic, M., Kojovic, V., Duisin, D., Stanojevic, D., Vujovic, S., Milosevic, A., et al., 2014. An overview of neovaginal reconstruction options in male to female rranssexuals. Scientific World Journal 2014. Available at https://doi.org/10.1155/2014/638919.

Blasdel, G., Kloer, C., Parker, A., Shakir, N., Zhao, L.C., Bluebond-Langner, R., 2023. Genital hypoplasia before gender-affirming vaginoplasty: does the robotic peritoneal flap method create equivalent vaginal canal outcomes? Plastic and Reconstructive Surgery 151 (4), 867–874. Available at https://doi.org/10.1097/PRS.0000000000010011.

Bordas, N., Stojanovic, B., Bizic, M., Szanto, A., Djordjevic, M.L., 2021. Metoidioplasty: surgical options and outcomes in 813 cases. Frontiers in Endocrinology 12. Available at https://doi.org/10.3389/fendo.2021.760284.

Boskey, E.R., Taghinia, A.H., Ganor, O., 2019. Association of surgical risk with exogenous hormone use in transgender patients: a systematic review. JAMA Surgery 154 (2), 109–115. Available at https://doi.org/10.1001/jamasurg.2018.4598.

Boskey, E.R., Jolly, D., Tabaac, A.R., Ganor, O., 2020. Behavioral health concerns and eligibility factors among adolescents and young adults seeking gender-affirming masculinizing top surgery. LGBT Health 7 (4), 182–189. Available at https://doi.org/10.1089/lgbt.2019.0213.

Boskey, E.R., Jolly, D., Kant, J.D., Ganor, O., 2023. Prospective evaluation of psychosocial changes after chest reconstruction in transmasculine and non-binary youth. Journal of Adolescent Health 73 (3), 503–509. Available at https://doi.org/10.1016/j.jadohealth.2023.04.029.

Bouman, M.B., Van Der Sluis, W.B., Buncamper, M.E., Özer, M., Mullender, M.G., Meijerink, W.J.H.J., 2016. Primary total laparoscopic sigmoid vaginoplasty in transgender women with penoscrotal hypoplasia: a prospective cohort study of surgical outcomes and follow-up of 42 patients. Plastic and Reconstructive Surgery 138 (4), 614–623. Available at https://doi.org/10.1097/PRS.0000000000002549.

Bränström, R., Pachankis, J.E., 2020. Reduction in mental health treatment utilization among transgender individuals after gender-affirming surgeries: a total population study. American Journal of Psychiatry 177 (8), 727–734. Available at https://doi.org/10.1176/appi.ajp.2019.19010080.

Bretschneider, C.E., Sheyn, D., Pollard, R., Ferrando, C.A., 2018. Complication rates and outcomes after hysterectomy in transgender men. Obstetrics and Gynecology 132 (5), 1265–1273. Available at https://doi.org/10.1097/AOG.0000000000002936.

Bruce, L., Khouri, A.N., Bolze, A., Ibarra, M., Richards, B., Khalatbari, S., et al., 2023. Long-term regret and satisfaction with decision following gender-affirming mastectomy. JAMA Surgery 158 (10), 1070–1077. Available at https://doi.org/10.1001/jamasurg.2023.3352.

Bultynck, C., Cosyns, M., T'Sjoen, G., Van Borsel, J., Bonte, K., 2021. Thyroplasty type III to lower the vocal pitch in trans men. Otolaryngology–Head and Neck Surgery 164 (1), 157–159. Available at https://doi.org/10.1177/0194599820937675.

Bungener, S.L., De Vries, A.L.C., Popma, A., Steensma, T.D., 2020. Sexual experiences of young transgender persons during and after gender-affirmative treatment. Pediatrics 146 (6). Available at https://doi.org/10.1542/peds.2019-1411.

Caprini, R.M., Oberoi, M.K., Dejam, D., Chan, C.H., Potemra, H.'ilani M.K., Morgan, K.B.J., et al., 2023. Effect of gender-affirming facial feminization surgery on psychosocial outcomes. Annals of Surgery 277 (5). Available at https://doi.org/10.1097/SLA.0000000000005472.

Casado, J.C., Rodríguez-Parra, M.J., Adrián, J.A., 2017. Voice feminization in male-to-female transgendered clients after Wendler's glottoplasty with vs. without voice therapy support. European Archives of Oto-Rhino-Laryngology 274 (4), 2049–2058. Available at https://doi.org/10.1007/s00405-016-4420-8.

Catani, G.S.do A., Catani, M.E.C., Nunes, M.C.de A., Nakagawa, É.Y., Ferreira, F.A., 2024. Efficacy of type III thyroplasty in vocal masculinization in trans men: a retrospective observational study. Journal of Voice. Available at https://doi.org/10.1016/j.jvoice.2024.09.015.

Coleman, E., Radix, A.E., Bouman, W.P., Brown, G.R., de Vries, A.L.C., Deutsch, M.B., et al., 2022. Standards of care for the health of transgender and gender diverse people, version 8. International Journal of Transgender Health 23 (1), S1. Available at https://doi.org/10.1080/26895269.2022.2100644.

Crabtree, L., Connelly, K.J., Guerriero, J.T., Battison, E.A.J., Tiller-Ormord, J., Sutherland, S.M., et al., 2024. A more nuanced story: pediatric gender-affirming healthcare is associated with satisfaction and confidence. Journal of Adolescent Health 75 (5), 772–779. Available at https://doi.org/10.1016/j.jadohealth.2024.06.016.

Cregten-Escobar, P., Bouman, M.B., Buncamper, M.E., Mullender, M.G., 2012. Subcutaneous mastectomy in female-to-male transsexuals: a retrospective cohort-analysis of 202 patients. Journal of Sexual Medicine 9 (12), 3148–3153. Available at https://doi.org/10.1111/j.1743-6109.2012.02939.x.

Dai, D., Charlton, B.M., Boskey, E.R., Hughes, L.D., Hughto, J.M.W., Orav, E.J., et al., 2024. Prevalence of gender-affirming surgical procedures among minors and adults in the US. JAMA Network Open 7 (6). Available at https://doi.org/10.1001/jamanetworkopen.2024.18814.

Das, R.K., Perdikis, G., Al Kassis, S., Drolet, B.C., 2023. Gender-affirming chest reconstruction among transgender and gender-diverse adolescents in the US from 2016 to 2019. JAMA Pediatrics 177 (1), 89–90. Available at https://doi.org/10.1001/jamapediatrics.2022.3595.

De Blok, C.J.M., Klaver, M., Wiepjes, C.M., Nota, N.M., Heijboer, A.C., Fisher, A.D., et al., 2018. Breast development in transwomen after 1 year of cross-sex hormone therapy: results of a prospective multicenter study. Journal of Clinical Endocrinology and Metabolism 103 (2), 532–538. Available at https://doi.org/10.1210/jc.2017-01927.

De Rosa, P., Kent, M., Regan, M., Purohit, R.S., 2024. Vaginal stenosis after gender-affirming vaginoplasty: a systematic review. Urology 186, 69–74. Available at https://doi.org/10.1016/j.urology.2024.02.005.

De Vries, A.L.C., McGuire, J.K., Steensma, T.D., Wagenaar, E.C.F., Doreleijers, T.A.H., Cohen-Kettenis, P.T., 2014. Young adult psychological outcome after puberty suppression and gender reassignment. Pediatrics 134 (4), 696–704. Available at https://doi.org/10.1542/peds.2013-2958.

Diaddigo, S.E., Asadourian, P.A., Lavalley, M.N., Marano, A.A., Rohde, C.H., 2024. Masculinizing chest reconstruction in adolescents and young adults: an analysis of national surgical quality improvement program data. Annals of Plastic Surgery 92 (2), 253–257. Available at https://doi.org/10.1097/SAP.0000000000003735.

Duffy, M.E., Henkel, K.E., Joiner, T.E., 2019. Prevalence of self-injurious thoughts and behaviors in transgender individuals with eating disorders: a national study. Journal of Adolescent Health 64 (4), 461–466. Available at https://doi.org/10.1016/j.jadohealth.2018.07.016.

Evans, C.L., Kamunya, R., Tibaijuka, G., 2020. Using helping mothers survive to improve intrapartum care. Pediatrics 146 (4). Available at https://doi.org/10.1542/peds.2019-3600.

Fisher, A.D., Castellini, G., Ristori, J., Casale, H., Cassioli, E., Sensi, C., et al., 2016. Cross-sex hormone treatment and psychobiological changes in transsexual persons: two-year follow-up data. The Journal of Clinical Endocrinology & Metabolism 101 (11), 4260–4269. Available at https://doi.org/10.1210/jc.2016-1276.

Frey, J.D., Poudrier, G., Chiodo, M.V., Hazen, A., 2016. A systematic review of metoidioplasty and radial forearm flap phalloplasty in female-to-male transgender genital reconstruction: is the "ideal" neophallus an achievable goal? Plastic and Reconstructive Surgery - Global Open 4 (12), e1131. Available at https://doi.org/10.1097/gox.0000000000001131.

Griffiths, S., Yager, Z., 2019. Gender, embodiment, and eating disorders. Journal of Adolescent Health 64 (4), 425–426. Available at https://doi.org/10.1016/j.jadohealth.2019.01.016.

Grimstad, F.W., Lee, J., 2024. Overview of gender-affirming surgery. Pediatric Radiology 54 (8), 1363–1370. Available at https://doi.org/10.1007/s00247-024-05874-z.

Haben, C.M., 2022. Masculinization laryngoplasty. Otolaryngologic Clinics of North America 55 (4), 757–765. Available at https://doi.org/10.1016/j.otc.2022.04.011.

Hassan, B., Zeitouni, F., Ascha, M., Sanders, R., Berger, Z., Fields, E., et al., 2024. Breast surgery in adolescents cisgender breast reduction versus transgender and nonbinary chest masculinization. Annals of Plastic Surgery 93 (2), 194–199. Available at https://doi.org/10.1097/SAP.0000000000003981.

Hein, I.M., De Vries, M.C., Troost, P.W., Meynen, G., Van Goudoever, J.B., Lindauer, R.J.L., 2015. Informed consent instead of assent is appropriate in children from the age of twelve: policy implications of new findings on children's competence to consent to clinical research ethics in public health, medical law, and health policy. BMC Medical Ethics 16 (1). Available at https://doi.org/10.1186/s12910-015-0067-z.

Herndon, J., Gupta, N., Davidge-Pitts, C., Imhof, N., Gonzalez, C., Carlson, S., et al., 2024. Genital surgery outcomes using an individualized algorithm for hormone management in transfeminine individuals. The Journal of Clinical Endocrinology & Metabolism 109 (11), 2774–2783. Available at https://doi.org/10.1210/clinem/dgae269.

James, S.E., Herman, J.L., Rankin, S., Keisling, M., Mottet, L., Anafi, M., 2016. The report of the 2015. U.S. Transgender Survey.

James, S.E., Herman, J.L., Durso, L.E., Heng-Lehtinen, R., 2024. Early Insights: A Report of the 2022. U.S. Transgender Survey.

Jarolím, L., Šedý, J., Schmidt, M., Naňka, O., Foltán, R., Kawaciuk, I., 2009. Gender reassignment surgery in male-to-female transsexualism: a retrospective 3-month follow-up study with anatomical remarks. The Journal of Sexual Medicine 6 (6), 1635–1644. Available at https://doi.org/10.1111/j.1743-6109.2009.01245.x.

Kilmer, L.H., Chou, J., Campbell, C.A., DeGeorge, B.R., Stranix, J.T., 2024. Gender-affirming surgery improves mental health outcomes and decreases antidepressant use in patients with gender dysphoria. Plastic and Reconstructive Surgery 154 (5), 1142–1149. Available at https://doi.org/10.1097/PRS.0000000000011325.

Kleinberg, D.L., Ruan, W., 2008. IGF-I, GH, and sex steroid effects in normal mammary gland development. Journal of Mammary Gland Biology and Neoplasia 13 (4), 353–360. Available at https://doi.org/10.1007/s10911-008-9103-7.

Lambert, A., Pratt, A., Conard, L.A.E., Grigg Dean, E., Page, E., Vaughn, L.M., et al., 2023. Supporting gender-related medical decision making for transgender and gender-diverse individuals: a scoping review. Transgender Health 8 (2), 113–123. Available at https://doi.org/10.1089/trgh.2021.0030.

Laspro, M., Hoffman, A., Chinta, S., Abdalla, J., Tran, D., Oh, C., et al., 2025. The effects of perioperative gender-affirming hormone therapy on facial feminization surgery adverse events, facial features addressed, and esthetic satisfaction: a multimodal analysis. Journal of Craniofacial Surgery 36 (2), 600–604. Available at https://doi.org/10.1097/SCS.0000000000010840.

Li, A.Y., Park, M.J., Fick, J., Ousterhout, D.K., Deschamps-Braly, J.C., 2024. Perioperative estrogen hormonal therapy does not increase venous thromboembolism risk in facial feminization surgery. Plastic and Reconstructive Surgery 154 (6), 1309–1315. Available at https://doi.org/10.1097/PRS.0000000000011437.

Mahfouda, S., Moore, J.K., Siafarikas, A., Hewitt, T., Ganti, U., Lin, A., et al., 2019. Gender-affirming hormones and surgery in transgender children and adolescents. The Lancet Diabetes and Endocrinology 7 (6), 484–498. Available at https://doi.org/10.1016/s2213-8587(18)30305-x.

Moghadam, S., Roca, Y., LaGuardia, J.S., Huang, K.X., Chin, M.G., Bedar, M., et al., 2024. Effect of duration of feminizing hormone therapy on facial fat volumes. Plastic and Reconstructive Surgery 154 (5), 1081–1088. Available at https://doi.org/10.1097/PRS.0000000000011200.

Morrison, S.D., Vyas, K.S., Motakef, S., Gast, K.M., Chung, M.T., Rashidi, V., et al., 2016a. Facial feminization: systematic review of the literature. Plastic and Reconstructive Surgery 137 (6), 1759–1770. Available at https://doi.org/10.1097/PRS.0000000000002171.

Morrison, S.D., Shakir, A., Vyas, K.S., Kirby, J., Crane, C.N., Lee, G.K., 2016b. Phalloplasty: a review of techniques and outcomes. Plastic and Reconstructive Surgery 138 (3), 594–615. Available at https://doi.org/10.1097/PRS.0000000000002518.

Murad, M.H., Elamin, M.B., Garcia, M.Z., Mullan, R.J., Murad, A., Erwin, P.J., et al., 2010. Hormonal therapy and sex reassignment: a systematic review and meta-analysis of quality of life and psychosocial outcomes. Clinical Endocrinology 72 (2), 214–231. Available at https://doi.org/10.1111/j.1365-2265.2009.03625.x.

Narayan, S.K., Hontscharuk, R., Danker, S., Guerriero, J., Carter, A., Blasdel, G., et al., 2021. Guiding the conversation—types of regret after gender-affirming surgery and their associated etiologies. Annals of Translational Medicine 9 (7), 605. Available at https://doi.org/10.21037/atm-20-6204 605.

Oles, N., Darrach, H., Landford, W., Garza, M., Twose, C., Park, C.S., et al., 2022. Gender affirming surgery: a comprehensive, systematic review of all peer-reviewed literature and methods of assessing patient-centered outcomes (part 1: breast/chest, face, and voice). Annals of Surgery 275 (1), e52. Available at https://doi.org/10.1097/sla.0000000000004728.

Olson-Kennedy, J., Warus, J., Okonta, V., Belzer, M., Clark, L.F., 2018. Chest reconstruction and chest dysphoria in transmasculine minors and young adults comparisons of nonsurgical and postsurgical cohorts. JAMA Pediatrics 172 (5), 431–436. Available at https://doi.org/10.1001/jamapediatrics.2017.5440.

Olson-Kennedy, J., Chan, Y.M., Rosenthal, S., Hidalgo, M.A., Chen, D., Clark, L., et al., 2019. Creating the trans youth research network: a collaborative research endeavor. Transgender Health 4 (1), 304–312. Available at https://doi.org/10.1089/trgh.2019.0024.

Papadopulos, N.A., Lellé, J.D., Zavlin, D., Herschbach, P., Henrich, G., Kovacs, L., et al., 2017. Quality of life and patient satisfaction following male-to-female sex reassignment surgery. Journal of Sexual Medicine 14 (5), 721–730. Available at https://doi.org/10.1016/j.jsxm.2017.01.022.

Park, R.H., Liu, Y.T., Samuel, A., Gurganus, M., Gampper, T.J., Corbett, S.T., et al., 2022. Long-term outcomes after gender-affirming surgery: 40-year follow-up study. Annals of Plastic Surgery 89 (4), 431–436. Available at https://doi.org/10.1097/SAP.0000000000003233.

Rabbitts, J.A., Aaron, R.V., Fisher, E., Lang, E.A., Bridgwater, C., Tai, G.G., et al., 2017. Long-term pain and recovery after major pediatric surgery: a qualitative study with teens, parents, and perioperative care providers. Journal of Pain 18 (7), 778–786. Available at https://doi.org/10.1016/j.jpain.2017.02.423.

Reisman, T., Goldstein, Z., Safer, J.D., 2019. A review of breast development in cisgender women and implications for transgender women. Endocrine Practice 25 (12), 1338–1345. Available at https://doi.org/10.4158/EP-2019-0183.

Ren, T., Galenchik-Chan, A., Erlichman, Z., Krajewski, A., 2024. Prevalence of regret in gender-affirming surgery a systematic review. Annals of Plastic Surgery 92 (5), 597–602. Available at https://doi.org/10.1097/SAP.0000000000003895.

Robinson, I.S., Carswell, J.M., Boskey, E., Agarwal, C.A., Brassard, P., Bélanger, M., et al., 2023. Gender-affirming surgery in adolescents and young adults: a review of ethical and surgical considerations. Plastic and Reconstructive Surgery 152 (4). Available at https://doi.org/10.1097/PRS.0000000000010325.

Roblee, C.V., Arteaga, R., Taritsa, I., Ascha, M., Weissman, J.P., Hackenberger, P., et al., 2024. Patient-reported and clinical outcomes following gender-affirming chest surgery: a comparison of binary and nonbinary transmasculine individuals. Plastic and Reconstructive Surgery - Global Open 12 (11), e6297. Available at https://doi.org/10.1097/GOX.0000000000006297.

Rogalska, M., Zielinski, M., Antkowiak, L., Kasperczuk, A., Misiolek, M., 2024. Impact of Wendler glottoplasty on acoustic measures and quality of voice in transgender women: a systematic review and meta-analysis. European Archives of Oto-Rhino-Laryngology 281 (7), 3345–3360. Available at https://doi.org/10.1007/s00405-024-08486-3.

Ross, M.B., Wesseling, S., Mullender, M., Kreukels, B.P.C., van de Grift, T., 2024. Expectations and experienced outcomes regarding gender-affirming surgeries: a pilot study. Transgender Health 9 (4), 326–338. Available at https://doi.org/10.1089/trgh.2022.0118.

Schoffer, A.K., Bittner, A.K., Hess, J., Kimmig, R., Hoffmann, O., 2022. Complications and satisfaction in transwomen receiving breast augmentation: short- and long-term outcomes. Archives of Gynecology and Obstetrics 305 (6), 1517–1524. Available at https://doi.org/10.1007/s00404-022-06603-3.

Slavit, D.H., Maragos, N.E., Lipton, R.J., 1990. Physiologic assessment of isshiki type III thyroplasty. Laryngoscope 100 (8), 844–848. Available at https://doi.org/10.1288/00005537-199008000-00009.

Snyder, L., Hebert, K.J., Horns, J.J., Schardein, J., McCormick, B.J., Downing, J., et al., 2023. Incidence and risk factors for postoperative venous thromboembolism after gender affirming vaginoplasty: a retrospective analysis of a large insurance claims database. Urology 182, 95–100. Available at https://doi.org/10.1016/j.urology.2023.08.041.

Stice, E., Shaw, H.E., 2002. Role of body dissatisfaction in the onset and maintenance of eating pathology: a synthesis of research findings. Journal of Psychosomatic Research 53 (5), 985–993. Available at https://doi.org/10.1016/S0022-3999(02)00488-9.

Swan, J., Phillips, T.M., Sanders, T., Mullens, A.B., Debattista, J., Brömdal, A., 2023. Mental health and quality of life outcomes of gender-affirming surgery: a systematic literature review. Journal of Gay & Lesbian Mental Health 27 (1), 2–45. Available at https://doi.org/10.1080/19359705.2021.2016537.

Tan, K.K.H., Byrne, J.L., Treharne, G.J., Veale, J.F., 2023. Unmet need for gender-affirming care as a social determinant of mental health inequities for transgender youth in Aotearoa/New Zealand. Journal of Public Health (United Kingdom) 45 (2). Available at https://doi.org/10.1093/pubmed/fdac131.

Tang, A., Hojilla, J.C., Jackson, J.E., Rothenberg, K.A., Gologorsky, R.C., Stram, D.A., et al., 2022. Gender-affirming mastectomy trends and surgical outcomes in adolescents. Annals of Plastic Surgery 88 (4), S325. Available at https://doi.org/10.1097/SAP.0000000000003135.

Thornton, S.M., Edalatpour, A., Gast, K.M., 2024. A systematic review of patient regret after surgery- a common phenomenon in many specialties but rare within gender-affirmation surgery. American Journal of Surgery 234, 68–73. Available at https://doi.org/10.1016/j.amjsurg.2024.04.021.

Tordoff, D.M., Wanta, J.W., Collin, A., Stepney, C., Inwards-Breland, D.J., Ahrens, K., 2022. Mental health outcomes in transgender and nonbinary youths receiving gender-affirming care. JAMA Network Open 5 (2). Available at https://doi.org/10.1001/jamanetworkopen.2022.0978.

Uffman, J.C., Willer, B.L., Mpody, C., Nafiu, O.O., Tobias, J.D., 2023. Characteristics of transgender and gender-diverse youth presenting for surgery in the United States. Anesthesia and Analgesia 137 (4), 882–886. Available at https://doi.org/10.1213/ANE.0000000000006618.

Vahabzadeh-Hagh, A.M., Walsh, E., Hamidi, V., McCowen, K., 2023. Voice outcomes from direct vocal fold testosterone injections, a case report. The Laryngoscope 133 (5), 1211–1213. Available at https://doi.org/10.1002/lary.30556.

Van Damme, S., Cosyns, M., Deman, S., Van den Eede, Z., Van Borsel, J., 2017. The effectiveness of pitch-raising surgery in male-to-female transsexuals: a systematic review. Journal of Voice 31 (2), 244.e1. Available at https://doi.org/10.1016/j.jvoice.2016.04.002.

van de Grift, T.C., Elaut, E., Cerwenka, S.C., Cohen-Kettenis, P.T., Kreukels, B.P.C., 2018. Surgical satisfaction, quality of life, and their association after gender-affirming surgery: a follow-up study. Journal of Sex and Marital Therapy 44 (2), 138–148. Available at https://doi.org/10.1080/0092623X.2017.1326190.

Van Der Meulen, I.S., Bungener, S.L., Van Der Miesen, A.I.R., Hannema, S.E., Kreukels, B.P.C., Steensma, T.D., et al., 2025. Timing of puberty suppression in transgender adolescents and sexual functioning after vaginoplasty. Journal of Sexual Medicine 22 (1), 196–204. Available at https://doi.org/10.1093/jsxmed/qdae152.

Veldhuis, J.D., Roemmich, J.N., Richmond, E.J., Rogol, A.D., Lovejoy, J.C., Sheffield-Moore, M., et al., 2005. Endocrine control of body composition in infancy, childhood, and puberty. Endocrine Reviews 26 (1), 114–146. Available at https://doi.org/10.1210/er.2003-0038.

Vrouenraets, L.J.J.J., de Vries, A.L.C., de Vries, M.C., van der Miesen, A.I.R., Hein, I.M., 2021. Assessing medical decision-making competence in transgender youth. Pediatrics 148 (6). Available at https://doi.org/10.1542/peds.2020-049643.

Webb, H., Free, N., Oates, J., Paddle, P., 2022. The use of vocal fold injection augmentation in a transmasculine patient unsatisfied with voice following testosterone therapy and voice training. Journal of Voice 36 (4), 588.e1. Available at https://doi.org/10.1016/j.jvoice.2020.08.011.

Weixel, T., Pellicane, M.J., Ciesla, J.A., Wildman, B., Medalie, D., 2025. Post-operative outcomes for transmasculine and nonbinary individuals who received top surgery as minors. International Journal of Transgender Health. Available at https://doi.org/10.1080/26895269.2025.2498751.

Wernick, J.A., Busa, S., Matouk, K., Nicholson, J., Janssen, A., 2019. A systematic review of the psychological benefits of gender-affirming surgery. Urologic Clinics of North America 46 (4), 475–486. Available at https://doi.org/10.1016/j.ucl.2019.07.002.

Wiepjes, C.M., Nota, N.M., de Blok, C.J.M., Klaver, M., de Vries, A.L.C., Wensing-Kruger, S.A., et al., 2018. The amsterdam cohort of gender dysphoria study (1972–2015): trends in prevalence, treatment, and regrets. Journal of Sexual Medicine 15 (4), 582–590. Available at https://doi.org/10.1016/j.jsxm.2018.01.016.

Wierckx, K., Gooren, L., T'Sjoen, G., 2014. Clinical review: breast development in trans women receiving cross-sex hormones. The Journal of Sexual Medicine 11 (5), 1240–1247. Available at https://doi.org/10.1111/jsm.12487.

Ziegler, A., Henke, T., Wiedrick, J., Helou, L.B., 2018. Effectiveness of testosterone therapy for masculinizing voice in transgender patients: a meta-analytic review. International Journal of Transgenderism 19 (1), 25–45. Available at https://doi.org/10.1080/15532739.2017.1411857.

Chapter 10

Mental healthcare for youth and families

Ashleigh Lin[1,3], Helen Morgan[2], Penelope Strauss[2,3], Yael Perry[2], and Blake S. Cavve[2]

[1]*Australian Research Centre in Sex, Health and Society, La Trobe University, Melbourne, VIC, Australia,* [2]*The Kids Research Institute Australia, Perth, WAS, Australia,* [3]*School of Population and Global Health, The University of Western Australia, Perth, WAS, Australia*

Learning objectives

- Briefly review the literature on the mental health of transgender and gender diverse (TGD) young people and their families.
- Describe the role of mental health providers in supporting TGD young people and their families.
- Summarize the literature on existing medical and psychosocial interventions to improve the mental health and well-being of TGD young people.
- Describe the role of the family in the mental health and well-being of TGD young people.
- Summarize existing interventions to improve the mental health and well-being of parents and caregivers.
- Highlight the strengths and weaknesses of research in this area and provide recommendations for future research on mental health interventions for TGD young people and their families.

Terminology

- **Young people** and **youth** are used to refer to children, adolescents, and young adults aged 26 years and younger.
- **Children** used alone refers to prepubescent young people.
- **Adolescents** used alone refers to young people under the age of 18 who are in puberty.
- **Young adults** refer to young people between the ages of 18 and 25.
- **Parent** is used to refer to anyone in the role of parenting and includes biological parents, foster parents, adoptive parents, other guardians, and other carers who function in the role of a parent.

Existing literature has largely approached the mental health experiences of transgender and gender diverse (TGD) young people and their families from a deficit perspective (Jacobsen and Devor, 2022; Vaughan and Rodriguez, 2014). Research establishing the preponderance of poor mental health is important in identifying the support needs of this population. However, negative framing of minority groups in relation to their health and well-being can contribute to pathologizing and patronizing discourses about a population. In this way, narratives that only present TGD youth in terms of poor mental health may inadvertently perpetuate transphobic attitudes (Vaughan and Rodriguez, 2014) and affect the self-perceptions, sense of hope, and health outcomes of TGD individuals as a result.

TGD youth and their parents regularly express the positive aspects of their identity. Findings from the Trans Pathways survey, which elicited responses from Australian TGD youth aged 14–25 years, identified four themes from the question "If you'd like, please tell us some positive aspects of being trans" (Amos et al., 2023). These were as follows: (1) connection and companionship with other TGD individuals, (2) emancipation from rigid gender role expectations, (3) gender euphoria, and (4) camaraderie and commiseration with other marginalized groups. TGD young people further

Advances in Research on Pediatric Gender Affirming Healthcare. DOI: https://doi.org/10.1016/B978-0-443-29019-0.00001-4
© 2026 Elsevier Inc. All rights are reserved, including those for text and data mining, AI training, and similar technologies.

imagined a future with a more progressive understanding of gender and TGD people, including a world in which TGD people could exist without society imposing restrictive expectations about gender on them.

Another positive factor that may prove useful, especially in assessing medical interventions, is the concept of "gender euphoria." This emerging concept is broadly understood to represent positive gender-related emotions, including both joy at specific transition milestones and a consistent sense of calmness, relief, and authenticity or "rightness" in transition (Ashley and Ells, 2018; Beischel et al., 2022; Bradford et al., 2021; Jacobsen and Devor, 2022). Emerging research indicates that gender euphoria and gender dysphoria are better considered interrelated attributes rather than different sides of one scale (Austin et al., 2022).

Nonbinary young adults (aged 18–23 years) described that being gendered correctly through the use of names and pronouns, and not being forced into "a false [gender] binary," as resulting in feelings of joy, affirmation, and support (Cosgrove, 2021). Qualitative interviews from a sample of Australian families (TGD youth aged 9–14) explored experiences of gender euphoria in youth, including themes that euphoria emerges through feelings of safety and comfort, affirming care, and hope or promise for the future (Skelton et al., 2024).

One study specifically examined the association between gender euphoria and mental health through a large survey study in Australia (Grant et al., 2024a). Although participants included adults over the age of 18, young people aged 18–24 years had the highest rates of experiencing gender euphoria at 39.3%. Nonbinary participants were least likely to report experiences of gender euphoria. Participants who reported current gender euphoria reported lower levels of psychological distress and were less likely to report recent suicidal ideation.

It is our firm belief that a more balanced view of the mental health and well-being of TGD people is required in the future so that we can celebrate TGD identity. Future research can better serve TGD people by going beyond deficit and distress-based models of TGD identities (Bradford and Catalpa, 2018; Jacobsen and Devor, 2022). Identifying the strengths and protective factors for TGD youth not only presents a more balanced picture by counteracting narratives of struggle and dysphoria but may also provide powerful factors to leverage in potential interventions to improve health and well-being (Jacobsen and Devor, 2022).

Mental health and well-being

It is well established that TGD young people experience higher rates of mental health concerns than their cisgender heterosexual peers and their sexuality diverse cisgender peers (Hill et al., 2021). An Australian study showed that three in four TGD young people have ever been diagnosed with depression and/or anxiety, and almost half of TGD young people have attempted suicide in their lifetime (Strauss et al., 2020). Similarly, high rates of mental health concerns in TGD young people have been reported in the United States (Price-Feeney et al., 2020), the United Kingdom (Bachmann and Gooch, 2018), and Aotearoa New Zealand (Tan et al., 2020). TGD youth also experience higher rates of eating disorders than the general population (Kerr et al., 2024), often in attempts to suppress menstruation and alter body shape (Coelho et al., 2019; Romito et al., 2021).

Some studies have shown a higher prevalence of traits associated with autism among TGD youth, and having autism is associated with additional mental health challenges (Mahfouda et al., 2019). While symptom prevalence is higher, research has not conclusively established higher rates of autism diagnoses among TGD youth, nor higher rates of gender dysphoria among autistic youth (Turban and van Schalkwyk, 2018). It is more likely that inadequate screening measures and overlapping experiences have confounded prevalence results (Thrower et al., 2020). Symptoms and experiences related to autism nonetheless warrant attention and intervention as needed.

Importantly, high rates of poor mental health are not inherently related to being TGD but rather the way in which TGD people are treated within society. Minority stress theory, originally developed for lesbian, gay and bisexual (LGB) people (Meyer, 1995), and later adapted for TGD populations (Testa et al., 2015, 2017), explains how people who are part of a minority group experience psychological stressors (e.g., discrimination, homophobia, transphobia) and can internalize these stressors (known as self-stigma), leading to poor mental health and increased suicidality.

Intersectionality is also an important consideration when understanding mental health experiences of certain populations. Intersectionality refers to the ways in which multiple identities (e.g., gender, race, sexuality, disability) overlap and intersect to compound health inequities (Medina-Martínez et al., 2021). The impact of minority stress on mental health suggests the possibility for effective educational interventions at a societal level. An example of this is the "We Are Allies" campaign in Canada (We Are Allies, 2025). Such campaigns are likely to result in improving the well-being of TGD people generally through education, which fosters understanding and inclusion. Additional research is needed to identify the ways in which TGD young people can be better supported and empowered by health systems, schools, families, and the broader society.

Gender affirmation as a mental health intervention

At an individual level, affirming a TGD young person's gender identity is vitally important for mental health and well-being. Social transition can help young people feel acknowledged and validated (Skelton et al., 2024). Parents have described observing parental affirmation and social transition as improving their child's well-being and lifting a weight from their own shoulders (Horton, 2023). TGD children who are supported by their parents to live openly and express their identified gender show improvements in demeanor and well-being (Durwood et al., 2017), no elevation in depressive symptoms, and only slightly elevated anxiety relative to population norms (Olson et al., 2016). However, social transition is not an exclusively positive experience, as shown in a community sample of TGD young people 14–21 years where social transition was associated with increased risk for verbal harassment, potentially due to increased visibility (Grant et al., 2024b).

Medical and legal affirmation have been associated with lower psychological distress and fewer anxiety symptoms, and legal affirmation has been further associated with less suicidal ideation. A self-report study of Brazilian youth aged 16–24 years found that having accessed multiple domains of (social, legal, and medical/surgical) gender affirmation was associated with fewer depression and anxiety symptoms, greater sense of pride and positivity about gender identity, and greater feeling of being socially accepted (Fontanari et al., 2020). Social gender affirmation by parents was also associated with a decrease in depression symptoms, although the relationship with anxiety symptoms was more complex.

There is now a growing literature on the mental health benefits of medical affirmation for adolescents, with most of this literature emerging from specialized pediatric gender services. Although mental health improvement is not meant to be the primary intended goal of medical affirmation, improved mental health is nevertheless often an additional welcomed outcome of medications that aim to align physical appearance with gender identity. Additionally, influences on mental health and mental health outcomes are multifactorial and, therefore, interpretation of results regarding medical interventions and mental health status must also account for additional variables impacting positive, negative, neutral, or unclear changes to mental health after initiation of gender-affirming medical options.

Overall, the evidence suggests that pubertal suppression either stabilizes or improves mental health, most likely because it alleviates the distress associated with the body developing in a way that does not align with gender identity. In the Netherlands, pubertal suppression was shown to reduce depression and other emotional and behavioral problems and improve general functioning, although it did not impact anxiety symptoms (De Vries et al., 2011). Mixed findings on the association between anxiety symptoms and pubertal suppression were also highlighted in a recent review, although notably, results demonstrated no negative associations or impacts on anxiety (Croteau et al., 2025).

A qualitative study found that young people and parents perceive access to pubertal suppression as a relief, suggesting that rather than improving mental health, pubertal suppression may be more closely associated with the prevention of worsening mental health (Horton, 2024). Furthermore, those with access to pubertal suppression also often have family support, and research has shown that TGD youth who are socially transitioned and have parental support experience relatively low symptomology and stable mental health across all stages of medical affirmation (Wittlin et al., 2025). There is some evidence that pubertal suppression is associated with lower rates of suicidality and self-harm (van der Miesen et al., 2020), although it is difficult to quantify the mediating role of family support.

Exploring mental health at hormone readiness assessment, one study found that TGD youth who initiated pubertal suppression (at sexual maturity rating stage 2 or 3) were found to have fewer mental health difficulties, including anxiety, depression, and suicidal ideation, than peers who had been through an incongruent puberty (McGregor et al., 2024). A study from the Netherlands similarly found that adolescents treated with pubertal suppression had internalizing symptoms comparable to their cisgender peers, despite much higher symptom levels pretreatment (van der Miesen et al., 2020). Psychosocial functioning has also been shown to improve with pubertal suppression (Costa et al., 2015). On the contrary, another study showed no changes in internalizing or externalizing symptoms associated with pubertal suppression, although no deterioration in mental health was noted either (Carmichael et al., 2021). This, however, was a small sample (N=44) of TGD youth who were mostly postpubertal, suggesting that the timing of pubertal suppression may be a crucial factor in considering mental health outcomes.

Gender-affirming hormones (e.g., estrogen and testosterone) are associated with improvements in mental health and quality of life. There is now a considerable body of evidence showing that gender-affirming hormones reduce gender dysphoria, anxiety, depression, and suicidality, in addition to improving quality of life and general well-being (Tordoff et al., 2022; Lavender et al., 2023; Olsavsky et al., 2023; Achille et al., 2020; Aldridge et al., 2021; Allen et al., 2019; Chen et al., 2023; Kuper et al., 2020). In young adults a recent waitlist control study of transmasculine individuals (median age: 22.5 years) demonstrated that those who received testosterone 3 months before the "waitlist" group showed reductions in depression and suicidality (Nolan et al., 2023). Additional large-scale research among TGD adults in the

United States has demonstrated significantly better mental health outcomes among those who had access to affirming hormones in adolescence compared to those who only gained access to hormones as adults (Turban et al., 2022).

Due to the nature of surgical interventions being primarily accessed by adults, there are fewer studies of gender-affirming surgeries and mental health in young people, but among the limited body of research, positive outcomes have been demonstrated (Boskey et al., 2023; Mehringer et al., 2021; Olson-Kennedy et al., 2018; Weigert et al., 2013).

There is no evidence to support the mental health benefits of psychological therapy in lieu of medical gender affirmation. The term "gender exploratory therapy" is sometimes misused to refer to psychotherapy to explore gender identity, which may appear innocuous at face value. Instead, this type of therapy is meant to discourage gender affirmation and has been likened to conversion therapy, also known as gender identity change efforts (Ashley, 2023). There is similarly no evidence for the benefits of withholding medical care or attempting to change a young person's gender identity. Psychotherapy administered on an indefinite basis under the guise of readiness assessment may be experienced as particularly coercive and damaging. Indeed, young people who have experienced interventions that aim to change gender identity show higher levels of suicidality than those who are not subjected to such interventions (Green et al., 2020).

Future research

The provision of gender-affirming medical intervention for children continues to be a highly contested space, with the political and social discourse ever-increasing. While robust research supports the benefits of medical care, future research should continue to rigorously measure the mental health (and other) impacts of pubertal suppression and gender-affirming hormones in adolescents. There is a clear need for longer-term follow-up in adulthood and a large multicenter collaboration to ensure adequate sample sizes. This is already underway in the United States, Canada, Europe, and Australia. There is also a need to better understand the mental health impacts of gender-affirming surgeries in youth.

One area where research on mental health and well-being is lacking is that of young people who reidentify with their birth-registered sex. Following a social, medical, surgical, or legal transition, some may come to cease identifying as TGD, experience feelings of regret, and/or stop treatment. There is substantial social and media interest in individuals who cease gender-affirming medical treatment and/or reidentify with their birth-registered sex (sometimes referred to as "detransition" or "retransition"). Data from Australia demonstrates that this number is very small, with only 5% of referrals to a pediatric gender service closed due to children reidentifying as their birth-registered sex (Cavve et al., 2024). As most of these young people reidentified while on the waitlist or during multidisciplinary assessment, this corresponded to 1% of all patients who had initiated medical treatment at this clinic reidentifying with their birth-registered sex. Similar rates have been found in the United States (Boskey et al., 2025).

There are a number of internal factors (such as developments in self-understanding, religious and political beliefs, unresolved dysphoria) and external factors (e.g., pressure from family, friends, and partners, societal stigma, and employment difficulties) that individuals have identified as connected to (previous or ongoing) shifts in identity or gender-affirmation wishes (Durwood et al., 2022; Vandenbussche, 2022; MacKinnon et al., 2022a,b; Turban et al., 2021). Many of these factors are likely to impact mental health, stability, and interpersonal relationships. In addition, the developing concept of detransphobia points to the compounding marginalization and stigma experienced by some people seeking to cease medical or social gender affirmation (MacKinnon et al., 2022a,b). Expectations and/or experiences of stigmatization may also contribute to withdrawal from social networks, lesbian, gay, bisexual, trans, queer + (LGBTQ+) communities, and mental health professionals (Vandenbussche, 2022; Gelly et al., 2025; MacKinnon et al., 2022a,b), meaning that many navigate these experiences without previous sources of support. However, self-reported outcomes following detransition may also include increased self-acceptance and improved well-being (Maragos et al., 2025) and do not necessarily entail regret (MacKinnon et al., 2022a,b). Furthermore, assumptions of regret as intrinsically linked with an evolution of identity can be experienced as stigmatizing to some (Gelly et al., 2025; MacKinnon et al., 2022a,b).

The current evidence base regarding the mental health and support needs of people who cease gender-affirming medical treatment and/or reidentify with their birth-registered sex is based largely upon a few in-depth qualitative projects with older adolescents and young adults participating in a few large North American research initiatives, including (De)trans Discourses (Sansfaçon et al., 2023; Gelly et al., 2025), De/Retrans Canada (MacKinnon et al., 2022a,b), and DARE (MacKinnon et al., 2025). One notable exception includes a secondary study within a larger longitudinal study of gender development, which included interviews of 15 young people (mean age: 11 years) who disengaged from social gender affirmation and provided some evidence that, with support, this can be undertaken without distress, rejection, or regret, but the study does not examine mental health in detail (Durwood et al., 2022).

There is an urgent need for research exploring the gender journey of children and younger adolescents with dynamic gender experiences and what mental health supports they and their families require. This needs to be conducted in an empathetic manner that avoids shame and recognizes that dynamic or nonlinear gender transition trajectories are possible. Researchers also need to be particularly clear in their terminology and definitions of "detransition," and the conflation of detransition/retransition with regret. For example, many participants who self-identify with the label "detransitioners" due to their experiences of treatment cessation or shifts in identity still identify as TGD and may have positive or mixed emotions about treatment and results (MacKinnon et al., 2022a,b). There is no current research to indicate the extent to which interventions adapted for TGD people serve those with past experiences of transition.

Psychosocial interventions to improve mental health

While there is no evidence that psychosocial interventions can reduce gender dysphoria itself, TGD youth may benefit from support for mental health for a variety of other reasons, including coping with stigma and discrimination, managing mental health concerns independent of gender, including complex and severe mental health concerns, and additional coping strategies for managing dysphoria while seeking appropriate affirmation steps (Huit et al., 2024). Minority stress theory and related research underpin the foundations of affirmative, evidence-based LGBTQ-focused interventions (Pachankis et al., 2024; Hollinsaid et al., 2023). Additionally, the psychological mediation framework, originally conceptualized by Hatzenbuehler (2009), has garnered empirical attention in the literature concerning TGD individuals' health (Sarno et al., 2020; Scandurra et al., 2018; Tucker et al., 2019). This framework proposes that the association between minority stress and psychopathology is mediated by cognitive, emotional, and interpersonal psychological processes. As minority stress theory and supporting empirical research indicate a need for cognitive, affective, and behavioral treatment objectives (Pachankis, 2015), contemporary evidence-based treatments addressing mental health disparities in TGD people have been predominantly underpinned by cognitive-behavioral approaches (Austin et al., 2018; Knutson et al., 2021; Lucassen et al., 2021; Riach, 2021).

Studies investigating the effectiveness and feasibility of gender-affirming psychological interventions are lacking in comparison with a strong body of research detailing the importance of good-quality, evidence-based, psychologically affirmative clinical practice (Chang and Singh, 2018; Ehrensaft, 2016; Nealy, 2017; Singh and Dickey, 2017). Almost all published psychological interventions have included TGD young people as a part of broader LGBTQ-focused interventions. A systematic review of 22 studies investigating LGBTQ-focused, empirically supported affirmative psychological interventions reported findings from eight studies which included TGD people only, with an additional two studies disaggregating findings for TGD participants (Expósito-Campos et al., 2023). Among these studies, four had youth samples (Austin et al., 2018; Lucassen et al., 2021; Stevens et al., 2020), and only one was a randomized controlled trial (RCT), presented within a doctoral thesis (Martin, 2019).

Interventions included a series of interactive online video tutorials focused on queer youth empowerment (QueerVIBE) (Martin, 2019), a face-to-face group affirmative cognitive behavioral therapy (CBT) program for TGD youth (AFFIRM) (Austin et al., 2018), an online CBT-based serious game SPARX (Lucassen et al., 2021), and a community-based model of care involving coordinated case management, CBT, sexual health, education, housing, and employment services (iTEAM) (Stevens et al., 2020). Overall, findings from TGD participants were mixed. Two studies noted improvements in psychological distress (Martin, 2019) and depressive symptoms (Austin et al., 2018), while the other two found no improvement in mental health symptoms (Lucassen et al., 2021; Stevens et al., 2020). However, other improvements were noted across the studies in terms of positive TGD identity and self-esteem. These findings should be considered in light of some significant methodological limitations. All but one study (Martin, 2019) was uncontrolled pre–post trial. Furthermore, two studies had very small sample sizes (Austin et al., 2018; Stevens et al., 2020), while the other two had more substantial samples but very high attrition (Lucassen et al., 2021; Martin, 2019).

A recent review of psychosocial support interventions for children and adolescents experiencing gender dysphoria or incongruence (Heathcote et al., 2024) identified six additional relevant studies evaluating interventions delivering individual or group psychosocial support within specialty gender services (Allen et al., 2021; Costa et al., 2015; Davidson et al., 2019), community-based programs (Hollinsaid et al., 2020), and residential settings (Silveri et al., 2021). These programs utilized a range of evidence-based approaches, including psychoeducation, CBT, systemic therapy, dialectical behavior therapy, and motivational enhancement therapy. One study evaluated the effectiveness of an online group Mindful Self-Compassion for Teens (MSC-T) (Bluth et al., 2023). Of the identified programs, three were designed specifically for young people engaged with specialty gender services, and one was adapted for TGD youth. Two interventions were delivered without modification for this population. Across studies, outcomes were mixed, reporting either improvements or no significant changes in outcomes, including depression, anxiety, general mental health, emotion dysregulation, and suicidality, as well as quality of life and functioning. Importantly, none observed adverse effects.

However, all were rated as low quality, with study designs, including four pre–post evaluations, two longitudinal cohorts, and one secondary analysis.

Currently, specialized psychosocial interventions tailored for young people in gender services remain limited to the settings in which they were developed. Beyond this, only AFFIRM and SPARX interventions are available outside of the research context, and only in limited global regions. Only AFFIRM and SPARX have been adapted specifically for TGD youth. SPARX-T, an adapted version of SPARX designed with and for TGD young people, is currently being trialed in an RCT in Australia. An RCT of an online CBT group-based minority stress intervention is also underway (Chinsen et al., 2024).

Research strengths and weaknesses

This is a developing area of research with many more findings likely to be available in the coming years. There has been a promising increase in focus on using an RCT design to test the efficacy of psychosocial interventions for TGD people (Expósito-Campos et al., 2023), although this has largely been in adult populations, and controlled trials in youth remain limited (Heathcote et al., 2024). Of note, existing psychosocial intervention studies in TGD youth have largely used standardized measures with sound psychometric properties to assess outcomes, although many have not been validated in TGD youth populations, and there has been substantial heterogeneity in study aims, intervention content, delivery, and focus, as well as outcome measures. Multifaceted target outcomes for interventions and standardized measures should be goals for future interventions.

Many interventions to date have been (and continue to be) designed for LGBTQ youth generally, with findings from TGD and cisgender sexuality diverse participants often not disaggregated. This disregards the key psychosocial and cultural differences among these subpopulations that may be pivotal to health risk behaviors and treatment needs (Smalley et al., 2016). Additionally, recent studies have shown crucial differences between binary and nonbinary TGD individuals (Bradford and Catalpa, 2018; Catalpa et al., 2020), indicating a need for heterogeneous consideration of subpopulations in TGD research.

In general, sample sizes are small, and therefore generalizability is limited (Heathcote et al., 2024; Expósito-Campos et al., 2023). Attrition rates are commonly high, particularly for digital interventions (Craig et al., 2021; Lucassen et al., 2021), and analysis does not routinely investigate differences between those withdrawing from and those completing studies, limiting the validity of results (Expósito-Campos et al., 2023). Most studies do not include follow-up evaluation, limiting the ability to assess enduring changes. Given that minority stress can occur across the lifespan (Chen et al., 2010), longer interventions with timely follow-up processes are indicated. It is worth noting that some studies do not distinguish between the findings of young adults (18–25 years) and older adults (Israel et al., 2021), limiting our specific understanding of the impact of these interventions on young people. As there are developmental and social differences between different age groups, with young people thought to experience a higher level of minority stress than their adult counterparts (Klein and Golub, 2016; Toomey et al., 2013), it is vital for study designs to measure youth-specific outcomes distinctly. Although there has been some increase in psychosocial interventions for TGD adolescents (Lucassen et al., 2021), there remains an absence of evidence for appropriate and effective interventions to improve the mental health of children under the age of 12. This should be a priority given that some families are supportive of gender exploration or affirmation of youth from before school age (Alegría, 2018; Gregor et al., 2015; Horton, 2023).

Future research

There is a clear need for TGD youth-specific psychosocial interventions to be developed and rigorously tested using RCT designs. These must consider the developmental and social contexts, and the understanding that risk and protective factors differ between TGD children, adolescents, and young adults. Careful consideration should be given to the primary and secondary outcomes, and the alignment of outcome measures between researchers in this field would be optimal. More measures that are developed, validated, and normed specifically for TGD youth (and in partnership with them) would be a useful addition to the field. We require a clearer understanding of how intervention outcomes differ for binary and nonbinary TGD youth, and how outcomes are impacted by intersectional identities. We would argue that, given the high rates of autism in TGD youth and the association between neurodivergence and mental well-being (Mahfouda et al., 2019; Strauss et al., 2021), it must be an important consideration in the design of any psychosocial intervention.

TGD young people face multiple barriers to accessing healthcare, over and above those experienced by their sexuality diverse counterparts (Hill et al., 2021; McDermott et al., 2016). The development of digital interventions represents a significant opportunity to address such inequalities. However, attrition rates in digital intervention studies have been particularly high (Craig et al., 2021; Lucassen et al., 2021). Consequently, digital psychological interventions are particularly lacking in evidence for efficacy, and the mechanisms for retaining adolescent participants throughout studies need to be urgently explored.

Finally, once the evidence base has been more clearly established in this field, efforts to improve the translation and accessibility of empirically supported psychosocial interventions beyond research and isolated clinical contexts will be critical. Implementation designs will be required to determine the feasibility of delivering and scaling up interventions with fidelity, with particular attention given to adaptations required for contexts such as geographical location, service delivery setting, and resourcing.

Family systems

Families are an important developmental context for children and adolescents generally, and, for TGD young people, they are a domain that can confer both risk and resilience, particularly once the young person 'comes out' (Simons et al., 2013; Travers et al., 2012). In the context of the challenges inherent in navigating a cisnormative world, parental support of TGD youth is strongly correlated with protective factors across a range of outcomes, such as lower rates of depression, anxiety, and suicidal ideation and behaviors, higher quality of life, and safer sex practices when compared with TGD youth experiencing ambivalent, low, or no parental support (Bauer et al., 2015; Johns et al., 2018; Olson et al., 2016; Simons et al., 2013; Veale et al., 2017; Watson et al., 2017; Wilson et al., 2012).

Conversely, parental rejection and ambivalence are related to negative outcomes such as increased self-harm, suicidality, risk-taking behaviors, and higher levels of mental health difficulties such as depression (Di Fini et al., 2025; Day et al., 2017; Fisher et al., 2018). A 2012 study found that 57% of TGD youth who reported experiencing parental rejection had attempted suicide in comparison with just 4% who reported experiences of parental acceptance (Travers et al., 2012). Parental rejection also has other serious, far-reaching impacts for TGD youth. Those who have experienced parental rejection are significantly overrepresented across the areas of homelessness, sexual exploitation, and the out-of-home/foster care system (Bateman, 2015; Choi et al., 2015; McCormick et al., 2018). Conversely, TGD youth who are supported by their parents to live openly and express their identified gender show improvements in demeanor and well-being (Durwood et al., 2017), no elevation in depressive symptoms, and only slightly elevated anxiety relative to population norms (Olson et al., 2016).

Consequently, parental support has been posited as the strongest protective factor against negative health outcomes for TGD young people (Bariola et al., 2015; Veale et al., 2017). Parental support is even more critical for TGD young people who often face barriers in accessing safe and affirming support and health services (Kerr et al., 2022; Telfer et al., 2018), culminating in parents being required to become the consistent and key support, and essential in facilitating contact for gender-affirming medical care (Riggs and Due, 2015). More commonly, however, young people report experiences of parental ambivalence or outright rejection once they disclose their gender identity, which negatively impacts their mental health (Caldarera et al., 2021; Clark et al., 2014; McDermott et al., 2016; Simons et al., 2013; Travers et al., 2012).

Mental health and well-being of parents

Parental development research indicates that parents learn and grow exponentially from their parent–child interactions (Dillon, 2002). These interactions may enhance personal aspects such as awareness of values, emotional sensitivity, and cognitive flexibility (Dillon, 2002). Additionally, supporting a child's development can facilitate the evolution of individual parental factors such as self-acceptance, open-mindedness, and personal authenticity (Luvmour, 2011). Family systems theory posits that the family functions as an emotional system with interdependent members that cannot be fully understood independent of the context of that system (Cox and Paley, 2003; Minuchin, 1985). Consequently, having a TGD child is especially pertinent within the family system.

That most parents report intense and complex emotional reactions, particularly in the early stages of becoming aware of their child's TGD identity, is well-documented. Reactions include anxiety and fear for the child's future, sadness, disbelief, belief that it is a temporary phase, loss of hope, and loneliness (Barron and Capous-Desyllas, 2017; Coolhart et al., 2018; Di Ceglie and Thümmel, 2006; Katz-Wise et al., 2017; Kuvalanka et al., 2018; Kuvalanka et al., 2014; Riley et al., 2011). Grief and loss have been reported to be dominant experiences (Norwood, 2013; Pepper and Brill, 2008; Wahlig, 2015). Another common theme is that of worry for their child's safety (Kuvalanka et al., 2014).

Some parents immediately affirm their child's gender and can express their unconditional support, love, acceptance, and pride (Alegría, 2018; Kuvalanka et al., 2014; Wong and Drake, 2018). More commonly, however, parents report a complex combination of these emotions that shift and change over time (Birnkrant and Przeworski, 2017; Katz-Wise et al., 2017; Kuvalanka et al., 2018). One qualitative study on the impact of having a TGD child on parental psychological well-being found that experiences of psychological well-being were multidimensional, characterized by numerous mental health threats (e.g., relationship ruptures with nonaffirming family members and friends, feelings of isolation, and anxiety related to their child's present and future well-being) and concurrent positive experiences of personal growth (e.g., increased open-mindedness, self-reflection, and confidence to live with greater authenticity) (Wells et al., 2022). There was also a focus on the importance of support and acceptance from others, and various coping strategies (e.g., creating family safety, self-care, and active emotion-focused strategies).

Parents of TGD children can also experience feelings of anxiety, stress, depression, hypervigilance, social isolation, increased substance use, sleep problems, caretaker fatigue, and relational strain with a partner (Barron and Capous-Desyllas, 2017; Bull and D'Arrigo-Patrick, 2018; Hidalgo and Chen, 2019; Sansfaçon et al., 2015; Wren, 2002). They can be impacted by stigma and rejection from their social networks and by healthcare professionals (Riggs and Due, 2015; Gray et al., 2016), leading to questioning of their parenting of their TGD child (Di Ceglie and Thümmel, 2006; Kuvalanka et al., 2014; Malpas, 2011). Stressors like stigma and rejection have been posited as a form of parental minority stress (Hidalgo and Chen, 2019).

On the contrary, many studies clearly document the positive aspects of parenting a TGD child. For example, a narrative literature review on parental reactions to having a TGD child found references to positive outcomes for parents in 16 studies out of 32 studies (Abreu et al., 2019). These included relational benefits (such as improved parent–child communication and connectedness), affirmation of values (e.g., unconditional love), newfound meaning through becoming activists, and the development of new personal narratives.

More recently, studies exploring parental experiences of having a TGD child have reflected increasingly positive narratives. For example, qualitative interviews demonstrated that supporting a child through social transition and seeing the child's increased happiness and ability to participate in life are affirmative parental experiences (Horton, 2023). In another study of parental narratives, confronting core beliefs, stepping into leadership and advocacy roles, and experiencing personal growth were all reported benefits for parents supporting a nonbinary child (Bull et al., 2022). Improved quality of the parent–child relationship has also been identified as a common experience following parental growth toward activism (Schlehofer et al., 2021). A recent qualitative study on the positive aspects of parenting a TGD child found similar results, with five major themes identified: (1) personal growth; (2) experiencing the strengths of the child; (3) improved aspects of the parent–child relationship; (4) positive change in the child; and (5) experiencing social connection and support (Morgan et al., 2023).

Mental health and well-being of siblings

Research concerning the impact of having a TGD child in the family has focused predominantly on the TGD child and parent well-being (Westwater et al., 2019). Given that a large majority of young people have siblings, and family systems theory indicates one individual's change inevitably impacts the entire system (Cox and Paley, 2003), understanding the impact of having a TGD sibling is crucial to the development of resources and support for siblings. Existing knowledge, resources, and support regarding siblings of TGD youth (henceforth, siblings) are scant (Norwood, 2013; Parker and Davis-McCabe, 2021; Wheeler et al., 2019). This limited literature indicates that siblings may experience grief (Norwood, 2013), bullying (Gregor et al., 2015), fear (Wheeler et al., 2019), relational strain, and divided loyalty due to familial tensions, isolation, suppression of negative emotions, and stress (Parker and Davis-McCabe, 2021). Siblings also report elevated rates of anxiety and depression when compared with the general population (Gibson et al., 2021).

Siblings identify a lack of access to accurate information and formal targeted support for themselves as pivotal to negative impacts on their mental health and ability to cope with relational challenges (Parker and Davis-McCabe, 2021). Knowledge acquisition regarding gender diversity and sibling support can result in improved sibling well-being, health outcomes, and improved familial relationships (Wheeler et al., 2019). Taken together, studies indicate siblings are inevitably impacted by having a TGD child in the family and appear to undertake significant personal and relational journeys. Consequently, researchers have called for dedicated research and resource development for improved sibling health outcomes (Klepper et al., 2024; Parker and Davis-McCabe, 2021; Wheeler et al., 2019). A research study is currently underway (in Perth, Australia), which is focused on better understanding the experiences and needs of siblings with the aim of developing dedicated sibling resources.

Interventions for families

Best practice for mental health providers in providing care to parents and carers should be individualized and comprehensive, assist in facilitating decision-making, address caregiver concerns from a risk/benefit perspective, and understand the complexity of emotions (Buckloh et al., 2022). However, research has focused primarily on group interventions for parents and carers. Worldwide, there have been several group-based interventions for parents of TGD youth. These studies indicate certain program elements are useful in facilitating parental understanding, support, and acceptance. Peer-support groups have been the most utilized format for parental support and are shown to confer positive benefits (Di Ceglie and Thümmel, 2006; Hillier and Torg, 2019; Lawlis et al., 2020; Menvielle and Rodnan, 2011). Such groups facilitate connection between parents undergoing similar experiences, normalize the parental journey, and reduce feelings of isolation in a cost-effective manner. Discussion in these groups centers around parental reactions to a child's disclosure, concerns about safety and the child's future, anxiety, uncertainty, loss and grief, transitional processes, and how to navigate challenging situations (Menvielle and Hill, 2010; Riley et al., 2013).

Notwithstanding the gains parents experience in these groups, they lack the structured, evidence-based approach and standardized evaluation that clinician-facilitated group parenting interventions for other populations afford (Malpas et al., 2022). Three recent clinician-led group programs underpinned by evidence-based approaches have been trialed (Caldarera et al., 2021; Dangaltcheva et al., 2021; Morgan et al., 2024). First, a 12-month program utilizing psychoanalytic group therapy and psychoeducational approaches was run for 14 parents of TGD children aged 8–17 years engaged with an Italian pediatric gender identity service (Caldarera et al., 2021). Participants completed a qualitative short form about the main program themes at 6 months and 12 months, and a semistructured feedback questionnaire about their experience at 12 months. Findings demonstrated that parents developed an evolved perspective of gender diversity and the needs of their children over time. They reported reduced loneliness and uncertainty regarding gender identity development and increased confidence and ability to communicate with their child. Perceived mechanisms for change included peer connection and the opportunity for mutual learning.

Connect, a trauma-informed, attachment-based group program developed for parents of at-risk adolescents aged 12–18 years in Canada, has been adapted to better meet the needs of parents of TGD youth (Dangaltcheva et al., 2021). The 10-week program utilizes reflective exercises, role-plays, and therapeutic discussion to build core skills such as reflective functioning, sensitivity, and dyadic affect regulation. The effectiveness of the program was measured through qualitative analysis of the transcribed group discussions and a posttreatment survey and semistructured feedback interview at the last session. Parents in the group (N=20) overwhelmingly reported that their relationship with their child had improved, all had applied their learning frequently or somewhat frequently, and parents stated that they had felt welcomed, respected, and safe in the program. Recommendations for changes included additional support for managing children's mental health crises, more time for peer connection, and a parallel group mirroring the program for their children.

A shorter, 4-week group program providing therapeutic support and resources for parents (N=16) has been developed and piloted for use with parents engaged with a pediatric gender diversity service in Australia, with the goals of improving parent–child relationship quality and increasing parental understanding, confidence, and support (Morgan et al., 2024). Preintervention and postintervention measures assessed parental knowledge of gender diversity and confidence in supporting their child, levels of negative emotional states, and parent–child relationship quality. Qualitative feedback primarily explored parents' subjective experience of the group. Findings indicate greater reported understanding, confidence, and support; improved levels of parental well-being; and high program acceptability. Peer connection, receiving good-quality information, and having a safe therapeutic space were cited as the most valuable program elements. This intervention has now been funded as a multisite RCT in Australia to begin in 2026.

Other interventions have been online. One recently trialed online intervention aimed at increasing transaffirming behaviors in parents through the provision of three modules, including psychoeducational content, educational videos, and interactive quizzes (Matsuno and Israel, 2021). The program content and structure were informed by psychological theories of behavior change and relevant research, and feedback was sought through focus groups with parents of TGD youth and experts in TGD mental health. Twenty-seven parents or primary caregivers of children aged 10–18 years were recruited online, predominantly via peer-support networks for parents of LGBTQIA+ children. Mixed-methods analyses indicated high acceptability and suggested improvements, such as making the intervention more tailored to participants (Matsuno and Israel, 2021).

The preliminary results from these programs indicate that peer support, evidence-based approaches, and psychoeducation are successful foundations for facilitating positive outcomes for parents of TGD young people. Taken together, these studies highlight several effective aspects of parent interventions in line with a recent systematic review of 32 articles that provided outcome data or clinical recommendations for family-based interventions with TGD youth and their families (Malpas et al., 2022). The review highlights that effective clinical practices include psychoeducation, providing space for family members to express their reactions regarding their child's gender identity, emphasizing the protectiveness of familial acceptance, utilizing multiple therapeutic modalities, providing opportunities for allyship and advocacy, connecting families to community and peer support, and centering intersectional approaches and concerns, including barriers to accessing such support.

More recently, a scoping literature review, which aimed to identify key components of psychosocial interventions focused on improving parental support of TGD youth, found four key strategies across the five reviewed studies (Vasen et al., 2025). Program components that appeared key to improving parental support for a child included access to (1) a confidential, nonjudgmental setting in which to process negative emotions, (2) information and education on gender diversity, (3) community support from other parents, and (4) resources for improving parent–child communication skills. Taken together, these reviews conclude that future research should examine the efficacy of family-based interventions by collecting quantitative data to determine impact, using validated measures in doing so, and translating findings into scalable interventions that effect enduring impact.

Future research

Parental support is strongly associated with protective factors across a range of child health outcomes for TGD young people. When parents strongly support their child's gender identity, the positive impacts are multifold, as evidenced by more positive health and well-being outcomes across several child domains, multifaceted positive parental growth, and enhanced parent–child relationship quality. In turn, such improvements can only benefit the wider family system.

Throughout the literature, improvements in the quality of future research with the families of TGD youth are indicated. To date, research samples have been characterized by a lack of diversity. Often, parent samples include those whose children are under the care of a pediatric gender service, resulting in findings that are unlikely to be generalizable to the wider parent population. Further research is also needed to better understand the experiences of cisgender male caregivers, as most parent research has elicited the perspectives of cisgender female caregivers to date, as well as parents from culturally diverse backgrounds, parents who do not identify as supportive of their child's gender identity, parents of children who are not engaged with gender-affirming health or support services, and those parenting nonbinary children. While parents are pivotal to children's well-being, others in the family system can also be highly important in promoting protective factors for supporting TGD young people's well-being. We are not aware of any interventions that have been developed to support the well-being of siblings of TGD youth, or the wider family system, although community organizations often have peer-support groups either online or face-to-face.

Many parents express that a lack of access to good-quality information, feeling overwhelmed and isolated, and fear for their child's future are barriers to supporting their child (Sansfaçon et al., 2023a,b; Riley et al., 2013; Wagner and Armstrong, 2020). There is a clear need for information, access to reliable research and guidelines, support from professionals, and contact with peers undergoing similar experiences to aid them during this journey (Riggs and Due, 2015; Pullen Sansfaçon et al., 2023a,b; Riley et al., 2013; Wren, 2002). There is a clear and pressing need for interventions to be developed that promote parental understanding and support (Matsuno and Israel, 2021; Di Fini et al., 2025; Israel et al., 2021; Parker and Davis-McCabe, 2021). Interventions targeting rejecting and mixed parental attitudes have the capacity to significantly improve both family functioning and the TGD child's mental health and well-being (Johnson et al., 2020). A combination of evidence-based approaches, good-quality information, and peer support appears to be a critical component of such interventions. The major limitation of the interventions to date is the lack of empirical evidence for their efficacy. Future studies should always incorporate an evaluation component and, optimally, use an RCT design. Furthermore, it is essential that interventions are developed collaboratively with parents to leverage the unique expertise of those with lived experience.

Research should also consider how interventions and resources can be effectively delivered outside of face-to-face group settings to maximize accessibility. For example, a recent research study in Australia utilized a codesign approach to develop a digital resource hub in collaboration with parents and community partner organizations. Transforming Families (https://transformingfamilies.org.au/) hosts a suite of resources and information to help parents better understand and support their child. It is the first website designed for and with Australian parents of TGD children and offers a significant opportunity to promote parental understanding, support, and acceptance of a child's gender identity through a widely accessible medium. Parents reported high satisfaction with and acceptability of the site (Morgan et al., 2026).

Finally, there is intrinsic value in balancing the existing dominant narrative of loss and hardship by highlighting the constructive and positive gains that so many parents of TGD young people report among the complexity of their journey. Raising awareness of positive parental experiences within a gender-affirmative framework and clinical approaches can facilitate increased parental acceptance, support, and well-being, which, in turn, can strengthen the family system and increase vital protective factors for child health and well-being outcomes.

Conclusion

It is well established that TGD young people face more mental health challenges than their cisgender LGB and heterosexual peers because of the discrimination and marginalization they face. Gender affirmation is likely to improve mental health and well-being related to gender dysphoria; however, psychological interventions may be required to support the overall mental health of TGD youth. Notably, mental health challenges cannot typically be ascribed to gender dysphoria alone, so additional supportive strategies are needed to address experiences that are unrelated to dysphoria. Additionally, while mental health can improve with social transition and the introduction of medical affirmation strategies, TGD youth and families need to maintain realistic expectations about the outcomes of interventions. Most existing interventions are developed for LGBTQ young people more generally and lack a consideration of the key psychosocial and cultural differences between LGB cisgender and TGD youth that may be pivotal to health risk behaviors and treatment needs. There remains a distinct lack of evidence-based psychological interventions to support TGD youth specifically.

Considering the family system within which a young person exists is essential, both in relation to the mental health and well-being of the TGD young person themselves and that of the parents and siblings. Parents experience a range of emotions when their child comes out as TGD, and this can have a substantial impact on their own mental health. Peer support has been shown to be a useful resource. However, there is a need for supporting parents by developing and disseminating trusted, accessible, evidence-based resources that simultaneously provide good-quality information, psychoeducation, peer-support access, and positive representation of gender diversity. This is a developing area of research that needs more attention. Additionally, the mental health and well-being of siblings and other extended family members of TGD youth remains poorly explored, and subsequently, there is no evidence for what can be done to best support them.

Research on the mental health and well-being of TGD youth and families can be strengthened by placing a greater focus on topics that have to date been neglected, such as the needs of young people who reidentify as their birth-registered sex and the siblings of TGD youth. There is also a need for the development and dissemination of psychological interventions specifically designed for and with TGD young people. RCT designs and rigorous evaluation are necessary. Although digital interventions hold promise, the challenges of attrition need to be addressed. Further research is also needed on the strength and resilience of TGD young people, and how this can be harnessed to protect against poor mental health. Affirmative research approaches should depathologize the TGD experience and celebrate the diversity of TGD people as a healthy expression of gender.

It is imperative that all research on the mental health of TGD young people and their families and the development of any intervention to improve it be authentically codesigned with the target population. This is key to ensuring that research remains relevant, safe, and acceptable. In this respect, interventions codesigned with TGD people represent a promising future for the field.

References

Abreu, R.L., Rosenkrantz, D.E., Ryser-Oatman, J.T., Rostosky, S.S., Riggle, E.D.B., 2019. Parental reactions to transgender and gender diverse children: a literature review. Journal of GLBT Family Studies 15 (5), 461–485. Available at https://doi.org/10.1080/1550428X.2019.1656132.

Achille, C., Taggart, T., Eaton, N.R., Osipoff, J., Tafuri, K., Lane, A., et al., 2020. Longitudinal impact of gender-affirming endocrine intervention on the mental health and well-being of transgender youths: preliminary results. International Journal of Pediatric Endocrinology 2020 (1). Available at https://doi.org/10.1186/s13633-020-00078-2.

Aldridge, Z., Patel, S., Guo, B., Nixon, E., Pierre Bouman, W., Witcomb, G.L., et al., 2021. Long-term effect of gender-affirming hormone treatment on depression and anxiety symptoms in transgender people: a prospective cohort study. Andrology 9 (6), 1808–1816. Available at https://doi.org/10.1111/andr.12884.

Allen, L.R., Watson, L.B., Egan, A.M., Moser, C.N., 2019. Well-being and suicidality among transgender youth after gender-affirming hormones. Clinical Practice in Pediatric Psychology 7 (3), 302–311. Available at https://doi.org/10.1037/cpp0000288.

Amos, N., Lim, G., Buckingham, P., Lin, A., Liddelow-Hunt, S., Mooney-Somers, J., et al., 2023. Rainbow Realities: In-Depth Analyses of Large-Scale LGBTQA+ Health and Wellbeing Data in Australia. La Trobe University.

Aramburu Alegría, C., 2018. Supporting families of transgender children/youth: parents speak on their experiences, identity, and views. International Journal of Transgenderism 19 (2), 132–143. Available at https://doi.org/10.1080/15532739.2018.1450798.

Ashley, F., 2023. Interrogating gender-exploratory therapy. Perspectives on Psychological Science 18 (2), 472–481. Available at https://doi.org/10.1177/17456916221102325.

Ashley, F., Ells, C., 2018. In favor of covering ethically important cosmetic surgeries: facial feminization surgery for transgender people. The American Journal of Bioethics 18 (12), 23–25. Available at https://doi.org/10.1080/15265161.2018.1531162.

Austin, A., Craig, S.L., D'Souza, S.A., 2018. An AFFIRMative cognitive behavioral intervention for transgender youth: preliminary effectiveness. Professional Psychology: Research and Practice 49 (1), 1–8. Available at https://doi.org/10.1037/pro0000154.

Austin, A., Papciak, R., Lovins, L., 2022. Gender euphoria: a grounded theory exploration of experiencing gender affirmation. Psychology & Sexuality 13 (5), 1406–1426. Available at https://doi.org/10.1080/19419899.2022.2049632.

Bachmann, C.L., Gooch, B., 2018. LGBT in Britain: Trans report. https://www.stonewall.org.uk/resources/lgbt-britain-trans-report-2018

Bariola, E., Lyons, A., Leonard, W., Pitts, M., Badcock, P., Couch, M., 2015. Demographic and psychosocial factors associated with psychological distress and resilience among transgender individuals. American Journal of Public Health 105 (10), 2108–2116. Available at https://doi.org/10.2105/AJPH.2015.302763.

Barron, C., Capous-Desyllas, M., 2017. Transgressing the gendered norms in childhood: understanding transgender children and their families. Journal of GLBT Family Studies 13 (5), 407–438. Available at https://doi.org/10.1080/1550428X.2016.1273155.

Bateman, W., 2015. LGBT Youth homelessness: A UK National Scoping of Cause, Prevelence, Response and Outcome. Albert Kennedy Trust.

Bauer, G.R., Scheim, A.I., Pyne, J., Travers, R., Hammond, R., 2015. Intervenable factors associated with suicide risk in transgender persons: a respondent driven sampling study in Ontario, Canada. BMC Public Health 15 (1). Available at https://doi.org/10.1186/s12889-015-1867-2.

Beischel, W.J., Gauvin, S.E.M., van Anders, S.M., 2022. A little shiny gender breakthrough": community understandings of gender euphoria. International Journal of Transgender Health 23 (3), 274–294. Available at https://doi.org/10.1080/26895269.2021.1915223.

Birnkrant, J.M., Przeworski, A., 2017. Communication, advocacy, and acceptance among support-seeking parents of transgender youth. Journal of Gay and Lesbian Mental Health 21 (2), 132–153. Available at https://doi.org/10.1080/19359705.2016.1277173.

Bluth, K., Lathren, C., Clepper-Faith, M., Larson, L.M., Ogunbamowo, D.O., Pflum, S., 2023. Improving mental health among transgender adolescents: implementing mindful self-compassion for teens. Journal of Adolescent Research 38 (2), 271–302. Available at https://doi.org/10.1177/07435584211062126.

Boskey, E.R., Jolly, D., Kant, J.D., Ganor, O., 2023. Prospective evaluation of psychosocial changes after chest reconstruction in transmasculine and non-binary youth. Journal of Adolescent Health 73 (3), 503–509. Available at https://doi.org/10.1016/j.jadohealth.2023.04.029.

Boskey, E.R., Scheffey, K.L., Pilcher, S., Barerra, E.P., McGregor, K., Carswell, J.M., et al., 2025. A retrospective cohort study of transgender adolescents' gender-affirming hormone discontinuation. Journal of Adolescent Health 76 (4), 584–591. Available at https://doi.org/10.1016/j.jadohealth.2024.11.002.

Bradford, N.J., Catalpa, J.M., 2018. Social and psychological heterogeneity among binary transgender, non-binary transgender and cisgender individuals. Psychology & Sexuality 10 (1), 69–82. Available at https://doi.org/10.1080/19419899.2018.1552185.

Bradford, N.J., Rider, G.N., Spencer, K.G., 2021. Hair removal and psychological well-being in transfeminine adults: associations with gender dysphoria and gender euphoria. Journal of Dermatological Treatment 32 (6), 635–642. Available at https://doi.org/10.1080/09546634.2019.1687823.

Buckloh, L.M., Poquiz, J.L., Alioto, A., Moyer, D.N., Axelrad, M.E., 2022. Best practices in working with parents and caregivers of transgender and gender diverse youth. Clinical Practice in Pediatric Psychology 10 (3), 325–335. Available at https://doi.org/10.1037/cpp0000442.

Bull, B., D'Arrigo-Patrick, J., 2018. Parent experiences of a child's social transition: moving beyond the loss narrative. Journal of Feminist Family Therapy 30 (3), 170–190. Available at https://doi.org/10.1080/08952833.2018.1448965.

Bull, B., Byno, L., D'Arrigo, J., Robertson, J., 2022. Parents of non-binary children: stories of understanding and support. Journal of Feminist Family Therapy 34 (1-2), 125–152. Available at https://doi.org/10.1080/08952833.2022.2029331.

Caldarera, A.M., Davidson, S., Vitiello, B., Baietto, C., 2021. A psychological support group for parents in the care of families with gender diverse children and adolescents. Clinical Child Psychology and Psychiatry 26 (1), 64–78. Available at https://doi.org/10.1177/1359104520963372.

Carmichael, P., Butler, G., Masic, U., Cole, T.J., de Stavola, B.L., Davidson, S., et al., 2021. Short-term outcomes of pubertal suppression in a selected cohort of 12 to 15 year old young people with persistent gender dysphoria in the UK. PLoS One 16 (2). Available at https://doi.org/10.1371/journal.pone.0243894.

Catalpa, J.M., McGuire, J.K., Fish, J.N., Rider, G.N., Bradford, N.J., Berg, D., 2020. Predictive Validity of the Genderqueer Identity Scale (GQI): Differences Between Genderqueer, Transgender and Cisgender Sexual Minority Individuals. Informa UK Limited, pp. 187–196. Available at https://doi.org/10.4324/9781003015888-17.

Cavve, B.S., Bickendorf, X., Ball, J., Saunders, L.A., Thomas, C.S., Strauss, P., et al., 2024. Reidentification with birth-registered sex in a western australian pediatric gender clinic cohort. JAMA Pediatrics 178 (5), 446–453. Available at https://doi.org/10.1001/jamapediatrics.2024.0077.

Chang, S.C., Singh, A.A., 2018. A Clinician's Guide to Gender-affirming Care: Working with Transgender and Gender Nonconforming Clients. New Harbinger Publications.

Chen, E.C., Androsiglio, R., Ng, V., 2010. Minority stress and health of lesbian, gay, and bisexual individuals: a developmental-contextual perspective. In: Ponterotto, J.G., Casas, J.M., Suzuki, L.A., Alexander, C.M. (Eds.), Handbook of multicultural counseling, third ed. Sage Publications, Inc., pp. 531–544.

Chinsen, A., Cronin, T. J., Pace, C. C., Tollit, M. A., & Pang, K. C. (2024). Evaluation of a codesigned group cognitive–behavioural therapy intervention for trans young people (TAG TEAM): protocol for a feasibility trial and a subsequent pilot RCT. BMJ Open, 14(1), e076511.

Chen, D., Berona, J., Chan, Y.M., Ehrensaft, D., Garofalo, R., Hidalgo, M.A., et al., 2023. Psychosocial functioning in transgender youth after 2 years of hormones. New England Journal of Medicine 388 (3), 240–250. Available at https://doi.org/10.1056/NEJMoa2206297.

Choi, S.K., Wilson, B.D.M., Shelton, J., Gates, G.J., 2015. Serving our youth 2015: The needs and experiences of lesbian, gay, bisexual, transgender, and questioning youth experiencing homelessness.

Clark, T.C., Lucassen, M.F.G., Bullen, P., Denny, S.J., Fleming, T.M., Robinson, E.M., et al., 2014. The health and well-being of transgender high school students: results from the New Zealand adolescent health survey (youth'12). Journal of Adolescent Health 55 (1), 93–99. Available at https://doi.org/10.1016/j.jadohealth.2013.11.008.

Coelho, J.S., Suen, J., Clark, B.A., Marshall, S.K., Geller, J., Lam, P.Y., 2019. Eating disorder diagnoses and symptom presentation in transgender youth: a scoping review. Current Psychiatry Reports 21 (11). Available at https://doi.org/10.1007/s11920-019-1097-x.

Coolhart, D., Ritenour, K., Grodzinski, A., 2018. Experiences of ambiguous loss for parents of transgender male youth: a phenomenological exploration. Contemporary Family Therapy 40 (1), 28–41. Available at https://doi.org/10.1007/s10591-017-9426-x.

Cosgrove, D., 2021. I am allowed to be myself": a photovoice exploration of non-binary identity development and meaning-making. Journal of Gay and Lesbian Social Services 33 (1), 78–102. Available at https://doi.org/10.1080/10538720.2020.1850385.

Costa, R., Dunsford, M., Skagerberg, E., Holt, V., Carmichael, P., Colizzi, M., 2015. Psychological support, puberty suppression, and psychosocial functioning in adolescents with gender dysphoria. The Journal of Sexual Medicine 12 (11), 2206–2214. Available at https://doi.org/10.1111/jsm.13034.

Cox, M.J., Paley, B., 2003. Understanding families as systems. Current Directions in Psychological Science 12 (5), 193–196. Available at https://doi.org/10.1111/1467-8721.01259.

Craig, S.L., Leung, V.W.Y., Pascoe, R., Pang, N., Iacono, G., Austin, A., et al., 2021. Affirm online: utilising an affirmative cognitive–behavioural digital intervention to improve mental health, access, and engagement among LGBTQA+ youth and young adults. International Journal of Environmental Research and Public Health 18 (4), 1–18. Available at https://doi.org/10.3390/ijerph18041541.

Croteau, T.A., Gelech, J., Morrison, M.A., Morrison, T.G., 2025. Psychological and physical health outcomes associated with gender-affirming medical care for transgender and gender-diverse youth: a critical review. Healthcare (Switzerland) 13 (14). Available at https://doi.org/10.3390/healthcare13141659.

Dahlgren Allen, S., Tollit, M. A., McDougall, R., Eade, D., Hoq, M., & Pang, K. C. (2021). A waitlist intervention for transgender young people and psychosocial outcomes. Pediatrics, 148(2).

Dangaltcheva, A., Booth, C., Moretti, M.M., 2021. Transforming connections: a trauma-informed and attachment-based program to promote sensitive parenting of trans and gender non-conforming youth. Frontiers in Psychology 12. Available at https://doi.org/10.3389/fpsyg.2021.643823.

Davidson, S., Morrison, A., Skagerberg, E., Russell, I., & Hames, A. (2019). A therapeutic group for young people with diverse gender identifications. Clinical Child Psychology and Psychiatry, 24(2), 241-257.

Day, J.K., Fish, J.N., Perez-Brumer, A., Hatzenbuehler, M.L., Russell, S.T., 2017. Transgender youth substance use disparities: results from a population-based sample. Journal of Adolescent Health 61 (6), 729–735. Available at https://doi.org/10.1016/j.jadohealth.2017.06.024.

De Vries, A.L.C., Steensma, T.D., Doreleijers, T.A.H., Cohen-Kettenis, P.T., 2011. Puberty suppression in adolescents with gender identity disorder: a prospective follow-up study. Journal of Sexual Medicine 8 (8), 2276–2283. Available at https://doi.org/10.1111/j.1743-6109.2010.01943.x.

Di Ceglie, D., Thümmel, E.C., 2006. An experience of group work with parents of children and adolescents with gender identity disorder. Clinical Child Psychology and Psychiatry 11 (3), 387–396. Available at https://doi.org/10.1177/1359104506064983.

Di Fini, G., Civilotti, C., Bolognino, A., Einaudi, G., Molo, M., Veglia, F., et al., 2025. The burden of the past: a systematic review of childhood trauma and mental health in transgender and gender nonconforming individuals. European Journal of Investigation in Health, Psychology and Education 15 (9). Available at https://doi.org/10.3390/ejihpe15090183.

Dillon, J.J., 2002. The role of the child in adult development. Journal of Adult Development 9 (4), 267–275. Available at https://doi.org/10.1023/A:1020286910678.

Durwood, L., McLaughlin, K.A., Olson, K.R., 2017. Mental health and self-worth in socially transitioned transgender youth. Journal of the American Academy of Child and Adolescent Psychiatry 56 (2), 116. Available at https://doi.org/10.1016/j.jaac.2016.10.016.

Durwood, L., Kuvalanka, K.A., Kahn-Samuelson, S., Jordan, A.E., Rubin, J.D., Schnelzer, P., et al., 2022. Retransitioning: the experiences of youth who socially transition genders more than once. International Journal of Transgender Health 23 (4), 409–427. Available at https://doi.org/10.1080/26895269.2022.2085224.

Ehrensaft, D., 2016. The Gender Creative Child, Pathways for Nurturing and Supporting Children Who Live Outside Gender Boxes. The Experiment, New York.

Expósito-Campos, P., Pérez-Fernández, J.I., Salaberria, K., 2023. Empirically supported affirmative psychological interventions for transgender and non-binary youth and adults: a systematic review. Clinical Psychology Review 100. Available at https://doi.org/10.1016/j.cpr.2022.102229.

Fisher, C.B., Fried, A.L., Desmond, M., Macapagal, K., Mustanski, B., 2018. Perceived barriers to HIV prevention services for transgender youth. LGBT Health 5 (6), 350–358. Available at https://doi.org/10.1089/lgbt.2017.0098.

Fontanari, A.M.V., Vilanova, F., Schneider, M.A., Chinazzo, I., Soll, B.M., Schwarz, K., et al., 2020. Gender affirmation is associated with transgender and gender nonbinary youth mental health improvement. LGBT Health 7 (5), 237–247. Available at https://doi.org/10.1089/lgbt.2019.0046.

Gelly, M.A., Atgé-Delbays, S., Sansfaçon, A.P., 2025. They're unable to see my decision to detransition for what it is": how detrans youth perceive and receive discourses on detransition. Sexuality Research and Social Policy 22 (4), 1748–1764. Available at https://doi.org/10.1007/s13178-024-01079-z.

Gibson, D.J., Glazier, J.J., Olson, K.R., 2021. Evaluation of anxiety and depression in a community sample of transgender youth. JAMA Network Open 4 (4). Available at https://doi.org/10.1001/jamanetworkopen.2021.4739.

Grant, R., Amos, N., Cook, T., Lin, A., Hill, A., Carman, M., et al., 2024. From euphoria to wellbeing: correlates of gender euphoria and its association with mental wellbeing among transgender adults. International Journal of Transgender Health 26 (4), 1111–1119. Available at https://doi.org/10.1080/26895269.2024.2324100.

Grant, R., Amos, N., Lin, A., Cook, T., Hill, A.O., Pang, K., et al., 2024b. Mental health and wellbeing outcomes associated with social, medical, and legal gender affirmation among trans young people in Australia. International Journal of Transgender Health. Available at https://doi.org/10.1080/26895269.2024.2366881.

Gray, S.A.O., Sweeney, K.K., Randazzo, R., Levitt, H.M., 2016. Am I doing the right thing?": pathways to parenting a gender variant child. Family Process 55 (1), 123–138. Available at https://doi.org/10.1111/famp.12128.

Green, A.E., Price-Feeney, M., Dorison, S.H., Pick, C.J., 2020. Self-reported conversion efforts and suicidality among us lgbtq youths and young adults, 2018. American Journal of Public Health 110 (8), 1221–1227. Available at https://doi.org/10.2105/AJPH.2020.305701.

Gregor, C., Hingley-Jones, H., Davidson, S., 2015. Understanding the experience of parents of pre-pubescent children with gender identity issues. Child and Adolescent Social Work Journal 32 (3), 237–246. Available at https://doi.org/10.1007/s10560-014-0359-z.

Hatzenbuehler, M.L., 2009. How does sexual minority stigma "get under the skin"? A psychological mediation framework. Psychological Bulletin 135 (5), 707–730. Available at https://doi.org/10.1037/a0016441.

Heathcote, C., Taylor, J., Hall, R., Jarvis, S.W., Langton, T., Hewitt, C.E., et al., 2024. Psychosocial support interventions for children and adolescents experiencing gender dysphoria or incongruence: a systematic review. Archives of Disease in Childhood 109 (Suppl 2). Available at https://doi.org/10.1136/archdischild-2023-326347.

Hidalgo, M.A., Chen, D., 2019. Experiences of gender minority stress in cisgender parents of transgender/gender-expansive prepubertal children: a qualitative study. Journal of Family Issues 40 (7), 865–886. Available at https://doi.org/10.1177/0192513X19829502.

Hill, A.O., Lyons, A., Jones, J., McGowan, I., Carman, M., Parsons, M., et al., 2021. Writing Themselves In 4: The health and wellbeing of LGBTQA+ young people in Australia. National report, monograph series number 124. Australian Research Centre in Sex, Health and Society, La Trobe University, Melbourne.

Hillier, A., Torg, E., 2019. Parent participation in a support group for families with transgender and gender-nonconforming children: "being in the company of others who do not question the reality of our experience. Transgender Health 4 (1), 168–175. Available at https://doi.org/10.1089/trgh.2018.0018.

Hollinsaid, N. L., Weisz, J. R., Chorpita, B. F., Skov, H. E., Price, M. A., & Research Network On Youth Mental Health. (2020). The effectiveness and acceptability of empirically supported treatments in gender minority youth across four randomized controlled trials. Journal of Consulting and Clinical Psychology, 88(12), 1053. Available at https://doi.org/10.1177/21677026231159050

Horton, C., 2023. I never wanted her to feel shame": parent reflections on supporting a transgender child. Journal of LGBT Youth 20 (2), 231–247. Available at https://doi.org/10.1080/19361653.2022.2034079.

Horton, C., 2024. Experiences of puberty and puberty blockers: insights from trans children, trans adolescents, and their parents. Journal of Adolescent Research 39 (1), 77–103. Available at https://doi.org/10.1177/07435584221100591.

Huit, T.Z., Coyne, C., Chen, D., 2024. State of the science: gender-affirming care for transgender and gender diverse youth. Behavior Therapy 55 (6), 1335–1347. Available at https://doi.org/10.1016/j.beth.2024.02.010.

Israel, T., Choi, A.Y., Lin, Y.J., Matsuno, E., Goodman, J.A., Kary, K.G., et al., 2021. Reducing internalized transnegativity: randomized controlled trial of an online intervention. Psychology of Sexual Orientation and Gender Diversity 8 (4), 429–439. Available at https://doi.org/10.1037/sgd0000447.

Jacobsen, K., Devor, A., 2022. Moving from gender dysphoria to gender euphoria: trans experiences of positive gender-related emotions. Bulletin of Applied Transgender Studies 1 (1-2), 119–143.

Johns, M.M., Beltran, O., Armstrong, H.L., Jayne, P.E., Barrios, L.C., 2018. Protective factors among transgender and gender variant youth: a systematic review by socioecological level. Journal of Primary Prevention 39 (3), 263–301. Available at https://doi.org/10.1007/s10935-018-0508-9.

Johnson, K.C., LeBlanc, A.J., Deardorff, J., Bockting, W.O., 2020. Invalidation experiences among non-binary adolescents. Journal of Sex Research 57 (2), 222–233. Available at https://doi.org/10.1080/00224499.2019.1608422.

Katz-Wise, S.L., Budge, S.L., Orovecz, J.J., Nguyen, B., Nava-Coulter, B., Thomson, K., 2017. Imagining the future: perspectives among youth and caregivers in the trans youth family study. Journal of Counseling Psychology 64 (1), 26–40. Available at https://doi.org/10.1037/cou0000186.

Kerr, L., Jones, T., Fisher, C.M., 2022. Alleviating gender dysphoria: a qualitative study of perspectives of trans and gender diverse people. Journal of Health Services Research and Policy 27 (1), 4–13. Available at https://doi.org/10.1177/13558196211013407.

Kerr, J.A., Paine, J., Thrower, E., Hoq, M., Mollica, C., Sawyer, S.M., et al., 2024. Prevalence of eating disorder symptoms in transgender and gender diverse adolescents presenting for gender-affirming care. Journal of Adolescent Health 74 (4), 850–853. Available at https://doi.org/10.1016/j.jadohealth.2023.11.396.

Klein, A., Golub, S.A., 2016. Family rejection as a predictor of suicide attempts and substance misuse among transgender and gender nonconforming adults. LGBT Health 3 (3), 193–199. Available at https://doi.org/10.1089/lgbt.2015.0111.

Klepper, M., Bosse, J., Flores, D.D., Sanders, R., 2024. Sibling support for transgender and gender-diverse youth: a scoping review. Journal of Adolescent Health 75 (4), 543–552. Available at https://doi.org/10.1016/j.jadohealth.2024.05.014.

Knutson, D., Kertz, S., Chambers-Baltz, S., Christie, M.B., Harris, E., Perinchery, R., 2021. A pilot test of a text message-based transgender and nonbinary affirmative cognitive-behavioral intervention for anxiety and depression. Psychology of Sexual Orientation and Gender Diversity 8 (4), 440–450. Available at https://doi.org/10.1037/sgd0000438.

Kuper, L.E., Stewart, S., Preston, S., Lau, M., Lopez, X., 2020. Body dissatisfaction and mental health outcomes of youth on gender-affirming hormone therapy. Pediatrics 145 (4). Available at https://doi.org/10.1542/peds.2019-3006.

Kuvalanka, K.A., Weiner, J.L., Mahan, D., 2014. Child, family, and community transformations: findings from interviews with mothers of transgender girls. Journal of GLBT Family Studies 10 (4), 354–379. Available at https://doi.org/10.1080/1550428X.2013.834529.

Kuvalanka, K.A., Allen, S.H., Munroe, C., Goldberg, A.E., Weiner, J.L., 2018. The experiences of sexual minority mothers with trans* children. Family Relations 67 (1), 70–87. Available at https://doi.org/10.1111/fare.12226.

Lavender, R., Shaw, S., Maninger, J.-K., Butler, G., Carruthers, P., Carmichael, P., et al., 2023. Impact of hormone treatment on psychosocial functioning in gender-diverse young people. LGBT Health 10 (5), 382–390. Available at https://doi.org/10.1089/lgbt.2022.0201.

Lawlis, S.M., Butler, P., Middleman, A., 2020. Evaluating transgender youth and parent interest and preferences regarding support groups. Global Pediatric Health 7. Available at https://doi.org/10.1177/2333794x20954680 2333794X2095468.

Lucassen, M.F.G., Stasiak, K., Fleming, T., Frampton, C., Perry, Y., Shepherd, M., et al., 2021. Computerized cognitive behavioural therapy for gender minority adolescents: analysis of the real-world implementation of SPARX in New Zealand. Australian and New Zealand Journal of Psychiatry 55 (9), 874–882. Available at https://doi.org/10.1177/0004867420976846.

Luvmour, J., 2011. Developing together: parents meeting children's developmental imperatives. Journal of Adult Development 18 (4), 164–171. Available at https://doi.org/10.1007/s10804-010-9111-x.

MacKinnon, K.R., Gould, W.A., Ashley, F., Enxuga, G., Kia, H., Ross, L.E., 2022a. De) transphobia: examining the socio-politically driven gender minority stressors experienced by people who detransitioned. Bulletin of Applied Transgender Studies 1 (3-4), 235–259.

MacKinnon, K.R., Kia, H., Salway, T., Ashley, F., Lacombe-Duncan, A., Abramovich, A., et al., 2022b. Health care experiences of patients discontinuing or reversing prior gender-affirming treatments. JAMA Network Open. Available at https://doi.org/10.1001/jamanetworkopen.2022.24717 E2224717.

MacKinnon, K.R., Khan, N., Newman, K.M., Gould, W.A., Marshall, G., Salway, T., et al., 2025. Introducing novel methods to identify fraudulent responses (sampling with sisyphus): web-based LGBTQ2S+ mixed-methods study. Journal of Medical Internet Research 27. Available at https://doi.org/10.2196/63252.

Mahfouda, S., Panos, C., Whitehouse, A.J.O., Thomas, C.S., Maybery, M., Strauss, P., et al., 2019. Mental health correlates of autism spectrum disorder in gender diverse young people: evidence from a specialised child and adolescent gender clinic in Australia. Journal of Clinical Medicine 8 (10). Available at https://doi.org/10.3390/jcm8101503.

Malpas, J., 2011. Between pink and blue: a multi-dimensional family approach to gender nonconforming children and their families. Family Process 50 (4), 453–470. Available at https://doi.org/10.1111/j.1545-5300.2011.01371.x.

Malpas, J., Pellicane, M.J., Glaeser, E., 2022. Family-based interventions with transgender and gender expansive youth: systematic review and best practice recommendations. Transgender Health 7 (1), 7–29. Available at https://doi.org/10.1089/trgh.2020.0165.

Maragos, A.C., Brown, C., Allen, L.R., 2025. Lived experiences: exploring detransition narratives. International Journal of Transgender Health. Available at https://doi.org/10.1080/26895269.2025.2458696.

Martin, S., 2019. Developing and evaluating Queervibe: an online intervention to empower trans and non-binary youth. Anglia Ruskin Research Online (ARRO). Thesis. Available at https://hdl.handle.net/10779/aru.23763150.v1.

Matsuno, E., Israel, T., 2021. The parent support program: development and acceptability of an online intervention aimed at increasing supportive behaviors among parents of trans youth. Journal of GLBT Family Studies 17 (5), 413–431. Available at https://doi.org/10.1080/1550428X.2020.1868369.

McCormick, A., Schmidt, K., Terrazas, S., 2018. Foster Family Acceptance. Informa UK Limited, pp. 166–188. Available at https://doi.org/10.4324/9781315717159-14.

McDermott, E.S., Hughes, E., Rawlings, V.E., 2016. Queer Future Final Report: Understanding Lesbian, Gay, Bisexual And Trans (LGBT) Adolescents' Suicide, Self-harm and Help-seeking Behaviour. . Available at https://www.queerfutures.co.uk/wp-content/uploads/2016/06/Queer-Futures-Final-Report.pdf.

McGregor, K., McKenna, J.L., Williams, C.R., Barrera, E.P., Boskey, E.R., 2024. Association of pubertal blockade at tanner 2/3 with psychosocial benefits in transgender and gender diverse youth at hormone readiness assessment. Journal of Adolescent Health 74 (4), 801–807. Available at https://doi.org/10.1016/j.jadohealth.2023.10.028.

Medina-Martínez, J., Saus-Ortega, C., Sánchez-Lorente, M.M., Sosa-Palanca, E.M., García-Martínez, P., Mármol-López, M.I., 2021. Health inequities in lgbt people and nursing interventions to reduce them: a systematic review. International Journal of Environmental Research and Public Health 18 (22). Available at https://doi.org/10.3390/ijerph182211801.

Mehringer, J.E., Harrison, J.B., Quain, K.M., Shea, J.A., Hawkins, L.A., Dowshen, N.L., 2021. Experience of chest dysphoria and masculinizing chest surgery in transmasculine youth. Pediatrics 147 (3). Available at https://doi.org/10.1542/PEDS.2020-013300.

Menvielle, E., Hill, D.B., 2010. An affirmative intervention for families with gender-variant children: a process evaluation. Journal of Gay and Lesbian Mental Health 15 (1), 94–123. Available at https://doi.org/10.1080/19359705.2011.530576.

Menvielle, E.J., Rodnan, L.A., 2011. A therapeutic group for parents of transgender adolescents. Child and Adolescent Psychiatric Clinics of North America 20 (4), 733–743. Available at https://doi.org/10.1016/j.chc.2011.08.002.

Meyer, I.H., 1995. Minority stress and mental health in gay men. Journal of Health and Social Behavior 36 (1), 38–56. Available at https://doi.org/10.2307/2137286.

Minuchin, P., 1985. Families and individual development: provocations from the field of family therapy. Child Development 56 (2), 289–302. Available at https://doi.org/10.1111/j.1467-8624.1985.tb00106.x.

Morgan, H., Raab, D., Lin, A., Strauss, P., Perry, Y., 2023. Knowledge is power: trans young people's perceptions of parental reactions to their gender identity, and perceived barriers and facilitators to parental support. LGBTQ+ Family: An Interdisciplinary Journal 19 (1), 35–53. Available at https://doi.org/10.1080/27703371.2022.2135660.

Morgan, H., van Hall, H.W., Moore, J.K., Lin, A., Perry, Y., 2024. A pilot group program for parents of trans young people at a specialized pediatric gender diversity service. LGBTQ+ Family: An Interdisciplinary Journal. Available at https://doi.org/10.1080/27703371.2024.2347495.

Morgan, H., Strauss, P., Lin, A., Moss, H., Perry, Y., 2026. Transforming families: evaluation of a digital resource to enhance parental support for trans children. International Journal of Transgender Health 1–12. Available at https://doi.org/10.1080/26895269.2026.2620480.

Nealy, E.C., 2017. Trans Kids and Teens: Pride, Joy, and Families in Transition. WW Norton & Company.

Nolan, B.J., Zwickl, S., Locke, P., Zajac, J.D., Cheung, A.S., 2023. Early access to testosterone therapy in transgender and gender-diverse adults seeking masculinization: a randomized clinical trial. JAMA Network Open 6 (9). Available at https://doi.org/10.1001/jamanetworkopen.2023.31919 E2331919,.

Norwood, K., 2013. Grieving gender: trans-identities, transition, and ambiguous loss. Communication Monographs 80 (1), 24–45. Available at https://doi.org/10.1080/03637751.2012.739705.

Olsavsky, A.L., Grannis, C., Bricker, J., Chelvakumar, G., Indyk, J.A., Leibowitz, S.F., et al., 2023. Associations among gender-affirming hormonal interventions, social support, and transgender adolescents' mental health. Journal of Adolescent Health 72 (6), 860–868. Available at https://doi.org/10.1016/j.jadohealth.2023.01.031.

Olson, K.R., Durwood, L., Demeules, M., McLaughlin, K.A., 2016. Mental health of transgender children who are supported in their identities. Pediatrics 137 (3). Available at https://doi.org/10.1542/peds.2015-3223.

Olson-Kennedy, J., Warus, J., Okonta, V., Belzer, M., Clark, L.F., 2018. Chest reconstruction and chest dysphoria in transmasculine minors and young adults comparisons of nonsurgical and postsurgical cohorts. JAMA Pediatrics 172 (5), 431–436. Available at https://doi.org/10.1001/jamapediatrics.2017.5440.

Pachankis, J.E., 2015. A transdiagnostic minority stress treatment approach for gay and bisexual men's syndemic health conditions. Archives of Sexual Behavior 44 (7), 1843–1860. Available at https://doi.org/10.1007/s10508-015-0480-x.

Pachankis, J.E., Hatzenbuehler, M.L., Klein, D.N., Bränström, R., 2024. The role of shame in the sexual-orientation disparity in mental health: a prospective population-based study of multimodal emotional reactions to stigma. Clinical Psychological Science 12 (3), 486–504. Available at https://doi.org/10.1177/21677026231177714.

Parker, E., Davis-McCabe, C., 2021. The sibling experience: growing up with a trans sibling. Australian Journal of Psychology 73 (2), 188–199. Available at https://doi.org/10.1080/00049530.2021.1882269.

Pepper, R., Brill, S., 2008. The Transgender Child. Cleis Press.

Price-Feeney, M., Green, A.E., Dorison, S., 2020. Understanding the mental health of transgender and nonbinary youth. Journal of Adolescent Health 66 (6), 684–690. Available at https://doi.org/10.1016/j.jadohealth.2019.11.314.

Pullen Sansfaçon, A., Robichaud, M.J., Dumais-Michaud, A.A., 2015. The experience of parents who support their children's gender variance. Journal of LGBT Youth 12 (1), 39–63. Available at https://doi.org/10.1080/19361653.2014.935555.

Pullen Sansfaçon, A., Gelly, M.A., Gravel, R., Medico, D., Baril, A., Susset, F., et al., 2023a. A nuanced look into youth journeys of gender transition and detransition. Infant and Child Development 32 (2). Available at https://doi.org/10.1002/icd.2402.

Pullen Sansfaçon, A., Planchat, T., Gelly, M.A., Baril, A., Susset, F., Millette, M., 2023b. Understanding the experiences of youth who have discontinued a gender transition: provider perspectives. Psychology of Sexual Orientation and Gender Diversity 12 (1), 154–164. Available at https://doi.org/10.1037/sgd0000644.

Riach, J., 2021. The effectiveness of a transaffirmative cognitive behavioural therapy group-based intervention to help transgender individuals suffering from depression. . Available at https://repository.usfca.edu/diss/548.

Riggs, D.W., Due, C., 2015. Support experiences and attitudes of Australian parents of gender variant children. Journal of Child and Family Studies 24, 1999–2007.

Riley, E.A., Sitharthan, G., Clemson, L., Diamond, M., 2011. The needs of gender-variant children and their parents: a parent survey. International Journal of Sexual Health 23 (3), 181–195. Available at https://doi.org/10.1080/19317611.2011.593932.

Riley, E.A., Sitharthan, G., Clemson, L., Diamond, M., 2013. Recognising the needs of gender-variant children and their parents. Sex Education 13 (6), 644–659. Available at https://doi.org/10.1080/14681811.2013.796287.

Romito, M., Salk, R.H., Roberts, S.R., Thoma, B.C., Levine, M.D., Choukas-Bradley, S., 2021. Exploring transgender adolescents' body image concerns and disordered eating: semi-structured interviews with nine gender minority youth. Body Image 37, 50–62. Available at https://doi.org/10.1016/j.bodyim.2021.01.008.

Sarno, E.L., Newcomb, M.E., Mustanski, B., 2020. Rumination longitudinally mediates the association of minority stress and depression in sexual and gender minority individuals. Journal of Abnormal Psychology 129 (4), 355–363. Available at https://doi.org/10.1037/abn0000508.

Scandurra, C., Bochicchio, V., Amodeo, A.L., Esposito, C., Valerio, P., Maldonato, N.M., et al., 2018. Internalized transphobia, resilience, and mental health: applying the psychological mediation framework to Italian transgender individuals. International Journal of Environmental Research and Public Health 15 (3). Available at https://doi.org/10.3390/ijerph15030508.

Schlehofer, M.M., Cortez-Regan, L., Bush, D., 2021. Experiences of parent-advocates of trans and gender non-conforming youth. Journal of GLBT Family Studies 17 (5), 450–469. Available at https://doi.org/10.1080/1550428X.2021.1902447.

Silveri, M. M., Schuttenberg, E. M., Schmandt, K., Stein, E. R., Rieselbach, M. M., Sternberg, A., ... & Boger, K. D. (2021). Clinical outcomes following acute residential psychiatric treatment in transgender and gender diverse adolescents. JAMA Network Open, 4(6), e2113637-e2113637.

Simons, L., Schrager, S.M., Clark, L.F., Belzer, M., Olson, J., 2013. Parental support and mental health among transgender adolescents. Journal of Adolescent Health 53 (6), 791–793. Available at https://doi.org/10.1016/j.jadohealth.2013.07.019.

Singh, A., dickey, lore m, 2017. Affirmative Counseling and Psychological Practice with Transgender and Gender Nonconforming Clients. American Psychological Association. Available at https://doi.org/10.1037/14957-000.

Skelton, S., Riggs, D.W., Pullen Sansfacon, A., Katz-Wise, S.L., Arora, M., Thibeault, C.A., 2024. It just feels really nice when people call me by my name': accounts of gender euphoria among Australian trans young people and their parents. Journal of Gender Studies 33 (4), 470–482. Available at https://doi.org/10.1080/09589236.2023.2285984.

Smalley, K.B., Warren, J.C., Barefoot, K.N., 2016. Differences in health risk behaviors across understudied LGBT subgroups. Health Psychology 35 (2), 103–114. Available at https://doi.org/10.1037/hea0000231.

Stevens, S., Haverly, K., Powell, C., 2020. Improvements in self-acceptance for LGBTQ+ and straight allied youth and young adults enrolled in an affirming system of care program. Children and Youth Services Review 118, 105382. Available at https://doi.org/10.1016/j.childyouth.2020.105382.

Strauss, P., Cook, A., Winter, S., Watson, V., Wright Toussaint, D., Lin, A., 2020. Associations between negative life experiences and the mental health of trans and gender diverse young people in australia: findings from Trans Pathways. Psychological Medicine 50 (5), 808–817. Available at https://doi.org/10.1017/S0033291719000643.

Strauss, P., Cook, A., Watson, V., Winter, S., Whitehouse, A., Albrecht, N., et al., 2021. Mental health difficulties among trans and gender diverse young people with an autism spectrum disorder (ASD): findings from trans pathways. Journal of Psychiatric Research 137, 360–367. Available at https://doi.org/10.1016/j.jpsychires.2021.03.005.

Tan, K.K.H., Ellis, S.J., Schmidt, J.M., Byrne, J.L., Veale, J.F., 2020. Mental health inequities among transgender people in aotearoa New Zealand: findings from the counting ourselves survey. International Journal of Environmental Research and Public Health 17 (8). Available at https://doi.org/10.3390/ijerph17082862.

Telfer, M., Kelly, F., Feldman, D., Stone, G., Robertson, R., Poulakis, Z., 2018. Transgender adolescents and legal reform: how improved access to healthcare was achieved through medical, legal and community collaboration. Journal of Paediatrics and Child Health 54 (10), 1096–1099. Available at https://doi.org/10.1111/jpc.14124.

Testa, R.J., Habarth, J., Peta, J., Balsam, K., Bockting, W., 2015. Development of the gender minority stress and resilience measure. Psychology of Sexual Orientation and Gender Diversity 2 (1), 65–77. Available at https://doi.org/10.1037/sgd0000081.

Testa, R.J., Michaels, M.S., Bliss, W., Rogers, M.L., Balsam, K.F., Joiner, T., 2017. Suicidal ideation in transgender people: gender minority stress and interpersonal theory factors. Journal of Abnormal Psychology 126 (1), 125–136. Available at https://doi.org/10.1037/abn0000234, http://doi.org/10.1037/abn0000240 Available from.

Thrower, E., Bretherton, I., Pang, K.C., Zajac, J.D., Cheung, A.S., 2020. Prevalence of autism spectrum disorder and attention-deficit hyperactivity disorder amongst individuals with gender dysphoria: a systematic review. Journal of Autism and Developmental Disorders 50 (3), 695–706. Available at https://doi.org/10.1007/s10803-019-04298-1.

Toomey, R.B., Ryan, C., Diaz, R.M., Card, N.A., Russell, S.T., 2013. Gender-nonconforming lesbian, gay, bisexual, and transgender youth: school victimization and young adult psychosocial adjustment. Psychology of Sexual Orientation and Gender Diversity 1 (S), 71–80. Available at https://doi.org/10.1037/2329-0382.1.s.71.

Tordoff, D.M., Wanta, J.W., Collin, A., Stepney, C., Inwards-Breland, D.J., Ahrens, K., 2022. Mental health outcomes in transgender and nonbinary youths receiving gender-affirming care. JAMA Network Open 5 (2). Available at https://doi.org/10.1001/jamanetworkopen.2022.0978.

Travers, R., Bauer, G., Pyne, J., Bradley, K., Gale, L., Papadimitriou, M., 2012. Impacts of strong parental support for trans youth. https://transpulseproject.ca/wp-content/uploads/2012/10/Impacts-of-Strong-Parental-Support-for-Trans-Youth-vFINAL.pdf

Tucker, R.P., Testa, R.J., Reger, M.A., Simpson, T.L., Shipherd, J.C., Lehavot, K., 2019. Current and military-specific gender minority stress factors and their relationship with suicide ideation in transgender veterans. Suicide and Life-Threatening Behavior 49 (1), 155–166. Available at https://doi.org/10.1111/sltb.12432.

Turban, J.L., van Schalkwyk, G.I., 2018. Gender dysphoria" and autism spectrum disorder: is the link real? Journal of the American Academy of Child and Adolescent Psychiatry 57 (1). Available at https://doi.org/10.1016/j.jaac.2017.08.017.

Turban, J.L., Loo, S.S., Almazan, A.N., Keuroghlian, A.S., 2021. Factors leading to "detransition" among transgender and gender diverse people in the United States: a mixed-methods analysis. LGBT Health 8 (4), 273–280. Available at https://doi.org/10.1089/lgbt.2020.0437.

Turban, J.L., King, D., Kobe, J., Reisner, S.L., Keuroghlian, A.S., 2022. Access to gender-affirming hormones during adolescence and mental health outcomes among transgender adults. PLoS One 17 (1). Available at https://doi.org/10.1371/journal.pone.0261039.

van der Miesen, A.I.R., Steensma, T.D., de Vries, A.L.C., Bos, H., Popma, A., 2020. Psychological functioning in transgender adolescents before and after gender-affirmative care compared with cisgender general population peers. Journal of Adolescent Health 66 (6), 699–704. Available at https://doi.org/10.1016/j.jadohealth.2019.12.018.

Vandenbussche, E., 2022. Detransition-related needs and support: a cross-sectional online survey. Journal of Homosexuality 69 (9), 1602–1620. Available at https://doi.org/10.1080/00918369.2021.1919479.

Vasen, E., Poteat, T., Maman, S., 2025. A review of psychosocial interventions to improve parental support of transgender youth. Transgender Health. Available at https://doi.org/10.1089/trgh.2024.0139.

Vaughan, M.D., Rodriguez, E.M., 2014. LGBT strengths: incorporating positive psychology into theory, research, training, and practice. Psychology of Sexual Orientation and Gender Diversity 1 (4), 325–334. Available at https://doi.org/10.1037/sgd0000053.

Veale, J.F., Peter, T., Travers, R., Saewyc, E.M., 2017. Enacted stigma, mental health, and protective factors among transgender youth in Canada. Transgender Health 2 (1), 207–216. Available at https://doi.org/10.1089/trgh.2017.0031.

Wagner, L.D., Armstrong, E., 2020. Families in transition: the lived experience of parenting a transgender child. Journal of Family Nursing 26 (4), 337–345. Available at https://doi.org/10.1177/1074840720945340.

Wahlig, J.L., 2015. Losing the child they thought they had: therapeutic suggestions for an ambiguous loss perspective with parents of a transgender child. Journal of GLBT Family Studies 11 (4), 305–326. Available at https://doi.org/10.1080/1550428X.2014.945676.

Watson, R.J., Veale, J.F., Saewyc, E.M., 2017. Disordered eating behaviors among transgender youth: probability profiles from risk and protective factors. International Journal of Eating Disorders 50 (5), 515–522. Available at https://doi.org/10.1002/eat.22627.

We Are Allies, 2025. Available from: http://www.weareallies.ca.

Weigert, R., Frison, E., Sessiecq, Q., Al Mutairi, K., Casoli, V., 2013. Patient satisfaction with breasts and psychosocial, sexual, and physical well-being after breast augmentation in male-to-female transsexuals. Plastic and Reconstructive Surgery 132 (6), 1421–1429. Available at https://doi.org/10.1097/01.prs.0000434415.70711.49.

Wells, L., Gringart, E., Perry, Y., Morgan, H., Lin, A., 2022. Psychological wellbeing amongst parents of trans children: an in-depth qualitative investigation. LGBTQ+ Family: An Interdisciplinary Journal 18 (5), 495–514. Available at https://doi.org/10.1080/27703371.2022.2125471.

Westwater, J.J., Riley, E.A., Peterson, G.M., 2019. What about the family in youth gender diversity? A literature review. International Journal of Transgenderism 20 (4), 351–370. Available at https://doi.org/10.1080/15532739.2019.1652130.

Wheeler, N.L., Langton, T., Lidster, E., Dallos, R., 2019. Understanding more about how young people make sense of their siblings changing gender identity: how this might affect their relationships with their gender-diverse siblings and their experiences. Clinical Child Psychology and Psychiatry 24 (2), 258–276. Available at https://doi.org/10.1177/1359104519830155.

Wilson, E.C., Iverson, E., Garofalo, R., Belzer, M., 2012. Parental support and condom use among transgender female youth. Journal of the Association of Nurses in AIDS Care 23 (4), 306–317. Available at https://doi.org/10.1016/j.jana.2011.09.001.

Wittlin, N.M., Gallagher, N.M., Atwood, Olson, K.R., 2025. Mental health during medical transition in a US and Canadian sample of early socially transitioned transgender youth. Journal of Adolescent Health 76 (2), 228–237. Available at https://doi.org/10.1016/j.jadohealth.2024.10.023.

Wong, W., Drake, S.J., 2018. A qualitative study of transgender children with early social transition: parent perspectives and clinical implications. PEOPLE: International Journal of Social Sciences 3 (2), 1970–1985. Available at https://doi.org/10.20319/pijss.2017.32.19701985.

Wren, B., 2002. I can accept my child is transsexual but if I ever see him in a dress I'll hit him': dilemmas in parenting a transgendered adolescent. Clinical Child Psychology and Psychiatry 7 (3), 377–397. Available at https://doi.org/10.1177/1359104502007003006.

Part III

Additional topics of interest

Chapter 11

Fertility and fertility preservation

Hayley K. Baines[1], Leena Nahata[2], Evelyn Mok-Lin[3], and Norah M. van Mello[4]

[1]*Department of Pediatric Endocrinology, Oregon Health & Science University, Portland, OR, United States,* [2]*Department of Pediatrics, The Ohio State University College of Medicine, Columbus, OH, United States,* [3]*Department of Obstetrics, Gynecology and Reproductive Sciences, University of California, San Francisco, CA, United States,* [4]*Department of Obstetrics & Gynaecology, Reproduction and Development, Amsterdam University Medical Center, Center of Expertise on Gender Dysphoria, Amsterdam, Netherlands*

Learning objectives

- Review recent research on fertility and fertility preservation among transgender and gender diverse (TGD) youth, including research on baseline knowledge, attitudes, and beliefs of patients, families, and health professional teams, fertility counseling, rates of fertility preservation, and outcomes related to fertility preservation procedures.
- Provide an overview of current fertility preservation options based on pubertal status and highlight the unique features of fertility preservation techniques and options for TGD youth seeking gender-affirming care, including clinical, developmental, and ethical considerations.
- Outline the limited original research in pediatric fertility preservation for TGD youth, incorporating relevant literature among adults.
- Discuss ongoing and future research directions, including the development of decision aids and pediatric fertility preservation protocols specific to TGD youth.

Key terms and definitions

- **Fertility** refers to reproductive potential, or the ability to have biologically related children.
- **Fertility preservation** is the process of storing gametes or reproductive tissue for potential later use for the purpose of having a biological child.

Reproductive health, including bodily autonomy, access to reproductive information, and the freedom to make choices about reproduction and family planning, is a fundamental human right (Sexual and reproductive health and rights, 2025). Unfortunately, many historically marginalized groups face inequities in access to reproductive health services and family planning support, including transgender and gender diverse (TGD) individuals (Allen et al., 2025). TGD people, including adolescents, have a right to understand their own fertility potential, to receive information on the potential impacts of gender-affirming medical and surgical interventions on fertility, and to make choices about their reproductive health and family planning. Although fertility and fertility preservation have been historically limited and even denied to TGD people, contemporary clinical practice and research have begun to prioritize reproductive health, resulting in a growing body of literature on fertility, including among TGD youth.

Early case reports and case series discussing the implications of gender-affirming medical interventions on reproductive health and fertility focused on the effects of testosterone on ovarian morphology in adults (Futterweit and Deligdisch, 1986). An early medical assumption was that fertility was sacrificed as a result of gender-affirming hormones (GAH), and

Advances in Research on Pediatric Gender Affirming Healthcare. DOI: https://doi.org/10.1016/B978-0-443-29019-0.00010-5
© 2026 Elsevier Inc. All rights are reserved, including those for text and data mining, AI training, and similar technologies.

the loss of reproductive potential was even required for the transition of legal documentation in some countries (Lawrence et al., 1996; Rodriguez-Wallberg et al., 2023). However, literature soon followed discussing TGD adults becoming biological parents following varying durations of GAH, specifically adult transgender men becoming pregnant following testosterone (Light et al., 2014).

There are no medical interventions indicated for TGD children before the onset of puberty. Youth must reach stage 2 of the Sexual Maturity Rating (SMR) scale (also known as Tanner stages), demonstrating hypothalamic–pituitary gonadal axis activation before considering gender-affirming medical interventions, such as gonadotropin releasing hormone agonists (GnRHa), testosterone, and estradiol (Coleman et al., 2022; Hembree et al., 2009). The most recent World Professional Association of Transgender Healthcare (WPATH) Standards of Care, version 8 (SOC8), outlines the importance of fertility in several chapters, including Reproductive Health, Hormone Therapy, and Adolescents. The authors state, "we recommend healthcare professionals working with transgender and gender diverse adolescents requesting gender-affirming medical or surgical treatments inform them, before initiating treatment, of the reproductive effects, including the potential loss of fertility and available options to preserve fertility within the context of the youth's stage of pubertal development" (Coleman et al., 2022). Despite this clear recommendation, there is also recognition that research in this area continues to evolve and to follow this recommendation, clinicians should stay informed on the latest research developments.

Two systematic reviews provide a synthesis of the research on fertility and fertility preservation in TGD youth, including adolescents and young adults (Baram et al., 2019; Warton and McDougall, 2022). From these reviews, several psychosocial and medical themes have been identified (Baram et al., 2019). Psychosocial themes that exist in the literature include the desire for biological children among TGD youth, attitudes, beliefs, and knowledge of patients, parents, and providers, counseling about fertility and referral for fertility preservation, fertility preservation utilization rates, and barriers to access. Medical themes include the impact of gender-affirming medical interventions on fertility, medical options for fertility preservation, and preservation outcomes. Ethical issues have also been identified in the literature, including the particular age and developmental stage at which TGD youth are considering their own fertility, the role of parents and caregivers in decision-making, and mental health complexities (Warton and McDougall, 2022).

Overview of pediatric fertility preservation

People assigned female at birth are typically born with immature, primordial ovarian follicles. With the maturation of the hypothalamic–pituitary–ovarian axis during puberty and throughout reproductive adulthood, gonadotropin-dependent folliculogenesis progresses to ovulation of mature oocytes with each menstrual hormonal cycle. The ability to preserve oocytes becomes available in mid to late puberty and is accomplished via ovarian stimulation protocols that include gonadotropin injections, resulting in temporary rises in estrogen levels. Typically, transvaginal ultrasounds are needed in preparation for the transvaginal oocyte aspiration procedure itself (Moravek, 2019; Finlayson et al., 2016). TGD youth with ovaries may not be comfortable with the idea of increasing estrogen levels or transvaginal procedures based on underlying levels of dysphoria related to genital anatomy (Finlayson et al., 2016). Transabdominal or transrectal ultrasounds can be considered along with oocyte retrieval with sedation to minimize distress when possible (Barrett et al., 2022). At this time, there is no information on the quality or future pregnancy success rates of cryopreserved unfertilized oocytes. Genetic testing and other predictors of viability are only available postfertilization at the embryo stage.

Ovarian tissue cryopreservation (OTC), or taking partial ovarian cortical biopsy or whole ovary for cryopreservation, is no longer considered experimental in the United States based on success of live births following cryopreservation, in vitro fertilization (IVF) after autotransplantation, and embryo transfer to gestational parent (Donnez and Dolmans, 2017; Ethics Committee of the American Society for Reproductive Medicine. Electronic address: ASRM@asrm.org, 2018; Nahata et al., 2020b). This is not necessarily a shared perspective internationally, and OTC has not been studied specifically in TGD youth with ovaries (Nahata et al., 2020b). Historically, OTC was considered the only option for early pubertal youth and continues to be the only fertility preservation option that has resulted in live births from prepubertal adolescents in the general population (Demeestere et al., 2015; Matthews et al., 2018). These cases represent patient populations facing fertility preservation decision-making in the setting of gonadotoxic medical interventions known to lead to fertility loss in the short term, such as oncological treatment. In recent years, it has been discovered that mature oocytes can be retrieved in pubertal youth on GnRHa (Rothenberg et al., 2019), although the quality of these oocytes is not yet known. TGD youth with ovaries may therefore have opportunities to consider broader fertility preservation options after initiation of gender-affirming medical treatments, depending on the individualized timeline of medical and surgical interventions.

People assigned male at birth are typically born with immature testicles that do not produce sperm. Therefore youth with testicles typically must advance to SMR stages 3–4, in some reports as early as stage 2, to reach spermatogenesis and potential sperm cryopreservation via masturbatory semen collection (de Nie et al., 2022b). A testicular volume of at least

10 mL is often utilized as a positive predictor of spermatogenesis (Peri et al., 2021; Yau et al., 2024). If masturbation is not possible but spermatogenesis has been reached, some youth may choose to proceed with prostate massage, or electroejaculation techniques or procedures such as sperm aspiration or microsurgical sperm extraction (Moravek, 2019).

Testicular sperm extraction (TESE) is an accepted method of obtaining sperm when an individual is not able to participate in obtaining masturbatory collection or is unable to ejaculate for medical reasons. As some youth with testicles may not have experience with masturbation or may experience significant dysphoria when thinking about or interacting with their genitals, TESE can reduce the risk of distress during fertility preservation (Barrera et al., 2023). Sperm has been isolated via TESE from individuals who have reached SMR stages 2–3 (Hagenäs et al., 2010). In a retrospective single-institution case series, 10 TGD youth with testicles (age: 12–16 years, SMR stages 3–5) successfully achieved sperm preservation via TESE (Barrera et al., 2023). TESE was completed at the same time as histrelin GnRHa implant placement. The study was limited by a lack of semen analysis regarding sperm quality parameters. For TGD youth with testicles in the earliest stages of puberty, testicular tissue cryopreservation (TTC) remains the only fertility preservation option and is primarily offered in oncology populations, but remains experimental as there have not been any reports of human live births using this technology to date (Moravek, 2019).

Ethical considerations

Research on the ethical considerations of fertility preservation in TGD youth is largely limited to case reports with accompanying discussions of beneficence, nonmaleficence, autonomy, informed consent, and justice (Schurman et al., 2018; McDougall et al., 2018; Varkey, 2021). While fertility counseling is critical to support informed decision-making and autonomy, the balance between beneficence and nonmaleficence may be more challenging in pediatric TGD populations than in adults who have access to established fertility preservation options and are developmentally closer to envisioning future parenthood. Providers and families also need to consider the impact of delaying or not initiating gender-affirming medical interventions for reasons of fertility potential alone, given robust evidence of the mental health benefits of affirming treatments.

Age and pubertal development can also limit fertility preservation options, and whether or not the available options are considered standard of care, or only available on a research protocol basis without demonstrated human live birth outcomes (McDougall et al., 2018). For example, OTC and TTC are both surgical fertility preservation interventions, but only OTC has led to mature oocytes and live births after autotransplantation, whereas TTC has not yet led to mature sperm or live births in humans (Moravek, 2019). Some health professionals continue to express concern that OTC and TTC could potentially limit future fertility potential by removing gonadal tissue from healthy patients (Nahata et al., 2020b). These procedures are often only offered through research-level protocols at academic medical centers in urban areas and may require shipment of gonadal tissue for ongoing storage (Valli-Pulaski et al., 2019; Winkelman et al., 2018). For all TGD youth, fertility preservation options should be considered at each point of decision-making when exploring gender-affirming medical and surgical options, such as at the time of considering GnRHa, before initiating estrogen or testosterone, and before gonadectomy.

From the perspective of autonomy, there may be disagreements between TGD youth and parents or caregivers regarding the necessity of fertility preservation before initiating gender-affirming medical care, which may require ongoing medical and psychosocial counseling to reach a shared decision (Nahata et al., 2018). Depending on local legislation, there is significant variation in the age of majority for legal consent or medical consent, which directly impacts TGD's youth autonomy in decision-making. For example, early published practice guidelines featured the threshold of 16 years before starting GAH (Hembree et al., 2009). This originates from the 2002 Dutch protocol for gender-affirming care, as 16 years was the age of majority in the Netherlands at the time, and the Dutch protocol is largely considered foundational evidence for pediatric gender-affirming care (Hembree et al., 2009, 2017). The more recent WPATH SOC8 removes definitive age parameters from practice guidelines in favor of a more individualized approach (Coleman et al., 2022).

Finally, the cost of fertility preservation is a justice consideration (Schurman et al., 2018). There is significant variation in access to pediatric fertility preservation from state to state in the United States, with most states not offering any health insurance benefit, regardless of indication (Kyweluk et al., 2019; Sax et al., 2020). Other countries may cover fertility preservation costs as a partial to full benefit depending on health insurance and benefit infrastructure (Brik et al., 2019). Both health professionals and families recognize cost as a barrier to pursuing fertility preservation for TGD youth (Harris et al., 2020; Nahata et al., 2017; Tishelman et al., 2019). One study used qualitative interviews to investigate ethical issues regarding fertility preservation among TGD youth and their parents (Harris et al., 2020). These personal narratives demonstrated fears of financial burden, highlighting the importance of appropriate informed consent and economic justice (Harris et al., 2020). Health professionals have also been interviewed to consider the nonmaleficence of early gender-affirming medical interventions that may impact future fertility potential, but the study sample (*N*=21) was too small to reach consensus, and broader discussion groups are needed (Vrouenraets et al., 2015).

Research methods and future directions

Research methods addressing ethical considerations of pediatric fertility and reproductive potential in TGD youth include case reports (Nahata et al., 2018), retrospective case series (Barrera et al., 2023), and qualitative interviews (Harris et al., 2020). To better address the ethical considerations regarding fertility preservation for pediatric TGD populations, research in this area would benefit from more (1) community engagement across the spectrum of study design and implementation, and (2) prospective, longitudinal studies to assess both reproductive (e.g., impact of medical interventions and fertility preservation on conception, live births, health of offspring) and psychosocial (e.g., satisfaction, regret, quality of life) outcomes.

Knowledge, attitudes, beliefs, and decision-making

Pediatric gender-affirming care research has long demonstrated a lack of understanding of the baseline needs and desires of TGD youth, their families, and even their health professionals, regarding fertility and fertility preservation (Nahata et al., 2018, 2019; Johnson et al., 2016; Chen et al., 2019a). Therefore there is an ongoing investigation to elucidate baseline knowledge, attitudes, and beliefs regarding fertility and fertility preservation in TGD youth and their families (Chen et al., 2018b; Chiniara et al., 2019; Halloran et al., 2023; Persky et al., 2020).

Most research on beliefs, attitudes, and knowledge related to fertility has been conducted with youth or adults actively seeking care within a specialty gender-affirming care clinic, or by surveying adults who previously received care as adolescents about their retrospective experiences. Several measurement tools have been developed to explore knowledge and attitudes related to fertility among this population. The Yale Gender Program Reproductive Knowledge and Experiences Survey (YPGP-RKES) is a 22-item measure adapted from oncofertility literature to assess the knowledge, attitudes, and beliefs about fertility specifically among TGD youth, and it can be found in the Appendix of the article about the development of the measure (Morrison et al., 2020). The Transgender Youth Fertility Attitudes Questionnaire (TYFAQ) is a 16-item measure with parallel versions for youth and parents that was developed through a participatory process, and it is freely available from the authors (Strang et al., 2018). The TYFAQ has also been adapted into a 22-item measure that expands on the original measure to include questions about gender identity and contraception (Nahata et al., 2020a).

Among one cohort of TGD youth (*N*=69, mean age ± SD=16.2 ± 2.0 years) who were administered the YPGP-RKES, 39.1% endorsed that it was important to them to have a child one day, 23.2% were interested in biological children, and 33.3% were unsure about having a child, with 37.6% reporting it was stressful to think about having a child one day (Halloran et al., 2023). In another cohort (*N*=44, mean age ± SD=16.3 ± 1.8 years) who were administered an adapted version of the TYFAQ, many (63.6%) participants thought their feelings about having children could change in the future, or were unsure about their parenthood goals (Nahata et al., 2020a). In a mixed cohort of TGD youth and adults seeking care (*N*=46 adults, *N*=11 youth, median age of 15 years with interquartile range of 13–15 years), 30% of the participants, all of whom were adults, expressed interest in a future biologically related child (Walton et al., 2023). Of the youth participants, only 3% were willing to consider delaying GAH to pursue fertility preservation. Finally, among a cohort of TGD adults who had initiated GAC as adolescents (*N*=89), 34% (*n*=30) reported they recognized a future desire for children at the start of gender-affirming care compared to 56% (*n*=50) who did not recognize this until adulthood, at the time of the study (de Nie et al., 2024).

The TYFAQ is a unique measure in that it can compare youth and parent attitudes across parallel measures. The initial pilot study following its development (*N*=25 TGD youth, *N*=26 parents) demonstrated that young people and their parents generally share similar beliefs and attitudes related to fertility and fertility preservation (Strang et al., 2018). In other words, while youth and parents may initially appear to disagree about the importance of these conversations, underlying values and attitudes are more likely to be relatively similar within family units. The only question that revealed a difference, which was nonsignificant after statistical correction, asked about the likelihood that the patient might change their mind about biological children in the future. While parents endorsed this item more strongly, nearly half of the youth agreed that their feelings might change, and another one-fifth were unsure. Many youth and parents agreed that it was important to learn about the effects of GAH on fertility and about fertility options, and few endorsed strong feelings about the importance of future children being biologically related.

Across several studies a large majority of youth and parents endorse knowledge of the potential impacts of GAH on future fertility (Morrison et al., 2020; Halloran et al., 2023; Nahata et al., 2020a; Strang et al., 2018). A small number of studies have also tested knowledge directly with accuracy questions. In one study of TGD adolescents and young adults aged 12–22 years, young people demonstrated more knowledge of gender specific topics, while lacking knowledge of more basic general reproduction topics (Morrison et al., 2020). This study also found that youth tend to overestimate the ability of their providers to predict the effects of GAH on fertility. In another study of baseline knowledge of TGD youth regarding fertility and reproduction, the mean percentage correct was 71.9% (Halloran et al., 2023). Health professional needs-assessment surveys aim to understand gender-affirming care providers' baseline knowledge and priorities in clinical tool development regarding pediatric fertility and fertility preservation (Johnson et al., 2016).

Decision-making

Multiple research methods have been utilized to explore TGD youth and family perspectives on decision-making related to fertility and fertility preservation. Early research in the field included the Trans Youth Fertility Study (TYFS), which conducted semistructured interviews among a small cohort of TGD youth and young adults (N=18, mean age: 18.4 ± 2.3 years, range: 15–24 years) (Kyweluk et al., 2018). Results revealed that patients often have goals to decouple reproductive biology from their gender and that patients feel limited by existing reproductive technology options. Factors impacting decision-making included individual identity, personal experiences, and family relationships. Another similar qualitative study revealed varied desires for parenthood among this patient population (Chen et al., 2019b). Semistructured interviews also revealed common perceived barriers to decision-making around fertility, including experiences of gender dysphoria, cost, and shortcomings of fertility counseling from healthcare providers. According to the results of semistructured interviews with TGD youth, as well as a small cohort of their parents, cost and other access barriers were similarly identified as concerns across youth and parents (Boguszewski et al., 2022). Finally, interviews focused on future family goals and barriers to fertility preservation among TGD youth (N=23, mean age: 16.9 ± 1.6 years) similarly demonstrated a desire for accurate information regarding the effects of GAH on fertility and fertility preservation options, extended resources for accessing fertility preservation, particularly financial resources, and respect and autonomy around the decision-making process (Kerman et al., 2021).

Some investigators are performing needs assessments to obtain data for ongoing projects, including designing and creating decision aids (Persky et al., 2020; Kolbuck et al., 2020; Strang et al., 2018). One needs assessment of TGD youth (N=64, mean age ± SD=16.8 ± 2.1 years) and their parents (N=46, mean age ± SD=48.3 ± 6.4 years) showed that 69% of youth and 63% of their parents thought they, or their child, was capable of making an educated decision regarding fertility preservation (Persky et al., 2020). This led investigators to recommend a qualitative study to help delineate decision-making around desire and goals for planning for biological children (Persky et al., 2020). Another needs assessment focused on clinicians and researchers across different aspects of pediatric gender-affirming care and used the Delphi procedure to create themes and desired content areas for a future fertility decision aid (Kolbuck et al., 2020).

Research methods and future directions

A strength of research on knowledge, beliefs, and attitudes about fertility is that there are a high number of nonretrospective studies, either cohort surveys (Johnson et al., 2016; Halloran et al., 2023; Nahata et al., 2020a; Walton et al., 2023) or qualitative semistructured interviews (Kyweluk et al., 2018; Chen et al., 2019b; Boguszewski et al., 2022; Kerman et al., 2021). In one study a Delphi procedure was used, interviewing multidisciplinary gender-affirming healthcare providers to determine content areas and learning objectives for development of a fertility-related decision support tool (Kolbuck et al., 2020). Perhaps most importantly, this area of research has relatively more real-time input directly from TGD youth, their parents or caregivers, and healthcare providers due to semistructured interviews and qualitative research models.

Based on having a better understanding of attitudes, beliefs, and knowledge, researchers are moving toward designing clinical tools, such as fertility preservation-related decision aids, to assist TGD youth, families, and health professionals through fertility preservation counseling (Chen et al., 2019a; Persky et al., 2020; Chen et al., 2019b; Kolbuck et al., 2020; Strang et al., 2018).

Fertility implications before and after gender-affirming medical interventions

Fertility counseling for TGD youth is challenging due to the ever-evolving body of knowledge on the long-term effects of the medical interventions considered by adolescents. While there is emerging data, ranging from case reports to retrospective cohort studies, in adult literature, these studies are only starting to include young adults who initiated gender-affirming medical treatments in their youth.

Animal models

Initial mouse models of testosterone effects on the ovary focused on characterizing testosterone suppression of the hypothalamic–pituitary gonadal axis and morphology changes that aligned with polycystic ovaries previously described in adult transgender men (Kinnear et al., 2019). To demonstrate the effects of short-term testosterone on female mice's fertility potential a specific mouse model was designed to include testosterone treatment for 6 weeks to postpubertal female mice and reviewed the response to gonadotropic stimulation while receiving testosterone (Bartels et al., 2021). Because the female mice had fertilizable eggs with serum testosterone levels comparable to male mice, researchers concluded that there may be potential for TGD individuals to trial ovarian stimulation for fertility preservation without pausing testosterone (Bartels et al., 2021).

More recently a model for female mice was designed to more closely mimic what TGD adolescents with ovaries may seek with a combination of GnRHa and testosterone in an effort to understand effects on future fertility (Godiwala et al., 2023). This mouse model showed that mouse ovaries exposed to GnRHa and testosterone could respond to gonadotropins and mature their oocytes, leading to live offspring. One limitation of the mouse model study was that compared to the specific protocol, human adolescent gender-affirming medical care timelines vary substantially, suggesting longer-term studies are needed to understand the effects of duration of therapy on oocyte potential (Godiwala et al., 2023).

Mouse models of estradiol effects on the testes have used varying estradiol doses with estradiol implant and found changes in gonadal histology of the seminiferous tubules and epididymis (Pfau et al., 2024). Though mature sperm were found, motility was decreased (Kinnear et al., 2019; Pfau et al., 2024). Ex vivo spermatogenesis in rat and human cytological analysis of isolated seminiferous tubules has also been demonstrated, which may have application for future fertility preservation options (Perrard et al., 2016).

Pubertal suppression

Pubertal suppression with GnRHa is the first medical intervention available for TGD youth once puberty begins. As GnRH is a pulsatile hypothalamic hormone stimulating pituitary gonadotropin release, agonist therapy prevents follicle-stimulating hormone (FSH) and luteinizing hormone release. Therefore end-organ gonadal sex steroid production is paused. Both WPATH and Endocrine Society practice guidelines outline GnRHa as the standard of care for pubertal suppression in TGD youth with evidence of SMR stage 2 or higher (Coleman et al., 2022; Hembree et al., 2017). As GnRHa act at the level of the hypothalamic–pituitary hormone regulation, there is no direct action on gonadal tissue, and the effect on fertility is considered temporary or reversible if discontinued. Youth with ovaries who start GnRHa before reaching menarche have immature ovarian follicles. Overall, GnRHa prevent ongoing ovarian maturation. This must be taken into consideration when TGD youth and caregivers are seeking pubertal suppression and balanced with potential future family planning and gender embodiment goals.

There are a few key factors to consider for TGD youth with testicles seeking pubertal suppression. Spermatogenesis typically occurs in later stages of puberty (de Nie et al., 2022b; Peri et al., 2021), whereas there may be an indication to start GnRHa management as early as SMR stage 2 before spermatogenesis is reached. Fertility preservation options may therefore be limited by stage of pubertal development, testicular volume, and ability to produce a sperm sample. Later in life, there may be options for TGD individuals with testicles to restore spermatogenesis after cessation of hormonal treatment, but this necessitates increasing testosterone levels, which could lead to undesired and irreversible physical changes (Sterling and Garcia, 2020). Therefore TGD youth with testicles should be counseled on their options for fertility preservation based on their pubertal status, including allowing pubertal progression for sperm maturation for fertility preservation before medical interventions.

Spironolactone

Spironolactone is an androgen receptor antagonist sometimes prescribed for TGD youth with testicles seeking to reduce testosterone-driven physical effects. It does not prevent the production of testosterone, so it may not be as effective as GnRHa in preventing ongoing androgen effects of testosterone. Preventing the action of testosterone, spironolactone reduces the frequency of spontaneous or desired erections, which can impact sexual function (Caminos-Torres et al., 1977). While this may be a goal of gender-affirming care for some individuals, this may limit capacity for sperm cryopreservation (Conflitti et al., 2026). Therefore it is generally recommended that spironolactone be held for weeks or months before attempting sperm collection via ejaculation, and standardized evidence-based protocols are currently in development.

Menstrual cycle suppression

Youth who are accessing GAH and have begun menstrual cycles have more options for menstrual management aside from GnRHa. Many TGD youth with ovaries will have more access to progesterone-only menstrual suppression and contraception, compared to GnRHa, given limited insurance coverage and local governmental restrictions. Some older youth may still prefer to access GnRHa if they have difficulty suppressing cycles with progesterone-only options, or if there is particular distress from ongoing chest tissue development. Progesterone-only options include norethindrone and norethindrone acetate daily pills, medroxyprogesterone injections, etonogestrel subdermal implant, and intrauterine devices. If oocyte cryopreservation is desired, progesterone management may be continued up until oocyte stimulation, or may be used as a bridge to GnRHa to prevent menses while awaiting fertility preservation. Progesterone-containing intrauterine devices (IUDs) can remain in place throughout the oocyte cryopreservation process (Friedenthal et al., 2017).

Testosterone

Early case reports studied testosterone effects on ovarian morphology and found thickening of the tunica albuginea and basal membrane changes similar to polycystic ovarian syndrome (PCOS) (Amirikia et al., 1986). It was largely assumed that TGD individuals with ovaries could not obtain pregnancy while taking gender-affirming testosterone, and that, essentially, testosterone damaged the ovarian tissue, preventing future fertility options. This assumption was proven incorrect when TGD individuals with prior testosterone use experienced spontaneous pregnancies (Light et al., 2014; Hassan et al., 2022).

Several outcomes related to fertility potential following gender-affirming testosterone use have been identified in TGD adults at the time of oopherectomy. In one study, TGD adults did not have a higher prevalence of PCOS morphology following testosterone compared to cisgender controls (Ikeda et al., 2013). In another report, approximately one-third of adults (33%, n=17 out of 52 participants, age: 20–28 years) continued to ovulate despite achieving amenorrhea while taking testosterone (Asseler et al., 2024). An analysis of ovarian tissue following 1 year of testosterone in 35 TGD adults (mean age: 24.30 ± 6.15 years) found cortical follicle distribution "surprisingly normal" (De Roo et al., 2017). However, one study found that ovarian tissue culture following testosterone (range: 18 months–10 years of treatment) demonstrated a higher prevalence of reduced follicle growth activation, poor follicle health, and increased DNA damage in transgender men (N=8, mean age: 27.6 years, range: 20–34 years) compared to ovarian cortical samples of cisgender age-matched controls (Bailie et al., 2023). The largest cohort study to date examining adolescent and young adult oocyte preservation after testosterone exposure reported that there was no difference in oocyte cryopreservation outcomes after exposure to testosterone (Cromack et al., 2024). Future research is still needed to assess testosterone therapy's impacts on oocyte quality and successful pregnancies.

Some studies show concern that long-term testosterone use can lead to impaired IVF due to lower numbers of mature oocytes and embryo maturation (Christodoulaki et al., 2023; Lierman et al., 2021; Schwartz et al., 2023). This evidence may have implications for youth starting GAH and understanding the potential timeline for fertility preservation opportunity before the long-term permanent impact of testosterone on gonadal tissue. This also underlines the importance of offering repeated fertility counseling throughout the reproductive life of TGD people, in addition to the need for further longitudinal studies as more youth age into adulthood and explore family planning.

Estradiol

Similar to early assumptions regarding testosterone, estradiol was expected to lead to infertility in TGD individuals with testicles. Studies reviewing the histological effects of estradiol on the testis have largely been done on removed testicular tissue at the time of orchiectomy in adulthood following varying durations of estradiol use (de Nie et al., 2022b). In an analysis of testicular cellular morphology in adults (age: 26–52 years) after 1–6 years of estradiol, both Sertoli and Leydig cells were impacted by estradiol, as were the seminiferous tubules (Leavy et al., 2017). This led to the early conclusion that estradiol severely impaired spermatogenesis, exacerbated by medical comanagement with androgen antagonists or GnRHa (Leavy et al., 2017). Cyproterone acetate seems to have a particular negative effect on adult testicular tissue (de Nie et al., 2023).

There have been case reports of successful restoration of spermatogenesis in TGD adults after long-term gender-affirming estradiol; however, this may take several months (Adeleye et al., 2019b) to more than a year to achieve (de Nie et al., 2023). In a longitudinal study of nine adults who discontinued estrogen for reproductive purposes, spermatogenesis was reversed for six (de Nie et al., 2023). Participants with recovery of viable spermatozoa started estrogen as young as age 18 years and were prescribed estrogen for as many as 18 years before discontinuing. Continued research is needed to determine estradiol's impact on testicular tissue and sperm production, alongside questions about duration of treatment, impacts of concurrent treatments, and viability of pregnancies from cryopreserved samples.

Research methods and future directions

Research investigating fertility and reproductive health implications of gender-affirming medical interventions is limited by minimal inclusion of TGD youth, or even young adults who initiated GAH in youth. Prospective mouse model studies (Kinnear et al., 2019; Bartels et al., 2021; Godiwala et al., 2023) have used adult mice to model potential effects of GAH and GnRHa on gonadal tissue, but direct applicability to TGD youth at varying stages of endogenous puberty is unclear. Of the human research available, most studies are retrospective studies using cisgender peer control groups (Ikeda et al., 2013; Bailie et al., 2023; Adeleye et al., 2019a). One study collected data prospectively in a descriptive observational study (De Roo et al., 2017). Case reports in adults may be starting to demonstrate the ability to preserve oocytes without discontinuation of GAH, including evidence of a healthy live birth (Stark and Mok-Lin, 2022; Greenwald et al., 2022). Existing studies suggest a permanent negative impact of estradiol on testicular tissue, but some promising case reviews indicate the possibility for restored spermatogenesis.

There are research opportunities in investigating oocyte cryopreservation protocols in TGD youth before initiating GAH, as well as additional cohorts who have already started testosterone at defined time intervals, such as 2 and 5 years. Expanding research to early pubertal TGD youth could investigate obtaining oocytes or ovarian cortical tissue after starting GnRHa and testosterone, to determine whether fertility preservation is even a necessary consideration at a young age. Additionally, studies confirming at what point in pubertal development fertility preservation is viable for youth with testicles are necessary, while continuing to expand into longitudinal research to assess the permanency of estradiol's impacts on testicular tissue and the ability to restore spermatogenesis.

Outcomes of fertility preservation

Existing research in fertility and fertility preservation for TGD youth largely focuses on education and understanding the underlying risks of gender-affirming medical interventions on fertility potential. There are limited available data regarding individuals who completed fertility preservation as youth, either before or after initiating any gender-affirming interventions. The majority of literature focusing on outcomes of fertility preservation in TGD individuals includes only adults who had reached reproductive maturity before starting GAH.

Oocyte preservation

Fertility research in TGD youth with ovaries mainly consists of either case reports or case series of youth who have completed fertility preservation, or describes the feasibility of ovarian stimulation and oocyte retrieval. Oocyte preservation procedures are more invasive than sperm preservation, requiring hormonal therapy preparation for over 1–2 weeks and then surgical intervention to access the ovaries. General fertility preservation research has demonstrated the overall safety and effectiveness of ovarian stimulation for youth (Manuel et al., 2020). Oocyte preservation rates are low in TGD youth with ovaries (Nahata et al., 2017; Pang et al., 2020).

In a case series of five TGD youth (age: 14–18 years) undergoing ovarian stimulation and oocyte preservation before gender-affirming medical treatments, all participants had mature oocytes cryopreserved (mean: 14.2 oocytes, range: 8–25), and no procedural complications were reported (Chen et al., 2018a). In a review of patients seen over a 3-year period, 26 referrals were received for TGD youth 21 years or younger, and 10 were seen for fertility preservation consultation, with five patients (19%) completing oocyte cryopreservation (Insogna et al., 2020). This case series included a 15-year-old patient undergoing oocyte cryopreservation with GnRHa implant in place. All patients tolerated stimulation, monitoring, and oocyte retrieval, and all patients had mature oocytes preserved (Insogna et al., 2020). In a case report of a single TGD youth who underwent oocyte cryopreservation at age 16 years while continuing GnRHa therapy the patient experienced depression and brief suicidal thoughts and was distressed by vaginal bleeding following oocyte retrieval and chest development from the ovarian stimulation hormones, which eventually regressed (Rothenberg et al., 2019). Four mature oocytes were preserved, and the patient initiated testosterone as planned following retrieval.

In a longer retrospective case series of fertility preservation referral outcomes over 12 years among TGD youth and young adults (*N*=44, median age: 17 years, range: 10–24 years), 35 participants completed fertility preservation consultation, with 15 (43%) declining fertility preservation and 20 (57%) completing oocyte preservation (Barrett et al., 2022). Reasons for declining fertility preservation after consultation included no interest in biological offspring, concern for dysphoria with fertility preservation, and desire to start testosterone. Two participants discontinued their testosterone before fertility preservation. A median of 22 mature oocytes (range: 5–59) were preserved per person (Barrett et al., 2022). Complications included four cases of mild to moderate ovarian hyperstimulation syndrome, including both participants who had started and paused testosterone.

In a review comparing TGD individuals (*n*=13) who had received testosterone to those who had not, and to age-matched cisgender female controls, successful pregnancies were achieved following assisted reproductive strategies using testosterone-exposed oocytes (Adeleye et al., 2019a). Overall, the number of oocytes obtained per cycle and peak estradiol levels were lower in TGD individuals with a history of testosterone use, but there was no difference in oocyte maturity rate. Therefore, even if testosterone may impact overall follicle count, there may not be a significant impact on the quality of oocytes and potential for maturation. This study included youth as young as 14 years, but no adolescents had received testosterone before study participation. The median age of participants with a history of testosterone was 26.9 years (interquartile range: 20.9–36.5), with a median testosterone use of 46 months and discontinuation of testosterone for 6 months (range: 1–13 months) before ovarian stimulation (Adeleye et al., 2019a). Another matched retrospective cohort study similarly compared TGD youth and adults (*N*=26) to cisgender adult women (Leung et al., 2019). The age range of the TGD group was 14–39 years, with younger patients planning oocyte cryopreservation only. All TGD participants discontinued testosterone before fertility preservation, and there were no statistically significant differences in the number of oocytes retrieved. Seven TGD adult couples had fresh or frozen embryo transfers, and all achieved live births (Leung et al., 2019).

A study from the Netherlands compared response in ovarian stimulation between young adult TGD individuals (*N*=24, median age: 22.3 years, interquartile range: 21.1–26.0) who had started testosterone compared to those who had not (Asseler et al., 2023). Seven participants (29.2%) had prior testosterone use, median: 23.9 months, and three participants (12.5%) had prior GnRHa use. All individuals who had started testosterone GAH paused it for 3 months before ovarian stimulation. Ovarian stimulation complications included five cases of ovarian hyperstimulation syndrome, two requiring hospitalization, and one ovarian torsion requiring emergency oophorectomy (Asseler et al., 2023). There were no statistically significant differences between those with prior testosterone use versus testosterone naïve participants in the incidence of complications, peak estradiol levels, or the number of oocytes retrieved or preserved. Investigators observed lower recombinant FSH doses needed in the prior testosterone group (Asseler et al., 2023).

A study from Israel focused on TGD youth specifically (*n*=9, median age: 16.4 ± 1.1 years), comparing fertility preservation outcomes in TGD youth to cisgender female peers undergoing fertility preservation due to cancer diagnosis and plan for gonadotoxic oncologic treatment (*n*=39, median age: 15.5 ± 1.3 years) (Amir et al., 2020). No one in the TGD cohort had initiated gender-affirming medical interventions before fertility preservation. Investigators observed that higher doses of gonadotropins were prescribed to the oncology cohort compared to TGD youth, but there were no significant differences in outcomes. The TGD participants collected a mean of 25.6 ± 12.9 mature oocytes compared to 18.8 ± 11.2 mature oocytes in the oncology cohort (Amir et al., 2020). In another study, fertilization rates and preimplantation embryo development were reviewed in adult TGD individuals (*n*=8) taking, then pausing, testosterone, and compared IVF outcomes to cisgender female peers (*n*=34) (Israeli et al., 2022). Adjusting for age differences the fertilization rate via conventional IVF was similar between the two groups, as was the number of embryos cryopreserved (Israeli et al., 2022).

Recent case reports have also challenged standard ovarian stimulation medication protocol use for TGD individuals (Stark and Mok-Lin, 2022; Martin et al., 2021; Gale et al., 2021). While conventional recommendations included discontinuation of testosterone 3–6 months or more before ovarian stimulation, there are now case reports of adults with successful oocyte preservation without discontinuing testosterone (Stark and Mok-Lin, 2022; Gale et al., 2021). One case report included alternative ovarian stimulation protocols, consisting of a patient receiving random start GnRH antagonist plus letrozole with leuprolide acetate trigger, leading to 35 oocytes retrieved and 23 metaphase II oocytes preserved. A second patient received a luteal phase start GnRH antagonist with leuprolide acetate trigger, leading to 14 oocytes retrieved and 9 oocytes preserved (Stark and Mok-Lin, 2022). Another case report described using letrozole to maintain lower estradiol levels while a TGD 15-year-old underwent ovarian stimulation with peak estradiol 510 pg/mL and cryopreserved 22 mature oocytes (Martin et al., 2021).

Beyond measuring the material success rate of oocyte preservation and alternative protocols for fertility preservation, research has also investigated psychosocial outcomes regarding fertility preservation decisions. One study compared TGD individuals to cisgender peers going through fertility preservation, with a small number of TGD individuals considering oocyte preservation (*n*=16, mean age: 25.5 years, range: 15.4–39.9 years) (Sundaram et al., 2024). No statistical differences were found in anxiety, depression, life satisfaction, or decisional regret between those who pursued oocyte preservation (*n*=10) and those who did not (*n*=6). Given the cross-sectional nature of the study, small sample size, and lack of analysis focused on youth specifically, more research is needed to determine whether there is a relationship between fertility preservation and psychosocial outcomes, particularly among youth.

Sperm preservation

In both pediatric and adult fertility preservation, there are higher rates of pursuing and completing sperm preservation compared to oocyte preservation (Nahata et al., 2017; Segev-Becker et al., 2020), and there are unique questions to be investigated for TGD youth with testes. In one academic medical center, sperm quality for TGD youth (n=18, mean age: 15.8 ± 1.6 years) was compared with that of cisgender youth diagnosed with cancer (n=27, mean age: 16.4 ± 1.8 years), all of whom had reached SMR 4 or 5 (Dilday et al., 2022). Overall, the TGD youth had lower sperm quality parameters before starting any gender-affirming medical interventions compared to the oncology group, but none of the differences reached statistical significance, resulting in still-viable samples. It was concluded that sperm cryopreservation by ejaculation was feasible for TGD youth (Dilday et al., 2022).

Among 23 TGD youth referred to urology for fertility preservation from a specialty pediatric gender clinic, eight participants (35%) completed sperm cryopreservation (age: 15.3 ± 1.4 years, mean SMR stage: 4.5 [range: 3–5], testicular volume: 18.5 ± 2.5 mL) compared to 15 participants (65%) who did not complete fertility preservation (age: 14.3 ± 1.7 years, mean SMR stage: 4.3 [range: 2–5], testicular volume: 16.6 ± 4.7 mL) (Yau et al., 2024). One patient preserved sperm at the SMR stage 3 testicular volume. Some patients did not complete fertility preservation due to a refusal to masturbate. Both groups included a mixture of participants who had and had not initiated GnRHa or GAH at the time of fertility preservation referral.

Research from Israel evaluated semen parameters from adolescents (N=25, mean age: 15.3 ± 1.97 years, age at preservation: 16.2 ± 1.38 years, SMR stage: 3–5) who underwent fertility preservation before initiating gender-affirming medical treatments (Amir et al., 2022). Twenty-one participants (83%) had at least one abnormal semen parameter, including one participant with azoospermia and seven (28%) with oligospermia. This study also investigated the rate of motile sperm recovery through the freeze-thaw process. Cohort rate of recovery was 35%, lower than 50% average for the general population (Amir et al., 2022). The same research group compared outcomes of fertility preservation among TGD adolescents and adults who had initiated gender-affirming medical interventions to those who had not, as well as a cisgender male control group, and found the lowest sperm quality and quantity among TGD participants who had pursued medical interventions (Barda et al., 2023).

Research from the Netherlands similarly evaluated semen parameters of TGD individuals greater than 16 years who completed fertility preservation (de Nie et al., 2020, 2022a). Quality parameters were lower than the WHO general population data, which was used by the Israeli research group for initial control data (Amir et al., 2022; de Nie et al., 2020, 2022a). In contrast to the outcomes found in Israel, this study did not find a significant association between impaired semen parameters and prior exposure to gender-affirming medical interventions (Amir et al., 2022; de Nie et al., 2022a).

Research from the United States further evaluated the age of consultation and outcomes of TGD adolescents and young adults with testicles referred for fertility preservation (Barnard et al., 2019). During the 3-year study period, 11 patients were referred for sperm cryopreservation (median age at initial presentation for gender-affirming medical treatment was 17 years, range: 15–24 years; median age at fertility preservation referral was 19.5 years, range: 16–24). Ten patients completed sperm cryopreservation, eight of whom had no history of medical interventions at the time of preservation. One patient had received 6 months of GnRHa, held the medication for 3 months, and semen analysis showed 12 sperm, of which only two were motile. A second attempt, 5 months after holding GnRHa, showed a normal semen analysis. The second patient had received spironolactone and estradiol for 26 months. Multiple attempts of sperm cryopreservation 2–4 months after holding medications showed persistent azoospermia (Barnard et al., 2019). In another study from the United States, individuals who had never used gender-affirming medical interventions (including adolescents, n=18) had statistically higher percent sperm motility, concentration, and total motile counts compared to groups who had previously pursued medical interventions (n=3) and those currently prescribed affirming interventions (n=7) (Adeleye et al., 2019b).

In addition to fertility preservation through ejaculation, researchers seek to understand the efficacy of alternative methods for fertility preservation. A case report of a 16-year-old patient demonstrated that early and late spermatids could be preserved via TESE at the time of orchiectomy following years of GnRHa and estrogen management (Adeleye et al., 2023). Research from Australia found that 17 of 25 participants (68%) without a history of gender-affirming medical interventions retrieved sperm via biopsy, and testicular volume of at least 10 mL was the most significant positive predictor of sperm retrieval success (Peri et al., 2021). Of the six participants who had already received GnRHa, one had sperm retrieved via testicular biopsy.

Research methods and future directions

The majority of research on the outcomes of fertility preservation consists of retrospective cohort studies (Peri et al., 2021; Valli-Pulaski et al., 2019; Asseler et al., 2023; Amir et al., 2020; Israeli et al., 2022; Dilday et al., 2022; Barnard et al., 2019;

Amir et al., 2022; Barda et al., 2023) demonstrating the feasibility of fertility preservation in TGD patient populations. These studies are supplemented by case reports (Rothenberg et al., 2019; Adeleye et al., 2023) and case series (Barrett et al., 2022; Chen et al., 2018a; Insogna et al., 2020), including those that are starting to demonstrate nonstandardized ovarian stimulation protocols for TGD individuals. Only one prospective cohort study (Valli-Pulaski et al., 2019) has been conducted, demonstrating the lack of prospective studies following TGD youth through their fertility preservation experiences compared to adults (Armuand et al., 2017). These reports focus on older adolescents and young adults, with very few studies including early pubertal youth.

The largest study on fertility preservation outcomes to date is in the process of coordinating a standardized research protocol for TTC across multiple academic centers (Valli-Pulaski et al., 2019). As TTC remains experimental, participants donate 25% of the collected tissue to research, and 75% is stored for potential future use. TGD youth who have not achieved spermarche do not have any other options for fertility preservation and can participate in this research. An OTC protocol has also been opened to offer a preservation option for TGD youth not advanced through puberty enough to preserve oocytes (Ausavarungnirun and Orwig, 2024).

Limitations in current outcomes data highlight that gametes are being preserved without fertilization, as most youth do not have reproductive partners, and therefore future live birth potential is not clearly understood. When TGD adults are preserving fertility, they are sometimes presenting with their reproductive partner, so embryo quality can be assessed (Israeli et al., 2022). Live births from preserved gametes of TGD youth have yet to be reported. This provides a vast potential for research to follow TGD youth through deciding to preserve fertility through returning as adults to access their gametes and proceed through thawing, fertilization, and embryo transfer, as well as pregnancy and live birth outcomes.

Larger scope studies should also explore patient experience and mental health outcomes when TGD youth return to access their preserved gametes in adulthood. Ongoing research is currently investigating mental health outcomes in TGD individuals who complete fertility preservation compared to those who do not (Sundaram et al., 2024). This should include investigating mental health outcomes in TGD individuals who preserved fertility in their youth and then returned to use gametes as adults. Potential outcomes include mental health for individuals who preserved gametes that led to live births compared to cohorts that decided not to use gametes for family building, and to those who returned to utilize gametes but did not achieve the desired family building results.

Fertility counseling and rates of fertility preservation

Overall, rates of pediatric fertility preservation in the United States are low and pediatric oncology literature demonstrates that young people may experience regrets related to missed opportunities to preserve fertility and that timely counseling about options is important as patients and families progress through medical decision-making (Nahata et al., 2017; Hudson et al., 2018; Stein et al., 2014; Klosky et al., 2015). Internationally, rates of pediatric fertility preservation may be higher than in the United States (Brik et al., 2019; Pang et al., 2020; Segev-Becker et al., 2020; Mattelin et al., 2022). Lack of health insurance benefit for pediatric fertility preservation in the United States is a recognized barrier to access (Vyas et al., 2021; Johnson et al., 2017), which is not an issue in many nations with national healthcare coverage that may include pediatric fertility preservation procedures and storage with partial to full coverage (Pang et al., 2020). Rates of fertility preservation referrals and completion are dependent on adequate counseling regarding the risks of certain gender-affirming medical and surgical interventions on fertility potential.

Rates of counseling, referral, and completion of fertility preservation have largely been studied through single-institution retrospective chart review. Early retrospective US data were concerning for low fertility preservation utilization rates among TGD youth. In one early study, only 2 patients (2.5%) attempted fertility preservation despite 72 patients (98.6%) having documented fertility counseling (Nahata et al., 2017). Of 66 TGD youth from another retrospective review, 78.8% (n=52) had documentation of fertility preservation counseling, 16.7% (n=11) were referred for fertility preservation, and only 3% (n=2) completed cryopreservation (Wakefield et al., 2019). Investigators noted it was more likely that fertility was discussed before starting estrogen or testosterone compared to pubertal suppression. Another early retrospective study found that 13 out of 105 TGD youth aged 14–20 years were referred for fertility preservation consultation, of which four patients completed sperm cryopreservation and one patient completed oocyte cryopreservation, totaling 4.8% of the study population (Chen et al., 2017).

A retrospective study that included mostly adults, but some youth (n=25, age: 15–29 years), investigated prior fertility counseling at the time of initial consultation for gender-affirming orchiectomy (Sineath et al., 2023). Results showed that preconsultation fertility counseling differed significantly by age group, with only 8% of the 15–29-year-old group reporting no prior fertility preservation discussions, compared to 50% of those older than 40 years (Sineath et al., 2023). Another surgical group examined a cohort that included youth and young adults (N=162, mean age ± SD=19.6 ± 4.4 years with a range of

15–33 years), all with documented fertility intentions and goals for either chest or genital surgeries (Mehra et al., 2022). The majority (94%) were already receiving GAH for a median duration of 14 months. Significantly more participants with testicles (*n*=6) than participants with ovaries (*n*=3) had already completed preservation before presenting for surgical consultation.

More recent research from the United States continues to demonstrate relatively low rates of fertility preservation despite high rates of fertility preservation counseling (Garborcauskas et al., 2023). Of 67 TGD youth ages 12–17 years from a more recent retrospective review, 55% (*n*=37) had documented discussions of the fertility implications of GAH, 33 patients were offered fertility preservation referrals, 5 pursued the referral, and 2 completed sperm cryopreservation, representing only 2.9% of the study cohort and 5.4% of those with documented discussions about fertility (Komorowski et al., 2023). Another recent study investigated whether having internal fertility preservation specialists in a multidisciplinary pediatric gender clinic would lead to higher referral rates and fertility preservation completion (Cooper et al., 2022). Although 31% (*n*=41) of reviewed patients (*N*=132) were referred for fertility preservation, 36% (*n*=15) completed consultation, and 22% completed fertility preservation procedures, or 6.8% of the study cohort. Results demonstrated that TGD youth with testicles are more likely to complete fertility preservation compared to peers with ovaries. However, the overall rates of fertility preservation completion were similar to prior reports.

Data published in other countries have shown higher fertility preservation rates. In the Netherlands a retrospective review of 35 patient records showed 91% (*n*=32) received fertility preservation counseling, and 12 patients (38%) attempted preservation, with 9 patients completing sperm cryopreservation (75%) (Brik et al., 2019). The other three patients did not complete fertility preservation due to an inability to produce sperm. Reviewed patients ranged in SMR stages 2–5, with a mean testicular volume of 15 ± 6 mL. In an Australian cohort of 102 TGD youth (mean age=15.6 years, range: 10.8–18.3 years) who received documented fertility counseling before initiating GAH, 62% (*n*=33) of youth with testicles attempted fertility preservation, with 22 (67%) preserving sperm via masturbatory sample and 11 (33%) via testicular biopsy (Pang et al., 2020). None of the TGD youth with ovaries attempted fertility preservation. Of the 11 youth with testicles who preserved via TTC, 5 had mature sperm and 6 had germ cells, and all samples were cryopreserved. These 11 youth were significantly younger than patients in most studies (mean age: 13.9 ± 1.5 years).

A retrospective review of patients in Israel showed that all pubertal TGD youth (*n*=96 of 107 patients reviewed) were offered fertility preservation before initiation of GAH, and of the 77 patients who pursued GAH, 31% (*n*=24) pursued internal fertility preservation referral. Of TGD youth with testicles, 100% completed sperm cryopreservation, and in youth with ovaries, 30% preserved oocytes (Segev-Becker et al., 2020). There was a noted association between younger age and increased likelihood of pursuing fertility preservation, but this did not reach statistical significance. Investigators theorized that cultural expectations and practices, both in Israel broadly and specifically within the gender clinic in which the review was conducted, may have contributed to higher fertility preservation rates compared to other international data (Segev-Becker et al., 2020).

Based on low utilization of fertility preservation, researchers have asked TGD youth, caregivers, and clinicians about their perspectives on optimizing delivery of counseling (Lai et al., 2021; Quain et al., 2021). A mixed-method study using an online survey and semistructured qualitative interviews investigated the preferences of TGD youth and their parents about who should provide fertility preservation counseling, and at what point in the gender-affirming journey (Quain et al., 2021). Sixty-four TGD youth (mean age: 17 years, range: 12–22 years) and 46 parents completed surveys, while 18 youth (mean age: 18 years, range: 15–24) and 13 parents completed interviews. Surveyed youth and parent groups had aligned preferences for gender clinic physicians, mental health professionals, or fertility specialists to provide fertility preservation counseling, whereas preferences on timing varied (Quain et al., 2021). Interview data added youth preference for parents to discuss fertility preservation information with them.

Qualitative research has also investigated the perspectives of healthcare professionals on fertility counseling for TGD youth (Lai et al., 2021). Semistructured individual interviews were conducted with a variety of pediatric gender-affirming health professionals (pediatricians, psychologists, psychiatrists, and gynecologists) in an Australian academic institution. These health professionals all worked directly with TGD youth clinically, but had varying lengths of experience doing so, from 2 months to 10 years (mean: 4.0 ± 3.0 years). Interview themes revealed a desire to prioritize a multidisciplinary practice model, utilize shared decision-making with young people, their caregivers, and clinicians, develop strategies for patient engagement, implement flexible individualized approaches, and engage in reflective practice among clinicians. Based on clinician experiences and perspectives, it was concluded that appropriately engaging with TGD youth can lead to them being "informed decision makers in their lives" (Lai et al., 2021).

Research methods and future directions

The vast majority of studies investigating fertility counseling and rates of referrals are retrospective cohort studies (Brik et al., 2019; Nahata et al., 2017; Pang et al., 2020; Segev-Becker et al., 2020; Wakefield et al., 2019; Sineath et al., 2023;

Komorowski et al., 2023; Garborcauskas et al., 2023; Cooper et al., 2022). This often allows a single institution to review its clinical experience over several years, but is not typically generalizable to other cohorts, nor necessarily a sustainable investigative practice. Demand for pediatric gender-affirming care clinical services is growing internationally, and small retrospective reviews will become more and more cumbersome as clinic cohorts grow exponentially (Segev-Becker et al., 2020). Two studies included qualitative semistructured individual interviews with TGD youth and parents, or clinicians (Lai et al., 2021; Quain et al., 2021). Developing prospective studies and utilizing validated shared decision-making tools are important next steps in optimizing fertility-related information sharing with TGD youth and their families.

Summary and future research directions

TGD young people have desires for future families and deserve access to fertility and family planning resources (Baram et al., 2019; Warton and McDougall, 2022). Research in pediatric TGD fertility and fertility preservation has focused on understanding the current state of fertility preservation in the patient population (Nahata et al., 2017), current clinical practices for fertility counseling (Hudson et al., 2018), and rates of referral and completion of fertility preservation (Pang et al., 2020; Cooper et al., 2022). Research outcomes have historically demonstrated low rates of fertility preservation, and subsequent research has therefore focused efforts on optimizing fertility counseling and education, and conducting needs assessments to better understand patient and parent attitudes and beliefs (Chen et al., 2018b; Nahata et al., 2020a). Ongoing research in this area is focusing on creating tools and decision aids to optimize fertility counseling for healthcare providers, patients, and patient support networks (Persky et al., 2020; Kolbuck et al., 2020). Future research is needed to evaluate decision-making outcomes once TGD youth reach adulthood and are able to reflect on past fertility preservation decisions.

Studies evaluating the effects of gender-affirming testosterone and estrogen have had mixed results, and it remains unclear exactly what level of affirming hormones and for what duration may be a tipping point to impairing fertility for TGD individuals. However, outcomes are generally positive, especially for oocyte health following testosterone use. While estradiol use has consistently been shown to negatively impact sperm, emerging research suggests that this impact may be reversible for some TGD individuals. Because research studies have not yet included TGD youth who have returned to utilize stored gametes for fertilization, embryo development, implantation, pregnancy, and birth, there is a dearth of outcome data aside from sperm analysis and oocyte cryopreservation numbers. Oocyte quality is difficult to assess before attempting fertilization. There is a lack of standardized protocols for how long someone may have to pause GAH before fertility preservation, or if pausing GAH is necessary (Stark and Mok-Lin, 2022).

Future research should focus on early pubertal TGD youth with limited options for fertility preservation based on gamete immaturity. Additional research is also needed to understand if premenarchal youth are able to respond similarly to ovarian stimulation after preceding directly from pubertal suppression to GAH, and whether GAH needs to be paused while pursuing fertility preservation. Existing case reports could be expanded into a prospective study including adolescents who have started testosterone in their youth and seek fertility preservation at older ages (Stark and Mok-Lin, 2022). Oocyte quality and fertilization success have not been reported from TGD youth gametes. If and when TGD youth return to their gametes, the medical and TGD community will gain knowledge from their experiences and outcomes that will inform future generations of TGD youth on the potential for future family planning.

References

Adeleye, A.J., Cedars, M.I., Smith, J., Mok-Lin, E., 2019a. Ovarian stimulation for fertility preservation or family building in a cohort of transgender men. Journal of Assisted Reproduction and Genetics 36 (10), 2155–2161. Available at https://doi.org/10.1007/s10815-019-01558-y.

Adeleye, A.J., Reid, G., Kao, C.N., Mok-Lin, E., Smith, J.F., 2019b. Semen parameters among transgender women with a history of hormonal treatment. Urology 124, 136–141. Available at https://doi.org/10.1016/j.urology.2018.10.005.

Adeleye, A.J., Stark, B.A., Jalalian, L., Mok-Lin, E., Smith, J.F., 2023. Evidence of spermatogenesis in the presence of hypothalamic suppression and low testosterone in an adolescent transgender female: a case report. Transgender Health 8 (1), 104–107. Available at https://doi.org/10.1089/trgh.2021.0034.

Allen, C.L., Muschialli, L., Nihlén, Å., Coates, A., Gonsalves, L.M., 2025. Barriers to sexual and reproductive health care faced by transgender and gender diverse people: a systematic review. Reproductive Health 22 (1). Available at https://doi.org/10.1186/s12978-025-02038-6.

Amir, H., Oren, A., Frishman, E.K., Sapir, O., Shufaro, Y., Becker, A.S., et al., 2020. Oocyte retrieval outcomes among adolescent transgender males. Journal of Assisted Reproduction and Genetics 37 (7), 1737–1744. Available at https://doi.org/10.1007/s10815-020-01815-5.

Amir, H., Perl, L., Barda, S., Lantsberg, D., Becker, A.S., Israeli, G., et al., 2022. Adolescent transgender females present impaired semen quality that is suitable for intracytoplasmic sperm injection even before initiating gender-affirming hormone treatment. Reproductive Sciences 29 (1), 260–269. Available at https://doi.org/10.1007/s43032-021-00561-y.

Amirikia, H., Savoy-Moore, R.T., Sundareson, A.S., Moghissi, K.S., 1986. The effects of long-term androgen treatment on the ovary. Fertility and Sterility 45 (2), 202–208. Available at https://doi.org/10.1016/S0015-0282(16)49155-7.

Armuand, G., Dhejne, C., Olofsson, J.I., Rodriguez-Wallberg, K.A., 2017. Transgender men's experiences of fertility preservation: a qualitative study. Human Reproduction 32 (2), 383–390. Available at https://doi.org/10.1093/humrep/dew323.

Asseler, J.D., Knieriem, J., Huirne, J.A., Goddijn, M., Verhoeven, M.O., van Mello, N.M., 2023. Outcomes of oocyte vitrification in trans masculine individuals. Reproductive Biomedicine Online 47 (1), 94–101. Available at https://doi.org/10.1016/j.rbmo.2023.03.007.

Asseler, J.D., del Valle, J.S., Chuva de Sousa Lopes, S.M., Verhoeven, M.O., Goddijn, M., Huirne, J.A.F., et al., 2024. One-third of amenorrheic transmasculine people on testosterone ovulate. Cell Reports Medicine 5 (3), 101440. Available at https://doi.org/10.1016/j.xcrm.2024.101440.

Ausavarungnirun, C., Orwig, K.E., 2024. Reproductive health in trans and gender-diverse patients: gonadal tissue cryopreservation in transgender and gender-diverse people. Reproduction (Cambridge, England) 168 (6). Available at https://doi.org/10.1530/REP-24-0253.

Bailie, E., Maidarti, M., Hawthorn, R., Jack, S., Watson, N., Telfer, E.E., et al., 2023. The ovaries of transgender men indicate effects of high dose testosterone on the primordial and early growing follicle pool. Reproduction and Fertility 4 (2). Available at https://doi.org/10.1530/raf-22-0102.

Baram, S., Myers, S.A., Yee, S., Librach, C.L., 2019. Fertility preservation for transgender adolescents and young adults: a systematic review. Human Reproduction Update 25 (6), 694–716. Available at https://doi.org/10.1093/humupd/dmz026.

Barda, S., Amir, H., Mizrachi, Y., Dviri, M., Yaish, I., Greenman, Y., et al., 2023. Sperm parameters in Israeli transgender women before and after cryopreservation. Andrology 11 (6), 1050–1056. Available at https://doi.org/10.1111/andr.13369.

Barnard, E.P., Priya Dhar, C., Rothenberg, S.S., Menke, M.N., Witchel, S.F., Montano, G.T., et al., 2019. Fertility preservation outcomes in adolescent and young adult feminizing transgender patients. Pediatrics 144 (3). Available at https://doi.org/10.1542/peds.2018-3943.

Barrera, E., Locks, R., Kremen, J., Yu, R., 2023. Fertility preservation in transfeminine adolescents using TESE at the time of Histrelin Acetate subcutaneous implant placement: a case series. Journal of Pediatric Urology 19 (5), 540. Available at https://doi.org/10.1016/j.jpurol.2023.02.006.

Barrett, F., Shaw, J., Blakemore, J.K., Fino, M.E., 2022. Fertility preservation for adolescent and young adult transmen: a case series and insights on oocyte cryopreservation. Frontiers in Endocrinology 13. Available at https://doi.org/10.3389/fendo.2022.873508.

Bartels, C.B., Uliasz, T.F., Lestz, L., Mehlmann, L.M., 2021. Short-term testosterone use in female mice does not impair fertilizability of eggs: implications for the fertility care of transgender males. Human Reproduction 36 (1), 189–198. Available at https://doi.org/10.1093/humrep/deaa282.

Boguszewski, K.E., Woods, S., Ducar, D.M., Taylor, J.F., 2022. Transgender and gender diverse fertility choices: supporting the decision-making process for adolescents and young adults. Journal of Pediatrics 240, 256. Available at https://doi.org/10.1016/j.jpeds.2021.10.013.

Brik, T., Vrouenraets, L.J.J.J., Schagen, S.E.E., Meissner, A., de Vries, M.C., Hannema, S.E., 2019. Use of fertility preservation among a cohort of transgirls in the Netherlands. Journal of Adolescent Health 64 (5), 589–593. Available at https://doi.org/10.1016/j.jadohealth.2018.11.008.

Caminos-Torres, R., Ma, L., Snyder, P.J., 1977. Gynecomastia and semen abnormalities induced by spironolactone in normal men. The Journal of Clinical Endocrinology and Metabolism 45 (2), 255–260.

Chen, D., Simons, L., Johnson, E.K., Lockart, B.A., Finlayson, C., 2017. Fertility preservation for transgender adolescents. Journal of Adolescent Health 61 (1), 120–123. Available at https://doi.org/10.1016/j.jadohealth.2017.01.022.

Chen, D., Bernardi, L.A., Pavone, M.E., Feinberg, E.C., Moravek, M.B., 2018a. Oocyte cryopreservation among transmasculine youth: a case series. Journal of Assisted Reproduction and Genetics 35 (11), 2057–2061. Available at https://doi.org/10.1007/s10815-018-1292-4.

Chen, D., Matson, M., Macapagal, K., Johnson, E.K., Rosoklija, I., Finlayson, C., et al., 2018b. Attitudes toward fertility and reproductive health among transgender and gender-nonconforming adolescents. Journal of Adolescent Health 63 (1), 62–68. Available at https://doi.org/10.1016/j.jadohealth.2017.11.306.

Chen, D., Kolbuck, V.D., Sutter, M.E., Tishelman, A.C., Quinn, G.P., Nahata, L., 2019a. Knowledge, practice behaviors, and perceived barriers to fertility care among providers of transgender healthcare. Journal of Adolescent Health 64 (2), 226–234. Available at https://doi.org/10.1016/j.jadohealth.2018.08.025.

Chen, D., Kyweluk, M.A., Sajwani, A., Gordon, E.J., Johnson, E.K., Finlayson, C.A., et al., 2019b. Factors affecting fertility decision-making among transgender adolescents and young adults. LGBT Health 6 (3), 107–115. Available at https://doi.org/10.1089/lgbt.2018.0250.

Chiniara, L.N., Viner, C., Palmert, M., Bonifacio, H., 2019. Perspectives on fertility preservation and parenthood among transgender youth and their parents. Archives of Disease in Childhood 104 (8), 739–744. Available at https://doi.org/10.1136/archdischild-2018-316080.

Christodoulaki, A., He, H., Zhou, M., Cardona Barberán, A., De Roo, C., Chuva De Sousa Lopes, S.M., et al., 2023. Characterization of ovarian tissue oocytes from transgender men reveals poor calcium release and embryo development, which might be overcome by spindle transfer. Human Reproduction 38 (6), 1135–1150. Available at https://doi.org/10.1093/humrep/dead068.

Coleman, E., Radix, A.E., Bouman, W.P., Brown, G.R., de Vries, A.L.C., Deutsch, M.B., et al., 2022. Standards of care for the health of transgender and gender diverse people, version 8. International Journal of Transgender Health 23 (1), S1. Available at https://doi.org/10.1080/26895269.2022.2100644.

Conflitti, A.C., Buonacquisto, A., Cicolani, G., Di Chiano, S., Ruberto, M., Pallotti, F., et al., 2026. Gender-affirming hormone therapy: effect on semen quality and use of fertility preservation. Andrology 14 (1), 322–330. Available at https://doi.org/10.1111/andr.70112.

Cooper, H.C., Long, J., Aye, T., Tian, X., 2022. Fertility preservation in transgender and non-binary adolescents and young adults. PLoS One 17 (3), e0265043. Available at https://doi.org/10.1371/journal.pone.0265043.

Craig Sineath, R., Guerre, M., Martin, L., Chouhan, J.D., 2023. Interest In and prior receipt of counseling on fertility preservation in transgender patients presenting for gender-affirming orchiectomy. Urology 182, 101–105. Available at https://doi.org/10.1016/j.urology.2023.07.018.

Cromack, S.C., Walter, J.R., Smith, K.N., Elvikis, J., Bazzetta, S.E., Goldman, K.N., 2024. Oocyte cryopreservation in transgender and gender-diverse individuals with or without prior testosterone use. Obstetrics and Gynecology 144 (6). Available at https://doi.org/10.1097/AOG.0000000000005749.

De Roo, C., Lierman, S., Tilleman, K., Peynshaert, K., Braeckmans, K., Caanen, M., et al., 2017. Ovarian tissue cryopreservation in female-to-male transgender people: insights into ovarian histology and physiology after prolonged androgen treatment. Reproductive Biomedicine Online 34 (6), 557–566. Available at https://doi.org/10.1016/j.rbmo.2017.03.008.

Demeestere, I., Simon, P., Dedeken, L., Moffa, F., Tsépélidis, S., Brachet, C., et al., 2015. Live birth after autograft of ovarian tissue cryopreserved during childhood. Human Reproduction 30 (9), 2107–2109. Available at https://doi.org/10.1093/humrep/dev128.

Dilday, E.A., Bukulmez, O., Saner, K., Lopez, X., Jarin, J., 2022. Sperm cryopreservation outcomes in transgender adolescents compared with adolescents receiving gonadotoxic therapy. Transgender Health 7 (6), 528–532. Available at https://doi.org/10.1089/trgh.2021.0037.

Donnez, J., Dolmans, M.M., 2017. Fertility preservation in women. New England Journal of Medicine 377 (17), 1657–1665. Available at https://doi.org/10.1056/NEJMra1614676.

Ethics Committee of the American Society for Reproductive Medicine. Electronic address: ASRM@asrm.org, 2018. Fertility preservation and reproduction in patients facing gonadotoxic therapies: an Ethics Committee opinion. Fertility and Sterility 110 (3), 380–386. Available at https://doi.org/10.1016/j.fertnstert.2018.05.034.

Finlayson, C., Johnson, E.K., Chen, D., Dabrowski, E., Gosiengfiao, Y., Campo-Engelstein, L., et al., 2016. Proceedings of the working group session on fertility preservation for individuals with gender and sex diversity. Transgender Health 1 (1), 99–107. Available at https://doi.org/10.1089/trgh.2016.0008.

Friedenthal, J., Maxwell, S.M., Willson, S., McCulloh, D.H., Grifo, J., Goldman, K.N., 2017. The progestin-containing intrauterine device (IUD) during ovarian stimulation and oocyte retrieval: should it stay or should it go? Fertility and Sterility 108 (3), e12. Available at https://doi.org/10.1016/j.fertnstert.2017.07.049.

Futterweit, W., Deligdisch, L., 1986. Histopathological effects of exogenously administered testosterone in 19 female to male transsexuals. The Journal of Clinical Endocrinology & Metabolism 62 (1), 16–21. Available at https://doi.org/10.1210/jcem-62-1-16.

Gale, J., Magee, B., Forsyth-Greig, A., Visram, H., Jackson, A., 2021. Oocyte cryopreservation in a transgender man on long-term testosterone therapy: a case report. F&S Reports 2 (2), 249–251. Available at https://doi.org/10.1016/j.xfre.2021.02.006.

Garborcauskas, G., Boskey, E.R., Guss, C.E., Grimstad, F.W., 2023. Retrospective review of sexual and reproductive health conversations during initial visits of adolescents seeking gender-affirming testosterone. Journal of Pediatric and Adolescent Gynecology 36 (1), 25–32. Available at https://doi.org/10.1016/j.jpag.2022.09.004.

Godiwala, P., Uliasz, T.F., Lowther, K.M., Kaback, D., Mehlmann, L.M., 2023. Puberty suppression followed by testosterone therapy does not impair reproductive potential in female mice. Endocrinology 164 (11). Available at https://doi.org/10.1210/endocr/bqad145.

Greenwald, P., Dubois, B., Lekovich, J., Pang, J.H., Safer, J., 2022. Successful in vitro fertilization in a cisgender female carrier using oocytes retrieved from a transgender man maintained on testosterone. AACE Clinical Case Reports 8 (1), 19–21. Available at https://doi.org/10.1016/j.aace.2021.06.007.

Hagenäs, I., Jørgensen, N., Rechnitzer, C., Sommer, P., Holm, M., Schmiegelow, K., et al., 2010. Clinical and biochemical correlates of successful semen collection for cryopreservation from 12-18-year-old patients: a single-center study of 86 adolescents. Human Reproduction 25 (8), 2031–2038. Available at https://doi.org/10.1093/humrep/deq147.

Halloran, J., Smidt, A.M., Morrison, A., Cron, J., Kallen, A.N., Olezeski, C.L., 2023. Reproductive and fertility knowledge and attitudes among transgender and gender-expansive youth: a replication and extension. Transgender Health 8 (4), 328–336. Available at https://doi.org/10.1089/trgh.2021.0080.

Harris, R.M., Kolaitis, I.N., Frader, J.E., 2020. Ethical issues involving fertility preservation for transgender youth. Journal of Assisted Reproduction and Genetics 37 (10), 2453–2462. Available at https://doi.org/10.1007/s10815-020-01873-9.

Hassan, A., Perini, J., Khan, Iyer, 2022. Transgender male: a case report and review of the literature. Case Rep Endocrinol 2022.

Hembree, W.C., Cohen-Kettenis, P., Delemarre-Van De Waal, H.A., Gooren, L.J., Meyer, W.J., Spack, N.P., et al., 2009. Endocrine treatment of transsexual persons: an endocrine society clinical practice guideline. Journal of Clinical Endocrinology and Metabolism 94 (9), 3132–3154. Available at https://doi.org/10.1210/jc.2009-0345.

Hembree, W.C., Cohen-Kettenis, P.T., Gooren, L., Hannema, S.E., Meyer, W.J., Murad, M.H., et al., 2017. Endocrine treatment of gender-dysphoric/gender-incongruent persons: an endocrine society*clinical practice guideline. Journal of Clinical Endocrinology and Metabolism 102 (11), 3869–3903. Available at https://doi.org/10.1210/jc.2017-01658.

Hudson, J., Nahata, L., Dietz, E., Quinn, G.P., 2018. Fertility counseling for transgender AYAs. Clinical Practice in Pediatric Psychology 6 (1), 84–92. Available at https://doi.org/10.1037/cpp0000180.

Ikeda, K., Baba, T., Noguchi, H., Nagasawa, K., Endo, T., Kiya, T., et al., 2013. Excessive androgen exposure in female-to-male transsexual persons of reproductive age induces hyperplasia of the ovarian cortex and stroma but not polycystic ovary morphology. Human Reproduction 28 (2), 453–461. Available at https://doi.org/10.1093/humrep/des385.

Insogna, I.G., Ginsburg, E., Srouji, S., 2020. Fertility preservation for adolescent transgender male patients: a case series. Journal of Adolescent Health 66 (6), 750–753. Available at https://doi.org/10.1016/j.jadohealth.2019.12.004.

Israeli, T., Preisler, L., Kalma, Y., Samara, N., Levi, S., Groutz, A., et al., 2022. Similar fertilization rates and preimplantation embryo development among testosterone-treated transgender men and cisgender women. Reproductive Biomedicine Online 45 (3), 448–456. Available at https://doi.org/10.1016/j.rbmo.2022.04.016.

Johnson, E.K., Chen, D., Gordon, E.J., Rosoklija, I., Holl, J.L., Finlayson, C., 2016. Fertility-related care for gender and sex diverse individuals: a provider needs-assessment survey. Transgender Health 1 (1), 197–201. Available at https://doi.org/10.1089/trgh.2016.0030.

Johnson, E.K., Finlayson, C., Rowell, E.E., Gosiengfiao, Y., Pavone, M.E., Lockart, B., et al., 2017. Fertility preservation for pediatric patients: current state and future possibilities. Journal of Urology 198 (1), 186–194. Available at https://doi.org/10.1016/j.juro.2016.09.159.

Kerman, H.M., Pham, A., Crouch, J.M., Albertson, K., Salehi, P., Inwards-Breland, D.J., et al., 2021. Gender diverse youth on fertility and future family: a qualitative analysis. Journal of Adolescent Health 68 (6), 1112–1120. Available at https://doi.org/10.1016/j.jadohealth.2021.01.002.

Kinnear, H.M., Constance, E.S., David, A., Marsh, E.E., Padmanabhan, V., Shikanov, A., et al., 2019. A mouse model to investigate the impact of testosterone therapy on reproduction in transgender men. Human Reproduction 34 (10), 2009–2017. Available at https://doi.org/10.1093/humrep/dez177.

Klosky, J.L., Simmons, J.L., Russell, K.M., Foster, R.H., Sabbatini, G.M., Canavera, K.E., et al., 2015. Fertility as a priority among at-risk adolescent males newly diagnosed with cancer and their parents. Supportive Care in Cancer 23 (2), 333–341. Available at https://doi.org/10.1007/s00520-014-2366-1.

Kolbuck, V.D., Sajwani, A., Kyweluk, M.A., Finlayson, C., Gordon, E.J., Chen, D., 2020. Formative development of a fertility decision aid for transgender adolescents and young adults: a multidisciplinary Delphi consensus study. Journal of Assisted Reproduction and Genetics 37 (11), 2805–2816. Available at https://doi.org/10.1007/s10815-020-01947-8.

Komorowski, A.S., Fisher, A.R., Jungheim, E.S., Lewis, C.S., Omurtag, K.R., 2023. Fertility preservation discussions, referral and follow-up in male-to-female and female-to-male adolescent transgender patients. Human Fertility 26 (5), 903–907. Available at https://doi.org/10.1080/14647273.2021.2015804.

Kyweluk, M.A., Sajwani, A., Chen, D., 2018. Freezing for the future: transgender youth respond to medical fertility preservation. International Journal of Transgenderism 19 (4), 401–416. Available at https://doi.org/10.1080/15532739.2018.1505575.

Kyweluk, M.A., Reinecke, J., Chen, D., 2019. Fertility preservation legislation in the United States: potential implications for transgender individuals. LGBT Health 6 (7), 331–334. Available at https://doi.org/10.1089/lgbt.2019.0017.

Lai, T.C., Davies, C., Robinson, K., Feldman, D., Elder, C.V., Cooper, C., et al., 2021. Effective fertility counselling for transgender adolescents: a qualitative study of clinician attitudes and practices. BMJ Open 11 (5), e043237. Available at https://doi.org/10.1136/bmjopen-2020-043237.

Lawrence, A.A., Shaffer, J.D., Snow, W.R., Chase, C., Headlam, bT., 1996. Health care needs of transgendered patients. JAMA: The Journal of the American Medical Association 276 (11), 874. Available at https://doi.org/10.1001/jama.1996.03540110028026.

Leavy, M., Trottmann, M., Liedl, B., Reese, S., Stief, C., Freitag, B., et al., 2017. Effects of elevated β-estradiol levels on the functional morphology of the testis - new insights. Scientific Reports 7 (1). Available at https://doi.org/10.1038/srep39931.

Leung, A., Sakkas, D., Pang, S., Thornton, K., Resetkova, N., 2019. Assisted reproductive technology outcomes in female-to-male transgender patients compared with cisgender patients: a new frontier in reproductive medicine. Fertility and Sterility 112 (5), 858–865. Available at https://doi.org/10.1016/j.fertnstert.2019.07.014.

Lierman, S., Tolpe, A., De Croo, I., De Gheselle, S., Defreyne, J., Baetens, M., et al., 2021. Low feasibility of in vitro matured oocytes originating from cumulus complexes found during ovarian tissue preparation at the moment of gender confirmation surgery and during testosterone treatment for fertility preservation in transgender men. Fertility and Sterility 116 (4), 1068–1076. Available at https://doi.org/10.1016/j.fertnstert.2021.03.009.

Light, A.D., Obedin-Maliver, J., Sevelius, J.M., Kerns, J.L., 2014. Transgender men who experienced pregnancy after female-to-male gender transitioning. Obstetrics and Gynecology 124 (6), 1120–1127. Available at https://doi.org/10.1097/AOG.0000000000000540.

Manuel, S.L., Moravek, M.B., Confino, R., Smith, K.N., Lawson, A.K., Klock, S.C., et al., 2020. Ovarian stimulation is a safe and effective fertility preservation option in the adolescent and young adult population. Journal of Assisted Reproduction and Genetics 37 (3), 699–708. Available at https://doi.org/10.1007/s10815-019-01639-y.

Martin, C.E., Lewis, C., Omurtag, K., 2021. Successful oocyte cryopreservation using letrozole as an adjunct to stimulation in a transgender adolescent after GnRH agonist suppression. Fertility and Sterility 116 (2), 522–527. Available at https://doi.org/10.1016/j.fertnstert.2021.02.025.

Mattelin, E., Strandell, A., Bryman, I., 2022. Fertility preservation and fertility treatment in transgender adolescents and adults in a Swedish region, 2013–2018. Human Reproduction Open 2022 (2). Available at https://doi.org/10.1093/hropen/hoac008.

Matthews, S.J., Picton, H., Ernst, E., Andersen, C.Y., 2018. Successful pregnancy in a woman previously suffering from β-thalassemia following transplantation of ovarian tissue cryopreserved before puberty. Minerva Ginecologica 70 (4), 432–435. Available at https://doi.org/10.23736/S0026-4784.18.04240-5.

McDougall, R.J., Gillam, L., Delany, C., Jayasinghe, Y., 2018. Ethics of fertility preservation for prepubertal children: should clinicians offer procedures where efficacy is largely unproven? Journal of Medical Ethics 44 (1), 27–31. Available at https://doi.org/10.1136/medethics-2016-104042.

Mehra, G., Boskey, E.R., Peters, C.J., Njubigbo, C., Charlton, B.M., Ganor, O., 2022. Assessing fertility intentions in patients presenting for gender-affirming surgery. LGBT Health 9 (5), 325–332. Available at https://doi.org/10.1089/lgbt.2021.0148.

Moravek, M.B., 2019. Fertility preservation options for transgender and gender-nonconforming individuals. Current Opinion in Obstetrics and Gynecology 31 (3), 170–176. Available at https://doi.org/10.1097/GCO.0000000000000537.

Morrison, A., Olezeski, C., Cron, J., Kallen, A.N., 2020. A pilot study to assess attitudes toward future fertility and parenthood in transgender and gender expansive adolescents. Transgender Health 5 (2), 129–137. Available at https://doi.org/10.1089/trgh.2019.0075.

Nahata, L., Tishelman, A.C., Caltabellotta, N.M., Quinn, G.P., 2017. Low fertility preservation utilization among transgender youth. Journal of Adolescent Health 61 (1), 40–44. Available at https://doi.org/10.1016/j.jadohealth.2016.12.012.

Nahata, L., Campo-Engelstein, L.T., Tishelman, A., Quinn, G.P., Lantos, J.D., 2018. Fertility preservation for a transgender teenager. Pediatrics 142 (3). Available at https://doi.org/10.1542/peds.2017-3142.

Nahata, L., Chen, D., Moravek, M.B., Quinn, G.P., Sutter, M.E., Taylor, J., et al., 2019. Understudied and under-reported: fertility issues in transgender youth—a narrative review. Journal of Pediatrics 205, 265–271. Available at https://doi.org/10.1016/j.jpeds.2018.09.009.

Nahata, L., Woodruff, T.K., Quinn, G.P., Meacham, L.R., Chen, D., Appiah, L.C., et al., 2020b. Ovarian tissue cryopreservation as standard of care: what does this mean for pediatric populations? Journal of Assisted Reproduction and Genetics 37 (6), 1323–1326. Available at https://doi.org/10.1007/s10815-020-01794-7.

Nahata, L., Chen, D., Quinn, G.P., Travis, M., Grannis, C., Nelson, E., et al., 2020a. Reproductive attitudes and behaviors among transgender/nonbinary adolescents. Journal of Adolescent Health 66 (3), 372–374. Available at https://doi.org/10.1016/j.jadohealth.2019.09.008.

de Nie, I., Meißner, A., Kostelijk, E.H., Soufan, A.T., Voorn-De Warem, I.A.C., den Heijer, M., et al., 2020. Impaired semen quality in trans women: prevalence and determinants. Human Reproduction 35 (7), 1529–1536. Available at https://doi.org/10.1093/humrep/deaa133.

de Nie, I., Mulder, C.L., Meißner, A., Schut, Y., Holleman, E.M., van der Sluis, W.B., et al., 2022b. Histological study on the influence of puberty suppression and hormonal treatment on developing germ cells in transgender women. Human Reproduction 37 (2), 297–308. Available at https://doi.org/10.1093/humrep/deab240.

de Nie, I., Asseler, J., Meißner, A., Voorn-de Warem, I.A.C., Kostelijk, E.H., den Heijer, M., et al., 2022a. A cohort study on factors impairing semen quality in transgender women. American Journal of Obstetrics and Gynecology 226 (3), 390. Available at https://doi.org/10.1016/j.ajog.2021.10.020 e1.

de Nie, I., van Mello, N.M., Vlahakis, E., Cooper, C., Peri, A., den Heijer, M., et al., 2023. Successful restoration of spermatogenesis following gender-affirming hormone therapy in transgender women. Cell Reports Medicine 4 (1), 100858. Available at https://doi.org/10.1016/j.xcrm.2022.100858.

de Nie, I., Asseler, J.D., Arnoldussen, M., Baas, S., de Vries, A.L.C., Huirne, J.A.F., et al., 2024. Reflecting on the importance of family building and fertility preservation: transgender people's experiences with starting gender-affirming treatment as an adolescent. Transgender Health 9 (4), 298–306. Available at https://doi.org/10.1089/trgh.2022.0135.

Pang, K.C., Peri, A.J.S., Chung, H.E., Telfer, M., Elder, C.V., Grover, S., et al., 2020. Rates of fertility preservation use among transgender adolescents. JAMA Pediatrics 174 (9), 890–891. Available at https://doi.org/10.1001/jamapediatrics.2020.0264.

Peri, A., Ahler, A., Gook, D., O'Connell, M.A., Bourne, H., Nightingale, M., et al., 2021. Predicting successful sperm retrieval in transfeminine adolescents after testicular biopsy. Journal of Assisted Reproduction and Genetics 38 (10), 2735–2743. Available at https://doi.org/10.1007/s10815-021-02293-z.

Perrard, M.-H., Sereni, N., Schluth-Bolard, C., Blondet, A., d'Estaing, S.G., Plotton, I., et al., 2016. Complete human and rat ex vivo spermatogenesis from fresh or frozen testicular tissue. Biology of Reproduction 95 (4), 89. Available at https://doi.org/10.1095/biolreprod.116.142802.

Persky, R.W., Gruschow, S.M., Sinaii, N., Carlson, C., Ginsberg, J.P., Dowshen, N.L., 2020. Attitudes toward fertility preservation among transgender youth and their parents. Journal of Adolescent Health 67 (4), 583–589. Available at https://doi.org/10.1016/j.jadohealth.2020.02.027.

Pfau, D.R., Schwartz, A.R., Dela Cruz, C., Padmanabhan, V., Moravek, M.B., Shikanov, A., 2024. A mouse model to investigate the impact of gender affirming hormone therapy with estradiol on reproduction. Advances in Biology 8 (9), e2300126. Available at https://doi.org/10.1002/adbi.202300126.

Quain, K.M., Kyweluk, M.A., Sajwani, A., Gruschow, S., Finlayson, C., Gordon, E.J., et al., 2021. Timing and delivery of fertility preservation information to transgender adolescents, young adults, and their parents. Journal of Adolescent Health 68 (3), 619–622. Available at https://doi.org/10.1016/j.jadohealth.2020.06.044.

Rodriguez-Wallberg, K., Obedin-Maliver, J., Taylor, B., Van Mello, N., Tilleman, K., Nahata, L., 2023. Reproductive health in transgender and gender diverse individuals: a narrative review to guide clinical care and international guidelines. International Journal of Transgender Health 24 (1), 7–25. Available at https://doi.org/10.1080/26895269.2022.2035883.

Rothenberg, S.S., Witchel, S.F., Menke, M.N., 2019. Oocyte cryopreservation in a transgender male adolescent. New England Journal of Medicine 380 (9), 886–887. Available at https://doi.org/10.1056/NEJMc1813275.

Sax, M.R., Pavlovic, Z., Decherney, A.H., 2020. Inconsistent mandated access to fertility preservation: a review of relevant state legislation. Obstetrics and Gynecology 135 (4), 848–851. Available at https://doi.org/10.1097/AOG.0000000000003758.

Schurman, J.V., Chen, D., Simons, L., 2018. Ethical considerations in fertility preservation for transgender youth: a case illustration. Clinical Practice in Pediatric Psychology 6 (1), 93–100. Available at https://doi.org/10.1037/cpp0000230.

Schwartz, A.R., Xu, M., Henderson, N.C., Dela Cruz, C., Pfau, D., Padmanabhan, V., et al., 2023. Impaired in vitro fertilization outcomes following testosterone treatment improve with washout in a mouse model of gender-affirming hormone treatment. American Journal of Obstetrics and Gynecology 229 (4), 419. Available at https://doi.org/10.1016/j.ajog.2023.07.013 e10419.

Segev-Becker, A., Israeli, G., Elkon-Tamir, E., Perl, L., Sekler, O., Amir, H., et al., 2020. Children and adolescents with gender dysphoria in israel: increasing referral and fertility preservation rates. Endocrine Practice 26 (4), 423–428. Available at https://doi.org/10.4158/ep-2019-0418.

Sexual and reproductive health and rights, 2025. Available at https://www.who.int/health-topics/sexual-and-reproductive-health-and-rights#tab=tab_1.

Stark, B.A., Mok-Lin, E., 2022. Fertility preservation in transgender men without discontinuation of testosterone. F and S Reports 3 (2), 153–156. Available at https://doi.org/10.1016/j.xfre.2022.02.002.

Stein, D.M., Victorson, D.E., Choy, J.T., Waimey, K.E., Pearman, T.P., Smith, K., et al., 2014. Fertility preservation preferences and perspectives among adult male survivors of pediatric cancer and their parents. Journal of Adolescent and Young Adult Oncology 3 (2), 75–82. Available at https://doi.org/10.1089/jayao.2014.0007.

Sterling, J., Garcia, M.M., 2020. Fertility preservation options for transgender individuals. Translational andrology and urology 9 (Suppl 2). Available at https://doi.org/10.21037/tau.2019.09.28.

Strang, J.F., Jarin, J., Call, D., Clark, B., Wallace, G.L., Anthony, L.G., et al., 2018. Transgender youth fertility attitudes questionnaire: measure development in nonautistic and autistic transgender youth and their parents. Journal of Adolescent Health 62 (2), 128–135. Available at https://doi.org/10.1016/j.jadohealth.2017.07.022.

Sundaram, V., Stark, B., Jaswa, E., Letourneau, J., Mok-Lin, E., 2024. Decision regret, and other mental health outcomes, following fertility preservation in the transgender individual compared to the cisgender woman. Journal of Assisted Reproduction and Genetics 41 (4), 1077–1085. Available at https://doi.org/10.1007/s10815-023-03013-5.

Tishelman, A.C., Sutter, M.E., Chen, D., Sampson, A., Nahata, L., Kolbuck, V.D., et al., 2019. Health care provider perceptions of fertility preservation barriers and challenges with transgender patients and families: qualitative responses to an international survey. Journal of Assisted Reproduction and Genetics 36 (3), 579–588. Available at https://doi.org/10.1007/s10815-018-1395-y.

Valli-Pulaski, H., Peters, K.A., Gassei, K., Steimer, S.R., Sukhwani, M., Hermann, B.P., et al., 2019. Testicular tissue cryopreservation: 8 years of experience from a coordinated network of academic centers. Human Reproduction 34 (6), 966–977. Available at https://doi.org/10.1093/humrep/dez043.

Varkey, B., 2021. Principles of clinical ethics and their application to practice. Medical Principles and Practice 30 (1), 17–28. Available at https://doi.org/10.1159/000509119.

Vrouenraets, L.J.J., Fredriks, A.M., Hannema, S.E., Cohen-Kettenis, P.T., De Vries, M.C., 2015. Early medical treatment of children and adolescents with gender dysphoria: an empirical ethical study. Journal of Adolescent Health 57 (4), 367–373. Available at https://doi.org/10.1016/j.jadohealth.2015.04.004.

Vyas, N., Douglas, C.R., Mann, C., Weimer, A.K., Quinn, M.M., 2021. Access, barriers, and decisional regret in pursuit of fertility preservation among transgender and gender-diverse individuals. Fertility and Sterility 115 (4), 1029–1034. Available at https://doi.org/10.1016/j.fertnstert.2020.09.007.

Wakefield, B.W., Boguszewski, K.E., Cheney, D., Taylor, J.F., 2019. Patterns of fertility discussions and referrals for youth at an interdisciplinary gender clinic. LGBT Health 6 (8), 417–421. Available at https://doi.org/10.1089/lgbt.2019.0076.

Walton, E., Abhari, S., Tangpricha, V., Futral, C., Mehta, A., 2023. Family planning and fertility counseling perspectives of gender diverse adults and youth pursuing or receiving gender affirming hormone therapy. Urology 171, 244–250. Available at https://doi.org/10.1016/j.urology.2022.10.007.

Warton, C., McDougall, R.J., 2022. Fertility preservation for transgender children and young people in paediatric healthcare: a systematic review of ethical considerations. Journal of Medical Ethics 48 (12), 1076–1082. Available at https://doi.org/10.1136/medethics-2021-107702.

Winkelman, W.D., Rosen, M.P., Mok-Lin, E., 2018. Fertility preservation services for women with newly diagnosed cancer. American Journal of Clinical Oncology: Cancer Clinical Trials 41 (10), 1031–1035. Available at https://doi.org/10.1097/COC.0000000000000420.

Yau, M., Malhotra, N.R., Steever, J., Baldisserotto, E., Lin, C., Fader, R., et al., 2024. Pubertal status at the time of fertility preservation in transgender girls. Endocrine Practice 30 (4), 356–359. Available at https://doi.org/10.1016/j.eprac.2024.01.008.

Chapter 12

Sexual development, health, and functioning

Theodore R. Burnes
Rossier School of Education, University of Southern California, Los Angeles, CA, United States

Learning objectives

- Summarize the research on sexual development, sexual functioning, and sexual health and well-being among transgender adolescents and young adults, both before and after various gender-affirming interventions.
- Describe historical, systemic, geographic, and individual barriers to research on sexual functioning and health.
- Highlight gaps in the research, including a need for more information on the benefits of gender-affirming healthcare on sexual experiences and outcomes, as well as sexual health literacy among transgender adolescents and young adults.
- Provide recommendations for community-driven research in the area of sexual health and functioning among transgender and gender diverse youth.

Key terms and definitions

- **Sex-positivity** includes openness, nonjudgmental attitudes, freedom, and liberation from antisex or sex-negative attitudes.
- **Sexual orientation** refers to a person's identity and behaviors related to romantic, sexual, or emotional attraction to others.
- **Sexual functioning** refers to the ability and opportunity to engage in desired actions of sexual interest and intimacy.
- **Sexual health** refers to the goals of preventing illness and maintaining health and happiness related to sexuality and sexual experiences.
- **Sexual well-being** includes the activities and practices aimed at achieving the goal of sexual health and happiness.

Sexual health and functioning, both before and after gender-affirming interventions, is an important component of providing competent healthcare to transgender and gender diverse (TGD) adolescents and young adults. TGD young people, like many young people seeking care, bring themes about their intimate relationships, sexual identity, and sexual health (including functioning) into their clinical encounters (Haley et al., 2019). More research on sexuality and sexual health is needed to provide accurate information, to inform comprehensive and competent education, and to slowly eradicate shame and stigma around talking about sexuality in the larger culture (Burnes et al., 2017).

Despite the articulated need to address sexuality and sexual health for TGD young people, clinicians and researchers often avoid these conversations and related interventions. This avoidance stems from, and cycles back into, research that avoids sexuality-related topics. Many clinical researchers and providers lack scientific knowledge about sexuality and receive little to no training outside of physiology (Verrastro et al., 2020). There are historical, systemic, geographic, and individual barriers to research and clinical care related to sexual health. TGD people face multiple obstacles to accessing quality sexuality healthcare, including outright discrimination and refusal of treatment, as well as a lack of relevant clinical understanding, gender-affirming practices, and cultural competence across various disciplines. Gender-segregated

Advances in Research on Pediatric Gender Affirming Healthcare. DOI: https://doi.org/10.1016/B978-0-443-29019-0.00012-9
© 2026 Elsevier Inc. All rights are reserved, including those for text and data mining, AI training, and similar technologies.

healthcare facilities further prevent many TGD patients from accessing appropriate care (White Hughto et al., 2018). The intimate nature of sexual and reproductive healthcare—such as screening and treatment for sexually transmitted infections (STIs) (Van Gerwen et al., 2020), screening for breast, cervical, and prostate cancers (de Blok et al., 2019), and contraception provision (Light et al., 2018)—makes these concerns especially acute (Kcomt, 2019), with greater impacts for TGD patients of color (Howard et al., 2019).

There is a dearth of research not only on the relationships between sexual health and gender transition, but also on experiences with minority stress and experiences of discrimination in healthcare environments. Patients with diverse gender identities and expressions seeking sexual and reproductive healthcare often fear that they will be treated in ways that are disrespectful or judgmental due to their gender identities or sexual practices (Sbragia and Vottero, 2020). These experiences are compounded by an incorrect conflation of sexual orientation and gender identity, resulting in research that does not assess any sexuality at all, given the inability to distinguish between these two different yet intersecting elements of identity. Other TGD patients may worry about interacting with clinical researchers and providers because aspects of their bodies do not conform to traditional gender norms. Some providers assume that their TGD patients do not need services such as pelvic exams or contraception, or that treating a variety of TGD patients is too complex. Not broaching the topic of sexuality with TGD patients perpetuates the invisibility of TGD sexualities, which in turn leads researchers to avoid these topics.

These dynamics contribute to significant disparities in sexual and reproductive health for TGD people, making researching these constructs particularly difficult methodologically. For example, a 2008 survey study indicated that one in three transgender people, and 48% of transgender men, have delayed or avoided preventive healthcare such as pelvic examinations or STI screening out of fear of discrimination or disrespect (Rachlin et al., 2008), but comparable studies with more recent data have yet to be published. While more recent survey data similarly indicate that many transgender men avoid annual pelvic exams (Obedin-Maliver and Care, 2016), these data are not accompanied by in-depth analyses of the reasons for this avoidance.

These concerns about sexual healthcare highlight the dearth of empirical, theoretical, and clinical research for TGD adolescents and young adults. Most existing literature on sexual wellness in TGD people is centered around the risk of HIV and living with HIV (Becasen et al., 2019; Marinelli et al., 2024), as well as other STIs (Bauer and Hammond, 2015). Furthermore, the impact of gender-affirming hormones (GAHs) and gender-affirming surgery (GAS) on sexual wellness in the general TGD population is not well described, and often based on research with ambiguous research designs and flawed methodologies. Such flawed methodologies include operationalizing sexuality-focused constructs using sex assigned at birth rather than gender identity, investigating gender-affirming interventions without assessing sexual pleasure, and using outdated measures that are not inclusive of the experiences of TGD people. With regard to research addressing the sexual health of TGD adolescents, there is an even greater lack of available information.

Sex-positivity

The construct of sex-positivity is a starting point when researching the multiple components of sexual health for TGD adolescents and young adults. Larger health systems often conflate sexuality with gender identity and expression, causing clinicians to sometimes inadvertently confabulate the two, creating historic and systemic barriers to sexual healthcare for TGD people. In contrast, current research identifies sex-positivity as a necessary best practice in clinical research and patient care (Kuzma et al., 2024). The term *sex-positivity* is used to describe openness, nonjudgmental attitudes, freedom, and liberation from antisex or sex-negative attitudes (Burnes, 2025). This approach encourages people to imagine what their world may have been like if they had lived free from sexual blame, judgment, or guilt. Sex-positivity is a framework that integrates the physical, emotional, cultural, intellectual, social, and spiritual aspects of sexual practice and sexual identity in positive, enriching ways (Syme et al., 2013). It highlights the strong connection between sexuality and diverse cultural identities and promotes the inclusion of dynamic sexual identities and fluidity, sexual satisfaction, and relational health.

There are some variations in the definition of sex-positivity, although there are many common core elements. An online survey asked 52 experts open-ended questions about how they define sex-positivity and sex-negativity (Ivanski and Kohut, 2017). One theme that emerged from these definitions was developing an understanding of sexuality that promotes support of personal sexual autonomy when there is regard for consent, health, and safety. Sex-positivity fosters a sense of entitlement to basic bodily autonomy, leading to a human sexuality that prioritizes confidence and joy (Nagoski, 2024). Other scholars have expanded the operationalization of the construct, suggesting that claiming and engagement with the erotic is an essential act of resistance and a way to stand against dehumanization and domination (Lorde, 1984).

Consequently, TGD people connecting with their erotic selves is a clear tool in promoting resilience and defying oppressive forces.

In contrast to sex-positivity, TGD adolescents and young adults may also have internalized sex-negative attitudes and belief systems. Sex-negative attitudes are those that understand, relay, and/or depict sex as abnormal, wrong, or shameful. For example, a patient might be told that they "have a problem" if they are having consensual sex with multiple people. A specific kind of sex negativity, erotophobia, is a cultural or religious fear, disdain, disgust, or devaluing of sensual or sexual pleasure, the erotic capacities of the mind and body, and nonprocreative sexual practices (Dawson, 2018).

Research investigating the sex-positive experiences of TGD young people is in its infancy (Anzani et al., 2024). Empirical considerations of sexual well-being with TGD patients include an ability to assess the presence or absence of gender dysphoria within assessment of and interventions on sexual health, as well as the shifts in attraction that may occur with GAH or GAS (Barbonetti, 2024). Unfortunately, research has historically focused on sexual attitudes alone (Hendrick and Hendrick, 1987; Fisher et al., 1988), and many measurement tools have not incorporated cultural and social evolutions about positive sexuality (Gerymski, 2021; Reisner et al., 2020).

Measuring sex-positivity as a continuum rather than a polarized dichotomy, and incorporating a wide range of sexual health behaviors and beliefs, will increase the validity of research in this domain. The Sex-Positivity Scale (Belous and Schulz, 2024) is a newer measurement tool that incorporates a range of beliefs, behaviors, and communication, but it has not been validated for adolescents or for TGD populations specifically. The Sex-Positivity Scale is a 26-item self-report measure, and it can be found in the Appendix of the published article. More sex-positive research among TGD young people is needed, taking into consideration intersecting identities and social locations.

Sexual identity and orientation

Sexual orientation refers to a person's identity and behaviors related to romantic, sexual, or emotional attraction to others. Research on sexual orientation and identity development among TGD youth and adults has been limited by historical heteronormative biases in the field of transgender health, in which patients were only considered "truly transgender" if they were also heterosexual (Reay, 2014). Until more recently, transgender people who acknowledged a nonheterosexual attraction, changes in sexual orientation, or diverse sexual activities may not have been granted access to gender-affirming medical care (Dickenson et al., 2023). More recent qualitative research supports a wide range of alternative perspectives, highlighting the various relational dynamics and intersectional identities that can occur across gender identity and sexual orientation (Skrzypczak et al., 2025; Baker, 2018; Hereth et al., 2020).

A systematic review of studies among TGD adults suggests a dynamic interplay between sexual orientation and gender identity at individual, relational, and systemic levels, with each aspect of identity playing a role in the development of the other (Skrzypczak et al., 2025). Compared to community samples of cisgender people, TGD adults are more likely to identify as a sexual minority and have heterogeneity in sexual orientation, adopting a wide range of labels and communities that are inclusive of diverse gender identities (Reisner et al., 2023). Many TGD people find traditional labels of sexual orientation to be limiting and outdated, reconstructing labels of sexual orientation identity and awareness that are more reflective of lived experience (Galupo et al., 2016). Prospective data also suggest that sexual orientation remains a constant developmental spectrum (Defreyne et al., 2021). While more limited, emerging research with TGD adolescents and young adults similarly demonstrates a high degree of heterogeneity, more expansive identification, and fluidity over the course of sexual identity development (Warwick et al., 2022).

While there is not always a cause or link between gender identity and sexual orientation, the relationship between these two multifaceted constructs is complex and dynamic. Qualitative research has shown that for some TGD young adults, sexual identity exploration facilitates gender identity development, but for others, clarifying gender identity is a necessary foundational step before exploration of sexual identity is possible (Hereth et al., 2020). Another qualitative study that included adolescents and young adults found that only a small portion of TGD youth identified as heterosexual, and that many reported changes in sexual identity both before and after gender-affirming medical treatments (Warwick et al., 2022).

Quantitative research among adolescents and young adults has demonstrated similar themes. In one study of binary and nonbinary transmasculine young people, less than one-third of participants identified as heterosexual/straight (Boskey and Ganor, 2022). Participants also reported attraction to people with a range of gender identities, and sexual orientation was not necessarily predictive of attraction or sexual activity, suggesting that identity, attraction, and behaviors may not always align. In another large study of US high school students, TGD adolescents were more likely to identify as gay or lesbian, queer, or asexual compared to their cisgender peers (Szoko et al., 2023). Among the TGD adolescents in this study, sexual orientation sometimes correlated with sexual activity, but not always.

An increasing number of TGD young people are identifying across the asexual spectrum, from having little to no sexual attraction to others to the presence of varying degrees of sexual attraction (Chen, 2020). Asexual people can enjoy caresses, kissing, cuddling, and forms of touch and eroticism that are not strictly sexual (Fielding, 2021). Fraysexuality occurs when, in comparison with asexuality, a person experiences sexual attraction with one or more other people until they begin to form an emotional connection, at which point the sexual attraction begins to disappear (Barron, 2023). Although asexuality may be an enduring sexual identity for some, other TGD youth may identify with asexuality due to the erasure of their sexuality as a result of their transgender status. Still others may identify as asexual temporarily, with shifts toward more expansive sexuality following gender-affirming interventions. Representation and inclusion of diverse sexual orientations, including asexuality, are needed to help young people clarify and engage with their own identity and attraction.

Sexual functioning

Critical scholar bell hooks (2010) wrote, "For sex is a danger zone. In sexual desire, in sexual acts, there is so much that is not equal" (bell hooks, 2010,p. 178). hooks' writing exemplifies how many TGD people may see various components of sexuality—including arousal, excitement, and orgasm—to be wrought with dynamics of marginalization and oppression. Patriarchal systems have defined sex, pleasure, and sexual functioning based on rigid, stereotyped gender roles, which are not inclusive of the spectrums of gender and sexual diversity.

Research and clinical care have been historically limited by these narrow conceptualizations of sexual functioning, especially for TGD people who may have more nuanced experiences of sexuality as a result of gender dysphoria, identity exploration, and the impact of gender-affirming care. TGD people may experience gender dysphoria that is centered on their own bodies during sexual experiences, while others may find sexual experiences affirming and empowering. These experiences can also shift over time as adolescents and adults experience more time affirmed in their gender and with the inclusion of gender-affirming medical interventions to relieve dysphoria.

Qualitative research has found that TGD adolescents and young adults define "sex" in terms of purpose, behavior, and consent (Warwick et al., 2022). A range of purposes was identified, including physical intimacy, sexual pleasure, orgasm, emotional intimacy, and romantic or spiritual connection. Behaviors might include penetrative intercourse, other interpersonal acts that involve genitals, or self-focused stimulation. When sexual acts include another person or persons, consent was also identified as a defining characteristic of sexual activity. Sexual functioning is therefore broadly defined here as the ability and opportunity to engage in desired actions of sexual interest and intimacy.

In addition to historical bias and patriarchal influence, research and clinical care related to sexual functioning are limited by noninclusive research methods and a lack of appropriate, validated research tools. For example, TGD participants may have sexual partners who identify as men, women, intersex, or some other identity. Most existing scales are focused on binary experiences and do not account for the variety of identities and practices of TGD people. The Male Sexual Function Index (MSFI) and Female Sexual Function Index (FSFI), for example, focus on binary experiences and do not take into account that body parts and functioning may be different before and after GAS (Barcelos et al., 2024). Although some scales have been adapted for use with transgender adults in clinical research, there are even fewer tools available to assess sexual functioning for TGD adolescents.

Despite these limitations, emerging research is beginning to demonstrate the various factors that can impact sexual functioning in TGD people. The impacts of GAH and GAS, for example, are highly variable and not fully understood (Holmberg et al., 2019). Previous research suggested that GAH can impact sexual arousal and libido in TGD adults. More recent research has demonstrated that the most well-established changes associated with GAH are initial changes to libido and increased long-term sexual satisfaction, likely through increased body satisfaction (Burns et al., 2022). A systematic review of studies of TGD adults found that GAH and GAS may help to improve sexual satisfaction (Mattawanon et al., 2021). Despite high rates of satisfaction, some TDG people nonetheless experience sexual dysfunctions following surgery due to stress about sexual performance or other sexual disturbances (Schardein and Nikolavsky, 2022; Wilder et al., 2023).

Most research to date on the impact of gender-affirming care on sexual functioning among TGD adolescents has come from a prominent clinical research group in the Netherlands at the Center for Expertise in Gender Dysphoria (Van Der Meulen et al., 2025a, 2025b; Bungener et al., 2020). TGD young adults aged 18–25 years who were referred to the center as adolescents and who received pubertal suppression, GAH, and at least one GAS (e.g., mastectomy, vaginoplasty, etc.) completed questionnaires on sexual behavior, functioning, and satisfaction at least 1 year postsurgery. Questionnaires were drawn from a large Dutch population study on sexual health and adapted based on clinical experience and patient feedback. Participants reported retrospectively on their experience before surgery and their current experiences. Results revealed that these TGD young adults had significant increases in almost all sexual activities compared to before surgery, particularly with regard to activities that involved their own genitals (Bungener et al., 2020). Transwomen in the study reported a greater increase in sexual intercourse, and transmen reported a greater increase in romantic relationships compared to one another, but there were no other differences based on gender. Most participants valued sex, love, and relationships and were satisfied with their current sexual functioning. A small number of participants also reported on negative sexual experiences. The results of this study highlight that various aspects of sexuality (e.g., desire, arousal, identity, sexual expression, sexual orientation, values, etc.) may shift with the introduction of gender-affirming social and medical interventions.

One consideration for sexual functioning that is unique to adolescents concerns the increasing availability of pubertal suppression and the potential long-term impacts of these interventions. In particular, there have been clinical research questions regarding adolescents assigned male at birth who, as a result of early pubertal suppression, halt the development of genitals and may never engage sexually with the anatomy that they were born with. To help clarify the impact of pubertal suppression the Dutch research group classified participants as having initiated pubertal suppression in early puberty (sexual maturity rating stage 2 or 3) or late puberty (sexual maturity rating stage 4 or 5) based on retrospective chart review in addition to the questionnaires described above (Van Der Meulen et al., 2025b). The study found that over 90% of participants experienced sexual desire postsurgery, with no difference between early and later suppression groups.

Before surgery, 58% of participants in the late suppression group reported a history of orgasm compared to only 19% in the early suppression group, consistent with what would be expected based on suppression timing. After surgery, nearly 80% of participants endorsed being able to achieve orgasm with no differences across the two suppression groups. Because several of the remaining 20% of participants reported that they had not yet tried to achieve orgasm, this number is expected to be even higher. Furthermore, these results demonstrate that a history of presurgical arousal and orgasm is not a prerequisite for sexual functioning after surgery. Some participants described sexual difficulties, the most common of which was pain, and there were similarly no differences based on timing of pubertal suppression (Van Der Meulen et al., 2025b). This is consistent with research showing that sexual functioning and satisfaction after vaginoplasty are high among adults (Kitic et al., 2025; Schardein and Nikolavsky, 2022), and suggests that early pubertal suppression is not a risk factor for sexual dysfunction later in life.

The Dutch research group also looked at early versus late pubertal suppression across both transmasculine and transfeminine participants and similarly found a high degree of sexual satisfaction, with no differences based on sex assigned at birth or pubertal suppression timing (Van Der Meulen et al., 2025a). This study found that the most common sexual problem among transmasculine participants was difficulty finding and initiating sexual content, with no significant difference based on timing of suppression. The most common sexual problem among transfeminine participants was orgasm, which was endorsed more often within the late suppression group. This last result was unexpected and not aligned with prior hypotheses and thus cannot be fully explained. However, it is possible that the ability to avoid some level of early genital dysphoria may actually facilitate sexual functioning later.

Recent qualitative research from the United States has demonstrated similar results among young adults who received gender-affirming care as adolescents (Finegan et al., 2025). According to the results of semistructured interviews and thematic analysis the primary effects of gender-affirming care on sexual functioning were (1) a decrease in erections, (2) a decrease in libido, and (3) an increase in sexual enjoyment and willingness to engage in sexual activities. Decreased erections were largely viewed as positive, having the effect of decreasing gender dysphoria and improving well-being. A decrease in libido was primarily observed among participants assigned male at birth, and this effect was seen as either positive or neutral.

It is recommended that researchers and clinicians decenter sexual functioning and instead center sexual pleasure as a more sex-positive focus. Researchers working with TGD young people are urged to deemphasize orgasm as the predominant means of sexual pleasure and understand that sexual well-being is not static, but a fluctuating state of being (Dickenson et al., 2023, pp. 174–175). Focusing less on genitalia as the sole focus of erotic pleasure is one step toward enhancing sexual satisfaction for TGD people (Fielding, 2021). Scholar Lucie Fielding introduces the idea of polymorphous perversity, or an embodied psychic disposition of being able to derive sensual, sexual, and erotic pleasure from just about any part of the body (Fielding, 2021). Fielding highlights the importance for TGD people to embrace a language that is comfortable for them when engaging with sexuality, such as names for certain body parts and articulating needs, desires, and sources of pleasure.

Sexual health and well-being

Clinical researchers have increasingly begun to understand the many ways that human sexuality intersects with physiology and the physical body, as well as the intersection between sexuality and overall health. Health is defined as a state of complete physical, mental, and social comfort and security, and not merely the absence of disease or sickness (WHO, 2006), and also includes the ability to adapt and manage physical, mental, and social challenges throughout life (Laprise, 2023). Whereas sexual health refers to a goal (i.e., people want to achieve sexual health), sexual well-being includes the activities and practices that people engage in so that they can eventually achieve that goal. In other words, sexual well-being refers to the process of living a lifestyle that is aimed at achieving complete or increased sexual health and happiness (Nelson and Gordon, 2020). Sexual well-being includes activities aimed at preventing illness, maintaining a healthy lifestyle, and making decisions that enable a healthy life.

Researchers investigating sexual health and well-being among TGD young people need to ensure that data collection methods accurately reflect their participants' experiences. TGD people may be less likely to speak about their experiences in qualitative interviews or focus groups due to the presence of dysphoria, and certain quantitative instruments may not incorporate the focus needed to accurately represent diverse experiences. Unfortunately, most available measurement tools are not gender inclusive, do not take into consideration potential gender diversity among partner(s), and only reflect a cisnormative understanding of sexual behaviors (Fornander et al., 2024). Well-designed research studies using developmentally appropriate measurement tools are needed to more accurately reflect the experiences of TGD youth

Differences in methodology across studies have resulted in mixed findings regarding the sexual health and well-being of TGD youth. For example, US data from the 2015 Healthy Kids Colorado Survey, which is a statewide survey of middle and high school students, indicated that TGD youth are more likely to engage in sexual experiences (Kattari et al., 2019) and experience a higher rate of both sexual victimization and dating conflict (Kattari et al., 2021) in comparison with their cisgender peers. In contrast a study of TGD adolescents seen in a specialty gender clinic in a similar region of the United States found that rates of dating and sexual behavior were significantly lower than national norms (Fornander et al., 2024). Another study of TGD youth seen in a gender clinic in the Netherlands similarly found that participants were overall less sexually experienced than their cisgender Dutch peers (Bungener et al., 2020). One reason for these differences may be that TGD youth seen within specialty gender clinics have more support and resources compared to the general population of TGD adolescents, who may experience more marginalization and minority stress.

The marginalization and abuse experienced by TGD adolescents and young adults not only create threats to their sexual well-being, but also increase health risks (Drabish and Theeke, 2022). Minority stress, which refers to the chronically high levels of stress faced by members of stigmatized minority groups such as LGBT people and people of color (Meyer, 2003), can impact sexual well-being for TGD young people. Minority stress may be caused by a number of factors, including lack of social support, lack of community, and lack of financial resources. Other sources of minority stress include interpersonal prejudice, discrimination, anticipated rejection, hiding or concealing an identity, and internalizing negative societal views (Matsuno et al., 2022; Meyer and Northridge, 2007; Riggle et al., 2008). Higher numbers of stress-level experiences, combined with the additive impact of the stress, can have a significant adverse impact on someone's lifetime (Riggle et al., 2008).

More research is needed to understand the relationship between minority stress and sexual health and well-being for TGD young people. The Gender Minority Stress and Resilience Scale (GMSR) (Testa et al., 2015) was developed for researchers to comprehensively assess resilience and minority stress, including gender-related discrimination, rejection, pride, and internalized transphobia. The GMSR was modified for use with adolescents (GMSR-A), and research found that minor adaptation of the existing adult measure resulted in high internal consistency and construct validity across nine subscales assessing domains of minority stress and resilience in a sample of TGD adolescents aged 12–17 years (Hidalgo et al., 2019, p. 278). Minority stress experiences are unfortunately widespread, with TGD people facing high rates of social and economic marginalization, as well as high rates of physical and sexual abuse (Thoma et al., 2021).

Research also indicates that TGD people are at high risk for HIV and other STIs compared to their cisgender peers (Stutterheim et al., 2021). According to a meta-analysis of 88 studies conducted between 2006 and 2017, laboratory-confirmed HIV prevalence was 14.1% for transgender women, 3.2% for transgender men, and 9.2% for transgender people overall (Becasen et al., 2019). By comparison, estimated HIV prevalence for all US adults during the same period was less than 0.5%. More recent data indicate that transgender people experience HIV infection at four times the national population level (Brown et al., 2024), with transgender women as well as Black and Latino/a/x/e TGD people at especially high risk for HIV and other STIs (Herman et al., 2024; Rivera et al., 2024). Transgender men also report high rates of unprotected sex (Badash et al., 2024; Maheux and Zhou, 2021; Van Gerwen et al., 2020) and are also at risk for STIs (Long et al., 2024), as well as unintended pregnancy if their partner(s) produce sperm (Silva et al., 2024).

Although most research on sexual health risks has been conducted in adult TGD populations (Agénor et al., 2021), qualitative research among TGD adolescents and young adults has shown similarly low rates of condom and contraception use (Warwick et al., 2022). Within this study, preventative and STI screening was endorsed by some participants, but several barriers were identified. In addition to time constraints and a lack of knowledge or access to healthcare resources a unique barrier for adolescents specifically was the fear of parents finding out about sexual activity.

Surveys focusing solely on experiences of victimization are important for understanding the TGD community's experiences with trauma and the implementation of prevention strategies. However, this singular focus is reductive (Burnes et al., 2026) and does not include an orientation toward sexual well-being as an expression of resilience. In addition, there is a paucity of research on how TGD young people explore essential components of sexual well-being, including giving and receiving consent, boundaries, and negotiating a continuum of intimacy with sexual and nonsexual intimate partners. More research using diverse methodologies that center on critical reflexivity and community-based participatory approaches is needed to better understand sexual health practices and risks for TGD adolescents.

Health literacy is another important factor in optimizing sexual health and well-being. In a large study of US high schools, sexual and gender minority students actually scored higher on knowledge of STIs and knowledge of sexual health rights compared to some cisgender students, and they scored similarly on health access self-efficacy and health communication (Marshall et al., 2025). However, students who identified as questioning their gender or sexuality scored significantly lower on all four scales. This may indicate that TGD adolescents who have a more established identity may also have access to more educational health resources than those who are still exploring their identity. TGD young people lack access to relevant sexual health information in part because sexual health education rarely addresses the bodies and identities of gender-diverse people. Sexual health education within school environments also varies widely, from a full course dedicated to human sexuality to a single lecture within a larger course or workshop focused on an unrelated topic (Burnes, 2017).

Qualitative research has found that sex education programs in schools include few LGBTQ+ themes and even less information relevant to TGD youth (Warwick et al., 2022). The results of this study showed that some adolescents are having conversations with their families, but that these conversations tend to be more superficial and less constructive than desired. Even in healthcare settings where adolescent sexual health should be a core topic of discussion, participants reported that discussions were highly limited, largely due to a lack of education among healthcare providers. TGD adolescents reported feeling more comfortable talking with specialty gender clinic providers and primary care providers with more past experience with TGD young people, although barriers nonetheless remained.

Sexual health education can be improved by including queer and gender-diverse perspectives and allowing the use of gender and sex continua to help learn about sexuality from a lens of pleasure, consent, and healthy communication (Burnes and Pascarella, 2025). Health educators can also empower TGD adolescents and young adults with the invitation to learn about their own sexual self-efficacy, connect to their bodies, explore their curiosities and boundaries, and evolve their gender identities and sexual orientations concurrently. Program evaluation research is needed to assess the benefits of sexuality education programs that intentionally include the sexual health and well-being of TGD individuals and to evaluate the effectiveness of these programs in helping learners engage in conversations about consent and personal sexuality-related values. These topics are understudied and underrepresented in sexual health education programs, especially those without a comprehensive sexual education framework (Dunleavy and Dodd, 2021).

Implications and recommendations for research

Sex-positivity, combined with foundational principles of social justice, can facilitate optimal patient care and clinical research for TGD young people. There are nonetheless limitations to the research available to date, which inform recommendations for future research and practice. These practical recommendations are offered to help other researchers to evaluate their own philosophies, research programs, and training to improve clinical research on sexual health and functioning and ultimately inform clinical care for TGD youth.

Recommendation 1: center sexuality-related values

In his landmark work, sexologist Dennis Dailey created a visual model to help individuals understand their sexual expression in terms of multiple interlocking circles that center sexuality-related values (Dailey, 1981). According to Dailey an exploration of values needs to be at the center of developing a healthy relationship to sex. Each of the circles in his model—*Sexual Health, Power and Sexualization*, *Sexual Identity*, *Sensuality*, and *Intimacy*—represents different domains of sexuality-related knowledge, attitudes, and skills. Together, five interlocking circles inform research questions and methodology related to the study of sexual health and functioning. Researchers can also use Dailey's work to ensure that their assessment tools accurately reflect different components of sexual attitudes, beliefs, values, and behaviors so as not to replicate reductionist data collection methods that focus solely on behavior for TGD youth.

By investigating the sexuality-based values of TGD young people, researchers can ensure that diversity of values—as well as beliefs, opinions, and behaviors—are reflected in their research programs. These values should also be incorporated into inclusive and validated measurement tools related to sexual health and functioning. Community-driven, participatory research designs will help to facilitate values-based research. A better understanding of the values of TGD youth will ensure that researchers know what constitutes sexual functioning, outside of cisnormative and heteronormative ideas of sexual activity, and it will inform sexual health education, ensuring appropriate content and instructional methods surrounding gender diversity, sexual pleasure, and consent.

Recommendation 2: consider unique sexual development trajectories

Future research should investigate the individual trajectories of sexual development experienced by TGD youth, including the personal desires, needs, and experiences that distinguish them from cisgender adolescents and TGD adults. By decentering cisnormative and heteronormative sexual practices, researcher can help to clarify the unique ways that TGD youth explore and develop sexual attraction and identity, create and sustain feelings of pleasure and well-being for themselves and their partner(s), and build intimacy and belonging within their communities. Future research in this area should include evaluation of sexuality education experiences, including how these experiences were helpful and/or harmful to sexual development and health. By acknowledging the unique sexual trajectory of individual TGD youth, researchers can help participants to value their unique sexuality (instead of internalizing shame based on unsaid societal and cultural "norms").

Recommendation 3: promote sexual health and well-being

Clinical research has historically focused solely on sexual behaviors and rates of STIs among TGD people, reinforcing the idea that these are the only aspects of sexual health and well-being that are valued. Health promotion and the use of routine preventative screening are an understudied area, even among TGD adults (Hoy-Ellis et al., 2022; Leone et al., 2023; Iwamoto et al., 2021; Schmidt and Rizzolo, 2017). Preventative screening is generally recommended for the body parts a patient has, regardless of that patient's gender identity, including breast, cervical, and prostate cancer screenings, in addition to sexual health and STI prevention.

More research is needed on health promotion and the benefits and barriers to preventative screening among TGD young people. Consideration for minority stress and resilience, using the GMSR (Testa et al., 2015) and GMSR-A (Hidalgo et al., 2019), for example, is a specific area of interest for clarifying predictors and barriers to sexual well-being among youth. By focusing on prevention, health, and resilience, researchers also communicate to TGD youth that their values and sexual well-being are important and valued by the medical community.

Recommendation 4: diversify research samples

Research on sexual health and functioning among TGD youth should take into consideration multiple dimensions of intersectional identity, both in research design and sampling. In addition to racial and cultural differences, researchers should intentionally recruit participants who reflect diverse sexual orientation identities, ability statuses, and geographic origins. Research would further benefit from representation of a diversity of relationship structures, including various forms of polyamory and relationship anarchy (Vaughan and Burnes, 2022).

National probability samples are one strategy for evaluating the wide range of experiences of TGD youth in the United States, including the empirical validation of measurement tools on sexual health and functioning. These samples should use random selection methods when possible to ensure diversity and accurate experiences across various research methodologies. For example, the use of data from the Transpop Study reflects diverse probability samples of transgender adults in the United States to provide evidence-based interventions and implications for policy research for adults (www.transpop.org; Krueger et al., 2020). There is an urgent need for an equivalent sample with TGD youth to ensure necessary and unbiased research with comparable youth samples. International research and cross-national collaborations will further improve the accuracy and utility of research outcomes.

Recommendation 5: reconstruct sexual safety

TGD youth are particularly at risk for sexual abuse (Messinger et al., 2022) and for engaging in commercial or survival sex work (Burnes and Dawson, 2023). When combined with experiences of harassment, victimization, and assaults, many TGD people do not feel a sense of safety in their own bodies (Fielding, 2021) or in exploring relationships. More research is needed on the experience of safety using a variety of methods, including trauma symptom inventories, qualitative work, and mixed methods approaches. This research will inform critical aspects of clinical care, including helping TGD youth identify what "safe enough" feels like (acknowledging that some patients may never feel 100% safe) and helping them to engage in "mini experiments" to question, gently expand, and possibly increase the boundaries of their safety over time.

Part of constructing sexual safety for TGD youth involves assisting them to build tools to cope with gender dysphoria and enhance sexual pleasure. Mixed-method research methods can be used to evaluate the links between gender dysphoria and sexual functioning, including sexual arousal and orgasm, from a trauma-informed lens. Research is needed to evaluate the effectiveness of various interventions, such as grounding and mindfulness skills, to decrease gender dysphoria during sexual experiences and to increase embodiment and pleasure (McLaughlin and Dawson, 2021). Finally, research is needed to clarify the benefits of gender-affirmation strategies and medical care on sexual experiences, intimacy, and relationships.

Recommendation 6: revise and empirically validate measurement tools

There are currently no available measurement tools that are empirically validated for evaluating sexual health and functioning in TGD youth. The above recommendations should also be incorporated into the development and validation of measurement tools, including centering sexuality-related values, considering unique sexual development trajectories, and encouraging sexual safety. Appropriate measurement should provide options for TGD young people to identify with multiple intersecting identities. Including a variety of sexual orientation identity labels, rather than expecting youth to conform to narrow, outdated labels, will result in more accurate outcomes. Tools that are specific to youth should consider developmentally appropriate knowledge levels and experiences (Fornander et al., 2024). Response options that reflect these developmental considerations and unique pathways of gender affirmation will further improve the quality of research outcomes.

Existing measures that are noninclusive of the spectrum of gender and sexual diversity, such as the FSFI and MSFI (Barcelos et al., 2024), should be revised to improve inclusivity, to center sexuality-related values, and to be developmentally appropriate before empirical validation. Other measures that were designed to be developmentally appropriate and gender- and sexuality-inclusive, such as the Sexual Behavior Questionnaire (Fornander et al., 2024), are in need of psychometric evaluation. The Sexual Behavior Questionnaire is a brief measure (four to seven items, depending on answers) of sexual activity specifically among TGD youth. It was created by medical and behavioral health experts based on their clinical experience and patient feedback (Fornander et al., 2024). In addition to quantitative measurement tools, qualitative tools that allow for the collection of more nuanced information will allow researchers the opportunity to build needed theory and to empirically validate constructs for further use with this population.

Conclusion

A research paradigm that centers sexuality-based values and focuses on sexual health and well-being for TGD youth will help to inform high-quality patient care. Future research should be grounded in sex-positivity and sexual liberation and employ multiple research methods, especially including community-based participatory methods (Burnes, 2022). By incorporating pleasure and values into research designs and interventions and moving away from outdated lenses that solely focus on identity labels and functioning, sexuality can shift to be a multifaceted experience that serves as a coping mechanism and source of joy for TGD youth.

References

Agénor, M., Murchison, G.R., Najarro, J., Grimshaw, A., Cottrill, A.A., Janiak, E., et al., 2021. Mapping the scientific literature on reproductive health among transgender and gender diverse people: a scoping review. Sexual and Reproductive Health Matters 29 (1). Available at https://doi.org/10.1080/26410397.2021.1886395.

Anzani, A., Rucco, D., Prunas, A., 2024. Exploring transgender and non-binary engagement with sex tech: a narrative review. Current Sexual Health Reports 16 (4), 237–250. Available at https://doi.org/10.1007/s11930-024-00395-0.

Badash, A., Grennan, D., Albrecht, J., 2024. Sexually transmitted diseases and HIV in transgender patients. Clinics in Dermatology 42 (2), 180–191. Available at https://doi.org/10.1016/j.clindermatol.2023.12.010.

Baker, W.B., 2018. Sexual and gender identities in transgender men: fluid and binary perspectives. Journal of Gay and Lesbian Mental Health 22 (3), 280–301. Available at https://doi.org/10.1080/19359705.2018.1458677.

Barbonetti, A., 2024. Hormones and sexuality: navigating the complex terrain of the interplay between endocrinology and sexual function. International Journal of Impotence Research 36 (4), 303–304. Available at https://doi.org/10.1038/s41443-024-00904-8.

Barcelos, T.M.R., de Oliveira Brito, L.G., Jorge, C.H., Franceschini, S.A., Troncon, J.K., Ferriani, R.A., et al., 2024. Evaluation of sexual function of transgender individuals. International Urogynecology Journal 35 (8), 1663–1671. Available at https://doi.org/10.1007/s00192-024-05857-6.

Barron, V., 2023. Amazing Ace, Awesome Aro: An Illustrated Exploration. Jessica Kingsley Publishers. Available at http://doi.org/10.5040/9781805017059.

Bauer, G.R., Hammond, R., 2015. Toward a broader conceptualization of trans women's sexual health. Canadian Journal of Human Sexuality 24 (1), 1–11. Available at https://doi.org/10.3138/cjhs.24.1-CO1.

Becasen, J.S., Denard, C.L., Mullins, M.M., Higa, D.H., Sipe, T.A., 2019. Estimating the prevalence of HIV and sexual behaviors among the US transgender population: a systematic review and meta-analysis, 2006–2017. American Journal of Public Health 109 (1). Available at https://doi.org/10.2105/AJPH.2018.304727.

Belous, C.K., Schulz, E.E., 2024. The sex positivity scale: a new way to measure sex positivity as a trait. Sexual and Relationship Therapy 39 (3), 944–962. Available at https://doi.org/10.1080/14681994.2022.2140136.

bell hooks, 2010. Teaching Critical Thinking Practical Wisdom. Routledge.

de Blok, C.J.M., Dreijerink, K.M.A., den Heijer, M., 2019. Cancer risk in transgender people. Endocrinology and Metabolism Clinics of North America 48 (2), 441–452. Available at https://doi.org/10.1016/j.ecl.2019.02.005.

Boskey, E.R., Ganor, O., 2022. Sexual orientation and attraction in a cohort of transmasculine adolescents and young adults. Transgender Health 7 (3), 270–275. Available at https://doi.org/10.1089/trgh.2020.0190.

Brown, E.E., Patel, E.U., Poteat, T.C., Mayer, K., Wawrzyniak, A.J., Radix, A.E., et al., 2024. Prevalence of sexually transmitted infections among transgender women with and without HIV in the Eastern and Southern United States. Journal of Infectious Diseases 229 (6), 1614–1627. Available at https://doi.org/10.1093/infdis/jiad605.

Bungener, S.L., De Vries, A.L.C., Popma, A., Steensma, T.D., 2020. Sexual experiences of young transgender persons during and after gender-affirmative treatment. Pediatrics 146 (6). Available at https://doi.org/10.1542/peds.2019-1411.

Burnes, T.R., 2017. Flying faster than the birds and the bees: Toward a sex-positive theory and practice in multicultural education. In: Gordon, R.K., Akutsu, T., McDermott, J.C., Lalas, J.W. (Eds.), Challenges Associated with Cross-Cultural and At-Risk Student Engagement. IGI Global Publishing, pp. 171–189.

Burnes, T.R., Singh, A.A., Witherspoon, R.G., 2017. Sex positivity and counseling psychology: an introduction to the major contribution. Counseling Psychologist 45 (4), 470–486. Available at https://doi.org/10.1177/0011000017710216.

Burnes, T.R., 2022. Practicalities and possibilities: PAR research in counseling with sex workers. Journal of Counseling Sexology & Sexual Wellness: Research, Practice, and Education 4 (1), 9–18. Available at https://doi.org/10.34296/04011056.

Burnes, T.R., Dawson, J.M., 2023. Essential clinical care for sex workers: a sex-positive handbook for mental health practitioners. North Atlantic Books.

Burnes, T.R., 2025. Engaging with human sexuality: diversity, well-being, and current perspectives. SAGE Publications.

Burnes, T.R., Pascarella III, J., 2025. Centering celebratory drag pedagogies in queer- and genderqueer-evasive K-12 educator preparation programs. International Journal of Qualitative Studies in Education 1–18. Available at https://doi.org/10.1080/09518398.2025.2572459.

Burnes, T.R., Estrellado, J.E., Singh, A., 2026. Chapter 12. Gender-affirming assessment and treatment of trauma with trans and nonbinary clients. In: Singh, A.A., McCullough, R. (Eds.), Affirmative Counseling and Psychological Practice With Transgender and Nonbinary Clients. American Psychological Association, Washington, DC, pp. 293–316. Available at https://doi.org/10.1037/0000471-013.

Burns, J.A., Beischel, W.J., van Anders, S.M., 2022. Hormone Therapy and Trans Sexuality: A Review. Psychology of Sexual Orientation and Gender Diversity 11 (1), 17–30. Available at https://doi.org/10.1037/sgd0000588.

Chen, A., 2020. Ace: What asexuality reveals about desire, society, and the meaning of sex. Massachusetts Beacon Press, pp. 210pgs.

Dailey, D., 1981. Sexual expression and aging. In: Berghorn, F., Schafer, D. (Eds.), The dynamics of aging. Westview Press, Boulder, CO, pp. 311–333.

Dawson, J., Burnes, T., 2018. Sexuality and clinical supervision. In: Burnes, T., Manese, J. (Eds.), Cases in Multicultural Supervision: New Lenses, Models, and Applications. Cognella, pp. 325–337.

WHO, 2006. Defining Sexual Health: Report of a Technical Consultation on Sexual Health. WHO. Available at http://www.who.int/reproductivehealth/topics/gender_rights/defining_sexual_health.pdf?ua=1.

Defreyne, J., Elaut, E., Den Heijer, M., Kreukels, B., Fisher, A.D., T'Sjoen, G., 2021. Sexual orientation in transgender individuals: results from the longitudinal ENIGI study. International Journal of Impotence Research 33 (7), 694–702. Available at https://doi.org/10.1038/s41443-020-00402-7.

Dickenson, J.A., Tebbe, E., Tellawi, G., 2023. A sexual wellbeing framework to address sexuality in therapy with transgender, nonbinary, and gender expansive clients. Routledge, United States Women and Therapy 46 (2), 171–198. Available at https://doi.org/10.1080/02703149.2023.2229684.

Drabish, K., Theeke, L.A., 2022. Health impact of stigma, discrimination, prejudice, and bias experienced by transgender people: a systematic review of quantitative studies. Issues in Mental Health Nursing 43 (2), 111–118. Available at https://doi.org/10.1080/01612840.2021.1961330.

Dunleavy, M., Dodd, S.J., 2021. The affirmative potential of sex-positive sex education with trans and gender expansive youth. Social Work Practice with Transgender and Gender Expansive Youth. Taylor and Francis, United States, pp. 83–97. Available at http://doi.org/10.4324/9780429297687-7.

Fielding, L., 2021. Trans Sex: Clinical Approaches to Trans Sexualities and Erotic Embodiments. Routledge, New York, pp. 1–250. Available at https://doi.org/10.4324/9780429318290.

Finegan, J.L., Marinkovic, M., Okamuro, K., Newfield, R.S., Anger, J.T., 2025. Experience with gender affirming hormones and puberty blockers (gonadotropin releasing hormone agonist): a qualitative analysis of sexual function. Journal of Sexual Medicine 22 (5), 945–950. Available at https://doi.org/10.1093/jsxmed/qdaf061.

Fisher, W.A., White, L.A., Byrne, D., Kelley, K., 1988. Erotophobia-erotophilia as a dimension of personality. Journal of Sex Research 25 (1), 123–151. Available at https://doi.org/10.1080/00224498809551448.

Fornander, M.J., Egan, A.M., Robertson, G.C., Moser, C.N., 2024. Self-reported sexual behavior of transgender youth. Journal of Pediatric and Adolescent Gynecology 37 (3), 336–341. Available at https://doi.org/10.1016/j.jpag.2023.12.003.

Galupo, M.P., Henise, S.B., Mercer, N.L., 2016. The labels don't work very well": transgender individuals' conceptualizations of sexual orientation and sexual identity. International Journal of Transgenderism 17 (2), 93–104. Available at https://doi.org/10.1080/15532739.2016.1189373.

Van Gerwen, O.T., Jani, A., Long, D.M., Austin, E.L., Musgrove, K., Muzny, C.A., 2020. Prevalence of sexually transmitted infections and human immunodeficiency virus in transgender persons: a systematic review. Transgender Health 5 (2), 90–103. Available at https://doi.org/10.1089/trgh.2019.0053.

Gerymski, R., 2021. Short sexual well-being scale – a cross-sectional validation among transgender and cisgender people. Health Psychology Report 9 (3), 276–287. Available at https://doi.org/10.5114/hpr.2021.102349.

Haley, S.G., Tordoff, D.M., Kantor, A.Z., Crouch, J.M., Ahrens, K.R., 2019. Sex education for transgender and non-binary youth: previous experiences and recommended content. Journal of Sexual Medicine 16 (11), 1834–1848. Available at https://doi.org/10.1016/j.jsxm.2019.08.009.

Hendrick, S., Hendrick, C., 1987. Multidimensionality of sexual attitudes. Journal of Sex Research 23 (4), 502–526. Available at https://doi.org/10.1080/00224498709551387.

Hereth, J., Pardee, D.J., Reisner, S.L., 2020. Gender identity and sexual orientation development among young adult transgender men sexually active with cisgender men: 'I had completely ignored my sexuality … that's for a different time to figure out. Culture, Health & Sexuality 22 (sup1), 31–47. Available at https://doi.org/10.1080/13691058.2019.1636290.

Herman, J.L., Salcedo, B., Chatham, K., Mahowald, L., Ortega, Q.V., Redfield, E., 2024. Para Mi Punto de Vista/From My Point of View: Results of the 2023 LA County Trans & Nonbinary Survey. The Williams Institute.

Hidalgo, M.A., Petras, H., Chen, D., Chodzen, G., 2019. The gender minority stress and resilience measure: psychometric validity of an adolescent extension. Clinical Practice in Pediatric Psychology 7 (3), 278–290. Available at https://doi.org/10.1037/cpp0000297.

Holmberg, M., Arver, S., Dhejne, C., 2019. Supporting sexuality and improving sexual function in transgender persons. Nature Reviews Urology 16 (2), 121–139. Available at https://doi.org/10.1038/s41585-018-0108-8.

Howard, S.D., Lee, K.L., Nathan, A.G., Wenger, H.C., Chin, M.H., Cook, S.C., 2019. Healthcare experiences of transgender people of color. Journal of General Internal Medicine 34 (10), 2068–2074. Available at https://doi.org/10.1007/s11606-019-05179-0.

Hoy-Ellis, C.P., Fredriksen-Goldsen, K.I., Kim, H.J., 2022. Utilization of recommended preventive health screenings between transgender and cisgender older adults in sexual and gender minority communities. Journal of Aging and Health. Available at https://doi.org/10.1177/08982643211068557.

Ivanski, C., Kohut, T., 2017. Exploring definitions of sex positivity through thematic analysis. The Canadian Journal of Human Sexuality 26 (3), 216–225. Available at https://doi.org/10.3138/cjhs.2017-0017.

Iwamoto, S.J., Grimstad, F., Irwig, M.S., Rothman, M.S., 2021. Routine screening for transgender and gender diverse adults taking gender-affirming hormone therapy: a narrative review. Journal of General Internal Medicine 36 (5), 1380–1389. Available at https://doi.org/10.1007/s11606-021-06634-7.

Kattari, S.K., Walls, N.E., Atteberry-Ash, B., Klemmer, C., Rusow, J.A., Kattari, L., 2019. Missing from the conversation: sexual risk factors across young people by gender identity and sexual orientation. International Journal of Sexual Health 31 (4), 394–406. Available at https://doi.org/10.1080/19317611.2019.1667938.

Kattari, S.K., Atteberry-Ash, B., Collins, C., Kattari, L., Harner, V., 2021. Increased prevalence, predictors, and in-group differences of forced sex and physical dating violence among trans/gender diverse youth. Social Sciences 10 (6), 236. Available at https://doi.org/10.3390/socsci10060236.

Kcomt, L., 2019. Profound health-care discrimination experienced by transgender people: rapid systematic review. Social Work in Health Care 58 (2), 201–219. Available at https://doi.org/10.1080/00981389.2018.1532941.

Kitic, A., Pozzo, V., Kitic, N., Atlan, M., Rausky, J., 2025. Impact of vaginoplasty on sexual health and satisfaction in transgender women. The Journal of Sexual Medicine 22 (3), 536–542. Available at https://doi.org/10.1093/jsxmed/qdae194.

Krueger, E.A., Divsalar, S., Luhur, W., Choi, S.K., Meyer, I.H., 2020. TransPop – U.S. Transgender Population Health Survey (Methodology and Technical Notes). The Williams Institute, Los Angeles, CA.

Kuzma, E.K., Ammerman, B.A., Lee, V., Baird, J., 2024. Re-envisioning youth sexual health care: supporting sex positivity in a digital world. Journal of Pediatric Health Care 38 (3), 310–322. Available at https://doi.org/10.1016/j.pedhc.2023.11.009.

Laprise, C., 2023. It's time to take a sustainable approach to health care in the face of the challenges of the 21st century. One Health 16, 100510. Available at https://doi.org/10.1016/j.onehlt.2023.100510.

Leone, A.G., Trapani, D., Schabath, M.B., Safer, J.D., Scout, N.F.N., Lambertini, M., et al., 2023. Cancer in transgender and gender-diverse persons: a review. JAMA Oncology 9 (4), 556–563. Available at https://doi.org/10.1001/jamaoncol.2022.7173.

Light, A., Wang, L.F., Zeymo, A., Gomez-Lobo, V., 2018. Family planning and contraception use in transgender men. Contraception 98 (4), 266–269. Available at https://doi.org/10.1016/j.contraception.2018.06.006.

Long, A., Vance, D., Fazeli, P., 2024. Sexual risk behaviors among the transgender and gender diverse population: a concept analysis. Nursing: Research and Reviews 14, 117–136. Available at https://doi.org/10.2147/nrr.s463639.

Lorde, A., 1984. Uses of the Erotic: The Erotic as Power. Sister Outsider. pp. 87–91.

Maheux, A.J., Zhou, Y., Thoma, B.C., Salk, R.H., Choukas-Bradley, S., 2021. Examining sexual behavior among U.S. transgender adolescents. Journal of Sex Research 58 (8), 1050–1060. Available at https://doi.org/10.1080/00224499.2020.1855408.

Marinelli, L., Cagnina, S., Bichiri, A., Magistri, D., Crespi, C., Motta, G., 2024. Sexual function of transgender assigned female at birth seeking gender affirming care: a narrative review. International Journal of Impotence Research 36 (4), 375–379. Available at https://doi.org/10.1038/s41443-023-00711-7.

Marshall, A., Mackson, G., Kaighobadi, F., Hoffman, N.D., Ciarleglio, A., Sandfort, T.G.M., 2025. Health literacy among sexual and gender-diverse adolescents in New York city. Health Education and Behavior 52 (3), 289–298. Available at https://doi.org/10.1177/10901981241307616.

Matsuno, E., Bricker, N.L., Collazo, E.N., Mohr, R., Balsam, K.F., 2022. The default is just going to be getting misgendered": minority stress experiences among nonbinary adults. Psychology of Sexual Orientation and Gender Diversity 11 (2), 202–214. Available at https://doi.org/10.1037/sgd0000607.

Mattawanon, N., Charoenkwan, K., Tangpricha, V., 2021. Sexual dysfunction in transgender people: a systematic review. Urologic Clinics of North America 48 (4), 437–460. Available at https://doi.org/10.1016/j.ucl.2021.06.004.

McLaughlin, A., Dawson, J.M., 2021. With Pleasure: Managing Trauma Triggers for More Vibrant Sex and Relationships. Chicago Review Press.

Messinger, A.M., Guadalupe-Diaz, X.L., Kurdyla, V., 2022. Transgender polyvictimization in the US transgender survey. Journal of Interpersonal Violence 37 (19-20), NP18810.

Van Der Meulen, I.S., Arnoldussen, M., Van Der Miesen, A.I.R., Hannema, S.E., Steensma, T.D., De Vries, A.L.C., et al., 2025a. Sexual satisfaction and dysfunction in transgender adults following puberty suppression treatment during adolescence. Journal of Sexual Medicine 22 (8), 1493–1503. Available at https://doi.org/10.1093/jsxmed/qdaf095.

Van Der Meulen, I.S., Bungener, S.L., Van Der Miesen, A.I.R., Hannema, S.E., Kreukels, B.P.C., Steensma, T.D., et al., 2025b. Timing of puberty suppression in transgender adolescents and sexual functioning after vaginoplasty. Journal of Sexual Medicine 22 (1), 196–204. Available at https://doi.org/10.1093/jsxmed/qdae152.

Meyer, I.H., 2003. Prejudice, social stress, and mental health in lesbian, gay, and bisexual populations: conceptual issues and research evidence. Psychological Bulletin 129 (5), 674–697. Available at https://doi.org/10.1037/0033-2909.129.5.674.

Meyer, I.H., Northridge, M.E., 2007. The Health of Sexual Minorities: Public Health Perspectives on Lesbian, Gay, Bisexual and Transgender Populations. Springer, New York, NY. Available at https://doi.org/10.1007/978-0-387-31334-4.

Nagoski, E., 2024. Come Together: The Science (and Art!) of Creating Lasting Sexual Connections. New York Times.

Nelson, K.M., Gordon, A.R., John, S.A., Stout, C.D., Macapagal, K., 2020. Physical sex is over for now": impact of COVID-19 on the well-being and sexual health of adolescent sexual minority males in the U.S. Journal of Adolescent Health 67 (6), 756–762. Available at https://doi.org/10.1016/j.jadohealth.2020.08.027.

Obedin-Maliver, J., 2016. Pelvic pain and persistent menses in transgender men. In: Deutsch, M.B. (Ed.), Guidelines for the Primary and Gender-Affirming Care of Transgender and Gender Nonbinary People. UCSF Gender Affirming Health Program, San Francisco, CA. Available at https://transcare.ucsf.edu/guidelines/pain-transmen.

Rachlin, K., Green, J., Lombardi, E., 2008. Utilization of health care among female-to-male transgender individuals in the United States. Journal of Homosexuality 54 (3), 243–258. Available at https://doi.org/10.1080/00918360801982124.

Reay, B., 2014. The Transsexual Phenomenon: A Counter-history. Journal of Social History 47 (4), 1042–1070. Available at https://doi.org/10.1093/jsh/shu018.

Reisner, S.L., Pletta, D.R., Potter, J., Deutsch, M.B., 2020. Initial psychometric evaluation of a brief sexual functioning screening tool for transmasculine adults: transmasculine sexual functioning index. Sexual Medicine 8 (3), 350–360. Available at https://doi.org/10.1016/j.esxm.2020.05.006.

Reisner, S.L., Choi, S.K., Herman, J.L., Bockting, W., Krueger, E.A., Meyer, I.H., 2023. Sexual orientation in transgender adults in the United States. BMC Public Health 23 (1). Available at https://doi.org/10.1186/s12889-023-16654-z.

Riggle, E.D.B., Whitman, J.S., Olson, A., Rostosky, S.S., Strong, S., 2008. The positive aspects of being a lesbian or gay man. Professional Psychology: Research and Practice 39 (2), 210–217. Available at https://doi.org/10.1037/0735-7028.39.2.210.

Rivera, A.V., Lopez, J.M., Braunstein, S.L., 2024. Recruiting black and latina/hispanic transgender women for HIV research: implementation of respondent-driven sampling and factors associated with peer recruitment. Transgender Health 9 (4), 339–347. Available at https://doi.org/10.1089/trgh.2021.0201.

Sbragia, J.D., Vottero, B., 2020. Experiences of transgender men in seeking gynecological and reproductive health care: a qualitative systematic review. JBI Evidence Synthesis 18 (9), 1870–1931. Available at https://doi.org/10.11124/JBISRIR-D-19-00347.

Schardein, J.N., Nikolavsky, D., 2022. Sexual Functioning of transgender females post-vaginoplasty: evaluation, outcomes and treatment strategies for sexual dysfunction. Sexual Medicine Reviews 10 (1), 77–90. Available at https://doi.org/10.1016/j.sxmr.2021.04.001.

Schmidt, E., Rizzolo, D., 2017. Disease screening and prevention for transgender and gender-diverse adults. JAAPA: Official Journal of the American Academy of Physician Assistants 30 (10), 11–16. Available at https://doi.org/10.1097/01.jaa.0000524709.87224.57.

Silva, G.C., Puccia, M.I.R., Barros, M.N.dos S., 2024. Transsexual men and pregnancy: an integrative literature review. Ciência & Saúde Coletiva 29 (4). Available at https://doi.org/10.1590/1413-81232024294.19612023en.

Skrzypczak, E., Grunt-Mejer, K., Ziółkowska, J., 2025. The meaning of sexual orientation in the lives of transgender people. Systematic review of qualitative studies. International Journal of Transgender Health 26 (3), 544–563. Available at https://doi.org/10.1080/26895269.2024.2376215.

Stutterheim, S.E., Van Dijk, M., Wang, H., Jonas, K.J., 2021. The worldwide burden of HIV in transgender individuals: an updated systematic review and meta-analysis. PLoS One 16 (12). Available at https://doi.org/10.1371/journal.pone.0260063.

Syme, M.L., Mona, L.R., Cameron, R.P., 2013. Sexual health and well-being after cancer: applying the sexual health model. Counseling Psychologist 41 (2), 268–285. Available at https://doi.org/10.1177/0011000012459970.

Szoko, N., Sequeira, G.M., Coulter, R.W.S., Kobey, J., Ridenour, E., Burnett, O., et al., 2023. Sexual orientation among gender diverse youth. Journal of Adolescent Health 72 (1), 153–155. Available at https://doi.org/10.1016/j.jadohealth.2022.08.016.

Testa, R.J., Habarth, J., Peta, J., Balsam, K., Bockting, W., 2015. Development of the gender minority stress and resilience measure. Psychology of Sexual Orientation and Gender Diversity 2 (1), 65–77. Available at https://doi.org/10.1037/sgd0000081.

Thoma, B.C., Rezeppa, T.L., Choukas-Bradley, S., Salk, R.H., Marshal, M.P., 2021. Disparities in childhood abuse between transgender and cisgender adolescents. APediatrics 148 (2). Available at https://doi.org/10.1542/peds.2020-016907.

Vaughan, M.D., Burnes, T.R., 2022. The Handbook of Consensual Non-monogamy Affirming Mental Health Practice. Rowman & Littlefield, Lanham, MD.

Verrastro, V., Saladino, V., Petruccelli, F., Eleuteri, S., 2020. Medical and health care professionals' sexuality education: state of the art and recommendations. International Journal of Environmental Research and Public Health 17 (7), 2186. Available at https://doi.org/10.3390/ijerph17072186.

Warwick, R.M., Araya, A.C., Shumer, D.E., Selkie, E.M., 2022. Transgender youths' sexual health and education: a qualitative analysis. Journal of Pediatric and Adolescent Gynecology 35 (2), 138–146. Available at https://doi.org/10.1016/j.jpag.2021.09.011.

White Hughto, J.M., Clark, K.A., Altice, F.L., Reisner, S.L., Kershaw, T.S., Pachankis, J.E., 2018. Creating, reinforcing, and resisting the gender binary: a qualitative study of transgender women's healthcare experiences in sex-segregated jails and prisons. International Journal of Prisoner Health 14 (2), 69–88. Available at https://doi.org/10.1108/IJPH-02-2017-0011.

Wilder, S., Shannon, B., Blasdel, G., Shakir, N., 2023. Sexual function outcomes following gender-affirming vaginoplasty: a literature review. Current Sexual Health Reports 15 (4), 301–306. Available at https://doi.org/10.1007/s11930-023-00377-8.

Chapter 13

Long-term satisfaction and risk of regret

Josh Tiller-Ormord
Randall Children's Hospital, Portland, OR, United States

Learning objectives

- Summarize the literature regarding long-term outcomes of pediatric gender-affirming healthcare, including psychosocial, satisfaction, and identity-related outcomes.
- Discuss internal and systemic biases that affect long-term outcome research on pediatric gender-affirming care (GAC).
- Identify and distinguish related long-term outcomes, such as discontinuation, detransition, reidentification, and regret.
- Describe current gaps in the literature for understanding the long-term outcomes of pediatric GAC and make recommendations for future research.

Key terms and definitions

- **Transition** refers to the process by which someone takes steps to change things about their life to be more aligned with their gender identity. This can include different types of transitions, including legal, social, and medical transitions.
- **Detransition** is broadly defined as the process of stopping or reversing pieces of transition after learning more throughout or after an initial transition process, often with the implication of reidentifying with the sex assigned at birth. However, stopping or reversing steps in transition is not always indicative of detransition. This term will specifically be used to describe research that uses this broadly defined term and to refer to individuals who personally identify as detransitioning. When possible, other, more precise terms will be used.
- **Retransition** refers to the process of transitioning either after a transition has been paused or to align with a different gender identity than the previous transition. Retransition does not necessarily include stopping or reversing any pieces of a previous transition but is typically associated with further gender identity development.
- **Reidentification** refers to a shift in how someone identifies after or during a transition, often used to describe identifying with the sex that was assumed at birth after a period of identifying as transgender or nonbinary. Frequently used terms in the research include "reidentification with sex assigned at birth" and "reidentification with birth-registered sex."

Understanding long-term outcomes, identifying benefits, and mitigating risk are important components of any healthcare intervention, and this is no less relevant to researchers and clinicians who work with transgender and gender diverse (TGD) youth. Unfortunately, politicization of gender-affirming care (GAC) in parts of the United States and Europe has negatively influenced the dialogue about long-term outcomes, interfering with ethical research and patient-centered care. At the time of this writing, at least 21 states in the United States have enacted a ban that prevents youth from accessing gender-affirming medical care, and another three states have enacted a ban that is currently unenforceable by court order (Redfield et al., 2024). Lawmakers passing these bans use misinformation and misinterpretation of scientific data to inaccurately claim that GAC is unsafe, experimental, and unnecessary, suggesting that young people will "grow out of" or "desist" from their diverse gender identity (McNamara et al., 2022). Excessive bans are rooted in political and religious motivation rather than an understanding of the literature and care for the TGD community (Preston and Lopez, 2024), but many researchers are nonetheless hesitant to produce research that could be misused. Research regarding

Advances in Research on Pediatric Gender Affirming Healthcare. DOI: https://doi.org/10.1016/B978-0-443-29019-0.00002-6
© 2026 Elsevier Inc. All rights are reserved, including those for text and data mining, AI training, and similar technologies.

reidentification with sex assigned at birth and posttransition regret is important to the field, but carries strong potential for misuse. Recommendations for ethical and informative research that minimizes the risk of misuse are therefore urgently needed.

Measuring patient-reported outcomes

Existing literature regarding long-term patient-reported outcomes of GAC has long included research examining the positive impacts of care, while more recent research has examined outcomes considered “negative” or undesirable. Among positive outcomes, mental and physical health outcomes are well-established, with GAC resulting in decreased depression, anxiety, gender dysphoria, and suicidality, particularly with pubertal suppression and gender-affirming hormone therapy (Mahfouda et al., 2019; Nguyen et al., 2018). More recent attempts to capture negative long-term outcomes have at times conflated multiple related but distinct concepts. For example, there is a growing body of research evaluating rates of discontinuation of GAC, often operationalized as ending or no longer picking up prescribed hormone therapy (Roberts et al., 2022; van der Loos et al., 2022). A potential danger in this measurement is conflating discontinuation of treatment with regret, detransitioning, or reidentifying with sex assigned at birth (Ashley, 2022; MacKinnon et al., 2023). Discontinuation of medical treatment should be viewed as a distinct outcome that is sometimes related to reaching transition goals, evolving gender identity, other health concerns, excessive cost, or other reasons (Crabtree et al., 2024; MacKinnon et al., 2024, 2025; Turban et al., 2018).

The process of detransitioning or retransitioning also has the potential to be conflated with regret, and research varies in the terms used, as well as how they are defined across studies. Definitions of detransition typically involve reversing components of gender transition and sometimes include discontinuing medical treatment specifically (MacKinnon et al., 2023; Turban et al., 2021; Vandenbussche, 2022). Often, but not always, detransition is used to describe someone who reidentifies with their sex assigned at birth after pursuing some form of gender affirmation. Retransition is a similar concept used to acknowledge that transition is not necessarily binary or linear, and it is sometimes used interchangeably with detransition (Gupta et al., 2023; Olson et al., 2022). The term retransition is less well-defined and is sometimes used to describe those who pause and reinitiate transition (MacKinnon et al., 2023).

Research outcomes can also be impacted by evolving terminology that remains subject to debate and reinterpretation. Terms are sometimes applied differently across cultures, geographic locations, and generational periods. For example, some experts advise against the use of the terms transition and detransition as they imply a linear process toward gender affirmation and an end goal that TGD people should strive for (National LGBTQIA+ Health Education Center, 2024; Turban et al., 2021). These terms are also often misused and misunderstood in the media (Turban et al., 2022b). Others argue detransition should be used in research due to its common use within the TGD community, and instead recommend that researchers work to destigmatize the concept of detransition (MacKinnon et al., 2023).

Retransition is defined here as a distinct outcome to represent individuals who engage in secondary transition. A secondary transition may occur either after a transition has been paused or because gender identity has evolved and changes or goals from a previous GAC no longer align with the current identity and expression. One example of this type of retransition is someone who previously sought GAC to align with a binary transgender identity and is now expressing a nonbinary identity with varying desired outcomes. Detransition is defined here as the process of discontinuing or reversing parts of transition to specifically realign with sex assigned at birth, with the acknowledgment that not everyone with this history will necessarily identify with this term or find it to be an appropriate reflection of their experience. The term detransition will also be used when it is the specific term used in research and in reference to individuals who identify with this term.

The final type of long-term outcome that will be summarized in this chapter is individual feelings toward transition and GAC. Investigators have historically examined feelings of satisfaction and regret regarding GAC, with recent research emerging on the mixed emotions in between (Crabtree et al., 2024; Narayan et al., 2021; Nieder et al., 2021; Pullen Sansfaçon et al., 2023). Regret is operationalized as a distinct outcome apart from reidentification, detransition, or discontinuation of care. TGD individuals may experience changes in gender identity throughout the transition process and often feel grateful for their care even in the context of discontinuation or detransition (Pullen Sansfaçon et al., 2023; Turban et al., 2018), or may feel regret about aspects of the transition while feeling grateful overall for the process and having no change in gender identity (Nieder et al., 2021; Vandenbussche, 2022).

Psychosocial and mental health outcomes

The measurement of psychosocial outcomes related to gender-affirming healthcare often involves following youth who receive GAC over time and tracking changes in mental health outcomes. The mental health benefits of specific gender-affirming interventions are outside of the scope of this chapter but will be summarized broadly, specifically highlighting overall research trends and limitations. Overall research offers strong evidence for the association between pediatric GAC and reductions in anxiety, depression, social distress, and gender dysphoria, as well as improvements in self-esteem and quality of life, among others (Mahfouda et al., 2019; Nguyen et al., 2018). This evidence is based on decades of high-quality cross-sectional and longitudinal research methods, with more recent research attempting to fill remaining research gaps by implementing randomization and exploring moderators and potential mechanisms of change.

While cross-sectional studies cannot establish causation or demonstrate changes in behavior over time, the overall body of cross-sectional research on pediatric GAC is strengthened by a diversity of study methods, targets of comparison, and sample characteristics. For example, one study evaluated psychosocial outcomes among four groups of German adolescents and young adults, and compared these outcomes to population norms (Becker-Hebly et al., 2021). The four groups were based on the services received (i.e., no medical treatment, puberty suppression only, gender-affirming hormones, or gender-affirming surgery), and the latter two groups were found to have similar psychosocial outcomes to the normed population. Other studies have compared similar groups to one another rather than population norms, with one study finding that Brazilian youth who had accessed more transition steps had lower anxiety and depression than youth who had accessed fewer steps (Fontanari et al., 2020). To establish long-term differences across groups, another study compared groups of adults based on recalled age of initiation of GAC and found that those with earlier access were the least likely to have experienced significant mental health challenges (Turban et al., 2022a). One notable limitation across these studies is that youth who are unable to access gender affirmation may be limited by other social variables, such as unsupportive parents, that may have also contributed to poorer mental health outcomes.

Early longitudinal research conducted in the Netherlands demonstrated decreased gender dysphoria and improved mental health and well-being over an average of 7 years with the receipt of multiple gender-affirming interventions (De Vries et al., 2014). Improvements in access to pediatric GAC have allowed researchers in other countries to implement similarly rigorous longitudinal methods, with increasingly long-term outcomes. At present, many longitudinal studies have been able to produce outcomes ranging from 6 months to 2 years (Achille et al., 2020; Becker-Hebly et al., 2021; Chelliah et al., 2024; Gadomski et al., 2023; Kuper et al., 2020; Tordoff et al., 2022), with the most recent data from the United States reaching up to 4 years following initiation of treatment (Chen et al., 2024). Longitudinal research has also facilitated the identification of mechanisms of change for improved mental health outcomes, most notably improved gender appearance congruence (Chen et al., 2023; Olson-Kennedy et al., 2025) and body satisfaction (Chelliah et al., 2024). Psychosocial factors that facilitate and hinder improvements over time can also be identified with this methodology. For example, parental support appears to be a moderating variable, where youth with lower levels of parental support demonstrate less improvement over the course of treatment (Chen et al., 2024).

Randomized control trial (RCT) studies on psychosocial outcomes are not well-suited for evaluating pediatric GAC due to the risks of withholding treatments with known benefits, high likelihood of attrition, and inability to blind condition arms, among others (Ashley et al., 2024). Researchers have attempted to overcome some of these limitations among adult TGD patients by randomizing patients into an accelerated pathway to care and comparing results to waitlist controls (Nolan et al., 2023). However, this would not be appropriate for transgender youth for whom multidisciplinary care and individualized timing are a core component of high-quality care (Coleman et al., 2022). Some researchers have used waitlist comparison groups among youth to most closely approximate RCT methodology while avoiding the inherent ethical and methodological limitations of the randomization process (Grannis et al., 2021). While cross-sectional in nature, this type of methodology minimizes biases introduced by other social factors, such as parent support, given that both groups have similar access to services.

Discontinuation or reversal of medical treatment

Research into the number of TGD youth who discontinue or reverse medical interventions represents an area of increasing interest. Within pediatric research, rates of discontinuation of gender-affirming hormone therapy range from 0.8% to 25.6% (Gupta et al., 2023; MacKinnon et al., 2024, 2025; Roberts et al., 2022; van der Loos et al., 2022). Researchers have also attempted to elucidate reasons for discontinuation of hormone use, but one limitation of this research has been a high risk of loss to follow-up among this population (Vandenbussche, 2022).

Research on the reversal of GAC (either reversal surgery or other interventions aimed specifically at reversing a medical change) in individuals receiving treatments during adolescence remains limited. One study on requests for reversal surgery reported an age range of 16–37 years at initial surgical consult, but did not specify whether surgeries occurred during adolescence for any participants (Jedrzejewski et al., 2023). Another study, specifically among individuals who had gender-affirming mastectomy as adolescents, found that while 2 of the 209 surgical patients expressed later regret, none underwent reversal surgery (Tang et al., 2022). The dearth of literature on reversal may be partially due to surgery being less common among TGD adolescents compared to adults, but more research is nonetheless warranted to understand surgical reversal as well as other interventions to reverse aspects of medical transition in this population.

One of the first studies to formally evaluate discontinuation of hormone therapy among youth used a retrospective chart review to identify patients within the US Veteran Affairs medical system with (1) two or more medical appointments with a TGD-related diagnostic code and (2) medical and pharmacy records showing at least two previous prescriptions for hormones (Roberts et al., 2022). The study only included patients who were spouses or children of veterans to minimize military regulation-related bias, and patients no longer receiving care in the system were excluded. Patients who had not obtained another prescription for hormones within 90 days of the most recent relevant encounter were categorized as having discontinued hormones. The results of the study showed 70.2% of the sample continued to receive hormones 4 years after initiation of hormone therapy. Of the 952 patients identified in the study, 39.1% were younger than 18 at the start of hormone therapy. For those who were below 18 when starting hormones, the continuation rate was 74.4%.

This study was able to evaluate a large sample of TGD adolescents and adults by using medical record information as a proxy for discontinuation of medical interventions among a population where all or most patients' medical and pharmacy records could be accessed by the researchers (Roberts et al., 2022). However, the continuation rate appeared lower than would be expected based on previous studies with adults (Hall et al., 2021). The study did not gather information directly from patients, limiting the conclusions that can be drawn about those categorized as having discontinued hormones. Furthermore, this study likely reflects broad care practices within a major medical organization, limiting generalizability based on specific clinic practices or standards of care (Coleman et al., 2022).

Another study using a similar approach reviewed the medical records of all patients seen within a major pediatric gender clinic in the Netherlands (van der Loos et al., 2022). The population in this study represented youth who accessed care according to a very specific clinic model, sometimes referred to as the "Dutch Protocol," where all patients starting hormone therapy first undergo a thorough diagnostic evaluation and begin a gonadotropin-releasing hormone agonist (GnRHa) to suppress the effects of endogenous estrogen or testosterone. All patients in this analysis received GnRHa, starting before age 18, for at least 3 months before beginning hormones. Of the 715 patients in their study, only 16 (2%) no longer had an active prescription for hormones in the database. The median duration of time since beginning hormones was 2.3 years for people assigned female at birth (AFAB) and 3.5 years for people assigned male at birth (AMAB). The study did not assess reasons for continuing or discontinuing treatment, but further data analysis showed no significant differences in discontinuation based on age or pubertal stage at the beginning of treatment.

While retrospective chart review was used for both studies, the study from the Netherlands (van der Loos et al., 2022) found a much lower rate of discontinuation at 2% than in the US-based study, at 25.6% (Roberts et al., 2022). This could be related to social or medical system differences between the Netherlands and the United States, such as the cost of healthcare. Differences could also be related to the evaluation of a specific clinic model, such as the Dutch Protocol, versus a broad medical system that may reflect differences in practice. Furthermore, one study evaluated a medical system associated with the military/military service, while the other evaluated a clinic serving a wider population. Additionally, the group in the Netherlands evaluated a shorter amount of time (2.3–3.5 years since beginning treatment) (van der Loos et al., 2022) than the group in the United States (4 years since beginning treatment) (Roberts et al., 2022). Neither study documented reasons for discontinuation of medical treatment, further limiting the interpretation of these differences.

One of the most recent large chart review studies on discontinuation of hormones examined the medical records of 1050 patients who accessed hormones in a large, pediatric gender clinic in the United States (Boskey et al., 2025). Outcomes of interest were whether or not the patient was taking hormones at the time of last communication, gender identity at discontinuation time, and reasons for discontinuation. The results of this study showed a low rate of discontinuation (4%) and an even lower rate of discontinuation due to reidentification with sex assigned at birth (0.5%). These results are consistent with other research showing that patients seen in pediatric medical clinics demonstrate low rates of discontinuation. The study is primarily limited by the method of discerning reasons for discontinuation, which was based on clinical documentation rather than gathering this information directly from patients.

One study of two academic centers in the United States assessed reasons for hormone discontinuation in addition to discontinuation rates (Gupta et al., 2023). In the first phase of this study, medical records were reviewed to identify individuals prescribed gender-affirming hormones as well as rates of continuation and discontinuation based on provider notes, chart review, and medication refills. If a patient disenrolled from the clinical center, their status at the time of their last visit acted as a proxy for hormone use. The review included 385 total participants, 121 of whom were younger than 18 when prescribed hormones. The average duration of time between starting hormones and the analysis in this study for the pediatric population was 24 months, and only one patient under 18 years old discontinued treatment (0.8%). Six patients discontinued treatment in total (1.6%). In phase 2 of the study the research team contacted patients directly to ask about reasons for discontinuation of treatment. Reasons included bullying from peers, financial barriers, and medical complications. The primary limitation of this component of the study was the low sample size. Additionally, the rate of discontinuation may differ for patients who sought care at another institution and discontinued later or who disenrolled from the institution due to financial barriers that also led to an inability to access the medication.

A recent survey study sought to recruit individuals who had received any gender-affirming medical treatment in the United States and Canada (MacKinnon et al., 2024, 2025). The survey was advertised broadly to Lesbian, Gay, Bisexual, Transgender, Queer, and more (LGBTQ+) individuals aged 15–29 rather than just to individuals who currently identify as TGD to increase the odds of recruiting individuals who discontinued GAC. Of the 720 participants who engaged in gender-affirming medical treatment, 121 (16.8%) responded to "Have you stopped or reversed any of these procedures or treatments?" with an affirmative answer. Of those 121 who stopped, 45 (37.2%) reported "Yes, but I wish I hadn't." Participants were able to select options from a list as to why they discontinued treatment in the survey, as well as the option to provide their own answer. The most common responses were health reasons, a change in gender identity, and cost. While 32% of those who discontinued treatment reported a change in gender identity as a reason, 80.2% reported an ongoing TGD identity. Endorsing a nonbinary identity at the time of the study was positively correlated with discontinuation of treatment. The study did not distinguish between those who started treatment in adolescence versus adulthood and only recorded whether participants had accessed certain GAC interventions (hormones or surgery) instead of the duration of treatment. Selection bias may have also played a role, as individuals who started and subsequently discontinued treatment due to reidentification with sex assigned at birth may be less likely to complete this study.

The Research-based, Ethical, Affirming, and Life-Saving (REAL) Story project is a 3-phase project currently underway at a large US pediatric gender clinic evaluating long-term outcomes, including discontinuation of treatment (Crabtree et al., 2024). The first of three phases included an anonymous survey distributed to a large cohort of individuals who had at least one visit in a pediatric gender clinic under 18 years of age. Of 141 survey respondents who had ever accessed gender-affirming hormones, 30 (21.3%) had ever stopped using hormones. Of those 30, 11 had since resumed hormones, and another 16 were considering resuming. Among the remaining three respondents with no plan to resume hormones, two (1.4%) reported discontinuing related to shifts in gender identity, and the third discontinued due to being satisfied with results. Other reasons for discontinuing hormones temporarily included external barriers (e.g., difficulty with injections, financial barriers), personal concerns (e.g., exploring fertility), and gender identity (e.g., selectively controlling the changes of hormones).

In summary, discontinuation of medical treatment has largely been studied using retrospective chart review, with more recent research using phone interviews and surveys to clarify reasons for discontinuation. Results suggest a high degree of variability among reasons for discontinuation, often including systemic barriers and social pressures. Studies have also shown that discontinuation can be temporary, and many individuals will resume treatments later. Research in this area is limited by the lack of prospective data and the apparent overrepresentation of individuals who are AFAB. Future research would benefit from longitudinal studies and qualitative research that may help to delineate reasons for discontinuation. These research methods could also ascertain whether the social experiences, medical sequelae, or gender journey related to continuing or discontinuing treatment differ based on the age at which treatment was initiated. Finally, long-term outcomes research that includes rates of reversal of medical interventions, including surgery, is also needed.

Detransition

Detransition generally refers to the process of discontinuing prior gender transition processes, often associated with reidentification with the sex assigned at birth. Exact rates of individuals who identify as having detransitioned are difficult to obtain. This is partially due to a lack of a consistent definition and uncertainty in existing methodologies (MacKinnon et al., 2023). Studies that follow a clinic sample may exclude detransitioners who are unsatisfied with care and go elsewhere for services or become lost to follow-up. Research on adults who detransition suggests that these individuals have multiple medical and psychological needs, but struggle to talk about these needs with providers due to real or perceived stigma (Vandenbussche, 2022). For this reason, individuals who identify as "detrans" (i.e., people who detransition) may be less inclined to engage in research or follow up with medical facilities where they received initial treatment. Additionally, online surveys that cater to LGBTQ+ or TGD individuals may not appear open to detrans individuals due to exclusionary language in the marketing.

One major study of 17,151 TGD adults in the United States found a detransition rate of 13.1% of those who sought any degree of social or medical gender affirmation (Turban et al., 2021). This study was a secondary analysis of data from the 2015 US Transgender Survey, which was a large-scale cross-sectional study that surveyed 27,715 TGD adults in total. The detransition rate was determined with a survey question asking, "Have you ever detransitioned? In other words, have you ever gone back to living as your sex assigned at birth, at least for a while?" Those who endorsed detransition were asked what reasons or influences contributed to that experience. Participants then were able to select multiple options from a list of pregenerated options that included external factors such as harassment or discrimination, and internal reasons such as transition being too difficult or a shift in gender identity. Results showed that 82.5% of those who detransitioned reported at least one external reason for detransitioning, and 15.9% reported at least one internal factor. Results also demonstrated that the younger cohorts were more likely to report reasons for detransition related to pressure from parents, community, or friends. Findings from this study highlight that detransition is not synonymous with regret and, in the majority of cases, is influenced by external factors.

A qualitative study explored reasons for detransition among 20 Canadian adolescents and young adults who detransitioned or discontinued transition (Pullen Sansfaçon et al., 2023). Within the sample, 13 participants underwent medical transition involving hormones and/or surgery, and the other 7 participants underwent social transition only. Participants were between the ages of 16 and 25, and three participants started medical transition before age 18. All but one of the participants were AFAB. The study involved a semistructured interview asking about the transition and detransition process. Results highlighted the nuances of transition, with some participants feeling grateful for the opportunity to transition despite ultimately choosing to detransition and others ultimately regretting the experience. Themes included feeling pressure to follow a certain transition path, mistaking other problems for gender dysphoria, understanding personal experiences differently now, and a lack of appropriate discussion or support about the decision to begin transition beforehand.

Some research on detransition has been criticized for risk of bias, particularly based on recruitment sampling and question administration. Recruitment among LGBTQ+ individuals may exclude people who do not actively identify as part of that community, and surveys with a pregenerated list of reasons for detransition may carry researcher bias based on predetermined beliefs about detransition (MacKinnon et al., 2024, 2025). One survey study of people who have detransitioned across the United States and Canada as part of the Detransition Analysis, Representation, and Exploration (DARE) study specifically attempted to overcome these limitations (MacKinnon et al., 2024, 2025). The research team recruited 962 participants with marketing materials that were tailored toward three different groups of people: TGD people who experienced shifts in identity, TGD individuals who were forced to detransition, and individuals who detransitioned and no longer identify as TGD. A total of 21 possible reasons for detransition were presented in the survey, grouped into four domains in the survey: mental health/psychological reasons, physical reasons, external reasons, and social/internal reasons. Analysis of this data is currently underway and is promising for collecting a broader group of detransitioners and a wider range of possible experiences.

Unfortunately, the concept of detransition has been politicized and misused by antitransgender groups to restrict access to care for TGD youth (McNamara et al., 2024). As a result, research has largely focused on reasons for and prevention of detransition, rather than the specific needs of individuals who have detransitioned. One study used structured interviews to evaluate the impact of public discourse on 25 detrans youth aged 16–25 years (Gelly et al., 2025). Qualitative analysis revealed that young people who have detransitioned feel misrepresented in the broader discourse and do not agree with detransition being seen as a solely negative outcome. They feel weaponized, unable to get their current needs met, and isolated from the potential community. This study is limited by a small sample size, but the results highlight the difficulty of reducing bias in research that is subject to politicization and misuse.

In summary, detransition has not historically been well-defined and is often conflated with other types of outcomes (Dolotina and Daniolos, 2023). Research has primarily used survey results and qualitative interviews to explore detransition experiences and has typically focused on factors influencing decisions. Future research should focus on the needs of individuals with this experience, using diverse methodologies and larger sample sizes. Standardized measurement tools could be useful for future research on detransition.

Reidentification

The concept of reidentification has often been conflated with detransition or desistance. The term reidentification is used here to refer to someone who identifies as transgender or nonbinary for a period of time, followed by identifying with the sex and/or gender they were assumed to be at birth, sometimes called "sex assigned at birth" or "sex registered at birth," depending on the country and birth record customs. Reidentification with the birth-registered sex may be temporary or long-term. For the purposes of this summary the term reidentification does not include individuals who demonstrate some gender nonconformity or gender exploration but never identify as transgender or nonbinary. Similarly, it does not include individuals who experience shifts in identity between binary transgender and nonbinary. While this definition refers specifically to a shift in identity irrespective of any particular transition steps or medical interventions, the intersection of reidentification and pediatric GAC remains an area of interest in long-term outcome research.

Given the history of conflating these concepts, few studies have systematically measured rates of reidentification. Some studies have implied information about reidentification without measuring it outright. For example, one study found that 80.2% of those who discontinued medical interventions endorsed a current TGD identity, implying that the other 19.8% of those who discontinued treatment reidentified with their sex assigned at birth, though it is not clearly stated as an outcome of medical treatment (MacKinnon et al., 2024, 2025).

One of the strongest US-based studies that explicitly examined reidentification as an outcome is a longitudinal study following youth who began social transition between the ages of 3 and 12, with an average of 5.4 years from social transition to the time of analysis (Olson et al., 2022). Participants were recruited as part of The TransYouth Project, which is a large longitudinal study following youth who socially transition in the United States and Canada (Gülgöz et al., 2019). For inclusion in the study, participants needed to have completed a "complete binary social transition," including changing pronouns, presumably in all settings. Requiring this level of social transition separates the study from earlier desistance literature that did not have stringent criteria for inclusion as a TGD participant (Ashley, 2022). At the same time the inclusion of only binary transgender children limits the generalizability of the findings, and these results may not therefore reflect the experiences of youth who initially identify as nonbinary.

Of the 317 youth in the study, 7.3% had some level of retransition to another identity after their initial transition, which included transition from binary transgender to nonbinary (Olson et al., 2022). A total of eight participants (2.5%) were using pronouns aligned with sex assigned at birth and were categorized as cisgender based on review of survey data and email communications with parents. Of those categorized as cisgender, none had accessed gender-affirming hormones, and only one had accessed puberty blockers. With an average age of 8.1 at the start of the study, more time is needed to determine how many of these youth will continue to access medications and whether gender identity continues to evolve. The TransYouth Project is ongoing, and therefore long-term results are expected over time.

Another of the strongest studies in this area examined reidentification by reviewing referral closures within the only pediatric gender clinic serving TGD youth in the state of Western Australia (Cavve et al., 2024). A comprehensive audit protocol identified 548 patients for whom a reason for referral closure could be determined. Results revealed 29 (5.3%) referrals were closed due to reidentification with sex registered at birth based on medical records, family correspondence, referral information, and other databases used to track patients over time. Of those who reidentified with sex registered at birth, 27 did so before or early in the assessment phase, before beginning any medical treatment. Of the remaining two patients, one accessed puberty blockers only and the other accessed puberty blockers followed by 1 month of affirming hormones before reidentification.

Another study recruited 92 TGD youth aged 15–21 from the community and followed them longitudinally over 2 years (Real and Russell, 2025). Participants were surveyed across four study waves, 9 months apart, and each participant was determined to be TGD or cisgender at each wave based on whether their current gender identity matched the sex assigned at birth using the two-step method of assessing gender identity. Identity trajectory patterns were categorized as follows: TGD throughout (n=32), cisgender to TGD (n=28), and TGD to cisgender (n=32). Notably, while 20 participants in the TGD throughout group had a history of hormone use (62.5%), only one participant in the TGD to cisgender group had a history of hormone use (3.1%). Similar to other studies, the small sample size was a limitation.

In summary, reidentification with sex assigned at birth has only recently been studied as an explicit outcome rather than being implied or assumed in studies of other long-term outcomes. Research to date suggests generally low rates of reidentification, more nuanced trajectories with periods of retransition, and especially low rates of reidentification following the initiation of GAC. This research is primarily limited by the length of follow-up and sample size. As longitudinal studies progress and more data become available over time, trajectories in identity from adolescence into adulthood will continue to be clarified. Future research would also benefit from more community-based samples and prospective studies, including TGD and cisgender youth.

Satisfaction, regret, and other emotions associated with care

One final area of long-term outcomes research includes the subjective experiences of satisfaction and regret (as well as all of the emotions in between). While there has been some historical hesitation to gather information on regret due to fear of politicization and misuse, there has been increasingly more interest in this topic from clinicians who want to better understand nuances in patient experiences to inform and improve care for youth and their families. As a result, there have been more intentional efforts in recent years to collect information about regret and to categorize it. Another shift over time has been from understanding satisfaction and regret as either binary choices or a linear spectrum to more recent research acknowledging the potential duality of these experiences and further recognizing that the spectrum of subjective experiences includes many more feelings than these two emotions. While satisfaction and regret can be related to social aspects of transition, the majority of research in this area focuses on specific medical interventions or on medical transition overall.

One comprehensive long-term outcomes study of TGD youth in the Netherlands treated with puberty blockers, hormones, and surgery included a brief questionnaire designed by the study team to evaluate satisfaction (De Vries et al., 2014). Patients were asked how satisfied they were with their appearance and with surgery on a numbered scale. Results revealed a high level of satisfaction (from fairly to very satisfied), and no patients indicated any regret. More recently, researchers in the Netherlands evaluated all patients (80% of whom were over 18 at first visit) seen in their Amsterdam clinic from 1972 to 2015 (Wiepjes et al., 2018). Regret was identified in 0.5% of all patients and categorized into three groups: social regret, true regret, and nonbinary identity. Regret was categorized as social when patients indicated no changes in their identity and cited other reasons, such as loss of relationships, whereas patients who indicated regret of diagnosis and treatment were categorized as having true regret. These results do not allow for comparison of rates of regret between those who began treatment in adolescence versus adulthood, and the categories are not well-operationalized for replication.

Another study similarly attempted to categorize types of regret based on a review of the literature and the results of a survey of surgeons from across the globe who received reports of regrets from patients (Narayan et al., 2021). Embracing a nonbinary identity was included in the category "true gender-related regret," along with changes in gender identity and misdiagnosis of gender dysphoria. "Social regret" remained a relevant category and included fear for safety due to judgment, difficulty in relationships, spiritual conflict, rejection from supports, and difficulties with employment. The third category identified was "medical regret," which includes complications due to surgeries and concern for health, as well as missing natal genitals. Survey results revealed regret rates per surgeon ranging from 0.2% to 0.3%, with the smallest percentage representing medical regret. The study was limited by relying on surgeons' reports of patients' disclosures and was subject to recall bias. Additionally, surgeons categorizing nonbinary individuals experiencing regret as "true regret" cases may be an indication of bias and not reflective of individual experiences, since patients were not asked directly.

A German research team attempted to better understand regret and reasons for dissatisfaction with medical interventions in their pediatric clinic by evaluating all patients in a prospective clinical cohort study using a relatively novel metric of tracking progress within their clinic using an individual treatment progress score (ITPS) (Nieder et al., 2021). The ITPS is a continuous score between 0% and 100% derived from the number of transition-related medical interventions the patient has received divided by the total number of interventions desired. Because treatment goals can change over time, the ITPS is frequently updated to reflect ongoing goals rather than relying solely on goals set at an initial visit. Patients were also surveyed using the Client Satisfaction Questionnaire (ZUF-8), which includes questions about satisfaction with the clinic and the care provided (Attkisson and Zwick, 1982). Results revealed a significant positive relationship between ITPS and satisfaction with transition-related care. One limitation of this methodology is that satisfaction with individual components of care was not ascertained. It is possible, for example, for a youth to feel satisfied with the overall clinic, but not with a particular aspect of their care. As with most survey research, this study is also limited by sampling bias, as some patients were not included due to not responding to the survey, refusing to participate, or having outdated contact information. Within the sample of 75 youth who were included, none who began medical treatments reported any regret.

A qualitative study of 20 detransitioned North American young adults provides additional insight into feelings related to care (Pullen Sansfaçon et al., 2023). Emotions discussed in the study included grief, gratitude, regret, happiness, and more. Some participants felt the initial transition helped them learn more about their gender identity in a way that would be impossible if they had never transitioned. Other participants reported feeling pressured into transitioning or regretting moving so quickly on transitioning, while also not addressing additional challenges.

In one study specific to gender-affirming mastectomies (also known as "top surgery"), a retrospective chart review was used to evaluate outcomes among 209 patients who underwent top surgery under the age of 18 years (Tang et al., 2022). Patients were aged 12–17 years at the time of referral, and the median postoperative follow-up time was 2 years. After this follow-up period, two patients (0.95%) expressed regret, although neither pursued reversal surgery during the next several years of the study, and it was not clear if it was "true regret" related to reidentification with sex assigned at birth. Both patients were nonbinary at the time of surgery and underwent surgery at age 16 years. One of the two patients was identified as being in an unsupportive environment. The study was limited by its retrospective design and lack of a standardized measure of regret. One notable finding from the study is that the referral rate over the course of the study period increased significantly, but rates of regret remained low and stable.

More longitudinal data on youth experiences of GAC is needed to understand subjective experiences with interventions over time. The REAL Story project is an ongoing study at a large pediatric gender clinic in the United States attempting to understand the full range of emotional experiences associated with receiving care in adolescence (Crabtree et al., 2024). TGD community members, working both within and outside of the clinic, were involved in the creation and implementation of the study design. The first phase involved an anonymous survey of patients who had at least one appointment in a pediatric clinic before age 18 years. The survey asked about the history of GAC, barriers to accessing care, and emotions associated with GAC. Respondents included 150 patients who were mostly satisfied and felt positive about their care, and results revealed that emotions such as grief, disappointment, and regret are rare but nuanced. Phases 2 and 3 are currently underway using a combination of retrospective chart review and qualitative interview methods, respectively. This combination of quantitative and qualitative work represents one of the first major studies evaluating satisfaction and regret in a sample where all patients sought care before age 18 years.

In summary, satisfaction and regret related to gender-affirming medical treatments are not mutually exclusive experiences, and TGD individuals may have a range of emotions associated with their care. Regret among those who accessed medical interventions in adolescence is very rare, occurring at similar rates in adult populations. Regret is nonetheless a variable of interest to clinicians who want to continue to improve care outcomes. Researchers have attempted to categorize regret, distinguishing between identity-related regret, socially-derived regret, and regret related to medical complications or expectations. Future research is needed to clarify the full range of experiences with GAC outside of satisfaction and regret alone, and this area of research would greatly benefit from standardized measures and prospective study designs.

Recommendations for future research

The studies discussed in this chapter represent a strong foundation of GAC research with particular strengths in short-term follow-up within medical systems. There are also novel research methods underway that will help clarify the experiences of detransition and the potential for regret. In addition to replicating and extending the timelines of these projects, additional recommendations include gathering more qualitative data about TGD experiences, increasing representation of diverse groups in research, conducting more community-based studies, and measuring outcomes in a way that allows for a full range of gender expression.

Longitudinal research

Gender clinics in the Netherlands, Germany, the United States, and Australia have published prospective cohort studies on clinical samples, and over time, these results will provide increasingly long-term outcomes. In particular, following these cohorts through adulthood will help to clarify the impact of receiving GAC at younger ages on a variety of experiences in adulthood. Additionally, longitudinal research among clinical samples would benefit from specifying any standardized practice criteria to clarify whether particular protocols result in different outcomes. Studies that specifically evaluate mechanisms of change and moderating factors will similarly help in understanding for whom and in what contexts pediatric GAC results in optimal outcomes.

Qualitative research

Some qualitative studies are currently underway, but more are needed to better understand the long-term experiences of TGD youth accessing GAC. Retrospective chart review and pregenerated survey questions are common research methods used in this area, but both are at risk of missing the authentic experience of study participants, particularly when TGD community members are not involved in study design. Qualitative research methods will provide more nuanced information on the true long-term experiences with pediatric GAC and will set a foundation for ongoing quantitative work. Researchers also recommend engaging in qualitative research to explore the language used by communities and to better understand how they define concepts such as detransition (MacKinnon et al., 2023). Rather than creating a definition based on scientific agreement, researchers can turn to TGD communities to help define appropriate targets of study.

Increasing representation

Researchers investigating TGD youth are at risk of overemphasizing Western, Educated, Industrialized, Rich, and Democratic (WEIRD) populations (Henrich et al., 2010). The sociopolitical climate of LGBTQ+ rights and access to healthcare limits the ability to replicate this research globally, and researchers should remain aware of the generalizability of their findings. Additionally, the research discussed in this chapter relied heavily on the experiences of individuals who are white, AFAB, and those with binary identities. Diversity and representation within the TGD community are important to understand, as the embodiment desires and outcomes can be different. Qualitative research can help clarify if experiential differences exist for those with different gender identities, as well as other experiences, such as neurodivergence. Autistic youth may have different experiences of gender that lead them to engage in care differently, which may require researchers to measure outcomes in a different way.

Community-based studies

The research summarized in this chapter has largely been conducted among clinical samples, often from major pediatric gender clinics. The generalizability of these findings is therefore limited, as results may exclude youth who are unable to access or who lose access to GAC at a major clinic due to financial barriers, unsupportive parents, distance from the clinic, political actions, and/or perceived stigma from providers. Some researchers have used surveys advertised in online forums and support groups of TGD or detransitioned youth (Fontanari et al., 2020; MacKinnon et al., 2024, 2025), which is one strategy for accessing nonclinic populations. More studies accessing youth this way or through community-based services could increase representation. A large research gap that currently exists is longitudinal research involving community-based samples.

Gender as a process, not a destination

When measuring long-term outcomes and seeking to clarify definitions, researchers should be mindful about potentially overemphasizing particular outcomes. The limited research on detransition, reidentification, and regret in particular represents areas with higher risks of personal bias. There is stigma surrounding these topics, partially due to politicalization and the use of these outcomes to justify restricting GAC for youth (MacKinnon et al., 2023; Preston and Lopez, 2024). There may also be a perception from non-TGD researchers and clinicians that detransition and/or reidentification represent unsuccessful outcomes. This aligns with theoretical notions that clinicians can somehow predict gender development over time and use that knowledge to prescribe or prohibit a treatment. As Dr. Baer Karrington a genderqueer medical provider, writes, “Cisgender identity does not invalidate transgender identity,” suggesting that someone who de- or retransitions may not reject or regret their experience of gender expansiveness and pursuit of embodiment goals (Karrington, 2022). When measuring GAC outcomes, researchers should be mindful to destigmatize discontinuation, detransition, and reidentification as negative outcomes and instead seek to understand how individuals experience these pathways and how they constitute a gender development process, which may require clinical assistance at various points.

References

Achille, C., Taggart, T., Eaton, N.R., Osipoff, J., Tafuri, K., Lane, A., et al., 2020. Longitudinal impact of gender-affirming endocrine intervention on the mental health and well-being of transgender youths: preliminary results. International Journal of Pediatric Endocrinology 2020 (1). Available at https://doi.org/10.1186/s13633-020-00078-2.

Ashley, F., 2022. The clinical irrelevance of “desistance” research for transgender and gender creative youth. Psychology of Sexual Orientation and Gender Diversity 9 (4), 387–397. Available at https://doi.org/10.1037/sgd0000504.

Ashley, F., Tordoff, D.M., Olson-Kennedy, J., Restar, A.J., 2024. Randomized-controlled trials are methodologically inappropriate in adolescent transgender healthcare. International Journal of Transgender Health 25 (3), 407–418. Available at https://doi.org/10.1080/26895269.2023.2218357.

Attkisson, C.C., Zwick, R., 1982. The client satisfaction questionnaire. Psychometric properties and correlations with service utilization and psychotherapy outcome. Evaluation and Program Planning 5 (3), 233–237. Available at https://doi.org/10.1016/0149-7189(82)90074-X.

Becker-Hebly, I., Fahrenkrug, S., Campion, F., Richter-Appelt, H., Schulte-Markwort, M., Barkmann, C., 2021. Psychosocial health in adolescents and young adults with gender dysphoria before and after gender-affirming medical interventions: a descriptive study from the Hamburg Gender Identity Service. European Child & Adolescent Psychiatry 30 (11), 1755–1767. Available at https://doi.org/10.1007/s00787-020-01640-2.

Boskey, E.R., Scheffey, K.L., Pilcher, S., Barerra, E.P., McGregor, K., Carswell, J.M., et al., 2025. A retrospective cohort study of transgender adolescents' gender-affirming hormone discontinuation. Journal of Adolescent Health 76 (4), 584–591. Available at https://doi.org/10.1016/j.jadohealth.2024.11.002.

Cavve, B.S., Bickendorf, X., Ball, J., Saunders, L.A., Thomas, C.S., Strauss, P., et al., 2024. Reidentification with birth-registered sex in a Western Australian pediatric gender clinic cohort. JAMA Pediatrics 178 (5), 446–453. Available at https://doi.org/10.1001/jamapediatrics.2024.0077.

Chelliah, P., Lau, M., Kuper, L.E., 2024. Changes in gender dysphoria, interpersonal minority stress, and mental health among transgender youth after one year of hormone therapy. Journal of Adolescent Health 74 (6), 1106–1111. Available at https://doi.org/10.1016/j.jadohealth.2023.12.024.

Chen, D., Berona, J., Chan, Y.M., Ehrensaft, D., Garofalo, R., Hidalgo, M.A., et al., 2023. Psychosocial functioning in transgender youth after 2 years of hormones. New England Journal of Medicine 388 (3), 240–250. Available at https://doi.org/10.1056/NEJMoa2206297.

Chen, D., Berona, J., Coyne, C.A., Chan, Y.-M., Ehrensaft, D., Hidalgo, M.A., et al., 2024. 9. Predictors of diverse mental health trajectories following 4 years of treatment with gender-affirming hormones. Journal of Adolescent Health 74 (3), S5. Available at https://doi.org/10.1016/j.jadohealth.2023.11.028.

Coleman, E., Radix, A.E., Bouman, W.P., Brown, G.R., de Vries, A.L.C., Deutsch, M.B., et al., 2022. Standards of care for the health of transgender and gender diverse people, version 8. International Journal of Transgender Health 23 (1), S1. Available at https://doi.org/10.1080/26895269.2022.2100644.

Crabtree, L., Connelly, K.J., Guerriero, J.T., Battison, E.A.J., Tiller-Ormord, J., Sutherland, S.M., et al., 2024. A more nuanced story: pediatric gender-affirming healthcare is associated with satisfaction and confidence. Journal of Adolescent Health 75 (5), 772–779. Available at https://doi.org/10.1016/j.jadohealth.2024.06.016.

De Vries, A.L.C., McGuire, J.K., Steensma, T.D., Wagenaar, E.C.F., Doreleijers, T.A.H., Cohen-Kettenis, P.T., 2014. Young adult psychological outcome after puberty suppression and gender reassignment. Pediatrics 134 (4), 696–704. Available at https://doi.org/10.1542/peds.2013-2958.

Dolotina, B., Daniolos, P.T., 2023. Dynamic gender identities and expressions. Child and Adolescent Psychiatric Clinics of North America 32 (4), 803–813. Available at https://doi.org/10.1016/j.chc.2023.05.002.

Fontanari, A.M.V., Vilanova, F., Schneider, M.A., Chinazzo, I., Soll, B.M.H., Schwarz, K., et al., 2020. Gender affirmation is associated with transgender and gender nonbinary youth mental health improvement. LGBT Health 7 (5), 237–247. Available at https://doi.org/10.1089/lgbt.2019.0046.

Gadomski, A.M., Scribani, M.B., Tallman, N., O'Bryan, J., Wolf-Gould, C., Wolf-Gould, C., 2023. Two-year follow-up study of health-related quality of life among transgender and gender expansive youth receiving gender-Affirming care. Transgender Health. Available at https://doi.org/10.1089/trgh.2022.0165.

Gelly, M.A., Atgé-Delbays, S., Sansfaçon, A.P., 2025. "They're unable to see my decision to detransition for what it is": how detrans youth perceive and receive discourses on detransition. Sexuality Research and Social Policy. Available at https://doi.org/10.1007/s13178-024-01079-z.

Grannis, C., Leibowitz, S.F., Gahn, S., Nahata, L., Morningstar, M., Mattson, W.I., et al., 2021. Testosterone treatment, internalizing symptoms, and body image dissatisfaction in transgender boys. Psychoneuroendocrinology 132. Available at https://doi.org/10.1016/j.psyneuen.2021.105358.

Gülgöz, S., Glazier, J.J., Enright, E.A., Alonso, D.J., Durwood, L.J., Fast, A.A., 2019. Similarity in transgender and cisgender children's gender development. Proceedings of the National Academy of Sciences of the United States of America 116 (49), 24480–24485. Available at https://doi.org/10.1073/pnas.1909367116.

Gupta, P., Patterson, B.C., Chu, L., Gold, S., Amos, S., Yeung, H., 2023. Adherence to gender affirming hormone therapy in transgender adolescents and adults: a retrospective cohort study. Journal of Clinical Endocrinology and Metabolism 108 (11), E1236. Available at https://doi.org/10.1210/clinem/dgad306.

Hall, R., Mitchell, L., Sachdeva, J., 2021. Access to care and frequency of detransition among a cohort discharged by a UK national adult gender identity clinic: retrospective case-note review. BJPsych Open 7 (6). Available at https://doi.org/10.1192/bjo.2021.1022.

Henrich, J., Heine, S.J., Norenzayan, A., 2010. The weirdest people in the world? Behavioral and Brain Sciences 33 (2-3), 61–83. Available at https://doi.org/10.1017/S0140525X0999152X.

Jedrzejewski, B.Y., Marsiglio, M.C., Guerriero, J., Penkin, A., Connelly, K.J., Berli, J.U., et al., 2023. Regret after gender-affirming surgery: a multidisciplinary approach to a multifaceted patient experience. Plastic and Reconstructive Surgery 152 (1), 206–214. Available at https://doi.org/10.1097/PRS.0000000000010243.

Karrington, B., 2022. Defining desistance: exploring desistance in transgender and gender expansive youth through systematic literature review. Transgender Health 7 (3), 189–212. Available at https://doi.org/10.1089/trgh.2020.0129.

Kuper, L.E., Stewart, S., Preston, S., Lau, M., Lopez, X., 2020. Body dissatisfaction and mental health outcomes of youth on gender-affirming hormone therapy. Pediatrics 145 (4). Available at https://doi.org/10.1542/peds.2019-3006.

MacKinnon, K.R., Expósito-Campos, P., Gould, W.A., 2023. Detransition needs further understanding, not controversy. BMJ (Clinical Research ed.). Available at https://doi.org/10.1136/bmj-2022-073584.

MacKinnon, K.R., Jeyabalan, T., Strang, J.F., Delgado-Ron, J.A., Lam, J.S.H., Gould, W.A., et al., 2024. Discontinuation of gender-affirming medical treatments: prevalence and associated features in a nonprobabilistic sample of transgender and gender-diverse adolescents and young adults in Canada and the United States. Journal of Adolescent Health 75 (4), 569–577. Available at https://doi.org/10.1016/j.jadohealth.2024.05.015.

MacKinnon, K.R., Khan, N., Newman, K.M., Gould, W.A., Marshall, G., Salway, T., et al., 2025. Introducing Novel Methods to Identify Fraudulent Responses (Sampling With Sisyphus): Web-Based LGBTQ2S+ Mixed-Methods Study. Journal of Medical Internet Research 27, e63252. Available at 10.2196/63252.

Mahfouda, S., Moore, J.K., Siafarikas, A., Hewitt, T., Ganti, U., Lin, A., et al., 2019. Gender-affirming hormones and surgery in transgender children and adolescents. The Lancet Diabetes and Endocrinology 7 (6), 484–498. Available at https://doi.org/10.1016/S2213-8587(18)30305-X.

McNamara, M., Lepore, C., Alstott, A., Kamody, R., Kuper, L., Szilagyi, N., et al., 2022. Scientific misinformation and gender affirming care: tools for providers on the front lines. Journal of Adolescent Health 71 (3), 251–253. Available at https://doi.org/10.1016/j.jadohealth.2022.06.008.

McNamara, M., McLamore, Q., Meade, N., Olgun, M., Robinson, H., Alstott, A., 2024. A thematic analysis of disinformation in gender-affirming healthcare bans in the United States. Social Science & Medicine 351, 116943. Available at https://doi.org/10.1016/j.socscimed.2024.116943.

Narayan, S.K., Hontscharuk, R., Danker, S., Guerriero, J., Carter, A., Blasdel, G., et al., 2021. Guiding the conversation—types of regret after gender-affirming surgery and their associated etiologies. Annals of Translational Medicine 9 (7), 605. Available at https://doi.org/10.21037/atm-20-6204 –605.

National LGBTQIA+ Health Education Center, 2024. A Program of the Fenway Institute: LGBTQIA+ Glossary of Terms for Health Care Teams. National LGBTQIA+ Health Education Center, Boston, MA. Available at https://www.lgbtqiahealtheducation.org/publication/glossary/.

Nguyen, H.B., Chavez, A.M., Lipner, E., Hantsoo, L., Kornfield, S.L., Davies, R.D., et al., 2018. Gender-affirming hormone use in transgender individuals: impact on behavioral health and cognition. Current Psychiatry Reports 20 (12). Available at https://doi.org/10.1007/s11920-018-0973-0.

Nieder, T.O., Mayer, T.K., Hinz, S., Fahrenkrug, S., Herrmann, L., Becker-Hebly, I., 2021. Individual treatment progress predicts satisfaction with transition-related care for youth with gender dysphoria: a prospective clinical cohort study. The Journal of Sexual Medicine 18 (3), 632–645. Available at https://doi.org/10.1016/j.jsxm.2020.12.010.

Nolan, B.J., Zwickl, S., Locke, P., Zajac, J.D., Cheung, A.S., 2023. Early access to testosterone therapy in transgender and gender-diverse adults seeking masculinization: a randomized clinical trial. JAMA Network Open 6 (9). Available at https://doi.org/10.1001/jamanetworkopen.2023.31919 E2331919.

Olson, K.R., Durwood, L., Horton, R., Gallagher, N.M., Devor, A., 2022. Gender identity 5 years after social transition. Pediatrics 150 (2). Available at https://doi.org/10.1542/peds.2021-056082.

Olson-Kennedy, J., Wang, L., Wong, C.F., Chen, D., Ehrensaft, D., Hidalgo, M.A., et al., 2025. Emotional health of transgender youth 24 months after initiating gender-affirming hormone therapy. Journal of Adolescent Health. Available at https://doi.org/10.1016/j.jadohealth.2024.11.014.

Preston, S., Lopez, X., 2024. Legislation, medicine, and politics: care for gender diverse youth. Current Opinion in Endocrinology, Diabetes & Obesity 31 (1), 43–52. Available at https://doi.org/10.1097/med.0000000000000845.

Pullen Sansfaçon, A., Gelly, M.A., Gravel, R., Medico, D., Baril, A., Susset, F., et al., 2023. A nuanced look into youth journeys of gender transition and detransition. Infant and Child Development 32 (2). Available at https://doi.org/10.1002/icd.2402.

Real, A.G., Russell, S.T., 2025. Transitions in gender identity among youth: trajectories of perceptions of family acceptance. The Journal of Adolescent Health: Official Publication of the Society for Adolescent Medicine 76 (5), 863–870.

Redfield, E., Conron, K.J., Mallory, C., 2024. The Impact of 2024 Anti-transgender Legislation on Youth. The Williams Institute.

Roberts, C.M., Klein, D.A., Adirim, T.A., Schvey, N.A., Hisle-Gorman, E., 2022. Continuation of gender-affirming hormones among transgender adolescents and adults. Journal of Clinical Endocrinology and Metabolism 107 (9). Available at https://doi.org/10.1210/clinem/dgac251 E3937.

Tang, A., Hojilla, J.C., Jackson, J.E., Rothenberg, K.A., Gologorsky, R.C., Stram, D.A., et al., 2022. Gender-affirming mastectomy trends and surgical outcomes in adolescents. Annals of Plastic Surgery 88 (4), S325. Available at https://doi.org/10.1097/SAP.0000000000003135.

Tordoff, D.M., Wanta, J.W., Collin, A., Stepney, C., Inwards-Breland, D.J., Ahrens, K., 2022. Mental health outcomes in transgender and nonbinary youths receiving gender-affirming care. JAMA Network Open 5 (2). Available at https://doi.org/10.1001/jamanetworkopen.2022.0978.

Turban, J.L., Carswell, J., Keuroghlian, A.S., 2018. Understanding pediatric patients who discontinue gender-affirming hormonal interventions. JAMA Pediatrics 172 (10), 903–904. Available at https://doi.org/10.1001/jamapediatrics.2018.1817.

Turban, J.L., Brady, C., Olson-Kennedy, J., 2022b. Understanding and supporting patients with dynamic desires for gender-affirming medical interventions. JAMA Network Open. Available at https://doi.org/10.1001/jamanetworkopen.2022.24722 E2224722,.

Turban, J.L., Loo, S.S., Almazan, A.N., Keuroghlian, A.S., 2021. Factors leading to "detransition" among transgender and gender diverse people in the United States: a mixed-methods analysis. LGBT Health 8 (4), 273–280. Available at https://doi.org/10.1089/lgbt.2020.0437.

Turban, J.L., King, D., Kobe, J., Reisner, S.L., Keuroghlian, A.S., 2022a. Access to gender-affirming hormones during adolescence and mental health outcomes among transgender adults. PLoS One 17 (1). Available at https://doi.org/10.1371/journal.pone.0261039.

van der Loos, M.A.T.C., Hannema, S.E., Klink, D.T., den Heijer, M., Wiepjes, C.M., 2022. Continuation of gender-affirming hormones in transgender people starting puberty suppression in adolescence: a cohort study in the Netherlands. The Lancet Child and Adolescent Health 6 (12), 869–875. Available at https://doi.org/10.1016/S2352-4642(22)00254-1.

Vandenbussche, E., 2022. Detransition-related needs and support: a cross-sectional online survey. Journal of Homosexuality 69 (9), 1602–1620. Available at https://doi.org/10.1080/00918369.2021.1919479.

Wiepjes, C.M., Nota, N.M., de Blok, C.J.M., Klaver, M., de Vries, A.L.C., Wensing-Kruger, S.A., et al., 2018. The Amsterdam cohort of gender dysphoria study (1972–2015): trends in prevalence, treatment, and regrets. Journal of Sexual Medicine 15 (4), 582–590. Available at https://doi.org/10.1016/j.jsxm.2018.01.016.

Chapter 14

Conclusion and future research directions

Jonna Finocchio and Natalie Nokoff
University of Colorado Anschutz, Aurora, CO, United States

Learning objectives

- Summarize the research in key areas of pediatric gender-affirming care, including gender identity development among transgender and gender diverse youth, risk, resilience, and barriers to care, behavioral and mental health, interdisciplinary healthcare models, and medical and surgical healthcare.
- Identify research gaps in key areas of pediatric gender-affirming care and make recommendations for future directions in research.

Key terms and definitions

- **Transgender health research** refers to historical and current research efforts related to improving the health and well-being of transgender and gender diverse (TGD) people of all ages.
- **Pediatric gender-affirming care** refers to medical, psychological, and social interventions aimed at improving the health and well-being of TGD children, adolescents, and young adults, as well as their families.
- **Community-based participatory research designs** include members of the community about and for which the research is being conducted at all levels of scientific design and implementation, from the development of research questions and priorities to the dissemination of research outcomes.

Since the turn of the century, transgender health research has evolved from a marginal academic topic to a rapidly expanding field shaped by heightened visibility of the transgender and gender diverse (TGD) community and sociopolitical debates. Multiple bibliometric studies have demonstrated a rapid rise in gender diversity-related research across health and social sciences. An analysis of all transgender health publications between 1900 and 2017 identified 5772 articles—nearly 60% of which were published between the years of 2005 and 2017 (Sweileh, 2018). Another scientometric study using the Scopus database searched for articles specific to TGD health and mental health, demonstrating a 4900% increase in less than 20 years (3 in 2004 to 150 in 2023) (Alikunju, 2024). Despite this surge in transgender scholarship, significant opportunities exist to further our understanding of the experiences of TGD youth as well as refine our methodological approaches in the process.

Risk, resilience, and access to healthcare

The World Professional Association for Transgender Health (WPATH) Standards of Care, Version 8 (SOC8) and the Endocrine Society Clinical Practice Guidelines recommend multidisciplinary or interdisciplinary care for TGD youth (Coleman et al., 2022; Hembree et al., 2017). In such models a team of providers from varying disciplines addresses a range of patient and family needs while refraining from emphasizing a particular treatment pathway (Hidalgo et al., 2013; Huit et al., 2024). In addition to medical care, this may involve behavioral healthcare, psychoeducation, social supports,

Advances in Research on Pediatric Gender Affirming Healthcare. DOI: https://doi.org/10.1016/B978-0-443-29019-0.00006-3
© 2026 Elsevier Inc. All rights are reserved, including those for text and data mining, AI training, and similar technologies.

and assistance with legal documentation. Care coordination and navigation, often embedded in multidisciplinary approaches, have been found to be particularly helpful to patients and families (Inwards-Breland et al., 2019). Such supports can help address school and community concerns, provide necessary resources, and connect families with similar lived experiences. Interventions should be grounded in a patient's own gender goals and adhere to international guidelines. Medical care is often provided by endocrinologists, adolescent medicine providers, or primary care providers, while behavioral health support is often provided by social workers or psychologists.

Despite the recommendation for multidisciplinary or interdisciplinary care, TGD youth often face numerous barriers to accessing both gender-affirming care and preventative healthcare. Studies evaluating the experiences of TGD youth have found numerous distinct impediments to care, including pervasive stigma and discrimination in the healthcare system, fear of entering such spaces due to being mocked or dismissed by staff, overt treatment denials based on gender identity, and confidentiality breaches (Chong et al., 2021; Pampati et al., 2021). Additionally, TGD youth are acutely aware of their provider's discomfort with their identity, and some report a sense of "moral opposition" on the part of the provider (Pampati et al., 2021).

One of the most commonly cited barriers to accessing care is a lack of knowledgeable providers, and many physicians agree with this sentiment. Prior research demonstrates significant deficiencies in medical providers' education and clinical preparedness related to transgender healthcare (Dubin et al., 2018; Korpaisarn and Safer, 2018; McPhail et al., 2016). Of medical students in the United States, 83% reported minimal or no education on transgender healthcare in residency (Altschuler and Mann, 2023). Subsequently, TGD youth have been found to utilize preventative healthcare services less than their cisgender peers (Rider et al., 2018) and experience health disparities such as chronic pain (Stocking et al., 2025), higher rates of obesity (Bishop et al., 2020; Linsenmeyer et al., 2021), and higher rates of disability (Smith-Johnson, 2022).

There are also unique barriers to accessing gender-affirming care providers. Many of the specialty multidisciplinary gender clinics in the United States are found in large metropolitan areas, creating access issues for many families relative to distance and cost (Gridley et al., 2016; Inwards-Breland et al., 2019; Sequeira et al., 2020; Strauss et al., 2022; Tollit et al., 2019). The relatively limited number of such programs has resulted in prolonged wait times for services, adding an additional layer of difficulty. At peak referral years, it was not uncommon for a specialty gender clinic to have wait times of 1–2 years (Allen et al., 2021; Lawson et al., 2024). TGD youth living in rural areas are noted to experience additional hardship in finding local affirming providers and accessing social supports due to location (Manns et al., 2023; Poquiz et al., 2021).

A robust body of literature highlights the innumerable risks faced by TGD people, many of which are amplified by social determinants of health. Previous research has explored the most common risk factors associated with poor physical and mental health outcomes in TGD youth, including discrimination, bullying, family rejection, social isolation, internalized stigma, and trauma (Bird et al., 2024; Delozier et al., 2020; Eisenberg et al., 2017; Johns et al., 2018; Newcomb et al., 2020; Ramos and Marr, 2023). While these risk factors are well-documented, protective factors that may mediate risk remain underexplored. Further clarification of potential protective factors is needed to facilitate the creation and implementation of effective interventions.

Newer research has turned its focus onto the resiliency of TGD youth. Historically, resilience research has positioned the capacity to "bounce back" as an individual-level trait (Puckett et al., 2024). This narrow focus is now seen as problematic in several ways. First, it does not consider the systems of oppression, power, and privilege that impact minoritized communities (Suslovic and Lett, 2024). It also assumes that adversity is temporal while placing the responsibility of overcoming such hardship on the shoulders of the impacted individual (McLean et al., 2008). A scoping review of studies published between 2010 and 2020 found that only five (of 33) specifically addressed whether resiliency factors buffered the effects of stigma-based stress for TGD individuals (Puckett et al., 2024). Additionally, results from these five studies were mixed and contradictory.

There are several promising protective factors for TGD youth that have emerged from recent research. In addition to being associated with improved self-esteem and mental health in TGD youth, there is evidence that positive family relationships are a buffer against risk (Johns et al., 2018; Tüzün et al., 2022). These relationships are also a predictor of primary care utilization in TGD youth (Taliaferro et al., 2019). This may be an indication that family support moderates the effects of the physical health disparities faced by the TGD community. In addition to families, peer support has also been found to promote improved well-being, sense of connectedness, and sense of purpose (Weinhardt et al., 2019). Finally, affirming school environments are shown to improve resiliency in TGD adolescents (Colvin et al., 2019), and having affirming adults within the school system is often indicative of improved functioning both inside and outside the school setting (Johns et al., 2018).

Key research gaps

Previous research evaluating risk and resiliency factors in TGD youth has largely prioritized distal stressors (e.g., bullying, discrimination, peer or family rejection) at the exclusion of proximal stressors (e.g., identity concealment, internalized transnegativity) and their interaction with protective and resilience processes (Delozier et al., 2020; Toomey, 2021). The term "resilience" is often vaguely defined in the literature, signaling the need for more precise, context-specific models that recognize joy, resistance, and pleasure rather than simply coping with harm (Robinson and Schmitz, 2021; Singh et al., 2014).

Consistent with broader trends in TGD research a key limitation of this body of work is a lack of sufficient sample diversity. Limited data exists on TGD youth of color, youth with disabilities, prepubertal children, nonbinary individuals, and youth living outside of North America or Europe (Olson-Kennedy et al., 2016; Stieglitz, 2010; Tankersley et al., 2021; Tyni et al., 2024). Additionally, much of the research assessing TGD risk and resilience remains characterized by cross-sectional designs and convenience sampling, limiting generalizability. The absence of longitudinal, nationally representative datasets is a critical gap in the evidence, preventing a more complex understanding of causal pathways and long-term impact.

Future directions for research

There are two theoretical frameworks that have been identified as a foundation for future research. The Gender Minority Stress and Resilience Framework (GMSR) is a synthesized framework for researchers conducting risk and resiliency research and one that explicitly situates health outcomes within systems of stigma, discrimination, and power (Hendricks and Testa, 2012). By linking minority stress to adverse outcomes, this framework counters deficit-based narratives by accounting for risk without implying inherent vulnerability. By embedding resilience within social contexts the GMSR addresses previous critiques of models focusing on individual-level resilience.

A second framework is the Transgender Resilience Intervention Model, which builds off the GMSR framework and emphasizes both group-level resilience factors (e.g., having role models, community belongingness, family acceptance, social support, activism) and individual-level factors (e.g., hope, self-acceptance, identity pride, self-worth, gender transition) (Matsuno and Israel, 2018). The aim of this model is to provide a framework for understanding how individual, group, and community interventions can target specific resilience factors. The use of the above frameworks is encouraged in future research to importantly shift the focus from risk to resiliency.

In addition to a renewed focus on protective factors and resilience, future research is advised to study how youth resist oppression, find joy, and enact situated agency—not just how they endure harm (Hillier et al., 2020; Robinson and Schmitz, 2021; Travers et al., 2022). Future work on TGD youth risk and resilience must move toward longitudinal, intersectional, and youth-led research that reduces harm, strengthens agency, and evaluates the protective mechanisms of affirming families, schools, communities, and healthcare systems (Toomey, 2021; Vaughan et al., 2021; Verguchten et al., 2023; Zeeman et al., 2017).

Gender identity development

Identity development research has been historically limited by cisnormative assumptions about gender, including that gender identity remains static and stable (Robinson, 2022). The majority of development literature is focused on (presumed) cisgender children, making the results far from generalizable to TGD children. Newer research has attempted to explore TGD identity development and reframe this development as a normative, dynamic process (Katz-Wise et al., 2017).

Previous research suggests that children can generally differentiate between male and female faces and voices as early as 6 months, and by 1 year can exhibit an understanding of their own assigned sex (Poulin-Dubois et al., 1994; Rhodes and Baron, 2019). By 18 months, many children begin to exhibit preferences for toys stereotypically associated with their assigned sex, as well as for friends who share their assigned sex (Solomon, 2016; Serbin et al., 2001). Middle childhood is marked by increasing flexibility in gender identity and preferences, with most children becoming less rigidly gender-conforming over time. Transgender children show similar developmental patterns to cisgender peers, maintaining stable gender identities while often becoming less bound by stereotypical preferences (Hässler et al., 2022). In adolescence, gender stereotyping generally declines and youth increasingly recognize both social and biological influences on gender, with transgender adolescents resembling cisgender peers in these trends but often placing greater importance on their gender identity (deMayo et al., 2022).

Results from the TransYouth Project (TYP), a longitudinal study following over 300 socially transitioned binary transgender children, have shown that children ages 3–5 years follow a similar trajectory as cisgender children regarding self-categorization of gender, gender-stereotyped behavior, and toy, peer, and clothing preferences (Fast and Olson, 2018; Olson and Gülgöz, 2018). A cohort of nonbinary and gender-nonconforming children was also recruited to the TYP to capture the experiences of youth who fall outside the binary (Rae et al., 2019). One of the major limitations of the above study and others focusing on TGD youth in early childhood is the exclusion of TGD individuals who realize their gender identity in adolescence or adulthood. A secondary analysis of the 2015 US Transgender Survey, assessing over 27,000 TGD adults, found that 40.8% of respondents reported realization of their TGD identity after age 10 (Turban et al., 2023b). Transgender women were more likely to report an understanding of their gender in childhood, while nonbinary individuals were more likely to report awareness in adolescence or later. Additionally, younger generations are more likely to identify with nonbinary identities than those in older generations (Puckett et al., 2022).

Key research gaps

Much of the existing research on TGD identity development stems from retrospective approaches with TGD adults. While offering rich insight, such research is limited by potential recall bias and cohort effects. The rapidly changing landscape of TGD healthcare further compounds these challenges, making it more difficult to differentiate cohort-specific influences and trajectories. Longitudinal data that follows TGD youth through various developmental milestones is needed. This work should include youth who initiate transition at multiple stages of childhood and adolescence, rather than restricting focus to the subgroup of early disclosers. Less is known about the early developmental pathways of TGD individuals disclosing their gender in adolescence or later.

An increase in the visibility of nonbinary-identified individuals calls for researchers to challenge how cisnormative ideals can pervade TGD research. Attempting to classify TGD youth in ways that reflect stereotypical "masculine" or "feminine" preferences is conceptually limited by conflating gender expression and interests with gender identity (Salinas-Quiroz et al., 2025). The developmental pathways demonstrated and reported by nonbinary youth indicate that identity development is not always linear or binary. Nonbinary individuals also have a unique and nuanced story to tell; those whose gender expression conform to binary expectations report being less subject to victimization and discrimination, but may also receive less inclusion and validation from others in the TGD community (Diamond, 2020). Further research is needed to better understand the distinct developmental needs of those with nonbinary identities.

Future directions for research

Although some have questioned or tried to identify the cause of TGD identities, researchers should cautiously examine the utility of such a research question. Previous attempts to determine the etiology of marginalized communities, such as the Lesbian, Gay, Bisexual, Transgender, and Queer (LGBTQ+) community or disability community, have been made with the goal of developing curative or preventative interventions. This is an outcome that many minoritized people are not interested in. As participants in a prior disability study noted, they would not wish to be "cured," rather, they "want to be accepted and appreciated, to have society make room for them as they are" (Dueker, 2002). Similarly, many TGD individuals experience true pride in their identities. In a qualitative study soliciting advice from TGD youth to other TGD youth, one participant stated, "It's not something you have to be ashamed of … there's nothing wrong with you" (Goffnett and Paceley, 2020).

Future research on gender identity formation is encouraged to move away from a stage-based view of development that views identity as evolving in a linear trajectory. Gender identity should be viewed as a fluid construct influenced by culture, context, experience, and environment. This research should examine the full range of subgroups within gender-diverse populations, including those with binary and nonbinary identities, those who disclose earlier in life and later in life, and those with intersectional identities.

Mental health

Previous research has consistently demonstrated that TGD individuals are disproportionately impacted by negative physical and mental health outcomes (Bariola et al., 2015; Chaudhary et al., 2024; Dickey and Budge, 2020). TGD individuals have been found to be disproportionately impacted by depression, anxiety, suicidal ideation and attempts, self-harm, posttraumatic stress disorder, and other mental health conditions (Glintborg et al., 2025; Reisner et al., 2016; Testa et al., 2017).

An unintended consequence of this essential research has been the normalization of a narrative that frames TGD people as chronically unwell or treatment-dependent. An aptly titled article called "Stop Raising Awareness Already" reminds researchers that raising awareness without subsequent action has the potential to be harmful, create backlash, and reach the wrong audience (Christiano and Neimand, 2017). Narratives that only focus on the negative experiences of being gender-diverse can exacerbate transphobic attitudes while also creating a sense of hopelessness among TGD people (Vaughan and Rodriguez, 2014).

Even limited engagement with TGD individuals makes clear that gender diversity is not universally experienced as distressing. Recent research has begun to explore the concept of gender euphoria, a term used to describe positive gender-related emotions and experiences (Jacobsen and Devor, 2022; Ashley and Ells, 2018; Beischel et al., 2022; Bradford et al., 2021). A study of TGD youth aged 9–14 years found that feelings of gender euphoria emerge through feelings of safety, access to affirming care, and hope or promise for the future (Skelton et al., 2024). Additionally, higher levels of gender euphoria have been found to reduce psychological distress and suicidal ideation (Grant et al., 2024). In addition to euphoria, TGD youth and young adults report numerous other positive aspects related to their identity, including connection and companionship, emancipation from rigid gender role expectations, and camaraderie with other marginalized groups (Amos et al., 2023). Future research must move beyond risk-only models and toward a strengths-based approach that aims to proactively bolster gender euphoria rather than solely mitigate mental health sequelae.

Minority stress as context

The minority stress model has been used as a framework for understanding poor mental health outcomes among TGD individuals. Initially created to describe the impact of negative experiences on the lesbian, gay and bisexual+ community, the minority stress model posits that the stressors faced by this population are unique, chronic, and socially-based (Meyer, 2003). As such, identified health and mental health disparities are not a reflection of inherent psychological issues. More recently, this model has been adapted to the unique stressors experienced by TGD people (Hendricks and Testa, 2012). Understanding the correlation between negative or "stressful" experiences and poor mental health outcomes helps to depathologize gender diversity. Within this model, stressors are defined as either distal (external) or proximal (internal). Distal stressors include things such as physical and sexual violence, discrimination, and harassment—all of which are experienced by TGD people at a disproportionate rate, and all of which are associated with mental health disorders, suicide, and substance abuse (Xavier et al., 2007; Beemyn and Rankin, 2011; Clements-Nolle et al., 2006; Collier et al., 2013; Factor and Rothblum, 2008).

Proximal stressors, while less studied, are also known to exacerbate negative mental health outcomes. Examples of proximal stressors may include the expectation of violence, discrimination, and rejection; internalized transphobia; and identity concealment (intentionally concealing a TGD identity to avoid intimidation) (Hendricks and Testa, 2012). It has also been proposed to add a new minority stress category focused on dysphoria stemming from both physical gender discrepancies as well as lack of interpersonal affirmation of gender identity (e.g., using the correct name and pronouns) (Toomey, 2021). Better understanding the direct, destructive impacts of minority stress is an important step in identifying protective factors. Effective interventions must consider these social determinants of health and approach minority stress and stigma as multilevel constructs that impact TGD youth on the individual, interpersonal, and structural levels (Hatzenbuehler and Pachankis, 2016).

Caregiver support and acceptance

Parent and caregiver support has been found to be one of the strongest protective factors against poor mental health outcomes in TGD youth (Bariola et al., 2015; Veale et al., 2017). In contrast, parental rejection or ambivalence is associated with increased self-harm, suicidality, risk-taking behavior, and other negative life circumstances such as homelessness, survival sex, sexual exploitation, and engagement with the foster system (Bateman, 2015; Choi et al., 2015; McCormick and Schmidt, 2018).

Parents report their own experiences with having a TGD-identified child, often involving complex feelings such as anxiety, sadness, loneliness, and grief and loss (Brill and Pepper, 2022; Norwood, 2013; Wahlig, 2015). However, half of all studies in a narrative literature review of parents found noted benefits from having a TGD child, including improved parent–child relationships, affirmation of values such as unconditional love, and watching their child flourish and participate in life (Horton, 2022; Abreu et al., 2019). A qualitative study centered on the positive aspects of parenting a TGD child found the following themes: personal growth, experiencing the strengths of the child, improved parent–child relationship, positive change in the child, and experiencing social connection and support (Morgan et al., 2023).

Research has also explored the benefits of peer support groups and other parent programs for caregivers of TGD youth. Peer support has demonstrated numerous benefits, such as facilitated connections, normalizing the parent journey, and reducing feelings of isolation (Di Ceglie and Thümmel, 2006; Hillier and Torg, 2019; Lawlis et al., 2020; Menvielle and Rodnan, 2011).

Social, medical, and legal affirmation

Extant research consistently demonstrates that social, medical, and legal affirmation are associated with improvements in mental health, elevated psychosocial functioning, and reduced gender dysphoria and suicidal ideation (Achille et al., 2020; Chen et al., 2023; Tordoff et al., 2022; Turban et al., 2023a). Research exploring the psychosocial impacts of gender-affirming medical interventions is limited but growing. Results from a prospective follow-up study suggest that pubertal suppression in TGD youth has a positive effect on overall mental health by reducing depressive symptoms and improving general functioning (De Vries et al., 2011). Questions remain as to whether pubertal suppression improves mental health on its own or merely serves as a prevention tool against worsening mental health (Horton, 2024). A study from the Netherlands found that TGD youth treated with pubertal suppression demonstrated similar mental health symptomology as their cisgender peers despite demonstrating poorer mental health pretreatment (van der Miesen et al., 2020). It is also well-documented that family support is a strong mediator. Socially transitioned TGD youth with the support of their families demonstrate stable mental health across all stages of medical affirmation (Wittlin et al., 2025).

Research has also repeatedly shown that initiation of gender-affirming hormones (GAH) is associated with improved quality of life and reductions in depression, anxiety, and suicidality (Achille et al., 2020; Allen et al., 2019; Chen et al., 2023; Kuper et al., 2018; Kuper et al., 2020; Lavender et al., 2023; Olsavsky et al., 2023; Tordoff et al., 2022). One study of TGD youth found that the initiation of GAH reduced the risk of an emergency department visit or inpatient hospitalization for suicidality by 43.6% compared to those never prescribed GAH or before GAH (Nunes-Moreno et al., 2025). There is also emerging evidence that the timing of these interventions matters. A study of TGD adults in the United States found that TGD individuals who accessed GAH as adolescents demonstrated better mental health outcomes than those who accessed GAH as adults (Turban et al., 2023b).

Intervention research

Relative to other fields of mental health research, there is a paucity of validated, evidence-based interventions and therapies developed for TGD youth. Despite this, there have been some promising developments in the last few years demonstrating the feasibility, acceptability, and effectiveness of interventions for youth mental health. The AFFIRM model is a group-based cognitive behavioral therapy (CBT) intervention for LGBTQ+ youth, which has demonstrated reductions in depressive symptoms among participants (Austin et al., 2018). It has also been adapted for parents and caregivers (AFFIRM Caregiver) to assist with increased caregiver support and affirmation (Austin et al., 2021). Releasing Internalized Stigma for Empowerment (RISE) is an online self-led intervention that helps TGD youth better understand and cope with gender-related stress (Israel et al., 2021).

Researchers have also begun to evaluate the efficacy of parent and caregiver support and therapies. Successful components of parent programs include: a nonjudgmental setting in which to process negative emotions, psychoeducation on gender diversity, community support from other parents, and resources for improving parent–child communication (Vasen et al., 2025). More recently, there have been several trials for clinician-led group therapy programs for parents grounded in evidence-based approaches (Caldarera et al., 2021; Dangaltcheva et al., 2021; Morgan et al., 2024).

While there are no mental health treatments validated exclusively for TGD youth, gender-affirming adaptations of existing therapies have shown efficacy with TGD youth. A 2022 systematic review of 22 studies of affirmative interventions showed consistent pre–post improvements in depression, anxiety, suicidality, coping, and self-esteem (Expósito-Campos et al., 2023). CBT, trauma-focused CBT, dialectical behavior therapy (DBT), and family therapies have all been validated for youth mental health and are supported by many years of randomized controlled trials in the general youth population. These interventions may require tailoring to meet the specific needs created by gender minority stressors.

Key research gaps

There are numerous methodological, structural, and political reasons for the insufficiency of validated, evidence-based treatment interventions for TGD youth. As a result, many of the identified interventions or proposals are situated as guidelines or theoretical models or based on expert opinion rather than trials (Dolotina and Turban, 2022; Holt et al., 2021).

Additionally, randomized controlled trials evaluating psychosocial outcomes are often considered unethical due to the nature of withholding treatments that are known to be beneficial (Ashley et al., 2024). Some researchers have attempted to approximate randomized control trials by using waitlist comparison groups (Chinsen et al., 2024; Allen et al., 2021), although these are imperfect substitutes. Studies with waiting list controls may overestimate treatment effects (Cunningham et al., 2013). Meta-analyses in people with anxiety disorders research have found that study participants randomized to waitlists may have worse outcomes than those in no-treatment groups (Patterson et al., 2016).

Given the robust evidence that family support serves as one of the strongest mediators of poor mental health outcomes, it is also important to highlight the gaps in family research. There is a substantial lack of underrepresentation of male caregivers in the current research base, as well as families from culturally diverse backgrounds. Psychosocial interventions for families have often included families accessing specialty pediatric gender care, and these results are not considered to be generalizable to the wider population, particularly as it inherently omits unsupportive families and underresourced families. The mental health and well-being of siblings of TGD youth are also underexplored (Klepper et al., 2024).

Future directions for research

Future research on the mental health of TGD youth, as well as accompanying interventions, should prioritize methodological rigor and address prior challenges such as small sample sizes, high attrition rates, and lack of long-term follow-up evaluating sustained outcomes (Expósito-Campos et al., 2023; Heathcote et al., 2024). There is a strong need for measures that are developed, validated, and normed for TGD youth specifically, and our understanding of the needs of the TGD community should not be dependent on the use of disaggregated LGBTQ+ data. Randomized control trials should be employed when possible and when deemed ethically appropriate.

Much of the previous research compares the mental health outcomes of TGD youth to their cisgender counterparts, but careful attention must also be paid to subgroups within the TGD community. Recent studies have found differences between nonbinary and binary TGD youth mental health; nonbinary individuals report higher levels of depression, anxiety, distress, and substance abuse (Burgwal et al., 2020; Hainey et al., 2025; Stanton et al., 2021). Better understanding the cause of these mental health discrepancies, as well as identifying effective interventions, is a much-needed addition to the evidence base.

Finally, a more inclusive approach to family research is also warranted, as existing studies have minimally included cisgender male caregivers, families of different cultural backgrounds, unsupportive parents, and parents of nonbinary children. Researchers are encouraged to challenge the dominant narrative of parental grief and loss while also examining the efficacy of family support interventions. Research should explore how to disseminate resources and tailor interventions for those with limited access to the larger gender centers, including those living in rural areas.

Disentangling the detransition literature

There are a limited but growing number of studies aiming to understand the rates of detransition or discontinuation of gender-affirming medical interventions and the needs of this subpopulation. A secondary data analysis of the 2015 US Transgender Survey sought to quantify the number of TGD adults who had ever reidentified with their assigned sex at birth (Turban et al., 2021). Results found that 13.1% of individuals endorsed having "gone back to living as [their] assigned sex, at least for a little while." Of those, the majority (82.5%) endorsed at least one external reason for stopping (such as discrimination or cost). The Research-Based, Ethical, Affirming, and Life-Saving (REAL) Story Project, currently underway at a pediatric gender center in the, found that 21.3% of 141 survey respondents reported having stopped GAH—although all but three patients had restarted GAH or were considering it. A retrospective chart review from a large, US pediatric gender center of 1050 patients found low rates of discontinuation of GAH (4%), with only 0.5% reidentifying with their assigned sex (Boskey et al., 2025). Another study of 720 individuals aged 15–29 years found that 16.5% of those who initiated some form of gender-affirming treatment later discontinued; of those, nearly 40% reported wishing they had not (MacKinnon et al., 2024). The most common reasons for stopping treatment were health concerns, a change in gender identity, and cost.

A qualitative study of youth and young adults in Canada who discontinued prior transition steps found that the reasons were highly nuanced, with some individuals feeling grateful for their experience with gender-affirming care and others feeling regret (Sansfaçon et al., 2023). Other resultant themes included feeling pressured to follow a predefined transition path and a lack of appropriate discussion about the decision to transition beforehand. Notably, this sample had a small sample size, with only 13 participants undergoing some form of medical transition, and only three participants started medical transition before the age of 18 years. More research is needed into individuals who make decisions regarding gender-affirming care as youth.

While the use of surveys has great potential to address limitations related to small sample sizes, researchers warn that surveys with pregenerated lists of potential rationales for stopping treatment carry a risk of researcher bias (MacKinnon et al., 2024). With this in mind the Detransition Analysis, Representation, and Exploration (DARE) study utilized a survey grouping 21 reasons for transition into four domains: mental health/psychological reasons, physical reasons, external reasons, and social/internal reasons (MacKinnon et al., 2024). Analysis is currently underway.

Key research gaps

One of the biggest limitations in research evaluating detransition is a lack of clearly agreed-upon terms to describe unique phenomena, and that these terms (e.g., detransition, retransition, reidentification, desistence) are often conflated with regret (Ashley, 2022; Dolotina and Daniolos, 2023; MacKinnon et al., 2023). Discontinuation of gender-affirming medical interventions, such as pubertal suppression and GAH, is its own distinct outcome that can be related to numerous factors beyond regret, such as achieving transition goals, evolving identity, cost, or health concerns (Crabtree et al., 2024; MacKinnon et al., 2024; Turban et al., 2018). This variable and often disparate use of terminology in research is a significant impediment to the field's ability to categorize this sample, and subsequently, the ability to provide ancillary supports.

Some researchers have advised against the use of words such as "transition" and "detransition" as they presuppose that TGD identities follow a linear trajectory and, in some cases, emphasize a belief that all TGD people should aspire to medical interventions to affirm their gender (Turban et al., 2022). Others posit that the terminology itself is not problematic, but rather the stigmatization of this language warrants change (MacKinnon et al., 2023). A recent qualitative study found that young people who have detransitioned do not agree with the current narrative around this concept, feel that they are being misrepresented, and report isolation from potential community (Gelly et al., 2025).

Future directions for research

There are several key limitations that need to be addressed in future research. It is critical for the field, with direct input from the community, to agree upon and operationalize the terms used in detransition and discontinuation research. Studies should also importantly shift away from the binary outcome categorization of "regret" versus "satisfaction" and toward a multidimensional spectrum of diverse experiences that cannot be meaningfully collapsed into a single category. Utilizing shared language that is clear and deliberate can prevent the conflation of experiences with regret. Longer follow-up periods and utilizing mixed-methods approaches may reduce bias while gaining qualitative insight into these complex experiences.

While pediatric gender centers are uniquely positioned to capture prospective and retrospective data, often from large patient populations, these studies have experienced high attrition for a variety of reasons, such as not returning for care after discontinuing a gender-affirming medication or discomfort discussing this topic with providers due to stigma around the concept of detransition (Vandenbussche, 2022). While retrospective chart review provides valuable data without the potential for losses to follow-up, it is also incomplete without direct feedback from the patient. Qualitative methods or mixed-methods designs may be useful in reducing observer and misclassification biases. Additionally, previous studies have had short follow-up windows, with most following youth for 6–24 months. Outcomes beyond 2 years remain undercharacterized, and studies that follow TGD youth longitudinally into adulthood are needed.

There is also a clear need for greater diversification of study samples. Individuals with access to specialty gender clinics are inherently better-resourced, and results may not be generalizable to individuals living in rural or low-resource communities. Also underrepresented in the current research are transfeminine youth, TGD youth of color, and nonbinary youth. All efforts should be made to intentionally recruit nonbinary and transfeminine youth and widen racial, ethnic, socioeconomic, and geographic representation. Finally, researchers must exercise great caution and consideration when identifying research questions and should collaborate and coset research priorities with TGD community members as often as possible and consider the impacts of research on the community.

Medical interventions

There is ongoing discussion about which outcome measurements are most appropriate for evaluating the effectiveness of gender-affirming medical interventions for TGD adolescents. Despite clear and consistent evidence of the physiological effectiveness of these treatments (i.e., successful prevention of unwanted pubertal changes using pubertal suppression and facilitating masculinizing and feminizing body changes using testosterone and estradiol), research to date has focused more heavily on the psychosocial outcomes associated with these treatments. A reflective thematic analysis of the literature cited in the Adolescent Chapter of the WPATH SOC8 identified 44 measures representing a range of outcomes, including gender dysphoria, body dissatisfaction, psychological functioning, social functioning, quality of life, and satisfaction with care (Oosthoek et al., 2024). The thematic analysis revealed four themes regarding perspectives on effectiveness, including (1) poorer functioning before medical treatment, (2) improvement following medical treatment, (3) achieving a consistent and often binary gender identity and presentation, and (4) absence of regret.

The narrative of ultimate gender stability and linear improvement in functioning is rooted in cisnormative bias and likely oversimplifies the complexity and mixed emotions that may be associated with affirming gender (Oosthoek et al., 2024). In fact, TGD adults who pursued care in adolescence often have positive feelings about their care and transition steps that are nonetheless accompanied by negative emotions, such as frustration with medical systems, disappointment with the extent of gender congruence that could be achieved, or even regret that some interventions could not be accessed earlier (Crabtree et al., 2024). While this body of research has been informative and nonetheless supports the provision of gender-affirming care for TGD adolescents, future research should continue to evaluate the utility and appropriateness of different measures of effectiveness, and in particular to involve TGD youth in the development of research questions and methods to produce research that directly meets the priorities of this community.

Pubertal suppression

The first article on pubertal suppression with gonadotropin-releasing hormone agonists (GnRHa) for TGD youth was published in 1996 (Gooren and Delemarre-van de Waal, 1996), and treatment became more widespread in the 2000s. WPATH and the Endocrine Society recommended GnRHa treatment for eligible youth starting at sexual maturity rating stage 2 of pubertal development, and assert that it is generally safe and effective (Coleman et al., 2022; Hembree et al., 2017). GnRHa treatment reliably suppresses gonadotropins and sex steroids, halting further development of unwanted secondary sex characteristics and allowing time for psychosocial development and clinical decision-making (Coleman et al., 2022; Hembree et al., 2017).

Initially, the use of GnRHa was intended as an extended diagnostic phase. Subsequent studies showed that GnRHa monotherapy is associated with lower anxiety, depression, and suicidal ideation compared to before treatment (Fisher et al., 2024; van der Miesen et al., 2020), and many adolescents report subjective relief from the cessation of unwanted pubertal changes (Carmichael et al., 2021). Limited evidence suggests no clear adverse effects on global cognitive ability, although some studies raise questions about executive functioning during prolonged GnRHa monotherapy and the need to account for confounders such as neurodiversity, anxiety, and educational context (Strang et al., 2022). High rates of continuation from GnRHa to GAH indicate that early treatment decisions are generally stable over time. Access barriers, consent processes, and variability in clinical pathways may influence outcomes from these interventions.

At baseline, before GnRHa treatment, many TGD youth (particularly those with a male sex assigned at birth) have lower bone mineral density (BMD) compared to their cisgender peers, likely reflecting differences in physical activity, nutrition, and psychosocial stressors (Lee et al., 2020). During GnRHa monotherapy, bone accrual slows relative to age-matched peers undergoing endogenous puberty, but longitudinal cohorts suggest that BMD generally normalizes after GAH, with the exception of the lumbar spine among those with a male sex at birth (Lee, 2023). Height velocity is generally maintained at prepubertal rates while on GnRHa monotherapy, and typically normalizes with the introduction of GAH. Final adult height usually aligns with genetic target height, although early suppression may modestly influence skeletal geometry toward affirmed-gender patterns (Boogers et al., 2022; van der Loos et al., 2021; Schulmeister et al., 2022). Body composition changes during GnRHa alone are variable; there are no significant changes in blood pressure (Taylor et al., 2024), and there are mixed results for insulin sensitivity (Klaver et al., 2020; Nokoff et al., 2021).

Key research gaps

Major gaps include the lack of large, prospective, long-term datasets extending from early pubertal suppression into later adulthood, when outcomes such as peak bone mass, fracture risk, cardiovascular disease, and reproductive health become more clinically meaningful. More research is needed on the impact of pubertal stage at initiation of GnRHa and age of subsequent GAH on adult bone health, skeletal structure, final height, and body composition in later adulthood. Neurocognitive outcomes and fertility remain underexplored, with insufficient data on short- and long-term cognitive effects and limited fertility preservation options for early pubertal youth, underscoring the need for both biomedical and qualitative research on decision-making. Finally, there remain significant barriers to access, and future research should evaluate how barriers to GnRHa access affect long-term well-being. Ethical and practical barriers limit randomized trials, underscoring the need for rigorous observational designs and harmonized outcome measures.

Future directions for research

Future work should prioritize multicenter, longitudinal cohorts with standardized measures to assess psychosocial measures, including quality of life, gender embodiment goals, appearance congruence, and mental health, as well as physical measures such as bone density, cardiometabolic health, and fertility outcomes. Studies are needed that follow individuals for a longer time into adulthood. Research is needed on the optimal timing and duration of GnRHa monotherapy, and the optimal timing for introduction of GAH. Dedicated neurocognitive studies using repeated measures and appropriate comparison groups are needed. Advances in fertility preservation technologies and applicability and outcomes for TGD individuals are needed. Finally, implementation and health equity research should address barriers to access, informed consent processes, and the psychosocial consequences of delayed or denied care, ensuring that future clinical guidance is grounded in robust evidence and inclusive of diverse patient experiences.

Testosterone and estradiol

GAH, including testosterone and estradiol, induce secondary sex characteristics that align with gender identity. Numerous longitudinal cohort and cross-sectional studies demonstrate that treatment with GAH is associated with reductions in depression, anxiety, suicidality, gender dysphoria, and body dissatisfaction, alongside improvements in quality of life, self-esteem, and overall well-being (De Vries et al., 2014; Chen et al., 2024, 2023; Olson-Kennedy et al., 2025). These benefits are shaped by contextual factors, particularly parental support, and appear to be mediated by increased gender appearance congruence and body satisfaction, which may enhance emotion regulation and executive functioning (Chen et al., 2024). Physical health parameters—including body mass index, glucose metabolism, blood pressure, liver enzymes, renal indices, and hematologic values—typically shift toward ranges observed in cisgender peers of the same affirmed gender after hormone therapy. Route of administration studies suggest comparable efficacy across testosterone formulations and potential advantages of transdermal estradiol for thromboembolic risk, bone health, and testosterone suppression (Cortez et al., 2024; King et al., 2024), although adolescent-specific comparative data remain limited.

Key research gaps

Key gaps include the absence of long-term prospective data extending into mid- and later adulthood, limited adolescent-specific comparisons of hormone delivery routes and dosing strategies, and an incomplete understanding of the neurocognitive effects of prolonged hormone exposure during adolescence. Evidence is sparse on individualized pathways of care, such as sustained low-dose regimens, partial or time-limited hormone use, and nonbinary embodiment goals. Physical health research has not consistently incorporated measures of minority stress, healthcare access, physical activity, nutrition, and socioeconomic context, despite strong evidence that these factors influence baseline health disparities. Finally, physical and mental health outcomes among adolescents who received pubertal blockade before initiation of GAH, compared to those who presented later in adolescence and were not treated with GnRHa, are needed.

Several large-scale studies across the United States and Europe have attempted to fill these gaps, including the Trans Youth Care—US Study (Olson-Kennedy et al., 2019), the Trans Youth Longitudinal Study (Kuper et al., 2020), the TransYouth Project (Wittlin et al., 2025), and the Trans Youth CAN! Study (Bauer et al., 2021). The majority of longitudinal studies have used clinic-based samples, although there are also examples from the community (Wittlin et al., 2025), youth seen in military hospitals (Hisle-Gorman et al., 2021), and large surveys (James et al., 2024).

Future directions for research

Prospective research should directly examine the timing of initiation, duration of exposure, and individualized dosing strategies, including low-dose and short-term pathways aligned with diverse embodiment goals. Future studies should prioritize large, multicenter longitudinal cohorts with standardized outcome measures of appearance congruence, gender embodiment goals, minority stress, quality of life, and mental health. Longer-term follow-up is needed to understand the possible impact of hormones on cardiometabolic risk, bone health, neurocognition, and whether those outcomes vary based on prior pubertal suppression or not. Finally, research framed within a minority stress and health equity lens will be essential to disentangle treatment effects from structural determinants of health and to inform ethical, patient-centered clinical guidance for GAH use in youth.

Other treatments

There are a broad range of gender-affirming medical and nonmedical interventions that extend beyond traditional hormone therapy and pubertal suppression, emphasizing the importance of individualized care for TGD youth, particularly those with nonbinary identities or specific embodiment goals. Goals may include steps around social transition, including using an affirmed name or pronouns. Nonmedical interventions may include chest binding, tucking, packing, use of body forms, voice training, and hair removal (Bradford et al., 2021; Julian et al., 2021; Russell and Abrams, 2019). Medical interventions, either instead of or concurrent with testosterone or estradiol, may include medications for menstrual suppression (progestin-only or combined estrogen-progestin) or antiandrogens (Alaniz et al., 2023). Medications may also be used to alleviate expected side effects such as hair loss or acne (Hollingshead et al., 2022). Together, these findings support a flexible model of care that prioritizes patient goals and lived experience over standardized or binary treatment pathways.

Key research gaps

Much of the current knowledge is derived from small samples, retrospective chart reviews, case reports, or adult populations, with relatively few prospective or controlled studies focused specifically on adolescents. Outcomes are inconsistently defined and measured. Research often prioritizes biomedical indicators while underassessing psychosocial outcomes such as quality of life, gender euphoria, or functional improvements in social participation. Moreover, many commonly used practices (including packing, tucking, shapewear, progesterone supplementation, and facial hair enhancement) have almost no systematic study despite their widespread use. The literature is further constrained by persistent binary assumptions about gender transition, with little data on those who identify as nonbinary.

Future directions for research

Future directions for research should focus on satisfaction and achievement of individual gender goals, including a focus on positive outcomes such as gender euphoria and identity affirmation, and broadening our understanding of what constitutes successful care. Additional research is needed on outcomes for nonbinary patients and the use of alternative hormone regimens, such as low-dose or a limited course of testosterone or estradiol. Safety and effectiveness data are needed for widely used but understudied practices such as binding, tucking, packing, hair modification, and adjunctive hormonal strategies. Research should incorporate mixed methods and participatory designs that center youth voices to better capture lived experience and real-world decision-making. Finally, investigations of cost, access, and implementation barriers are essential to ensure that beneficial interventions are equitably available.

Surgical interventions

While most professional organizations recommend deferring genital surgeries until the age of majority, other surgeries, such as chest masculinizing surgery, have been considered earlier, depending on the developmental stage of the adolescent. Some adolescents considering future surgery would benefit from support with ongoing preparation for surgery and recovery; and it is useful for clinicians to understand the range of potential procedures and considerations. Surgical options include chest ("top"), genital ("bottom"), facial, vocal, and body contouring procedures, with timing and selection individualized based on developmental stage, embodiment goals, prior medical treatments such as GAH or pubertal suppression, and local/regional age-based guidelines. Surgeries for people with a male sex at birth may include breast augmentation, orchiectomy, vulvoplasty, vaginoplasty, facial feminization surgery, glottoplasty (vocal surgery), and body contouring (Safa et al., 2019a). Surgeries for people with a female sex at birth may include mastectomy or chest masculinizing surgery, metoidioplasty, phalloplasty, hysterectomy, salpingo-oophorectomy, facial masculinization surgery, vocal fold augmentation, and thyroplasty (Safa et al., 2019b).

Across the literature, masculinizing chest surgery is the most commonly accessed procedure in adolescence (Handler et al., 2019) and is associated with marked reductions in chest dysphoria and improvements in gender congruence, mood, and social functioning (Olson-Kennedy et al., 2018). Studies in adults and emerging-adults show that gender-affirming surgery is linked to lower depression, anxiety, suicidality, and psychological distress, with high satisfaction, low regret, and improved quality of life, suggesting durable mental health benefits when care is accessible and appropriately timed (Murad et al., 2010; Park et al., 2022; Swan et al., 2023).

However, access to surgery remains limited. National surveys and health system data indicate that many youth desire surgery, yet only a small fraction undergo procedures due to structural barriers, including cost, insurance coverage, geographic availability of surgeons, and fear of losing family support (Abern et al., 2023; Uffman et al., 2023). Youth who do undergo surgery are disproportionately insured and less racially or ethnically diverse (Uffman et al., 2023).

Key research gaps

Much of the research on gender-affirming surgery relies on retrospective chart reviews, single-center samples, or adult cohorts, with relatively few prospective or longitudinal studies focused specifically on adolescents and young adults. Research is heavily US-based, limiting generalizability to international contexts. Outcome measures are inconsistent, and studies often prioritize surgical complications or satisfaction with less focus on broader domains such as functional recovery, long-term psychosocial adjustment, pain interference, and quality of life. Evidence is limited for the less common or more complex procedures in young adults, including genital surgeries, facial surgery, vocal surgery, and body contouring. Evidence is also limited for nonbinary patients.

Future directions for research

Immediate postoperative challenges such as pain, recovery time, temporary mood declines, or functional limitations may be especially salient for young adults balancing school, work, and independence, yet these experiences are rarely captured in research. More research is needed on outcomes based on individualized embodiment goals. Psychosocial supports, including family involvement, peer support, mental healthcare, and preparation for postoperative recovery, are critical determinants of outcomes but remain understudied. There is a need for larger, prospective, and longitudinal studies that track both medical and psychosocial outcomes over time, greater inclusion of adolescents, nonbinary youth, and diverse racial and socioeconomic groups, and more international collaboration to understand regional differences in access and practice. Standardized outcome tools, mixed-methods approaches, and patient-reported measures are needed to better capture satisfaction, gender congruence, recovery experiences, and quality of life. Research should also examine optimal timing of interventions, the long-term impact of early pubertal suppression on surgical needs, comparative risks and benefits of different techniques, and the effectiveness of perioperative supports such as counseling, peer networks, and pain management strategies.

Sexual health

Sexual development, health, and functioning are central, yet often neglected, components of comprehensive care for TGD adolescents and young adults. TGD adolescents and adults are more likely to identify as a sexual minority, have heterogeneity in their sexual orientation, or adopt a range of labels compared to cisgender adults, with fluidity in identification over time (Reisner et al., 2023; Warwick et al., 2022). Qualitative and quantitative studies indicate that sexual identity development may interact dynamically with gender identity development and may shift before and after gender-affirming interventions (Barbonetti, 2024). Furthermore, reported sexual orientation, attraction, and behaviors do not always align (Boskey and Ganor, 2022), and an increasing number of people are identifying as asexual (Chen, 2020). Some people may experience gender dysphoria around their bodies or sexual experiences, whereas others may find these affirming and empowering.

Some TGD people delay or avoid preventative healthcare due to fear of discrimination or disrespect (Jaffee et al., 2016; Rachlin et al., 2008). Most research on sexual wellness in TGD populations is focused on those living with HIV, or the risk of HIV or other sexually transmitted infections. TGD individuals have a higher prevalence of HIV infection, particularly black transgender women (Becasen et al., 2019). A sex-positive framework of sexual health among TGD individuals is needed, emphasizing autonomy, pleasure, consent, and normalization of sexual experience that moves beyond a focus on negative outcomes.

Emerging evidence also suggests that GAH and surgeries are often associated with increased sexual satisfaction and activity, mediated by reduced dysphoria and improved body congruence, although individual trajectories vary (Bungener et al., 2020; Mattawanon et al., 2021). More data are needed on the impact of pubertal blockade and sexual functioning and satisfaction, particularly for those with a male sex at birth. In one study, there were no differences in sexual desire postsurgery among individuals assigned male at birth with a history of early versus late pubertal suppression (Van Der Meulen et al., 2025).

Key research gaps

There are gaps in theory, measurement, and sampling related to the sexual health experiences of TGD young people. Most available instruments for sexual functioning and well-being were developed for cisgender, binary populations and fail to capture diverse bodies (including after GAH or gender-affirming surgery), practices, partners, and values among TGD people (Barcelos et al., 2024). There are currently no empirically validated, developmentally appropriate measures specifically designed for TGD adolescents. Much of the literature overemphasizes risk (e.g., HIV, victimization) while underrepresenting pleasure, resilience, intimacy, consent, and positive sexual development. There is a need for more research involving the sex-positive experiences of TGD people (Anzani et al., 2024). The Sex Positivity Scale is a measurement tool that may be used but has not been validated in adolescents or TGD individuals (Belous and Schulz, 2024). Existing studies are often cross-sectional, clinic-based, geographically limited, and insufficiently diverse with respect to race, socioeconomic status, disability, relationship structures, and geographic context. Research rarely integrates minority stress, intersectionality, or community perspectives, and longitudinal data on how sexual well-being evolves across adolescence and into adulthood remain scarce.

Research on the impact of early pubertal suppression, GAH, or gender-affirming surgery on sexual wellness is needed, particularly using rigorous methodology and outcomes that are known to be important to patients. This is particularly relevant for individuals assigned male at birth who had early pubertal suppression and the impact of limited genital growth on future gender-affirming surgery (particularly vaginoplasty) and future sexual functioning.

Future directions for research

The primary recommendations for research on sexual health among TGD individuals, and adolescents in particular, are to (1) center sexuality-related values, (2) consider unique sexual development trajectories, (3) promote sexual health and well-being, (4) diversify research samples, (5) reconstruct sexual safety, and (6) revise and empirically validate measurement tools.

Reproductive health and fertility preservation

Reproductive health counseling is an essential but often underdeveloped component of adolescent care. Current research in this area has focused on clinical practices for fertility counseling, referral rates to fertility specialists, and completion of fertility preservation procedures (Cooper et al., 2022; Hudson et al., 2018; Nahata et al., 2017; Pang et al., 2020). There is ongoing research focusing on the creation of tools and decision aids to optimize fertility counseling for providers, patients, and families (Kolbuck et al., 2020; Persky et al., 2020).

Before any medical interventions, adolescents in the later stages of puberty have viable fertility preservation options, including oocyte cryopreservation, ovarian tissue cryopreservation, sperm cryopreservation, and testicular sperm extraction. There are more limited data and options for individuals early in puberty or those treated with a GnRHa for pubertal suppression. A testicular volume of $\geq$10 mL has a 92% sensitivity and 71% specificity for predicting successful sperm retrieval via testicular biopsy (Peri et al., 2021). In one study the youngest patient with an ejaculate containing motile sperm was 12.2 years, and the smallest testicular volumes associated with motile spermatozoa in the ejaculate were 6–7 mL (Hagenäs et al., 2010). There is a case report of successful ovarian stimulation in a transgender male suppressed at sexual maturity rating stage 2 after GnRHa implant removal (Martin et al., 2021).

GAH does not uniformly eliminate fertility potential, as ovulation and spermatogenesis may persist or recover after treatment interruption, and there is a lack of data on what duration of GAH may impact fertility. There are a growing number of TGD individuals with a female sex assigned at birth who successfully complete oocyte cryopreservation on testosterone (Mortimer et al., 2025). Collectively, these findings challenge earlier assumptions that transition inevitably results in sterility and support the importance of offering fertility preservation counseling before and throughout medical care.

Fertility preservation rates among TGD adolescents are low, with only 5%–7% undergoing preservation despite 24%–50% expressing interest in future biological parenthood (Cooper et al., 2022). Barriers include procedural invasiveness, exacerbation of dysphoria (e.g., genital exams or hormone stimulation), costs and lack of insurance coverage, limited geographic access to specialists, uncertainty about future parenthood goals, and the desire not to delay gender-affirming medical treatment (Baram et al., 2019). Fertility preservation decisions occur within complex developmental, emotional, and practical contexts.

Key research gaps

Most research consists of retrospective chart reviews, small case series, and single-center reports, with very few prospective or longitudinal studies following youth over time. Outcomes are largely limited to immediate procedural metrics (e.g., number of oocytes or sperm retrieved). Long-term endpoints including fertilization success, pregnancy, live births, or adult satisfaction with earlier decisions are rarely measured. Early pubertal youth are especially understudied, and experimental approaches such as testicular tissue cryopreservation lack demonstrated human live births. Evidence regarding the effects of GAH duration, optimal timing of preservation, and whether GnRHa or GAH must be paused before fertility preservation remains inconsistent.

Conceptual and ethical gaps further complicate care. Decisions about fertility preservation intersect with autonomy, parental involvement, developmental readiness, and justice considerations such as financial access. Youth may experience tension between immediate gender affirmation and hypothetical future fertility, and some procedures may intensify dysphoria or emotional distress. Research also shows variability in baseline fertility knowledge among youth, families, and providers, and highlights the need for counseling approaches that are affirming, individualized, and developmentally appropriate. Importantly, many studies emphasize that fertility goals may evolve over time, underscoring the need for repeated conversations rather than one-time counseling before treatment initiation.

Future directions for research

Larger, prospective, and longitudinal studies that evaluate both reproductive outcomes (e.g., conception, live birth, offspring health) and psychosocial outcomes (e.g., satisfaction, regret, quality of life) are needed. Continued work to develop fertility-related decision aids is needed (Chen et al., 2019; Kolbuck et al., 2020). Future research should focus on early pubertal TGD youth with limited options for fertility preservation based on gamete immaturity. Additionally, outcomes are needed for fertility preservation on or after GAH alone, or concurrent with or after GnRHa treatment. Furthermore, standardized outcome measures and multicenter collaborations are needed to strengthen generalizability, optimize stimulation protocols that minimize dysphoria, and track live birth outcomes over time.

Conclusion

Since the early 2000s, research with TGD populations has expanded rapidly. Despite this growth, substantial gaps remain, particularly for longitudinal outcomes in TGD adolescents. Multidisciplinary, patient-centered care is recommended but often inaccessible due to stigma, lack of trained providers, geographic inequities, or long wait times. TGD youth face elevated mental health risks driven by minority stress, discrimination, and family rejection, yet emerging research highlights resilience, gender euphoria, and protective factors such as family support, affirming schools, peers, and healthcare environments.

Community-based participatory research designs are needed to understand and meet the priorities of TGD youth and their families. These research designs involve relevant stakeholders in every stage of research, which can improve recruitment effectiveness, optimize the experience of research participants, and ultimately empower underserved communities to shape the scientific agendas that are relevant to them (Katz-Wise et al., 2018). Participatory research designs are growing in their use and recognized value across fields of research related to specific communities. Researchers will need to educate ethics review boards, institutional stakeholders, and policy makers to continue improving community-based participatory research methods in this area (Shook et al., 2025).

Participation of TGD youth and their families and caregivers is especially needed in identifying the most appropriate outcomes to measure the effectiveness of gender-affirming care, as well as in the development of measurement tools (Oosthoek et al., 2024). Research based on existing measures has informed the field in many ways, but there is growing recognition that improved measurement based on TGD youth priorities is needed to advance research in this field. Patient-reported outcome measures that are developed with direct input from TGD youth, such as GENDER-Q Youth, are being developed (Kennedy et al., 2025, 2026). International and institutional collaboration is needed to develop measurements and produce research results that are robust, generalizable, and multiculturally relevant.

Participatory designs can also inform overall research priorities and initiatives. One study employed a Youth Advisory Board to help inform focus groups and a subsequent survey related to the specific priorities of TGD youth, including a focus group specifically for indigenous TGD youth (Clark et al., 2026). Study results demonstrated a need for accessible, up-to-date, quality health information on a variety of topics. Specific research priorities included as follows: (1) how to better prepare youth for medical appointments, talking with providers, and coping with needle phobia, (2) patient-reported outcomes related to gender-affirming medical interventions, and (3) evaluating the unique care needs and experiences of youth experiencing intersectional oppression, such as racism and ableism.

Overall, evidence consistently links social, medical, and legal affirmation to improved mental health and well-being for TGD individuals. Future research in the area of pediatric gender-affirming care must prioritize longitudinal, intersectional, patient-informed, and equity-focused approaches that move beyond risk to center resilience, agency, and quality of life.

AI disclosure

During the preparation of this work the author(s) used ChatGPT to outline key points for the conclusion chapter. After using this tool/service the author(s) reviewed and edited the content as needed and take(s) full responsibility for the content of the publication.

References

Abern, L., Diego, D., Krempasky, C., Cook, J., Maguire, K., 2023. Desire for gender-affirming surgery in a sample of transgender and gender diverse individuals in late adolescence assigned female at birth. Journal of Pediatric and Adolescent Gynecology 36 (5), 488–490. Available at https://doi.org/10.1016/j.jpag.2023.07.001.

Abreu, R.L., Rosenkrantz, D.E., Ryser-Oatman, J.T., Rostosky, S.S., Riggle, E.D.B., 2019. Parental reactions to transgender and gender diverse children: a literature review. Journal of GLBT Family Studies 15 (5), 461–485. Available at https://doi.org/10.1080/1550428X.2019.1656132.

Achille, C., Taggart, T., Eaton, N.R., Osipoff, J., Tafuri, K., Lane, A., et al., 2020. Longitudinal impact of gender-affirming endocrine intervention on the mental health and well-being of transgender youths: preliminary results. International Journal of Pediatric Endocrinology 2020 (1). Available at https://doi.org/10.1186/s13633-020-00078-2.

Alaniz, V.I., Sheeder, J.L., Whitmore, G.T., Wilde, M.D., Hutchens, K.J., Nokoff, N.J., et al., 2023. Menstrual suppression in adolescent and young adult transgender males. Journal of Pediatric and Adolescent Gynecology 36 (2), 116–121. Available at https://doi.org/10.1016/j.jpag.2022.10.007.

Alikunju, H., 2024. Unravelling the movements in the transgender health research of the past 20 years. Journal of Data Science, Informetrics, and Citation Studies 3 (3), 337–348. Available at https://doi.org/10.5530/jcitation.3.3.33.

Allen, L.R., Watson, L.B., Egan, A.M., Moser, C.N., 2019. Well-being and suicidality among transgender youth after gender-affirming hormones. Clinical Practice in Pediatric Psychology 7 (3), 302–311. Available at https://doi.org/10.1037/cpp0000288.

Allen, S.D., Tollit, M.A., McDougall, R., Eade, D., Hoq, M., Pang, K.C., 2021. A waitlist intervention for transgender young people and psychosocial outcomes. Pediatrics 148 (2). Available at https://doi.org/10.1542/peds.2020-042762.

Altschuler, R., Mann, A.K., 2023. The development of accessible and sustainable transgender health education for rural-serving medical providers in Appalachia. Rural and Remote Health 23 (2). Available at https://doi.org/10.22605/RRH7827.

Amos, N., Lim, G., Buckingham, P., Lin, A., Liddelow-Hunt, S., Mooney-Somers, J., et al., 2023. Rainbow realities: in-depth analyses of large-scale LGBTQA+ health and wellbeing data in Australia.

Anzani, A., Rucco, D., Prunas, A., 2024. Exploring transgender and non-binary engagement with sex tech: a narrative review. Current Sexual Health Reports 16 (4), 237–250. Available at https://doi.org/10.1007/s11930-024-00395-0.

Ashley, F., 2022. The clinical irrelevance of "desistance" research for transgender and gender creative youth. Psychology of Sexual Orientation and Gender Diversity 9 (4), 387–397. Available at https://doi.org/10.1037/sgd0000504.

Ashley, F., Ells, C., 2018. In favor of covering ethically important cosmetic surgeries: facial feminization surgery for transgender people. The American Journal of Bioethics 18 (12), 23–25. Available at https://doi.org/10.1080/15265161.2018.1531162.

Ashley, F., Tordoff, D.M., Olson-Kennedy, J., Restar, A.J., 2024. Randomized-controlled trials are methodologically inappropriate in adolescent transgender healthcare. International Journal of Transgender Health 25 (3), 407–418. Available at https://doi.org/10.1080/26895269.2023.2218357.

Austin, A., Craig, S.L., D'Souza, S.A., 2018. An AFFIRMative cognitive behavioral intervention for transgender youth: preliminary effectiveness. Professional Psychology: Research and Practice 49 (1), 1–8. Available at https://doi.org/10.1037/pro0000154.

Austin, A., Craig, S.L., Matarese, M., Greeno, E.J., Weeks, A., Betsinger, S.A., 2021. Preliminary effectiveness of an LGBTQ+ affirmative parenting intervention with foster parents. Children and Youth Services Review 127. Available at https://doi.org/10.1016/j.childyouth.2021.106107.

Baram, S., Myers, S.A., Yee, S., Librach, C.L., 2019. Fertility preservation for transgender adolescents and young adults: a systematic review. Human Reproduction Update 25 (6), 694–716. Available at https://doi.org/10.1093/humupd/dmz026.

Barbonetti, A., 2024. Hormones and sexuality: navigating the complex terrain of the interplay between endocrinology and sexual function. International Journal of Impotence Research 36 (4), 303–304. Available at https://doi.org/10.1038/s41443-024-00904-8.

Barcelos, T.M.R., de Oliveira Brito, L.G., Jorge, C.H., Franceschini, S.A., Troncon, J.K., Ferriani, R.A., et al., 2024. Evaluation of sexual function of transgender individuals. International Urogynecology Journal 35 (8), 1663–1671. Available at https://doi.org/10.1007/s00192-024-05857-6.

Bariola, E., Lyons, A., Leonard, W., Pitts, M., Badcock, P., Couch, M., 2015. Demographic and psychosocial factors associated with psychological distress and resilience among transgender individuals. American Journal of Public Health 105 (10), 2108–2116. Available at https://doi.org/10.2105/AJPH.2015.302763.

Bateman, W., 2015. LGBT Youth Homelessness: A UK National Scoping of Cause, Prevelence, Response and Outcome. Albert Kennedy Trust, London, UK.

Bauer, G., Churchill, S., Ducharme, J., 2021. Trans Youth CAN! Research TeamTrans Youth CAN! Gender Distress Scale (TYC-GDS).

Becasen, J.S., Denard, C.L., Mullins, M.M., Higa, D.H., Sipe, T.A., 2019. Estimating the prevalence of HIV and sexual behaviors among the US transgender population: a systematic review and meta-analysis, 2006–2017. American Journal of Public Health 109 (1), E1. Available at https://doi.org/10.2105/AJPH.2018.304727.

Beemyn, G., Rankin, S., 2011. The Lives of Transgender People. Columbia University Press.

Beischel, W.J., Gauvin, S.E.M., van Anders, S.M., 2022. "A little shiny gender breakthrough": community understandings of gender euphoria. International Journal of Transgender Health 23 (3), 274–294. Available at https://doi.org/10.1080/26895269.2021.1915223.

Belous, C.K., Schulz, E.E., 2024. The Sex Positivity Scale: a new way to measure sex positivity as a trait. Sexual and Relationship Therapy 39 (3), 944–962. Available at https://doi.org/10.1080/14681994.2022.2140136.

Bird, K., Arcelus, J., Matsagoura, L., O'Shea, B.A., Townsend, E., 2024. Risk and protective factors for self-harm thoughts and behaviours in transgender and gender diverse people: a systematic review. Heliyon 10 (5), e26074. Available at https://doi.org/10.1016/j.heliyon.2024.e26074.

Bishop, A., Overcash, F., McGuire, J., Reicks, M., 2020. Diet and physical activity behaviors among adolescent transgender students: school survey results. Journal of Adolescent Health 66 (4), 484–490. Available at https://doi.org/10.1016/j.jadohealth.2019.10.026.

Boogers, L.S., Wiepjes, C.M., Klink, D.T., Hellinga, I., Van Trotsenburg, A.S.P., Den Heijer, M., et al., 2022. Transgender girls grow tall: adult height is unaffected by GnRH Analogue And Estradiol Treatment. Journal of Clinical Endocrinology and Metabolism 107 (9), E3805. Available at https://doi.org/10.1210/clinem/dgac349.

Boskey, E.R., Ganor, O., 2022. Sexual orientation and attraction in a cohort of transmasculine adolescents and young adults. Transgender Health 7 (3), 270–275. Available at https://doi.org/10.1089/trgh.2020.0190.

Boskey, E.R., Scheffey, K.L., Pilcher, S., Barerra, E.P., McGregor, K., Carswell, J.M., et al., 2025. A retrospective cohort study of transgender adolescents' gender-affirming hormone discontinuation. Journal of Adolescent Health 76 (4), 584–591. Available at https://doi.org/10.1016/j.jadohealth.2024.11.002.

Bradford, N.J., Rider, G.N., Spencer, K.G., 2021. Hair removal and psychological well-being in transfeminine adults: associations with gender dysphoria and gender euphoria. Journal of Dermatological Treatment 32 (6), 635–642. Available at https://doi.org/10.1080/09546634.2019.1687823.

Brill, S., Pepper, R., 2022. The Transgender Child: A Handbook for Parents and Professionals Supporting Transgender and Nonbinary Children. Cleis Press.

Bungener, S.L., De Vries, A.L.C., Popma, A., Steensma, T.D., 2020. Sexual experiences of young transgender persons during and after gender-affirmative treatment. Pediatrics 146 (6). Available at https://doi.org/10.1542/peds.2019-1411.

Burgwal, A., Gvianishvili, N., Hård, V., Kata, J., Nieto, I.G., Orre, C., et al., 2020. Non-Binary and Genderqueer Genders Health Disparities between Binary and Non Binary Trans People: A Community-Driven Survey. Routledge, pp. 100–111.

Caldarera, A.M., Davidson, S., Vitiello, B., Baietto, C., 2021. A psychological support group for parents in the care of families with gender diverse children and adolescents. Clinical Child Psychology and Psychiatry 26 (1), 64–78. Available at https://doi.org/10.1177/1359104520963372.

Carmichael, P., Butler, G., Masic, U., Cole, T.J., de Stavola, B.L., Davidson, S., et al., 2021. Short-term outcomes of pubertal suppression in a selected cohort of 12 to 15 year old young people with persistent gender dysphoria in the UK. PLoS One 16 (2). Available at https://doi.org/10.1371/journal.pone.0243894.

Chaudhary, S., Hoffmann, J.A., Pulcini, C.D., Zamani, M., Hall, M., Jeffries, K.N., et al., 2024. Youth suicide and preceding mental health diagnosis. JAMA Network Open 7 (7). Available at https://doi.org/10.1001/jamanetworkopen.2024.23996.

Chen, A., 2020. Ace: What Asexuality Reveals about Desire, Society, and the Meaning of Sex. Beacon Press.

Chen, D., Kolbuck, V.D., Sutter, M.E., Tishelman, A.C., Quinn, G.P., Nahata, L., 2019. Knowledge, practice behaviors, and perceived barriers to fertility care among providers of transgender healthcare. Journal of Adolescent Health 64 (2), 226–234. Available at https://doi.org/10.1016/j.jadohealth.2018.08.025.

Chen, D., Berona, J., Chan, Y.M., Ehrensaft, D., Garofalo, R., Hidalgo, M.A., et al., 2023. Psychosocial functioning in transgender youth after 2 years of hormones. New England Journal of Medicine 388 (3), 240–250. Available at https://doi.org/10.1056/NEJMoa2206297.

Chen, D., Berona, J., Coyne, C.A., Chan, Y.-M., Ehrensaft, D., Hidalgo, M.A., et al., 2024. 9. Predictors of diverse mental health trajectories following 4 years of treatment with gender-affirming hormones. Journal of Adolescent Health 74 (3), S5. Available at https://doi.org/10.1016/j.jadohealth.2023.11.028.

Chinsen, A., Cronin, T.J., Pace, C.C., Tollit, M.A., Pang, K.C., 2024. Evaluation of a codesigned group cognitive-behavioural therapy intervention for trans young people (TAG TEAM): protocol for a feasibility trial and a subsequent pilot RCT. BMJ Open 14 (1). Available at https://doi.org/10.1136/bmjopen-2023-076511.

Choi, S.K., Wilson, B.D., Shelton, J., Gates, G.J., 2015. Serving our youth 2015: the needs and experiences of lesbian, gay, bisexual, transgender, and questioning youth experiencing homelessness.

Chong, L.S.H., Kerklaan, J., Clarke, S., Kohn, M., Baumgart, A., Guha, C., et al., 2021. Experiences and perspectives of transgender youths in accessing health care: a systematic review. JAMA Pediatrics 175 (11), 1159–1173. Available at https://doi.org/10.1001/jamapediatrics.2021.2061.

Christiano, C., Neimand, A., 2017. Stop raising awareness already. Stanford Social Innovation Review 15 (2), 34–41. Available at https://doi.org/10.48558/7MA6-J918, https://doi.org/10.48558/7MA6-J918 Available from.

Clark, D.B.A., Ji, D., Foran, E., Seburn, K., Temple, J., 2026. Addressing health information needs and research priorities of transgender, two-spirit, and nonbinary youth. International Journal of LGBTQ+ Youth Studies 23 (1), 186–202. Available at https://doi.org/10.1080/29968992.2025.2523244.

Clements-Nolle, K., Marx, R., Katz, M., 2006. Attempted suicide among transgender persons: the influence of gender-based discrimination and victimization. Journal of Homosexuality 51 (3), 53–69. Available at https://doi.org/10.1300/J082v51n03_04.

Coleman, E., Radix, A.E., Bouman, W.P., Brown, G.R., de Vries, A.L.C., Deutsch, M.B., et al., 2022. Standards of care for the health of transgender and gender diverse people, version 8. International Journal of Transgender Health 23 (sup1), S1. Available at https://doi.org/10.1080/26895269.2022.2100644.

Collier, K.L., Van Beusekom, G., Bos, H.M.W., Sandfort, T.G.M., 2013. Sexual orientation and gender identity/expression related peer victimization in adolescence: a systematic review of associated psychosocial and health outcomes. Journal of Sex Research 50 (3-4), 299–317. Available at https://doi.org/10.1080/00224499.2012.750639.

Colvin, S., Egan, J.E., Coulter, R.W.S., 2019. School climate & sexual and gender minority adolescent mental health. Journal of Youth and Adolescence 48 (10), 1938–1951. Available at https://doi.org/10.1007/s10964-019-01108-w.

Cooper, H.C., Long, J., Aye, T., Tian, X., 2022. Fertility preservation in transgender and non-binary adolescents and young adults. PLoS One 17 (3), e0265043. Available at https://doi.org/10.1371/journal.pone.0265043.

Cortez, S., Moog, D., Lewis, C., Williams, K., Herrick, C.J., Fields, M.E., et al., 2024. Effectiveness and safety of different estradiol regimens in transgender females: a randomized controlled trial. Journal of the Endocrine Society 8 (8). Available at https://doi.org/10.1210/jendso/bvae108.

Crabtree, L., Connelly, K.J., Guerriero, J.T., Battison, E.A.J., Tiller-Ormord, J., Sutherland, S.M., et al., 2024. A more nuanced story: pediatric gender-affirming healthcare is associated with satisfaction and confidence. Journal of Adolescent Health 75 (5), 772–779. Available at https://doi.org/10.1016/j.jadohealth.2024.06.016.

Cunningham, J.A., Kypri, K., McCambridge, J., 2013. Exploratory randomized controlled trial evaluating the impact of a waiting list control design. BMC Medical Research Methodology 13 (1). Available at https://doi.org/10.1186/1471-2288-13-150.

Dangaltcheva, A., Booth, C., Moretti, M.M., 2021. Transforming connections: a trauma-informed and attachment-based program to promote sensitive parenting of trans and gender non-conforming youth. Frontiers in Psychology 12. Available at https://doi.org/10.3389/fpsyg.2021.643823.

De Vries, A.L.C., Steensma, T.D., Doreleijers, T.A.H., Cohen-Kettenis, P.T., 2011. Puberty suppression in adolescents with gender identity disorder: a prospective follow-up study. Journal of Sexual Medicine 8 (8), 2276–2283. Available at https://doi.org/10.1111/j.1743-6109.2010.01943.x.

De Vries, A.L.C., McGuire, J.K., Steensma, T.D., Wagenaar, E.C.F., Doreleijers, T.A.H., Cohen-Kettenis, P.T., 2014. Young adult psychological outcome after puberty suppression and gender reassignment. Pediatrics 134 (4), 696–704. Available at https://doi.org/10.1542/peds.2013-2958.

Delozier, A.M., Kamody, R.C., Rodgers, S., Chen, D., 2020. Health disparities in transgender and gender expansive adolescents: a topical review from a minority stress framework. Journal of Pediatric Psychology 45 (8), 842–847. Available at https://doi.org/10.1093/jpepsy/jsaa040.

deMayo, B.E., Jordan, A.E., Olson, K.R., 2022. Gender development in gender diverse children. Annual Review of Developmental Psychology 4 (1), 207–229. Available at https://doi.org/10.1146/annurev-devpsych-121020-034014.

Di Ceglie, D., Thümmel, E.C., 2006. An experience of group work with parents of children and adolescents with gender identity disorder. Clinical Child Psychology and Psychiatry 11 (3), 387–396. Available at https://doi.org/10.1177/1359104506064983.

Diamond, L.M., 2020. Gender fluidity and nonbinary gender identities among children and adolescents. Child Development Perspectives 14 (2), 110–115. Available at https://doi.org/10.1111/cdep.12366.

Dickey, L.M., Budge, S.L., 2020. Suicide and the transgender experience: a public health crisis. American Psychologist 75 (3), 380–390. Available at https://doi.org/10.1037/amp0000619.

Dolotina, B., Daniolos, P.T., 2023. Dynamic gender identities and expressions: detransition and affirming non-linear gender pathways among transgender and gender diverse youth. Child and Adolescent Psychiatric Clinics of North America 32 (4), 803–813. Available at https://doi.org/10.1016/j.chc.2023.05.002.

Dolotina, B., Turban, J.L., 2022. A multipronged, evidence-based approach to improving mental health among transgender and gender-diverse youth. JAMA Network Open 5 (2). Available at https://doi.org/10.1001/jamanetworkopen.2022.0926.

Dubin, S.N., Nolan, I.T., Streed, C.G., Greene, R.E., Radix, A.E., Morrison, S.D., 2018. Transgender health care: improving medical students' and residents' training and awareness. Advances in Medical Education and Practice 9, 377–391. Available at https://doi.org/10.2147/AMEP.S147183.

Dueker, A.K., 2002. Unequal rights: discrimination against people with mental disabilities and the Americans with disabilities act. Journal of Health Politics, Policy and Law 27 (3), 519–522. Available at https://doi.org/10.1215/03616878-27-3-519.

Eisenberg, M.E., Gower, A.L., McMorris, B.J., Rider, G.N., Shea, G., Coleman, E., 2017. Risk and protective factors in the lives of transgender/gender nonconforming adolescents. Journal of Adolescent Health 61 (4), 521–526. Available at https://doi.org/10.1016/j.jadohealth.2017.04.014.

Expósito-Campos, P., Pérez-Fernández, J.I., Salaberria, K., 2023. Empirically supported affirmative psychological interventions for transgender and non-binary youth and adults: a systematic review. Clinical Psychology Review 100, 102229. Available at https://doi.org/10.1016/j.cpr.2022.102229.

Factor, R.J., Rothblum, E.D., 2008. A study of transgender adults and their non-transgender siblings on demographic characteristics, social support, and experiences of violence. Journal of LGBT Health Research 3 (3), 11–30. Available at https://doi.org/10.1080/15574090802092879.

Fast, A.A., Olson, K.R., 2018. Gender development in transgender preschool children. Child Development 89 (2), 620–637. Available at https://doi.org/10.1111/cdev.12758.

Fisher, A.D., Ristori, J., Romani, A., Cassioli, E., Mazzoli, F., Cocchetti, C., et al., 2024. Back to the future: is GnRHa treatment in transgender and gender diverse adolescents only an extended evaluation phase? Journal of Clinical Endocrinology and Metabolism 109 (6), 1565–1579. Available at https://doi.org/10.1210/clinem/dgad729.

Gelly, M.A., Atgé-Delbays, S., Sansfaçon, A.P., 2025. "They're unable to see my decision to detransition for what it is": how detrans youth perceive and receive discourses on detransition. Sexuality Research and Social Policy.

Glintborg, D., Møller, J.J.K., Rubin, K.H., Lidegaard, Ø., T'sjoen, G., Larsen, M.L.J.Ø., et al., 2025. Mental and physical health among danish transgender persons compared with cisgender persons. JAMA Network Open 8 (4). Available at https://doi.org/10.1001/jamanetworkopen.2025.7115.

Goffnett, J., Paceley, M.S., 2020. Challenges, pride, and connection: a qualitative exploration of advice transgender youth have for other transgender youth. Journal of Gay and Lesbian Social Services 32 (3), 328–353. Available at https://doi.org/10.1080/10538720.2020.1752874.

Gooren, L., Delemarre-van de Waal, H., 1996. The feasibility of endocrine interventions in juvenile transsexuals. Journal of Psychology & Human Sexuality 8 (4), 69–74. Available at https://doi.org/10.1300/j056v08n04_05.

Grant, R., Amos, N., Cook, T., Lin, A., Hill, A., Carman, M., 2024. From euphoria to wellbeing: correlates of gender euphoria and its association with mental wellbeing among transgender adults. International journal of transgender health 1–9.

Gridley, S.J., Crouch, J.M., Evans, Y., Eng, W., Antoon, E., Lyapustina, M., et al., 2016. Youth and caregiver perspectives on barriers to gender-affirming health care for transgender youth. Journal of Adolescent Health 59 (3), 254–261. Available at https://doi.org/10.1016/j.jadohealth.2016.03.017.

Hagenäs, I., Jørgensen, N., Rechnitzer, C., Sommer, P., Holm, M., Schmiegelow, K., et al., 2010. Clinical and biochemical correlates of successful semen collection for cryopreservation from 12-18-year-old patients: a single-center study of 86 adolescents. Human Reproduction 25 (8), 2031–2038. Available at https://doi.org/10.1093/humrep/deq147.

Hainey, K.J., Connolly, D.J., Thomson, R., Smalley, N., Campbell, D.D., Wells, V., et al., 2025. Mental health outcomes in transgender and nonbinary people: an umbrella review. JAMA Psychiatry 82 (11), 1142–1151. Available at https://doi.org/10.1001/jamapsychiatry.2025.1850.

Handler, T., Hojilla, J.C., Varghese, R., Wellenstein, W., Satre, D.D., Zaritsky, E., 2019. Trends in referrals to a pediatric transgender clinic. Pediatrics 144 (5). Available at https://doi.org/10.1542/peds.2019-1368.

Hässler, T., Glazier, J.J., Olson, K.R., 2022. Consistency of gender identity and preferences across time: an exploration among cisgender and transgender children. Developmental Psychology 58 (11), 2184–2196. Available at https://doi.org/10.1037/dev0001419.

Hatzenbuehler, M.L., Pachankis, J.E., 2016. Stigma and minority stress as social determinants of health among lesbian, gay, bisexual, and transgender youth: research evidence and clinical implications. Pediatric Clinics of North America 63 (6), 985–997. Available at https://doi.org/10.1016/j.pcl.2016.07.003.

Heathcote, C., Taylor, J., Hall, R., Jarvis, S.W., Langton, T., Hewitt, C.E., et al., 2024. Psychosocial support interventions for children and adolescents experiencing gender dysphoria or incongruence: a systematic review. Archives of Disease in Childhood 109 (2), s19. Available at https://doi.org/10.1136/archdischild-2023-326347.

Hembree, W.C., Cohen-Kettenis, P.T., Gooren, L., Hannema, S.E., Meyer, W.J., Murad, M.H., et al., 2017. Endocrine treatment of gender-dysphoric/gender-incongruent persons: an endocrine society clinical practice guideline. Endocrine Practice: Official Journal of the American College of Endocrinology and the American Association of Clinical Endocrinologists 23 (12), 1437. Available at https://doi.org/10.4158/1934-2403-23.12.1437.

Hendricks, M.L., Testa, R.J., 2012. A conceptual framework for clinical work with transgender and gender nonconforming clients: an adaptation of the minority stress model. Professional Psychology: Research and Practice 43 (5), 460–467. Available at https://doi.org/10.1037/a0029597.

Hidalgo, M.A., Ehrensaft, D., Tishelman, A.C., Clark, L.F., Garofalo, R., Rosenthal, S.M., et al., 2013. The gender affirmative model: what we know and what we aim to learn. Human Development 56 (5), 285–290. Available at https://doi.org/10.1159/000355235.

Hillier, A., Torg, E., 2019. Parent participation in a support group for families with transgender and gender-nonconforming children: "being in the company of others who do not question the reality of our experience" Transgender Health 4 (1), 168–175. Available at https://doi.org/10.1089/trgh.2018.0018.

Hillier, A., Kroehle, K., Edwards, H., Graves, G., 2020. Risk, resilience, resistance and situated agency of trans high school students. Journal of LGBT Youth 17 (4), 384–407. Available at https://doi.org/10.1080/19361653.2019.1668326.

Hisle-Gorman, E., Schvey, N.A., Adirim, T.A., Rayne, A.K., Susi, A., Roberts, T.A., et al., 2021. Mental healthcare utilization of transgender youth before and after affirming treatment. Journal of Sexual Medicine 18 (8), 1444–1454. Available at https://doi.org/10.1016/j.jsxm.2021.05.014.

Hollingshead, N., Hodax, J.K., Boos, M.D., 2022. Management of acne in transgender and gender diverse youth part 2: unique considerations and strategies in medical treatment. Pediatric Dermatology 39 (6), 870–875. Available at https://doi.org/10.1111/pde.15114.

Holt, N.R., Ralston, A.L., Hope, D.A., Mocarski, R., Woodruff, N., 2021. A systematic review of recommendations for behavioral health services for transgender and gender diverse adults: the three-legged stool of evidence-based practice is unbalanced. Clinical Psychology: Science and Practice 28 (2), 186–201. Available at https://doi.org/10.1037/cps0000006.

Horton, C., 2022. I never wanted her to feel shame": parent reflections on supporting a transgender child. Journal of LGBT Youth 19, 1–17.

Horton, C., 2024. Experiences of puberty and puberty blockers: insights from trans children, trans adolescents, and their parents. Journal of Adolescent Research 39 (1), 77–103. Available at https://doi.org/10.1177/07435584221100591.

Hudson, J., Nahata, L., Dietz, E., Quinn, G.P., 2018. Fertility counseling for transgender AYAs. Clinical Practice in Pediatric Psychology 6 (1), 84–92. Available at https://doi.org/10.1037/cpp0000180.

Huit, T.Z., Coyne, C., Chen, D., 2024. State of the science: gender-affirming care for transgender and gender diverse youth. Behavior Therapy 55 (6), 1335–1347. Available at https://doi.org/10.1016/j.beth.2024.02.010.

Inwards-Breland, D.J., Divall, S., Salehi, P., Crouch, J.M., Negaard, M., Lu, A., et al., 2019. Youth and parent experiences in a multidisciplinary gender clinic. Transgender Health 4 (1), 100–106. Available at https://doi.org/10.1089/trgh.2018.0046.

Israel, T., Choi, A.Y., Lin, Y.J., Matsuno, E., Goodman, J.A., Kary, K.G., et al., 2021. Reducing internalized transnegativity: randomized controlled trial of an online intervention. Psychology of Sexual Orientation and Gender Diversity 8 (4), 429–439. Available at https://doi.org/10.1037/sgd0000447.

Jacobsen, K., Devor, A., 2022. Moving from gender dysphoria to gender euphoria: trans experiences of positive gender-related emotions. Bulletin of Applied Transgender Studies 1 (1-2).

Jaffee, Shires, Stroumsa, 2016. Discrimination and delayed health care among transgender women and men: implications for improving medical education and health care delivery. Medical Care 54 (11), 1010–1016. Available at https://doi.org/10.1097/mlr.0000000000000583.

James, S.E., Herman, J.L., Durso, L.E., Heng-Lehtinen, R., 2024. Early insights: a report of the 2022 US transgender survey.

Johns, M.M., Beltran, O., Armstrong, H.L., Jayne, P.E., Barrios, L.C., 2018. Protective factors among transgender and gender variant youth: a systematic review by socioecological level. Journal of Primary Prevention 39 (3), 263–301. Available at https://doi.org/10.1007/s10935-018-0508-9.

Julian, J.M., Salvetti, B., Held, J.I., Murray, P.M., Lara-Rojas, L., Olson-Kennedy, J., 2021. The impact of chest binding in transgender and gender diverse youth and young adults. Journal of Adolescent Health 68 (6), 1129–1134. Available at https://doi.org/10.1016/j.jadohealth.2020.09.029.

Katz-Wise, S.L., Budge, S.L., Orovecz, J.J., Nguyen, B., Nava-Coulter, B., Thomson, K., 2017. Imagining the future: perspectives among youth and caregivers in the trans youth family study. Journal of Counseling Psychology 64 (1), 26–40. Available at https://doi.org/10.1037/cou0000186.

Katz-Wise, S.L., Ehrensaft, D., Vetters, R., Forcier, M., Austin, S.B. 2018. Family functioning and mental health of transgender and gender-nonconforming youth in the trans teen and family narratives project. The Journal of Sex Research, 55(4–5), 582–590.

Kennedy, S.L., Jack, S.M., Johnson, N., Couturier, J., Rae, C., Klassen, A.F., 2025. Prioritizing participant and research team emotional safety during data generation and analysis in qualitative patient-reported outcome measure research: development of a framework informed by the GENDER-Q youth study. Patient 18 (6), 597–608. Available at https://doi.org/10.1007/s40271-025-00756-4.

Kennedy, S.L., Jack, S.M., Johnson, N., Couturier, J., Rae, C., Klassen, A.F., 2026. Using a timeline activity in virtual concept elicitation interviews with youth: methods and lessons learned from an interpretive descriptive study to develop a patient-reported outcome measure (GENDER-Q youth). Patient 19 (1), 5–20. Available at https://doi.org/10.1007/s40271-025-00770-6.

King, H., Kelley, T.P., Shatzel, J.J., 2024. Gender-affirming hormone therapy in the transgender patient: influence on thrombotic risk. Hematology (US) 2024 (1), 652–663. Available at https://doi.org/10.1182/hematology.2024000592.

Klaver, M., De Mutsert, R., Van Der Loos, M.A.T.C., Wiepjes, C.M., Twisk, J.W.R., Den Heijer, M., et al., 2020. Hormonal treatment and cardiovascular risk profile in transgender adolescents. Pediatrics 145 (3). Available at https://doi.org/10.1542/peds.2019-0741.

Klepper, M., Bosse, J., Flores, D.D., Sanders, R., 2024. Sibling support for transgender and gender-diverse youth: a scoping review. Journal of Adolescent Health 75 (4), 543–552. Available at https://doi.org/10.1016/j.jadohealth.2024.05.014.

Kolbuck, V.D., Sajwani, A., Kyweluk, M.A., Finlayson, C., Gordon, E.J., Chen, D., 2020. Formative development of a fertility decision aid for transgender adolescents and young adults: a multidisciplinary Delphi consensus study. Journal of Assisted Reproduction and Genetics 37 (11), 2805–2816. Available at https://doi.org/10.1007/s10815-020-01947-8.

Korpaisarn, S., Safer, J.D., 2018. Gaps in transgender medical education among healthcare providers: a major barrier to care for transgender persons. Reviews in Endocrine and Metabolic Disorders 19 (3), 271–275. Available at https://doi.org/10.1007/s11154-018-9452-5.

Kuper, L.E., Adams, N., Mustanski, B.S., 2018. Exploring cross-sectional predictors of suicide ideation, attempt, and risk in a large online sample of transgender and gender nonconforming youth and young adults. LGBT Health 5 (7), 391–400. Available at https://doi.org/10.1089/lgbt.2017.0259.

Kuper, L.E., Stewart, S., Preston, S., Lau, M., Lopez, X., 2020. Body dissatisfaction and mental health outcomes of youth on gender-affirming hormone therapy. Pediatrics 145 (4). Available at https://doi.org/10.1542/peds.2019-3006.

Lavender, R., Shaw, S., Maninger, J.K., Butler, G., Carruthers, P., Carmichael, P., et al., 2023. Impact of hormone treatment on psychosocial functioning in gender-diverse young people. LGBT Health 10 (5), 382–390. Available at https://doi.org/10.1089/lgbt.2022.0201.

Lawlis, S.M., Butler, P., Middleman, A., 2020. Evaluating transgender youth and parent interest and preferences regarding support groups. Global Pediatric Health 7. Available at https://doi.org/10.1177/2333794X20954680.

Lawson, M.L., Gotovac, S., Couch, B., Gale, L., Vandermorris, A., Ghosh, S., et al., 2024. Pathways to care for adolescents attending a first hormone appointment at Canadian gender affirming medical clinics: a cross-sectional analysis from the trans youth CAN! study. Journal of Adolescent Health 74 (1), 140–147. Available at https://doi.org/10.1016/j.jadohealth.2023.07.021.

Lee, J.Y., 2023. Bone health in the transgender and gender diverse youth population. Current Osteoporosis Reports 21 (4), 459–471. Available at https://doi.org/10.1007/s11914-023-00799-2.

Lee, J.Y., Finlayson, C., Olson-Kennedy, J., Garofalo, R., Chan, Y.M., Glidden, D.V., et al., 2020. Low bone mineral density in early pubertal transgender/gender diverse youth: findings from the trans youth care study. Journal of the Endocrine Society 4 (9). Available at https://doi.org/10.1210/jendso/bvaa065.

Linsenmeyer, W.R., Katz, I.M., Reed, J.L., Giedinghagen, A.M., Lewis, C.B., Garwood, S.K., 2021. Disordered eating, food insecurity, and weight status among transgender and gender nonbinary youth and young adults: a cross-sectional study using a nutrition screening protocol. LGBT Health 8 (5), 359–366. Available at https://doi.org/10.1089/lgbt.2020.0308.

van der Loos, M.A.T.C., Hellinga, I., Vlot, M.C., Klink, D.T., den Heijer, M., Wiepjes, C.M., 2021. Development of hip bone geometry during gender-affirming hormone therapy in transgender adolescents resembles that of the experienced gender when pubertal suspension is started in early puberty. Journal of Bone and Mineral Research 36 (5), 931–941. Available at https://doi.org/10.1002/jbmr.4262.

MacKinnon, K.R., Jeyabalan, T., Strang, J.F., Delgado-Ron, J.A., Lam, J.S.H., Gould, W.A., et al., 2024. Discontinuation of gender-affirming medical treatments: prevalence and associated features in a nonprobabilistic sample of transgender and gender-diverse adolescents and young adults in Canada and the US. Journal of Adolescent Health 75 (4), 569–577. Available at https://doi.org/10.1016/j.jadohealth.2024.05.015.

MacKinnon, K.R., Expósito-Campos, P., Gould, W.A., 2023. Detransition needs further understanding, not controversy. BMJ (Clinical Research ed.) 381. Available at https://doi.org/10.1136/bmj-2022-073584.

Manns, P.Z., Buckman, C., Tumin, D., Mathai, A., Dendy, C., 2023. Rural perspectives on health services for transgender and gender diverse youth. Health Promotion Practice 24 (4), 609–611. Available at https://doi.org/10.1177/15248399221146805.

Martin, C.E., Lewis, C., Omurtag, K., 2021. Successful oocyte cryopreservation using letrozole as an adjunct to stimulation in a transgender adolescent after GnRH agonist suppression. Fertility and Sterility 116 (2), 522–527. Available at https://doi.org/10.1016/j.fertnstert.2021.02.025.

Matsuno, E., Israel, T., 2018. Psychological interventions promoting resilience among transgender individuals: transgender resilience intervention model (TRIM). The Counseling Psychologist 46 (5), 632–655. Available at https://doi.org/10.1177/0011000018787261.

Mattawanon, N., Charoenkwan, K., Tangpricha, V., 2021. Sexual dysfunction in transgender people: a systematic review. Urologic Clinics of North America 48 (4), 437–460. Available at https://doi.org/10.1016/j.ucl.2021.06.004.

McCormick, A., Schmidt, K., Terrazas, S., 2018. Foster family acceptance: understanding the role of foster family acceptance in the lives of LGBTQ youth. LGBTQ Youth in Foster Care. Routledge.

McLean, J., Maxwell, M., Platt, S., Harris, F., Jepson, R., 2008 Risk and protective factors for suicide and suicidal behaviour: a literature review.

McPhail, D., Rountree-James, M., Whetter, I., 2016. Addressing gaps in physician knowledge regarding transgender health and healthcare through medical education. Canadian Medical Education Journal 7 (2). Available at https://doi.org/10.36834/cmej.36785.

Menvielle, E.J., Rodnan, L.A., 2011. A therapeutic group for parents of transgender adolescents. Child and Adolescent Psychiatric Clinics of North America 20 (4), 733–743. Available at https://doi.org/10.1016/j.chc.2011.08.002.

Meyer, I.H., 2003. Prejudice, social stress, and mental health in lesbian, gay, and bisexual populations: conceptual issues and research evidence. Psychological Bulletin 129 (5), 674–697. Available at https://doi.org/10.1037/0033-2909.129.5.674.

van der Miesen, A.I.R., Steensma, T.D., de Vries, A.L.C., Bos, H., Popma, A., 2020. Psychological functioning in transgender adolescents before and after gender-affirmative care compared with cisgender general population peers. Journal of Adolescent Health 66 (6), 699–704. Available at https://doi.org/10.1016/j.jadohealth.2019.12.018.

Morgan, H., Raab, D., Lin, A., Strauss, P., Perry, Y., 2023. Knowledge is power: trans young people's perceptions of parental reactions to their gender identity, and perceived barriers and facilitators to parental support. LGBTQ+ Family: An Interdisciplinary Journal 19 (1), 35–53. Available at https://doi.org/10.1080/27703371.2022.2135660.

Morgan, H., van Hall, H.W., Moore, J.K., Lin, A., Perry, Y., 2024. A pilot group program for parents of trans young people at a specialized pediatric gender diversity service. LGBTQ+ Family: An Interdisciplinary Journal 20 (5), 417–431. Available at https://doi.org/10.1080/27703371.2024.2347495.

Mortimer, R.M., Walker, Z.W., Ashby, R.K., Ginsburg, E.S., Srouji, S.S., 2025. Ovarian stimulation and retrieval outcomes in transgender males maintained on testosterone: a case series. Journal of Assisted Reproduction and Genetics 42 (12), 4433–4437. Available at https://doi.org/10.1007/s10815-025-03699-9.

Murad, M.H., Elamin, M.B., Garcia, M.Z., Mullan, R.J., Murad, A., Erwin, P.J., et al., 2010. Hormonal therapy and sex reassignment: a systematic review and meta-analysis of quality of life and psychosocial outcomes. Clinical Endocrinology 72 (2), 214–231. Available at https://doi.org/10.1111/j.1365-2265.2009.03625.x.

Nahata, L., Tishelman, A.C., Caltabellotta, N.M., Quinn, G.P., 2017. Low fertility preservation utilization among transgender youth. Journal of Adolescent Health 61 (1), 40–44. Available at https://doi.org/10.1016/j.jadohealth.2016.12.012.

Newcomb, M.E., Hill, R., Buehler, K., Ryan, D.T., Whitton, S.W., Mustanski, B., 2020. High burden of mental health problems, substance use, violence, and related psychosocial factors in transgender, non-binary, and gender diverse youth and young adults. Archives of Sexual Behavior 49 (2), 645–659. Available at https://doi.org/10.1007/s10508-019-01533-9.

Nokoff, N.J., Scarbro, S.L., Moreau, K.L., Zeitler, P., Nadeau, K.J., Reirden, D., et al., 2021. Body composition and markers of cardiometabolic health in transgender youth on gonadotropin-releasing hormone agonists. Transgender Health 6 (2), 111–119. Available at https://doi.org/10.1089/trgh.2020.0029.

Norwood, K., 2013. Grieving gender: trans-identities, transition, and ambiguous loss. Communication Monographs 80 (1), 24–45. Available at https://doi.org/10.1080/03637751.2012.739705.

Nunes-Moreno, M., Furniss, A., Cortez, S., Davis, S.M., Dowshen, N., Kazak, A.E., et al., 2025. Mental health diagnoses and suicidality among transgender youth in hospital settings. LGBT Health 12 (1), 20–28. Available at https://doi.org/10.1089/lgbt.2023.0394.

Olsavsky, A.L., Grannis, C., Bricker, J., Chelvakumar, G., Indyk, J.A., Leibowitz, S.F., et al., 2023. Associations among gender-affirming hormonal interventions, social support, and transgender adolescents' mental health. Journal of Adolescent Health 72 (6), 860–868. Available at https://doi.org/10.1016/j.jadohealth.2023.01.031.

Olson, K.R., Gülgöz, S., 2018. Early findings from the trans youth project: gender development in transgender children. Child Development Perspectives 12 (2), 93–97. Available at https://doi.org/10.1111/cdep.12268.

Olson-Kennedy, J., Cohen-Kettenis, P.T., Kreukels, B.P.C., Meyer-Bahlburg, H.F.L., Garofalo, R., Meyer, W., et al., 2016. Research priorities for gender nonconforming/transgender youth: gender identity development and biopsychosocial outcomes. Current Opinion in Endocrinology, Diabetes, and Obesity 23 (2), 172–179. Available at https://doi.org/10.1097/MED.0000000000000236.

Olson-Kennedy, J., Warus, J., Okonta, V., Belzer, M., Clark, L.F., 2018. Chest reconstruction and chest dysphoria in transmasculine minors and young adults comparisons of nonsurgical and postsurgical cohorts. JAMA Pediatrics 172 (5), 431–436. Available at https://doi.org/10.1001/jamapediatrics.2017.5440.

Olson-Kennedy, J., Chan, Y.M., Rosenthal, S., Hidalgo, M.A., Chen, D., Clark, L., et al., 2019. Creating the trans youth research network: a collaborative research endeavor. Transgender Health 4 (1), 304–312. Available at https://doi.org/10.1089/trgh.2019.0024.

Olson-Kennedy, J., Wang, L., Wong, C.F., Chen, D., Ehrensaft, D., Hidalgo, M.A., et al., 2025. Emotional health of transgender youth 24 months after initiating gender-affirming hormone therapy. Journal of Adolescent Health 77 (1), 41–50. Available at https://doi.org/10.1016/j.jadohealth.2024.11.014.

Oosthoek, E.D., Stanwich, S., Gerritse, K., Doyle, D.M., de Vries, A.L.C., 2024. Gender-affirming medical treatment for adolescents: a critical reflection on "effective" treatment outcomes. BMC Medical Ethics 25 (1). Available at https://doi.org/10.1186/s12910-024-01143-8.

Pampati, S., Andrzejewski, J., Steiner, R.J., Rasberry, C.N., Adkins, S.H., Lesesne, C.A., et al., 2021. "We deserve care and we deserve competent care": qualitative perspectives on health care from transgender youth in the Southeast US. Journal of Pediatric Nursing 56, 54–59. Available at https://doi.org/10.1016/j.pedn.2020.09.021.

Pang, K.C., Peri, A.J.S., Chung, H.E., Telfer, M., Elder, C.V., Grover, S., et al., 2020. Rates of fertility preservation use among transgender adolescents. JAMA Pediatrics 174 (9), 890–891. Available at https://doi.org/10.1001/jamapediatrics.2020.0264.

Park, R.H., Liu, Y.T., Samuel, A., Gurganus, M., Gampper, T.J., Corbett, S.T., et al., 2022. Long-term outcomes after gender-affirming surgery: 40-year follow-up study. Annals of Plastic Surgery 89 (4), 431–436. Available at https://doi.org/10.1097/SAP.0000000000003233.

Patterson, B., Boyle, M.H., Kivlenieks, M., Van Ameringen, M., 2016. The use of waitlists as control conditions in anxiety disorders research. Journal of Psychiatric Research 83, 112–120. Available at https://doi.org/10.1016/j.jpsychires.2016.08.015.

Peri, A., Ahler, A., Gook, D., O'Connell, M.A., Bourne, H., Nightingale, M., et al., 2021. Predicting successful sperm retrieval in transfeminine adolescents after testicular biopsy. Journal of Assisted Reproduction and Genetics 38 (10), 2735–2743. Available at https://doi.org/10.1007/s10815-021-02293-z.

Persky, R.W., Gruschow, S.M., Sinaii, N., Carlson, C., Ginsberg, J.P., Dowshen, N.L., 2020. Attitudes toward fertility preservation among transgender youth and their parents. Journal of Adolescent Health 67 (4), 583–589. Available at https://doi.org/10.1016/j.jadohealth.2020.02.027.

Poquiz, J., Moser, C.N., Grimstad, F., Boman, K., Sonneville, S.A., Turpin, A., et al., 2021. Gender-affirming care in the midwest: reaching rural populations. Journal of Rural Mental Health 45 (2), 121–128. Available at https://doi.org/10.1037/rmh0000174.

Poulin-Dubois, D., Serbin, L.A., Kenyon, B., Derbyshire, A., 1994. Infants' intermodal knowledge about gender. Developmental Psychology 30 (3), 436–442. Available at https://doi.org/10.1037/0012-1649.30.3.436.

Puckett, J.A., Tornello, S., Mustanski, B., Newcomb, M.E., 2022. Gender variations, generational effects, and mental health of transgender people in relation to timing and status of gender identity milestones. Psychology of Sexual Orientation and Gender Diversity 9 (2), 165–178. Available at https://doi.org/10.1037/sgd0000391.

Puckett, J.A., Domínguez, S., Matsuno, E., 2024. Measures of resilience: do they reflect the experiences of transgender individuals? Transgender Health 9 (1), 1–13. Available at https://doi.org/10.1089/trgh.2022.0015.

Pullen Sansfaçon, A., Gelly, M.A., Gravel, R., Medico, D., Baril, A., Susset, F., et al., 2023. A nuanced look into youth journeys of gender transition and detransition. Infant and Child Development 32 (2). Available at https://doi.org/10.1002/icd.2402.

Rachlin, K., Green, J., Lombardi, E., 2008. Utilization of health care among female-to-male transgender individuals in the US. Journal of Homosexuality 54 (3), 243–258. Available at https://doi.org/10.1080/00918360801982124.

Rae, J.R., Gülgöz, S., Durwood, L., DeMeules, M., Lowe, R., Lindquist, G., et al., 2019. Predicting early-childhood gender transitions. Psychological Science 30 (5), 669–681. Available at https://doi.org/10.1177/0956797619830649.

Ramos, N., Marr, M.C., 2023. Traumatic stress and resilience among transgender and gender diverse youth. Child and Adolescent Psychiatric Clinics of North America 32 (4), 667–682. Available at https://doi.org/10.1016/j.chc.2023.04.001.

Reisner, S.L., Biello, K.B., Hughto, J.M.W., Kuhns, L., Mayer, K.H., Garofalo, R., et al., 2016. Psychiatric diagnoses and comorbidities in a diverse, multicity cohort of young transgender women: baseline findings from project LifeSkills. JAMA Pediatrics 170 (5), 481–486. Available at https://doi.org/10.1001/jamapediatrics.2016.0067.

Reisner, S.L., Choi, S.K., Herman, J.L., Bockting, W., Krueger, E.A., Meyer, I.H., 2023. Sexual orientation in transgender adults in the US. BMC Public Health 23 (1). Available at https://doi.org/10.1186/s12889-023-16654-z.

Rhodes, M., Baron, A., 2019. The development of social categorization. Annual Review of Developmental Psychology 1 (1), 359–386. Available at https://doi.org/10.1146/annurev-devpsych-121318-084824.

Rider, G.N., Mcmorris, B.J., Gower, A.L., Coleman, E., Eisenberg, M.E., 2018. Health and care utilization of transgender and gender nonconforming youth: a population-based study. Pediatrics 141 (3). Available at https://doi.org/10.1542/peds.2017-1683.

Robinson, B.A., 2022. Non-binary embodiment, queer knowledge production, and disrupting the cisnormative field: notes from a trans ethnographer. Journal of Men's Studies 30 (3), 425–445. Available at https://doi.org/10.1177/10608265221108204.

Robinson, B.A., Schmitz, R.M., 2021. Beyond resilience: resistance in the lives of LGBTQ youth. Sociology Compass 15 (12). Available at https://doi.org/10.1111/soc4.12947.

Russell, M.R., Abrams, M., 2019. Transgender and nonbinary adolescents: the role of voice and communication therapy. Perspectives of the ASHA Special Interest Groups 4 (6), 1298–1305. Available at https://doi.org/10.1044/2019_persp-19-00034.

Safa, B., Lin, W.C., Salim, A.M., Deschamps-Braly, J.C., Poh, M.M., 2019a. Current concepts in feminizing gender surgery. Plastic and Reconstructive Surgery 143 (5), 1081–1091. Available at https://doi.org/10.1097/PRS.0000000000005595.

Safa, B., Lin, W.C., Salim, A.M., Deschamps-Braly, J.C., Poh, M.M., 2019b. Current concepts in masculinizing gender surgery. Plastic and Reconstructive Surgery 143 (4), 857–871. Available at https://doi.org/10.1097/PRS.0000000000005453.

Salinas-Quiroz, F., Aral, T., Hillekens, J., Hölscher, S., Demos, J., 2025. "You're free from just a girl or a boy": nonbinary children's understanding of their gender. International Journal of Transgender Health 26 (2), 378–395. Available at https://doi.org/10.1080/26895269.2024.2351470.

Schulmeister, C., Millington, K., Kaufman, M., Finlayson, C., Kennedy, J.O., Garofalo, R., et al., 2022. Growth in transgender/gender-diverse youth in the first year of treatment with gonadotropin-releasing hormone agonists. Journal of Adolescent Health 70 (1), 108–113. Available at https://doi.org/10.1016/j.jadohealth.2021.06.022.

Sequeira, G.M., Ray, K.N., Miller, E., Coulter, R.W.S., 2020. Transgender youth's disclosure of gender identity to providers outside of specialized gender centers. Journal of Adolescent Health 66 (6), 691–698. Available at https://doi.org/10.1016/j.jadohealth.2019.12.010.

Serbin, L.A., Poulin-Dubois, D., Colburne, K.A., Sen, M.G., Eichestedt, J.A., 2001. Gender stereotyping in infancy: visual preferences for and knowledge of gender-stereortyped toys in the second year. International Journal of Behavioral Development 25 (1), 7–15. Available at https://doi.org/10.1080/01650250042000078.

Shook, A.G., Lucas, R., Liang, C., Tordoff, D.M., Skytta, J.A.F., Kientz, J., et al., 2025. Navigating challenges with IRB review for community-engaged research with transgender youth in the US. Ethics and Human Research 47 (6), 13–25. Available at https://doi.org/10.1002/eahr.70002.

Singh, A.A., Meng, S.E., Hansen, A.W., 2014. "I am my own gender": resilience strategies of trans youth. Journal of Counseling and Development 92 (2), 208–218. Available at https://doi.org/10.1002/j.1556-6676.2014.00150.x.

Skelton, S., Riggs, D.W., Pullen Sansfacon, A., Katz-Wise, S.L., Arora, M., Thibeault, C.A., 2024. 'It just feels really nice when people call me by my name': accounts of gender euphoria among Australian trans young people and their parents. Journal of Gender Studies 33 (4), 470–482. Available at https://doi.org/10.1080/09589236.2023.2285984.

Smith-Johnson, M., 2022. Transgender adults have higher rates of disability than their cisgender counterparts. Health Affairs 41 (10), 1470–1476. Available at https://doi.org/10.1377/hlthaff.2022.00500.

Solomon, J., 2016. Gender identity and expression in the early childhood classroom. YC Young Children 71 (3).

Stanton, A.M., Batchelder, A.W., Kirakosian, N., Scholl, J., King, D., Grasso, C., et al., 2021. Differences in mental health symptom severity and care engagement among transgender and gender diverse individuals: findings from a large community health center. PLoS One 16 (1). Available at https://doi.org/10.1371/journal.pone.0245872.

Stieglitz, K.A., 2010. Development, risk, and resilience of transgender youth. Journal of the Association of Nurses in AIDS Care 21 (3), 192–206. Available at https://doi.org/10.1016/j.jana.2009.08.004.

Stocking, S.Q., Webb, C.K., Miller, G.H., Thomeer, M.B., Goodin, B.R., Sorge, R.E., 2025. Understanding risk of chronic pain development and related mental health disparities among transgender people: a review of current literature and future directions. Journal of Pain 37. Available at https://doi.org/10.1016/j.jpain.2024.104681.

Strang, J.F., Chen, D., Nelson, E., Leibowitz, S.F., Nahata, L., Anthony, L.G., et al., 2022. Transgender youth executive functioning: relationships with anxiety symptoms, autism spectrum disorder, and gender-affirming medical treatment status. Child Psychiatry and Human Development 53 (6), 1252–1265. Available at https://doi.org/10.1007/s10578-021-01195-6.

Strauss, P., Winter, S., Waters, Z., Wright Toussaint, D., Watson, V., Lin, A., 2022. Perspectives of trans and gender diverse young people accessing primary care and gender-affirming medical services: findings from trans pathways. International Journal of Transgender Health 23 (3), 295–307. Available at https://doi.org/10.1080/26895269.2021.1884925.

Suslovic, B., Lett, E., 2024. Resilience is an adverse event: a critical discussion of resilience theory in health services research and public health. Community Health Equity Research & Policy 44 (3), 339–343. Available at https://doi.org/10.1177/2752535x231159721.

Swan, J., Phillips, T.M., Sanders, T., Mullens, A.B., Debattista, J., Brömdal, A., 2023. Mental health and quality of life outcomes of gender-affirming surgery: a systematic literature review. Journal of Gay and Lesbian Mental Health 27 (1), 2–45. Available at https://doi.org/10.1080/19359705.2021.2016537.

Sweileh, W.M., 2018. Bibliometric analysis of peer-reviewed literature in transgender health (1900 - 2017). BMC International Health and Human Rights 18 (1). Available at https://doi.org/10.1186/s12914-018-0155-5.

Taliaferro, L.A., Harder, B.M., Lampe, N.M., Carter, S.K., Rider, G.N., Eisenberg, M.E., 2019. Social connectedness factors that facilitate use of healthcare services: comparison of transgender and gender nonconforming and cisgender adolescents. Journal of Pediatrics 211, 172–178. Available at https://doi.org/10.1016/j.jpeds.2019.04.024.

Tankersley, A.P., Grafsky, E.L., Dike, J., Jones, R.T., 2021. Risk and resilience factors for mental health among transgender and gender nonconforming (TGNC) youth: a systematic review. Clinical Child and Family Psychology Review 24 (2), 183–206. Available at https://doi.org/10.1007/s10567-021-00344-6.

Taylor, J., Mitchell, A., Hall, R., Heathcote, C., Langton, T., Fraser, L., et al., 2024. Interventions to suppress puberty in adolescents experiencing gender dysphoria or incongruence: a systematic review. Archives of Disease in Childhood 109, s33. Available at https://doi.org/10.1136/archdischild-2023-326669.

Testa, R.J., Michaels, M.S., Bliss, W., Rogers, M.L., Balsam, K.F., Joiner, T., 2017. Suicidal ideation in transgender people: gender minority stress and interpersonal theory factors. Journal of Abnormal Psychology 126 (1), 125–136. Available at https://doi.org/10.1037/abn0000234, http://doi.org/10.1037/abn0000240 Available from.

Tollit, M.A., Feldman, D., McKie, G., Telfer, M.M., 2019. Patient and parent experiences of care at a pediatric gender service. Transgender Health 3 (1), 251–256. Available at https://doi.org/10.1089/trgh.2018.0016.

Toomey, R.B., 2021. Advancing research on minority stress and resilience in trans children and adolescents in the 21st century. Child Development Perspectives 15 (2), 96–102. Available at https://doi.org/10.1111/cdep.12405.

Tordoff, D.M., Wanta, J.W., Collin, A., Stepney, C., Inwards-Breland, D.J., Ahrens, K., 2022. Mental health outcomes in transgender and nonbinary youths receiving gender-affirming care. JAMA Network Open 5 (2). Available at https://doi.org/10.1001/jamanetworkopen.2022.0978.

Travers, A., Marchbank, J., Boulay, N., Jordan, S., Reed, K., 2022. Talking back: trans youth and resilience in action. Journal of LGBT Youth 19 (1), 1–30. Available at https://doi.org/10.1080/19361653.2020.1758275.

Turban, J.L., Carswell, J., Keuroghlian, A.S., 2018. Understanding pediatric patients who discontinue gender-affirming hormonal interventions. JAMA Pediatrics 172 (10), 903–904. Available at https://doi.org/10.1001/jamapediatrics.2018.1817.

Turban, J.L., Loo, S.S., Almazan, A.N., Keuroghlian, A.S., 2021. Factors leading to "detransition" among transgender and gender diverse people in the US: a mixed-methods analysis. LGBT Health 8 (4), 273–280. Available at https://doi.org/10.1089/lgbt.2020.0437.

Turban, J.L., Brady, C., Olson-Kennedy, J., 2022. Understanding and supporting patients with dynamic desires for gender-affirming medical interventions. JAMA Network Open 5 (7), e2224722. Available at https://doi.org/10.1001/jamanetworkopen.2022.24722.

Turban, J.L., King, D., Kobe, J., Reisner, S.L., Keuroghlian, A.S., 2023a. Correction: access to gender-affirming hormones during adolescence and mental health outcomes among transgender adults. PLoS One 18 (6), e0287283. Available at https://doi.org/10.1371/journal.pone.0287283.

Turban, J.L., Dolotina, B., Freitag, T.M., King, D., Keuroghlian, A.S., 2023b. Age of realization and disclosure of gender identity among transgender adults. Journal of Adolescent Health 72 (6), 852–859. Available at https://doi.org/10.1016/j.jadohealth.2023.01.023.

Tüzün, Z., Başar, K., Akgül, S., 2022. Social connectedness matters: depression and anxiety in transgender youth during the COVID-19 pandemic. The Journal of Sexual Medicine 19 (4), 650–660. Available at https://doi.org/10.1016/j.jsxm.2022.01.522.

Tyni, K., Wurm, M., Nordström, T., Bratt, A.S., 2024. A systematic review and qualitative research synthesis of the lived experiences and coping of transgender and gender diverse youth 18 years or younger. International Journal of Transgender Health 25 (3), 352–388. Available at https://doi.org/10.1080/26895269.2023.2295379.

Uffman, J.C., Willer, B.L., Mpody, C., Nafiu, O.O., Tobias, J.D., 2023. Characteristics of transgender and gender-diverse youth presenting for surgery in the US. Anesthesia and Analgesia 137 (4), 882–886. Available at https://doi.org/10.1213/ANE.0000000000006618.

Van Der Meulen, I.S., Arnoldussen, M., Van Der Miesen, A.I.R., Hannema, S.E., Steensma, T.D., De Vries, A.L.C., et al., 2025. Sexual satisfaction and dysfunction in transgender adults following puberty suppression treatment during adolescence. Journal of Sexual Medicine 22 (8), 1493–1503. Available at https://doi.org/10.1093/jsxmed/qdaf095.

Vandenbussche, E., 2022. Detransition-related needs and support: a cross-sectional online survey. Journal of Homosexuality 69 (9), 1602–1620. Available at https://doi.org/10.1080/00918369.2021.1919479.

Vasen, E., Poteat, T., Maman, S., 2025. A review of psychosocial interventions to improve parental support of transgender youth. Transgender Health. Available at https://doi.org/10.1089/trgh.2024.0139.

Vaughan, M.D., Rodriguez, E.M., 2014. LGBT strengths: incorporating positive psychology into theory, research, training, and practice. Psychology of Sexual Orientation and Gender Diversity 1 (4), 325–334. Available at https://doi.org/10.1037/sgd0000053.

Vaughan, E., Költő, A., Ravikumar, D., Kelly, C., Nic Gabhainn, S., 2021. Gender minority youth: shifting the paradigm from risk to resilience. European Journal of Public Health 31 (Supplement_3). Available at https://doi.org/10.1093/eurpub/ckab164.598.

Veale, J.F., Peter, T., Travers, R., Saewyc, E.M., 2017. Enacted stigma, mental health, and protective factors among transgender youth in Canada. Transgender Health 2 (1), 207–216. Available at https://doi.org/10.1089/trgh.2017.0031.

Verguchten, J., Demeyer, C., Vanoverbeke, I., Vandenbussche, O.K., Kerckhof, M., 2023. The right of transgender people to be resilient!. European Journal of Public Health 33 (Supplement_2). Available at https://doi.org/10.1093/eurpub/ckad160.903.

Wahlig, J.L., 2015. Losing the child they thought they had: therapeutic suggestions for an ambiguous loss perspective with parents of a transgender child. Journal of GLBT Family Studies 11 (4), 305–326. Available at https://doi.org/10.1080/1550428X.2014.945676.

Warwick, R.M., Araya, A.C., Shumer, D.E., Selkie, E.M., 2022. Transgender youths' sexual health and education: a qualitative analysis. Journal of Pediatric and Adolescent Gynecology 35 (2), 138–146. Available at https://doi.org/10.1016/j.jpag.2021.09.011.

Weinhardt, L.S., Xie, H., Wesp, L.M., Murray, J.R., Apchemengich, I., Kioko, D., et al., 2019. The role of family, friend, and significant other support in well-being among transgender and non-binary youth. Journal of GLBT Family Studies 15 (4), 311–325. Available at https://doi.org/10.1080/1550428X.2018.1522606.

Wittlin, N.M., Gallagher, N.M., Atwood, S., Olson, K.R., 2025. Mental health during medical transition in a US and Canadian sample of early socially transitioned transgender youth. Journal of Adolescent Health 76 (2), 228–237. Available at https://doi.org/10.1016/j.jadohealth.2024.10.023.

Xavier, J., Honnold, J.A., Bradford, J., 2007. The health, health-related needs, and lifecourse experiences of transgender Virginians: Virginia Transgender Health Initiative Study Statewide Survey Report. Virginia Department of Health, Division of Disease Prevention, Richmond, VA.

Zeeman, L., Aranda, K., Sherriff, N., Cocking, C., 2017. Promoting resilience and emotional well-being of transgender young people: research at the intersections of gender and sexuality. Journal of Youth Studies 20 (3), 382–397. Available at https://doi.org/10.1080/13676261.2016.1232481.

Author Index

Note: Page numbers followed by "*t*" refer to tables.

A

Aaen, E. K., 138
Aaron, M. S., 98–99
Abern, L., 131, 223–224
Abrams, M., 117–118
Abreu, R. L., 32, 61, 156, 217
Achenbach, T. M., 12, 106
Achille, C., 71, 83–84, 103*t*, 104–106, 151–152, 203, 218
Ackerman, K. E., 98–99
Acosta, W., 54
Adeleye, A. J., 175–179
Adkins, V., 33
Agénor, M., 192–193
Akgul, S., 115–116
Alaniz, V. I., 119–120, 223
Aldridge, Z., 151–152
Alfieri, T., 24
Alikunju, H., 213
Allen, B. J., 33
Allen, C. L., 169
Allen, L. R., 69, 71, 73, 103*t*, 104, 106, 151–154, 218
Allen, S. D., 5, 74, 214, 218–219
Altman, K., 138–139
Altschuler, R., 214
Ameri, K., 121
Amir, H., 177–179
Amirikia, H., 175
Amos, N., 114–115, 149–150
Andrzejewski, J., 43–44
Angelino, A. C., 33–34
Angus, L. M., 121
Anzani, A., 189, 225
Araujo, V., 31–32
Arcelus, J., 105*t*
Aristizabal, A., 140
Armuand, G., 178–179
Arnoldussen, M., 27, 88, 103–106, 103*t*
Ascha, M., 71, 132–135, 142*t*
Ashley, F., 31, 102, 114, 150, 152, 202–203, 207, 217–220
Asseler, J. D., 175, 177–179
Attkisson, C. C., 208
Ausavarungnirun, C., 179
Austin, A., 72, 150, 153, 218

B

Baams, L., 5
Bachmann, C. L., 150
Badash, A., 192
Bahr, C., 121
Bailie, E., 175–176
Baines, H. K., 98
Baker, W. B., 189
Band, I. C., 60
Bandura, A., 31
Bao, A. M., 29–30
Baram, S., 170, 181, 225
Barbonetti, A., 189, 224
Barcelos, T. M. R., 190, 195, 225
Barda, S., 178–179
Bariola, E., 155, 216–217
Barnard, E. P., 178–179
Baron, A., 23
Barr, S. M., 115
Barrera, E., 141, 171–172
Barrett, F., 170, 176–179
Barron, C., 155–156
Barron, V., 190
Bartels, C. B., 174, 176
Baskaran, C., 102
Bateman, W., 155, 217
Battison, E. A. J., 132, 136
Bauer, G. R., 12*t*, 13–14, 73–74, 104, 107, 155, 188
Baum. J., 22, 25
Becasen, J. S., 188, 192, 224
Becerra-Culqui, T. A., 72
Becker-Hebly, I., 83–84, 103*t*, 104, 203
Beemyn, G., 217
Beischel, W. J., 150, 217
Bell, B., 33–34
Bell, T., 117
Belous, C. K., 189, 225
Bene, N. C., 90
Berger, M. N., 33
Bernacki, J. M., 132, 134
Berrahou, I. K., 119–120
Bertelloni, C., 82
Bertrand, A. A., 141*t*
Bettcher, T. M., 113–114
Bird, K., 214
Birmaher, B., 106
Birnkrant, J. M., 155–156
Bishop, A., 42, 214
Bishop, M. D., 115
Bizic, M., 138*t*
Blakemore, J. E. O., 23–24
Blasdel, G., 138
Bluth, K., 153–154
Bockting, W. O., 43
Boerner, K. E., 42
Boguszewski, K. E., 173
Boogers, L. S., 11, 82, 85–88, 91, 221
Boos, M. D., 122
Borah, L., 6, 60
Bordas, N., 141
Borger, O., 99
Boskey, E. R., 33, 131, 133–135, 152, 189, 204, 219, 224
Boulware, S., 14
Bouman, M. B., 138
Bowman, S. J., 12*t*
Bradford, N. J., 118, 150, 154, 217, 223
Bradley, S. J., 4
Bragg, J. E., 75
Brandon-Friedman, R. A., 61
Branstrom, R., 133
Breland, D. J., 55, 58–59
Brennan, L., 52, 55
Bretschneider, C. E., 142*t*
Brezin, F., 14
Brik, T., 171–172, 179–181
Brill, S., 217
Bronfenbrenner, U., 46
Brown, C., 29–30, 32–34, 43, 45
Brown, E. E., 192
Bruce, L., 134
Bruno, J., 73
Buckloh, L. M., 8, 156–157
Budge, S. L., 28, 72
Bull, B., 156
Bullock, L. M., 43
Bultynck, C., 143
Bungener, S. L., 132, 190–192, 224
Burgener, K., 121–122
Burgwal, A., 219
Burinkul, S., 121
Burnes, T. R., 187–188, 193–195
Burns, J. A., 190
Bussey, K., 31–32

C

Caldarera, A. M., 155, 157, 218
Call, D. C., 42–45, 74
Caminos-Torres, R., 174
Cantu, A. L., 103*t*
Caprini, R. M., 133
Carlile, A., 73

Carlson, S. M., 98
Carmichael, P., 71, 83–84, 86, 151, 221
Casado, J. C., 140*t*
Catalpa, J. M., 150, 154
Catani, G. S., 143
Cavve, B. S., 13, 152, 207
Chadwick, K. A., 117
Chang, J., 117, 121
Chang, S. C., 72, 75
Chaudhary, S., 216
Chelliah, P., 103*t*, 104, 106, 203
Chen, A., 190, 224
Chen, D., 8, 12, 25–27, 44, 69, 71, 73, 88, 91–92, 103*t*, 104, 106, 151–152, 156, 172–173, 176, 178–179, 181, 203, 218, 222, 226
Chen, E. C., 154, 156
Chew, D., 102
Chiniara, L. N., 89, 172
Chinsen, A., 218–219
Chodzen, G., 42, 45, 68, 72
Choe, J., 122
Choi, S. K., 155
Chong, L. S. H., 214
Christiano, C., 217
Christodoulaki, A., 175
Chu, L., 122
Ciancia, S., 86
Clark, B. A., 8–9, 51–57
Clark, D. B. A., 226–227
Clark, T. C., 155
Clements, Z. A., 46
Clements-Nolle, K., 217
Coelho, J. S., 150
Cohen-Kettenis, P. T., 4, 12–13, 22–23, 69, 82–84, 103–104, 103*t*, 106–107
Coleman, E., 3, 6–8, 10, 68–69, 82, 84, 89–90, 98, 102, 132, 170–171, 174, 203–204, 213–214, 221
Collier, K. L., 217
Colvin, S., 43–44, 214
Condry, J., 31–32
Conn, B. M., 45
Connelly, K. J., 56–57, 98
Connolly, M. D., 72, 102
Cook, R. E., 32, 34
Cooley, S., 23
Coolhart, D., 155
Coolidge, F. L., 29
Cooper, H. C., 180–181, 225
Cortez, S., 98, 222
Cosgrove, D., 28, 150
Costa, R., 71, 83–84, 151, 153–154
Cousino, M. K., 8, 69
Cox, M. J., 155–156
Coyne, C. A., 6, 8, 45, 69, 72, 75
Crabtree, L., 122, 134, 202, 205, 208–209, 220–221
Craig, S. L., 4, 46, 154–155
Cregten-Escobar, P., 141*t*
Crenshaw, K., 46
Croteau, T. A., 151
Crowley, W. F., 82
Cunningham, J. A., 218–219

D

Dacakis, G., 117
Dai, D., 9, 131
Dailey, D., 193
Daley, T., 9, 122
Dangaltcheva, A., 157, 218
D'Arrigo-Patrick, J., 156
Das, R. K., 131
D'Augelli, A. R., 32
Davies, S., 117–118
Davis, T. E., 123
Dawson, J., 189, 194–195
Day, J. K., 155
De Blok, C. J. M., 101–102, 137, 187–188
de Bres, J., 33
De Chants, J. P., 115
de Graaf, N. M., 22
De Lara, L. D., 103*t*, 104
De Roo, C., 175–176
De Rosa, P., 138*t*
De Vries, A. L. C., 4, 11, 83–84, 103–104, 103*t*, 106, 139, 141–142, 151, 203, 208, 218, 222
Defreyne, J., 101, 189
DeGuzman, P. B., 59
Delemarre-van de Waal, H., 4
Delozier, A. M., 6, 42, 214–215
deMayo, B., 24–25, 31, 34–35, 215
Demeestere, I., 170
Dessens, A. B., 30
Devor, A., 149–150
D'hoore, L., 97, 102
Di Ceglie, D., 155–157, 218
Di Fini, G., 155, 158
Diaddigo, S. E., 135–136
Diamond, L. M., 28, 216
Diana, P., 115–116
Dickenson, J. A., 189, 191
Dickey, L. M., 216
Dijkman, B. A. M., 121
Dilday, E. A., 178–179
Dillon, J. J., 155
DiMatteo, M. R., 122
Dixon, M., 10
Doll, E., 98
Dolotina, B., 206, 218–220
Donaghy, O. J. E., 12*t*, 13
Donnez, J., 170
Dopp, A. R., 14
Doyle, D. M., 31, 97, 102
Drabish, K., 192
Drummond, K. D., 23
Dubin, S. N., 214
Dueker, A. K., 216
Duffy, M. E., 134–135
Dunleavy, M., 193
Durkin, K., 33
Durwood, L., 9, 23, 27, 71, 151–152, 155

E

Eaton, W. O., 23–24
Eaton, W. W., 106
Eccles, J. S., 33
Edwards-Leeper, L., 4, 69
Ehrensaft, D., 4, 31, 34, 153
Eisenberg, M. E., 54, 214
Eitel, K. B., 83
Elizabeth, P. H., 29
Ells, C., 150
Endendijk, J. J., 30, 32
Enright, E. A., 32
Erickson-Schroth, L., 122
Evans, C. L., 134
Expósito-Campos, P., 72, 218–219

F

Fabes, R. A., 23–24, 32–33
Factor, R. J., 217
Fagan, J., 23
Fagart, J., 121
Fagot, B. I., 23
Fairchild, K., 24
Fast, A. A., 22–24, 216
Feil, K., 42
Feinberg, L., 114
Fernández, R., 29
Fielding, L., 190–191, 194–195
Finegan, J. L., 191
Finlayson, C., 170
Fisch-Shvalb, N., 88
Fisher, A. D., 83–84, 86, 137, 221
Fisher, C. B., 155
Fisher, W. A., 189
Fivush, R., 32
Fletcher, D., 43
Flores, A. R., 5, 67–68
Fontanari, A. M. V., 42, 105, 105*t*, 203, 210
Foo, M. M., 59
Fornander, M. J., 192, 195
Frey, J. D., 138*t*, 141–142, 142*t*
Friedenthal, J., 175
Fuqua, J. S., 121–122

G

Gadomski, A. M., 203
Gale, J., 177
Galupo, M. P., 189
Gao, J. L., 118
Garborcauskas, G., 180–181
Gauthier, J., 114
Gelly, M. A., 206, 220
Gelman, S. A., 24, 32
Gerymski, R., 189
Gibson, D. J., 43, 156
Giles-Petersen, J., 4
Gill-Peterson, J., 35
Ginsberg, B. A., 118
Glintborg, D., 216
Glodowski, M. B., 89
Godiwala, P., 174, 176
Goffnett, J., 216
Goldberg, A. E., 60
Goldberg, R. F., 14
Goldenberg, T., 44, 51, 53, 68
Golombok, S., 24
Goodey, C., 22
Gooren, L., 4, 82, 221
Goossens, G. H., 99
Govier, E., 31

Grannis, C., 104–105, 105*t*, 203
Grant, R., 150–151, 217
Gray, S. A. O., 156
Green, A. E., 4, 105, 105*t*, 152
Green, R., 29
Greenwald, P., 176
Gregor, C., 154, 156
Gridley, S. J., 6, 55, 58–59, 68, 72–73, 214
Griffiths, S., 134–135
Grimstad, F. W., 140, 142
Gülgöz, S., 22–23
Gupta, P., 202–205
Gurrala, R. R., 101

H

Haben, C. M., 143
Hagenäs, I., 171, 225
Hainey, K. J., 219
Haley, S. G., 187
Hall, R., 204
Halloran, J., 172–173
Hancock, J., 118
Handler, T., 223
Hannema, S. E., 100
Haroz, E. E., 106
Harris, R. M., 171–172
Harrison, J. E., 4
Hassan, A., 175
Hassan, B., 131, 135
Hässler, T., 24, 215
Hastings, J., 74
Hatzenbuehler, M. L., 5, 153, 217
Heathcote, C., 153–154, 219
Hedrick, H. R., 59
Hein, I. M., 132
Hembree, W. C., 3, 6–8, 68, 82, 84, 89–90, 98, 101, 170–171, 174, 213–214, 221
Hendrick, S., 189
Hendricks, M. L., 5, 42, 44–45, 215, 217
Henrich, J., 209–210
Herbison, A. E., 81–82
Hereth, J., 189
Herman, J., 10–11, 67–68, 74, 192
Herndon, J., 135
Herrmann, L., 27
Heylens, G., 29
Hidalgo, M. A., 4, 43, 45, 68, 156, 192, 194, 213–214
Hill, A., 150, 154–157
Hillier, A., 156–157, 215, 218
Hines, M., 29–30, 34
Hisle-Gorman, E., 103*t*, 104, 222
Hodax, J. K., 6, 75, 85
Hollingshead, N., 122, 223
Hollinsaid, N. L., 153–154
Holmberg, M., 190
Holt, M. K., 11
Holt, N. R., 73, 218–219
Hood, K. E., 29
Hope, D. A., 75
Horowitz, L. M., 12, 106
Horton, C., 151, 154, 156, 217–218
Hou, L., 87–88
Howard, S. D., 187–188
Hoy-Ellis, C. P., 194
Hudson, J., 179, 181, 225
Hughes, L. D., 60–61
Huit, T. Z., 68, 72, 153, 213–214
Hunter, J., 5, 45, 72
Hussain, S. H., 121

I

Ikeda, K., 175–176
Ingprasert, S., 118
Insogna, I. G., 176, 178–179
Inwards-Breland, D. J., 8, 59, 68–70, 72–73, 75–76, 213–214
Irwin, D. E., 106
Israel, T., 46, 72, 75, 154, 157–158, 215, 218
Israeli, T., 177–179
Ivanski, C., 188–189
Iwamoto, S. J., 194

J

Jackson, E. F., 32, 34
Jacobsen, K., 149–150, 217
Jacques, K. P., 32
James, S. E., 114, 130, 133
Jarin, J., 99–101
Jarolim, L., 138*t*
Jedrzejewski, B. Y., 203–204
Joel, D., 29–30
Johns, M. M., 43–44, 74, 155, 214
Johnson, E. K., 172–173, 179
Johnson, H. E., 57–58
Johnson, K. C., 158
Jones, B. A., 12–13, 116
Joseph, T., 85–86
Julian, J. M., 116, 223
Juurlink, D. N., 121

K

Kahn, N. F., 11, 56–57
Kaltiala, R., 67–68, 103*t*
Kanj, R. V., 119–120
Karamanis, G., 4
Karrington, B., 23, 210
Kattari, S. K., 5, 74, 192
Katz-Wise, S. L., 10, 68, 155–156, 215, 226
Kauffman, R. P., 29
Kaur, M. N., 12*t*, 13
Kausar, N., 113–114
Kcomt, L., 187–188
Kearns, S., 56
Kelly, A., 82, 86
Kennedy, S. L., 226
Kerman, H. M., 173
Kerr, J. A., 150
Kerr, L., 155
Keuroghlian, A. S., 60
Khonina, M., 22
Kidd, K. M., 10–11, 55–56, 59, 61, 68
Killen, M., 23
Kilmer, L. H., 133
King, H., 98–99, 102, 222
King, W. M., 42
Kinnear, H. M., 174, 176
Kitic, A., 191
Klaver, M., 87–88, 91, 99–100, 221
Klein, A., 154
Kleinberg, D. L., 137
Klepper, M., 156, 219
Klink, D., 85–88
Klosky, J. L., 179
Knaus, S., 99
Knutson, D., 153
Kolbuck, V. D., 173, 181, 225–226
Komorowski, A. S., 180–181
Kornienko, O., 33
Korpaisarn, S., 214
Kosciw, J. G., 6
Kossor, D. C., 121
Kovacs, M., 106
Kozee, H. B., 12*t*, 13, 106–107
Kozlowska, K., 74
Kremen, J., 73
Krishna, K. B., 82, 86, 89
Kroenke, K., 12, 106
Krueger, E. A., 194
Kuper, L. E., 25–26, 34, 71, 73, 103*t*, 104, 151–152, 203, 218, 222
Kuvalanka, K. A., 155–156
Kuzma, E. K., 188
Kyweluk, M. A., 171–173

L

LaFleur, J., 14
Lahlou, N., 82
Lai, T. C., 89, 180–181
Lambert, A., 132
Lan, M., 32
Lange, T. M., 72
Laprise, C., 191–192
LaRocca, D., 6
Laspro, M., 135
Lavalley, M. N., 14
Lavender, R., 103*t*, 151–152, 218
Lawlis, S. M., 156–157, 218
Lawrence, A. A., 169–170
Lawson, M. L., 5, 11, 73–74, 214
Lazar, L., 89
Leavy, M., 175
Lee, J., 69, 72, 84–85, 140, 142, 221
Lee, M. K., 105–106, 105*t*
Lee, W. Y., 5, 61
Lefevor, G. T., 74
Leinbach, M. D., 23
Leone, A. G., 194
Lett, E., 43, 74
Leung, A., 177
Levy, J., 121
Li, A. Y., 135
Lierman, S., 175
Light, A., 169–170, 175, 187–188
Lindgren, T. W., 12–13, 106
Linsenmeyer, W., 42, 116, 214
Locke, E. R., 56–57
Long, A., 192
Lorde, A., 188–189
Lu, J., 84
Lucas, S., 28
Lucassen, M. F. G., 153–155

Luvmour, J., 155
Lykens, J. E., 114

M

MacKinnon, K. R., 152–153, 202–203, 205–207, 209–210, 219–220
Maheux, A. J., 192
Mahfouda, S., 102, 131, 150, 154, 202–203
Malpas, J., 156–157
Manns, P. Z., 73, 214
Manuel, S. L., 176
Mar, S. A., 60
Maragos, A. C., 152
Marinelli, L., 118, 188
Marshall, A., 193
Marshall, W. A., 81
Martin, C. E., 177, 225
Martin, C. L., 23–24, 32–33
Martin, D. J., 33
Martin, S., 153
Marwa, A., 85
Matsuno, E., 28, 46, 72, 75, 157–158, 192, 215
Mattawanon, N., 190, 224
Mattelin, E., 179
Matthews, S. J., 170
McCarthy, M. M., 30
McCauley, C. J., 89
McCormick, A., 155, 217
McCoy, R. T., 33–34
McDermott, E. S., 154–155
McDougall, R. J., 170–171, 181
McDowell, A., 6
McGregor, K., 83–84, 151
McGuire, J. K., 12–13, 22, 32, 106–107
McKleroy, V. S., 46
McLaughlin, A., 195
McLean, K. C., 43
McLemore, K. A., 115
McNamara, M., 14, 102, 201–202, 206
McPhail, D., 214
McPherson, L. E., 44
McQuillan, M. T., 45
Medico, D., 26–28
Medina-Martínez, J., 150
Mehra, G., 179–180
Mehringer, J. E., 152
Mejia-Otero, J. D., 83
Menvielle, E. J., 156–157, 218
Messinger, A. M., 194–195
Meyer, H. M., 73
Meyer, I. H., 42–44, 150, 192, 217
Meyer-Bahlburg, H. F. L., 30
Mikovits, J. C., 114
Miller, A. S., 15
Miller, C. L., 23
Miller-Jacobs, C., 10–11
Miller-Perusse, M., 45
Millington, K., 99–101, 121
Mills, M., 117
Miner, M., 98
Minuchin, P., 155
Misakian, A., 98
Moghadam, S., 131
Moghetti, P., 121
Monstrey, S., 115–116
Moravek, M. B., 170–171
Morgan, H., 156–158, 217–218
Morrison, A., 172–173
Morrison, S. D., 139, 142*t*
Mortimer, R. M., 225
Mosier-Mills, A., 71
Moussaoui, D., 89, 119–120
Moyer, D. N., 42
Murad, M. H., 133, 223
Myers, B., 117

N

Nagoski, E., 188–189
Nahata, L., 57, 89, 170–173, 176, 178–181, 225
Narayan, S. K., 134, 202, 208
Nasomyont, N., 84–85
Navabi, B., 85, 87–88, 91
Nealy, E. C., 153
Nelson, H. D., 102
Nelson, K. M., 191–192
Newcomb, M. E., 72, 214
Newhook, J. T., 21–22, 27
Newman, L. E., 4
Nguyen, H. B., 29–31, 97, 102, 202–203
Nie, I. D., 170–172, 174–176, 178
Nieder, T. O., 202, 208
Nightingale, K. J., 59, 98
Nokoff, N. J., 87–88, 91, 99–101, 221
Nolan, B. J., 151–152, 203
Nolan, I. T., 117
Nora Isacoff, N., 32
Norwood, K., 155–156, 217
Nota, N. M., 101
Notini, L., 88
Nugent, B., 33
Nunes-Moreno, M., 105, 105*t*, 218

O

Obedin-Maliver, J., 188
Ojeda, S. R., 81–82
Oles, N., 137*t*, 139*t*, 141*t*
Olezeski, C. L., 69
Olsavsky, A. L., 105, 105*t*, 151–152, 218
Olson, J., 26
Olson, K. R., 9, 11, 13, 22–24, 26–27, 32, 71, 115–116, 151, 155, 202, 207, 216
Olson-Kennedy, J., 11, 69, 100–101, 103*t*, 104, 116, 132–134, 152, 203, 215, 222–223
Onejeme, C., 118
Oosthoek, E. D., 11, 220–221, 226
Orenius, T., 122–123
O'Shea, J., 42–44, 47
Osman, A., 106

P

Pachankis, J. E., 5, 133, 153
Paley, B., 155–156
Pampati, S., 72–73, 214
Panagiotakopoulos, L., 89–90
Pang, K. C., 118, 176, 179–181, 225
Papadopulos, N. A., 138*t*
Park, R. H., 133, 223
Parker, E., 156, 158
Pasquino, A. M., 82, 89
Patel, K. T., 121
Patrick, K. L., 42–44, 47
Patterson, B., 218–219
Patton, G. C., 81
Pauly, I. B., 12–13, 106
Peitzmeier, S. M., 115–116
Pelusi, C., 98
Pepper, R., 155
Peri, A., 89, 170–171, 174, 178–179, 225
Perl, L., 100
Perrard, M. H., 174
Persky, R. W., 172–173, 181, 225
Peters, B. R., 90
Pfau, D. R., 174
Phulwani, P., 98–99
Pine-Twaddell, E., 83
Plant, E. A., 31–32
Pollitt, A. M., 44, 115
Poquiz, J., 73, 76, 214
Posner, K., 106
Poteat, V. P., 43–44
Poulin-Dubois, D., 23, 215
Preston, S., 60, 201–202, 210
Price-Feeney, M., 42
Puckett, J. A., 25–26, 43, 46, 56, 75, 214, 216
Pulice-Farrow, L., 46
Pullen Sansfaçon, A., 35, 158, 202, 206, 208

Q

Quain, K. M., 180–181
Quinn, S., 117

R

Rabbitts, J. A., 135
Rachlin, K., 188, 224
Rae, J. R., 23, 216
Rafferty, J., 7–8
Ramos, N., 214
Rapoport, S. K., 117–118
Ravindranath, O., 8–9
Real, A. G., 207
Reay, B., 189
Redfield, E., 73, 201–202
Reisman, T., 137
Reisner, S. L., 114, 189, 216, 224
Ren, T., 134
Reynolds, C. R., 106
Rhodes, M., 23, 215
Riach, J., 153
Richmond, B. O., 106
Rider, G. N., 42–44, 51–54, 101, 214
Riggle, E. D. B., 192
Riggs, D. W., 155–156, 158
Riley, E. A., 155–158
Riskind, R. G., 32, 34
Rivera, A. V., 192
Rizvi, S. L., 72
Roberge, S., 86
Roberts, C. M., 202–204
Robinson, B. A., 215
Robinson, I. S., 131–133
Roblee, C. V., 141
Rodriguez, M., 59–60

Rodriguez-Wallberg, K., 169–170
Rogalska, M., 139
Romito, M., 150
Rose, L. I., 121
Rosenfield, R. L., 81
Ross, M. B., 136
Rothenberg, S. S., 170, 176, 178–179
Rotondi, N. K., 114
Roy, M. K., 85
Ruan, W., 137
Rudman, L. A., 24
Rusow, J. A., 51–53, 55
Russell, M. R., 5, 117–118, 223
Russell, S. T., 6, 44, 71, 115
Ryan, C., 32

S

Safa, B., 90, 223
Salinas-Quiroz, F., 22, 28, 32–33, 35, 216
Salvetti, B., 9
Salway, T., 22
Samargia, L. A., 52
Sansfaçon, A. P., 152, 156, 158, 219–220
Sarkar, M., 43
Sarno, E. L., 45, 153
Sax, M. R., 171–172
Sbragia, J. D., 188
Scandurra, C., 43, 153
Schagen, S. E. E., 82–88
Schardein, J. N., 190–191
Schlehofer, M. M., 156
Schmidt, E., 194
Schoffer, A. K., 137*t*
Schulmeister, C., 82, 86, 221
Schwartz, A. R., 175
Schwartz, B. I., 119–120
Schwarz, K., 117
Seager van Dyk, I., 32
Segev-Becker, A., 178–181
Sequeira, G. M., 60, 72–74, 214
Serbin, L. A., 23–24, 215
Shaw, H. E., 134–135
Shi, Q., 52, 55
Shook, A. G., 226
Shuhart, C. R., 84, 86
Shumer, D. E., 82, 89–90
Signorella, M. L., 23–24
Silva, G. C., 192
Silveira, J. C. P., 53, 55
Simons, L., 155
Singh, A., 46, 153, 215
Singhal, V., 98–99
Sisk, C. L., 88
Skelton, S., 150–151, 217
Skrzypczak, E., 189
Slack, D., 98
Slavit, D. H., 143
Smalley, K. B., 154
Smiler, A. P., 24
Smith, C., 117
Smith, K. P., 121
Smith, Y. L. S., 103–104, 103*t*
Smith-Johnson, M., 214
Snyder, L., 135
Solomon, J., 23, 215
Sorbara, J. C., 25–27
Spencer, K. G., 8
Spielberger, C. D., 106
Spitzer, R. L., 12, 106
Spivey, L. A., 32
Stanton, A. M., 219
Staphorsius, A. S., 88
Stark, B. A., 176–177, 181
Steensma, T. D., 24, 27, 31
Stein, D. M., 179
Steinberg, L., 8–9
Sterling, J., 174
Stevens, S., 153
Stever, G. S., 33
Stewart, H. L., 119–120, 122
Stice, E., 134–135
Stieglitz, K. A., 215
Stocking, S. Q., 214
Stoffers, I. E., 99–101
Strang, J. F., 12*t*, 13, 88, 105–106, 105*t*, 172–173, 221
Strauss, P., 72–73, 106–107, 150, 154, 214
Strenth, C. R., 42
Stutterheim, S. E., 192
Suarez, N. A., 10–11, 22
Sumerwell, C., 98–99
Sundaram, V., 177, 179
Suslovic, B., 43, 214
Swaab, D. F., 29–30
Swan, J., 133, 223
Sweder, N., 28, 35
Sweileh, W. M., 213
Sylliboy, J. R., 33–34
Syme, M. L., 188
Szoko, N., 189

T

Ta, V., 98
Tack, L. J. W., 100–101
Taddio, A., 122
Taliaferro, L. A., 43, 52–53, 214
Tan, K. K. H., 134, 150
Tang, A., 131, 134–135, 203–204, 208
Tankersley, A. P., 42–44, 47, 74, 215
Tanner, J. M., 81
Taylor, J., 83–84, 88, 102, 221
Taylor, M. G., 24
Telfer, M., 155
Testa, R. J., 5, 42, 44–45, 72, 150, 192, 194, 215–217
Thoma, B. C., 192
Thomas, M., 33–34
Thoreson, N., 118
Thornton, S. M., 134
Thrower, E., 150
Tishelman, A. C., 171–172
Todd, K., 27, 114
Toland, M. K., 102
Tollit, M. A., 70, 72–73, 214
Toomey, R. B., 154, 215, 217
Topaz, E., 25–27
Tordoff, D. M., 5, 103*t*, 104, 134, 151–152, 203, 218
Tornello, S. L., 32, 34
Tornese, G., 83, 85, 91–92
Travers, A., 215
Travers, R., 155
Troy, A. S., 43
Tucker, R. P., 153
Turban, J. L., 4, 25–26, 73, 83–84, 90, 105–106, 105*t*, 150–152, 202–203, 205–206, 216, 218–220
Tüzün, Z., 42–43, 214
Twist, J., 22
Tyni, K., 215

U

Uffman, J. C., 131, 223–224

V

Vahabzadeh-Hagh, A. M., 143
Valentine, A., 100
Valli-Pulaski, H., 171, 178–179
Van Damme, S., 140*t*
van de Grift, T. C., 83, 86–88, 133
van der Loos, M. A. T. C., 84–87, 91, 202–204, 221
Van Der Meulen, I. S., 90, 138, 190–191, 224
van der Miesen, A. I. R., 83–84, 151, 218, 221
Van Gerwen, O. T., 187–188, 192
van Leerdam, T. R., 97, 102
Vance, S. R., 42, 56, 74–75
Vandenberg, N., 14
Vandenbussche, E., 152, 202–203, 205, 220
Vandereycken, W., 106
Varkey, B., 171
Vasen, E., 158, 218
Vaughan, E., 215
Vaughan, M. D., 149, 194, 217
Veale, J. F., 42–43, 72, 155, 217
Veldhuis, C. B., 33
Veldhuis, J. D., 137
Verguchten, J., 215
Verrastro, V., 187–188
Vestering, A., 102
Vincent, B., 74
Vincenzo, C. D., 71
Vlot, M. C., 85
Voss, R. V., 116
Vries, A. L. C., 69, 71
Vrouenraets, L. J. J. J., 90–91, 132, 171–172
Vyas, N., 179

W

Waagen, T., 24, 32–33
Wagner, L. D., 158
Wahlig, J. L., 155, 217
Wakefield, B. W., 179–181
Wallien, M. S. C., 23
Walton, E., 172–173
Ward, L., 28
Warton, C., 170, 181
Warwick, R. M., 69, 189–190, 192–193, 224
Waterhouse, P., 44
Watson, R. J., 155
Webb, H., 143
Weigert, R., 152

Weimer, A. K., 132, 134
Weinhardt, L. S., 43, 114, 214
Weixel, T., 133–134
Wells, L., 155–156
Werner, E. E., 43
Wernick, J. A., 133
Westwater, J. J., 156
Wheeler, N. L., 156
White Hughto, J. M., 187–188
Wiepjes, C. M., 134, 208
Wierckx, K., 137
Wilde, B., 121–122
Wilder, S., 190
Willemsen, L. A., 82, 86
Williams, C. R., 114
Williams, K. A., 52
Wilson, E. C., 155
Wilson, L. M., 97, 102
Winer, J., 100
Winkelman, W. D., 171
Wittlin, N. M., 24–25, 27, 42–44, 72, 103*t*, 104, 151, 218, 222
Wong, W., 155–156
Wren, B., 156, 158
Wylie, K., 6

X

Xerxa, Y., 31
Xiao, S. X., 32–33
Xu, W., 99–100

Y

Yager, Z., 134–135
Yau, M., 170–171, 178
Yeh, D., 53, 55
Yung, K., 117

Z

Zeeman, L., 215
Zhang, Q., 22
Zhou, J. N., 30–31
Ziegler, A., 143
Zolkoski, S. M., 43
Zosuls, K. M., 23–24

Subject Index

Note: Page numbers followed by "*t*" refer to tables.

A

Achenbach System of Empirically Based Assessment (ASEBA), 106
Alanine transaminase (ALT), 101
American Academy of Paediatrics (AAP), 7
American Society for Reproductive Medicine, 170
Appearance congruence, 13, 97
Ask Suicide Screening Questionnaire (ASQ), 106
Aspartate aminotransferase (AST), 101
Assigned female at birth (AFAB), 81, 98, 204
Assigned male at birth (AMAB), 81, 98, 204

B

Barriers, 51
Behavior Rating Inventory of Executive Function (BRIEF), 88
Behavioral healthcare, 8
Body composition and cardiometabolic health, 87–88
Body Image Scale (BIS), 12–13, 106
Body Image Scale—Gender Spectrum (BIS-GS), 106
Bone mineral density (BMD), 81–82
Brain development and cognition, 88
Building social supports and community connection, 75

C

Center for Epidemiological Studies Depression Scale (CESD-R), 106
Center for Expertise in Gender Dysphoria, 4
Center for Medicaid Services, 15
Center of Expertise on Gender Dysphoria (CEGD), 103–104
Child Behavior Checklist (CBCL), 106
Children's Depression Inventory (CDI), 106
Cisgender, 21
Cognitive behavioral therapy (CBT), 153, 218
Columbia-Suicide Severity Rating Scale (C-SSRS), 106
Community-based participatory research, 213
The Companionship Model for Youth (TCM-Y), 72
Complete androgen insensitivity syndrome (CAIS), 29
Congenital adrenal hyperplasia (CAH), 29
Cricothyroid approximation (CTA), 139
Cross-linked telopeptide of type I collagen (ICTP), 84
C-terminal telopeptide (s-CTX), 84
Cultural context, inclusivity, and equity, 74

D

Department of Health and Human Services, 15
Detransphobia, 152
Dialectical behavior therapy (DBT), 72
Disinformation, 14–15
Disparities
  clinical practice recommendations, 55
  research strengths and limitations, 54–55
DSM-5-TR, 4
DSM-IV criteria, 29
Dutch protocol, 4

E

Early pubertal suppression, 129
Embodiment, 113
Endocrine Society guidelines, 7
Estradiol, 175
  body satisfaction, 102–107
  cross-sectional research methods, 105–106, 105*t*
  dosing methods, 98–99
  future aspects, 107–108, 222
  key research gaps, 222
  longitudinal research methods, 103–105, 103*t*
  measurement tools, 105*t*, 106–107
  mental health, 102–107
  overview, 97–99
  physical health, body composition, and risk profiles, 99–102
  well-being, 102–107
European Network for the Investigation of Gender Incongruence (ENIGI), 11
Evidence base for holistic and interdisciplinary care
  medical and surgical interventions, 71
  psychological assessments and interventions, 71–72
  social interventions, 71

F

Facial feminization surgery (FFS), 133
Family systems
  future aspects, 158
  interventions for families, 156–158
  mental health and well-being of parents, 155–156
  mental health and well-being of siblings, 156
Female Sexual Function Index (FSFI), 190
Feminizing surgery, 129
Ferriman Gallwey (FGm), 118
Fertility, 169
  preservation, 169
Fertility and fertility preservation
  counseling, 179–181
  ethical considerations, 171
  fertility implications
    animal models, 174
    estradiol, 175
    future aspects, 176
    menstrual cycle suppression, 175
    pubertal suppression, 174
    spironolactone, 174
    testosterone, 175
  knowledge, attitudes, beliefs, and decision-making, 172–173
    decision-making, 173
    future aspects, 173
  oocyte preservation, 176–177
  overview, 170–172
  rates of fertility preservation, 179–181
  sperm preservation, 178
Fertility preservation, 89
Follicle-stimulating hormone (FSH), 81–82, 174

G

Gammaglutamyl transferase (GGT), 101
Gay-straight alliances (GSA), 43–44
Gender affirmation, mental health intervention
  future aspects, 152–153
  psychosocial interventions, 153–155
    future aspects, 154–155
    research strengths and weaknesses, 154
Gender Congruence and Life Satisfaction Scale (GCLS), 12–13
Gender development and developmental pathways
  active agents, children, 34
  biological factors
    gender-affirming medical interventions, 31
    heritability, 29
    prenatal hormonal exposure, 29–30
    transgender neural development, 30–31
  biological, psychological, and social processes, 34

childhood
adolescence, 24–25
ages 0-3 years, 23
early childhood, 23–24
middle childhood, 24
cisnormativity and bias, 22–23
common developmental pathways
first, 27
second, 27
third, 27–28
future aspects, 34–35
gender diverse children, 27–28
importance, 22
nonbinary, 28
overview, 22–25
social and environmental factors
cultural factors, 33–34
parents, 31–32
peers and other sources, 32–33
transactional model, 28
(trans)gender developmental milestones
disclosure, 26
medical transition, 27
realization, 26
social transition, 26–27
Gender diversity, 10–11
Gender dysphoria, 113
hormone management, 119–121
management
dermatologic effects, 122
needle phobia, 122–123
pill swallowing aversion, 122–123
sex characteristics appearance
chest binding, 115–116, 119*t*
facilitating facial hair, 118
hair removal, 118
packing, tucking and body forms, 116–117
voice training, 117–118
social and legal transition steps, 115
Gender embodiment desires and goals, 67, 97
Gender expression, 21
Gender identity, 3–4, 21
Gender Identity Acceptance, 13
Gender minority stress and resilience (GMSR), 44–45
Gender Minority Stress and Resilience Framework (GMSR), 215
Gender nonconforming, 21
Gender policing, 32–33
Gender Preoccupation and Stability Questionnaire (GPSQ), 13
Gender Self-Report (GSR), 13
Gender web, 34
Gender-affirming care (GAC), 7–8, 67, 82, 201–202
Gender-affirming care models, existing
adjunctive programming and collaboration, 70
initial assessment of care needs, 69
medical and surgical interventions, 70
psychosocial care and supportive interventions, 69–70
Gender-affirming healthcare. *See also* Transgender and gender diverse (TGD)
clinical guidelines, 6–7
clinical practice
behavioral healthcare, 8
informed assent, 8–9
informed consent, 8–9
legal affirmation, 8
medical and mental health supports, 9
social affirmation, 8
ethical research
epidemiology of gender diversity, 10–11
evaluation, 11
measurement, 11–13, 12*t*
outcomes, 11–13, 12*t*
evolution, 4–5
gender dysphoria, 4
gender incongruence, 4
gender-affirmative model, 4–5
guideline development, 13–14
minority stress, 5–6
mitigators, 6
sex and gender, 4
sociopolitical climate
disinformation, 14–15
future aspects, 15
misinformation, 14–15
United States, 14
structural transphobia, 5–6
systematic reviews, 13–14
Gender-affirming hormones (GAHs), 82, 97, 188, 218
Gender-affirming surgery (GAS), 129–130, 188
Generalized Anxiety Disorder-7 (GAD-7), 12
Glomerular filtration rate (GFR), 100–101
Gonadal suppression, 81
Gonadotropin-releasing hormone (GnRH), 81–82
Gonadotropin-releasing hormone agonists (GnRHa), 9, 98, 119, 204, 221
Grading of Recommendations, Assessment, Development, and Evaluation (GRADE), 7, 83

H

Healthcare and barriers
disparities
recommendations for clinical practice, 55
research strengths and limitations, 54–55
electronic consultation, 59–60
future aspects, 61–62
legal barriers, 60–61
pediatric gender-affirming care, access and barriers
recommendations for clinical practice, 58–59
research strengths and limitations, 58
political attacks, 60–61
telehealth, 51, 59–60
Health disparities, 41, 51
Holistic and interdisciplinary healthcare
cultural context, inclusivity, and equity, 74
evidence base, 70–72
existing gender-affirming care models
adjunctive programming and collaboration, 70
initial assessment of care needs, 69
medical and surgical interventions, 70
psychosocial care and supportive interventions, 69–70
interdisciplinary gender-affirming care, 68
navigating healthcare and addressing care barriers, 75–76
regional and global distinctions, 72–74
social supports and community connection, 75
tailoring inclusive care approaches, 74–75
Homeostatic model assessment for insulin resistance (HOMA-IR), 99–100
Hormone management
antiandrogens, 121
menstrual cycle suppression, 119–120
progesterone for individuals assigned male at birth, 121

I

Individual treatment progress score (ITPS), 208
Informed assent, 8–9
Informed consent, 8–9
Interdisciplinary gender-affirming care, 68
International Classification of Diseases-11, 4
Intrauterine device (IUD), 119–120

L

Laser-assisted voice adjustment (LAVA), 139
Legal affirmation, 8
Long-term satisfaction and risk of regret
community-based studies, 210
detransition, 201, 205–206
discontinuation, 203–205
gender as a process, not a destination, 210
increasing representation, 209–210
longitudinal research, 209
measuring patient-reported outcomes, 202
psychosocial and mental health outcomes, 202–203
qualitative research, 209
reidentification, 201, 206–207
reversal of medical treatment, 203–205
satisfaction, regret, and other emotions, 207–209

M

MacArthur competence assessment tool, 9
Male Sexual Function Index (MSFI), 190
Masculinizing surgery, 129
Medical and mental health supports, 9
Mental healthcare, youth and families
family systems
future aspects, 158
interventions for families, 156–158
mental health and well-being of parents, 155–156
mental health and well-being of siblings, 156
gender affirmation, mental health intervention, 151–153
future aspects, 152–153
psychosocial interventions, 153–155
mental health and well-being, 150

Mindful Self-Compassion for Teens (MSC-T), 153–154
Minnesota Student Survey, 52–54
Minority stress, 41
Misinformation, 14–15
Multidimensional Transgender and Nonbinary Resilience (MTNBR), 46–47
Multidisciplinary gender clinics, 6

N

National Health and Nutrition Examination Survey (NHANES), 99
Navigating healthcare and addressing care barriers, 75–76
Nonbinary, 21, 113
Nonmedical interventions, 113

O

Ovarian tissue cryopreservation (OTC), 170

P

Patient Health Questionnaire-9 (PHQ-9), 12, 106
Patient Reported Outcomes Measurement Information System (PROMIS), 106
Patient-reported outcome measure (PROM), 13
Patient-Reported Outcomes Measurement Information System (PROMIS), 133
Pediatric gender-affirming care, 3, 213
  clinical practice recommendations, 58–59
  research strengths and limitations, 58
Physical health, body composition, and risk profiles
  anthropometric measurement, 99
  blood pressure, 100
  body mass index, 99
  glucose metabolism, 99–100
  hematocrit, 100
  hemoglobin, 100
  hyperprolactinemia, 101
  lipids, 100
  long-term health risks, 101–102
    cancer, 101–102
    venous thromboembolism, 102
  renal function, 100–101
  serum creatinine, 100–101
Political attacks, 60–61
Polycystic ovarian syndrome (PCOS), 175
Pubertal suppression, 174, 221
  access, decision-making, and treatment continuation, 90–91
  body composition and cardiometabolic health, 87–88
  bone health, 84–86
  brain development and cognition, 88
  clinical outcomes, 83–90
  efficacy, 83
  fertility, 89
  fertility preservation, 89
  future aspects, 91–92
  gender-affirming care, 82
  height and skeletal structure, 86–87
  mental health and psychosocial outcomes, 83–84

Q

Quantitative computed tomography (QCT), 84

R

Randomized control trial (RCT), 203
Randomized controlled trial (RCT), 153
Releasing internalized stigma for empowerment (RISE), 72
Resilience, 41
Revised Children's Manifest Anxiety Scale (RCMAS-2), 106
Risk and resilience factors
  future aspects, 47–48
  gender minority stress and resilience (GMSR) model, 44–45
  Multidimensional Transgender and Nonbinary Resilience (MTNBR), 46–47
  resiliency and protective factors, 43–44
  risk factors, 41–42
  transgender resilience intervention model (TRIM), 46

S

Screen for Anxiety Related Emotional Disorders (SCARED), 106
Self-Injury Questionnaire-Treatment Related (SIQ-TR), 106
Sex assigned at birth, 21
Sexual and gender minority (SGM), 72
Sexual development, health, and functioning
  center sexuality-related values, 193–194
  consider unique sexual development trajectories, 194
  diversify research samples, 194
  promote sexual health and well-being, 194
  reconstruct sexual safety, 194–195
  revise and empirically validate measurement tools, 195
  sex orientation, 187, 189–190
  sex-positivity, 187–189
  sexual functioning, 187, 190–191
  sexual health, 191–193
  sexual identity, 189–190
  sexual well-being, 187, 191–193
Sexual maturity rating (SMR), 81, 132, 170
Sexually transmitted infections (STIs), 187–188
Social affirmation, 8
Spironolactone, 174
Standards of Care, Version 8 (SOC8), 6–7
Structural transphobia, 5–6
Suicidal Behavior Questionnaire-Revised (SBQ-R), 106
Surgical considerations, 89–90
Surgical interventions, in adolescence and young adults
  community acknowledgment and considerations, 130
  overview of GAS, 131
  surgical trends, 131–137
    age and timing of interventions, 131–133
    individualized surgical planning, 133
    postoperative care and support, 135–137
    preoperative considerations, 134–135
    psychosocial outcomes, 133–134
    satisfaction and risk of regret, 134
  transgender men and nonbinary people assigned female at birth, 140–143
  transgender women and nonbinary people assigned male at birth, 137–140

T

Tanner staging, 81
Telehealth, 51, 59–60
Testicular sperm extraction (TESE), 171
Testicular tissue cryopreservation (TTC), 171
Testosterone, 175
  body satisfaction, 102–107
  cross-sectional research methods, 105–106, 105*t*
  dosing methods, 98–99
  future aspects, 107–108, 222
  key research gaps, 222
  longitudinal research methods, 103–105, 103*t*
  measurement tools, 105*t*, 106–107
  mental health, 102–107
  overview, 97–99
  physical health, body composition, and risk profiles, 99–102
  well-being, 102–107
Transgender and gender diverse (TGD)
  detransition literature, 219–220
  gender identity development, 215–216
  medical interventions, 220–223
  mental health, 216–219
    caregiver support and acceptance, 217–218
    future aspects, 219
    intervention research, 218
    key research gaps, 218–219
    minority stress as context, 217
    social, medical, and legal affirmation, 218
  pubertal suppression. *See* Pubertal suppression
  reproductive health and fertility preservation, 225–226
  risk, resilience, and access to healthcare, 213–215
  sexual health, 224–225
  surgical interventions, 223–224
  testosterone and estradiol. *See* Estradiol. *See* Testosterone
  treatments, 223
Trans Woman Voice Questionnaire (TWVQ), 117
Trans Youth CAN! Gender Distress Scale (TYC-GDS), 13
Trans Youth CAN! Gender Positivity Scale (TYC-GPS), 13
Trans Youth Care—United States (TYCUS) study, 104
Trans Youth Fertility Study (TYFS), 173
Trans Youth Project (TYP), 11
Transgender and gender diverse, 3, 21
Transgender Congruence Scale (TCS), 13, 106–107
(Trans)gender developmental milestones
  disclosure, 26
  medical transition, 27
  realization, 26

social transition, 26–27
Transgender health research, 213
Transgender men and nonbinary people assigned female at birth
bottom surgery, 141–142, 142t
facial gender-affirming surgery, 142
gender-affirming vocal surgery, 143
top surgery, 140–141, 141t
Transgender resilience intervention model (TRIM), 46, 215
Transgender Self-Evaluation Questionnaire (TSEQ), 117
Transgender women and nonbinary people assigned male at birth
bottom surgery, 137–138, 138t
facial feminization surgery, 138–139, 139t
facial gender-affirming surgery, 138–139, 139t
gender-affirming body contouring, 140
gender-affirming vocal surgery, 139–140, 140t
top surgery, 137, 137t
Transgender Youth Fertility Attitudes Questionnaire (TYFAQ), 172
Transition, 113, 201
Transsexual Voice Questionnaire (TVQ^{MtF}), 117
TransYouth Project (TYP), 23, 216
Type 1 N-terminal propeptide (P1NP), 84
Type III procollagen peptide (P3NP), 84

U

University Medical Center Utrecht, 103–104
Utrecht Gender Dysphoria Scale (UGDS), 12–13, 106–107
Utrecht Gender Dysphoria Scale — Gender Spectrum (UGDS-GS), 106–107

V

Venous thromboembolism (VTE), 102, 135

W

Western, Educated, Industrialized, Rich, and Democratic (WEIRD), 209–210
World Professional Association for Transgender Health (WPATH), 6–7, 68, 82, 132, 213–214
World Professional Association of Transgender Healthcare (WPATH), 170

Y

Yale Gender Program Reproductive Knowledge and Experiences Survey (YPGP-RKES), 172
Youth Risk Behavior Survey (YRBS), 10–11
Youth Self-Report (YSR), 106